Praise for *Altered Genes,*

D0390316

"Without doubt, one of the most importan
shall urge everyone I know who cares about life on earth, and the future of their children, and children's children, to read it. It will go a long way toward dispelling the confusion and delusion that has been created regarding the genetic engineering process and the foods it produces. . . . Steven Druker is a hero. He deserves at least a Nobel Prize."

— **Jane Goodall, PhD, DBE,** UN Messenger of Peace
(*from the Foreword*)

"A fascinating book: highly informative, eminently readable, and most enjoyable. It's a real page-turner and an eye-opener."

— **Richard C. Jennings, PhD,** Department of History and Philosophy of Science, University of Cambridge, UK

"This incisive and insightful book is truly outstanding. Not only is it well-reasoned and scientifically solid, it's a pleasure to read – and a must-read. Through its masterful marshalling of facts, it dispels the cloud of disinformation that has misled people into believing that GE foods have been adequately tested and don't entail abnormal risk."

— **David Schubert, PhD,** molecular biologist and
Head of Cellular Neurobiology,
Salk Institute for Biological Studies

"*Altered Genes, Twisted Truth* is lucid, illuminating, and alarming. As a former New York City prosecutor, I was shocked to discover how the FDA illegally exempted GE foods from the rigorous testing mandated by federal statute. And as the mother of three young kids, I was outraged to learn how America's children are being callously exposed to experimental foods that were deemed abnormally risky by the FDA's own experts."

— **Tara-Cook Littman, JD**

"Steven Druker has written a great book that could well be a milestone in the endeavor to establish a scientifically sound policy on genetically engineered foods. The evidence is comprehensive and irrefutable; the reasoning is clear and compelling. No one has documented other cases of irresponsible behavior by government regulators and the scientific establishment nearly as well as Druker documents this one. His book should be widely read and thoroughly heeded."

— **John Ikerd, PhD,** Professor Emeritus of Agricultural and Applied Economics, University of Missouri – Columbia

"*Altered Genes, Twisted Truth* will stand as a landmark. It should be required reading in every university biology course."

— **Joseph Cummins, PhD,** Professor Emeritus of Genetics,
Western University, London, Ontario

"Steven Druker's meticulously documented, well-crafted, and spellbinding narrative should serve as a clarion call to all of us. In particular, his chapter detailing the deadly epidemic of 1989-90 that was linked with a genetically engineered food supplement is especially significant. I and my Mayo Clinic colleagues were active participants in the attempt to identify the cause of this epidemic. Druker provides a comprehensive analysis of all the evidence and also presents new findings from our work. Overall his discussion of this tragic event, as well as its ominous implications, is the most comprehensive, evenly-balanced and accurate account that I have read."

— **Stephen Naylor, PhD,** CEO and Chairman of MaiHealth Inc.
Professor of Biochemistry and Molecular Biology, & Pharmacology
Mayo Clinic (1991-2001)

"*Altered Genes, Twisted Truth* is very readable, thorough, logical and thought-provoking. Steven Druker exposes shenanigans employed to promote genetic engineering that will surprise even those who have followed the ag-biotech industry closely for years. I strongly recommend his book."

— **Belinda Martineau, PhD,** molecular biologist, a co-developer of the first genetically engineered whole food, and author of *First Fruit: The Creation of the Flavr Savr™ Tomato and the Birth of Biotech Foods*

"Steven Druker has done a beautiful job of weaving a compelling scientific argument into an engaging narrative that often reads like a detective story, and he makes his points dramatically and clearly. The examination of genetic engineering from the standpoint of software engineering is especially insightful, exposing how the former is more like a 'hackathon' than a careful, systematic methodology for revising complex information systems. I will recommend this book to my friends."

— **Thomas J. McCabe,** developer of the cyclomatic complexity software metric, a key analytic tool in computer programming employed throughout the world

"Based on over 30 years of teaching computer science at universities and on extensive experience as a programmer in private industry, I can state that Steven Druker has done an excellent job of demonstrating the recklessness of the current practices of genetic engineering in comparison to the established practices of software engineering. His book presents a striking contrast between the two fields, showing how software engineers progressively developed greater awareness of the inherent risks of altering complex information systems – and accordingly developed more rigorous procedures for managing them – while genetic technicians have largely failed to do either, despite the fact that the information systems they alter are far more complex, and far less comprehended, than any human-made system."

— **Ralph Bunker, PhD**

"Steven Druker has written one of the few books I have encountered, in my many years of public interest work, with the capacity to drive major change in a major issue. What Ralph Nader's *Unsafe at Any Speed* was to the auto industry and what Rachel Carson's *Silent Spring* was to synthetic pesticides, *Altered Genes, Twisted Truth* will be to genetically engineered food. It is profoundly penetrating, illuminating, and compelling, and it could stimulate a monumental and beneficial shift in our system of food production.

— **Joan Levin, JD, MPH**

"Druker's brilliant exposé catches the promoters of GE food red-handed: falsifying data, corrupting regulators, lying to Congress. He thoroughly demonstrates how distortions and deceptions have been piled one on top of another, year after year, producing a global industry that teeters on a foundation of fraud and denial. This book is sure to send shockwaves around the world."

— **Jeffrey M. Smith**, international bestselling author of
Seeds of Deception & *Genetic Roulette*

"*Altered Genes, Twisted Truth* reveals how the inception of molecular biotechnology ignited a battle between those committed to scientific accuracy and the public interest and those who saw genetic engineering's commercial potential. Steven Druker's meticulously researched book pieces together the deeply disturbing and tremendously important history of the intertwined science and politics of GMOs. Understanding this ongoing struggle is a key to understanding science in the modern world."

— **Allison Wilson, PhD,** molecular geneticist
Science Director, The Bioscience Resource Project

"*Altered Genes, Twisted Truth* is a remarkable work that may well change the public conversation on one of the most important issues of our day. If the numerous revelations it contains become widely known, the arguments being used to defend genetically engineered foods will be untenable."

— **Frederick Kirschenmann, PhD,** Distinguished Fellow, Leopold Center for Sustainable Agriculture, Iowa State University
Author of *Cultivating an Ecological Conscience*

"Steven Druker's exceptionally well-researched and well-written book elucidates the scientific facts about genetically engineered foods that the PR myths have been obscuring. It provides a unique and invaluable resource not only for concerned citizens, but for historians of science and technology as well. In a comprehensive and skillful manner, it demonstrates how the integrity of science was compromised as a highly influential community of biologists with special interests in genetic engineering muddled scientific truth in order to protect the image of bioengineered foods and to advance their growing partnerships with big business and government. Ultimately, the book reveals that what's at stake here is not only the safety of our food supply, but the future of science.

I am pleased that Steven made good use of the extensive firsthand information I shared about the unsavory behind-the-scenes machinations of biotech promoters in both scientific institutions and government agencies, and I am very impressed with the book as a whole – and expect that a large number of other scientists will be too."

— **Philip Regal, PhD,** Professor Emeritus, College of Biological Sciences, University of Minnesota

Altered Genes, Twisted Truth

How the Venture to Genetically Engineer Our Food Has Subverted Science, Corrupted Government, and Systematically Deceived the Public

STEVEN M. DRUKER

Clear River Press

Altered Genes, Twisted Truth

How the Venture to Genetically Engineer Our Food Has Subverted Science, Corrupted Government, and Systematically Deceived the Public

Steven M. Druker

Copyright © 2015 Steven M. Druker

Foreword © 2015 Jane Goodall

All Rights Reserved.
No part of this book may be used or reproduced in any manner whatsoever without prior written permission from the publisher, except for the inclusion of brief quotations in reviews.

Library of Congress Control Number: 2014951117

ISBN 978-0-9856169-1-5 (Hardcover)

ISBN 978-0-9856169-0-8 (Softcover)

Clear River Press
P.O. Box 520022
Salt Lake City, UT 84152
www.alteredgenestwistedtruth.com

Distributed to the book trade by Chelsea Green Publishing
85 North Main Street, Suite 120
White River Junction, VT 05001
www.chelseagreen.com

Cover design by George Foster
Book design by Lisa DeSpain
Original illustrations by Michael Albertsen

Printed in the USA on partially recycled paper.

Neither the author nor the publisher assumes responsibility for errors in internet addresses or for changes in the addresses after publication. Nor are they responsible for the content of websites they do not own.

Trademark Acknowledgements:
Roundup® Roundup Ready® and YieldGard® are registered trademarks of Monsanto Company.
Prozac® is a registered trademark of Eli Lilly and Company.
Flavr Savr™ is a registered trademark of Calgene and Monsanto Company.

Dedication

To the courageous scientists who have endeavored to uphold truth and scientific integrity regarding the risks of genetic engineering, especially those whose clarity of vision and power of expression inspired a wave of remedial action.

CONTENTS

Note: Appendices C and D and the Executive Summary are available online at:

http://alteredgenestwistedtruth.com/appendix-c/
http://alteredgenestwistedtruth.com/appendix-d/
http://alteredgenestwistedtruth.com/executive-summary/

FOREWORD

JANE GOODALL

I well remember how horrified I felt when I learned that scientists had succeeded in reconfiguring the genetics of plants and animals. The first genetically engineered (GE) plants were created in the 1980s, but I did not hear about them until the 1990s when they were first commercialized. It seemed a shocking corruption of the life forms of the planet, and it was not surprising that many people were as appalled as I was – and that these altered organisms became known as 'Frankenfoods'.

In fact, there were good science-based reasons to mistrust the new foods; yet GE crops have spread throughout North America and several other parts of the world. How has this come about? The answer to that question is to be found in Steven Druker's meticulously researched book. Several years in the making, it is a fascinating, if chilling story.

I did not realize what a formidable task the bioengineers faced as they struggled to introduce new genes into a variety of agricultural crops. Their intent was to make them produce toxins that would deter insect pests, or enable them to resist herbicides, and so on. A major challenge was the need to overcome the various defensive mechanisms of the plants themselves, which did their best to repel the alien material. Another was to compel the foreign genes to function in a cellular environment where they would ordinarily remain dormant. It is a testament to human persistence and ingenuity that the scientists finally succeeded!

But the reconfigured plants they eventually created were, as Druker explains in engaging detail, different in a variety of ways from their parents; and from the outset many qualified scientists expressed concerns about the safety of the new crops for both the environment and human and animal health. He further demonstrates that this very real difference between GE plants and their conventional counterparts is one of the basic truths that biotech proponents have endeavored to obscure. As part of the process, they portrayed the various concerns as merely the ignorant opinions of misinformed individuals – and derided them as not only unscientific, but anti-science. They then set to work to convince the public and government officials, through the dissemination of false information, that there was an overwhelming expert consensus, based on solid evidence, that the new foods were safe. Yet this, as Druker points out, was clearly not true.

As the chapters progress, we read how the advocates of genetic engineering have steadfastly maintained that the crops created by this radical technology are essentially similar to those from which they have been derived, that the process is splendidly exact, and that GE foods, therefore, are if anything *safer* than their traditionally bred 'parents' – when in fact, there's significant dissimilarity, the process is far from exact, and the risks are greater, especially the risk of creating unexpected toxins that are difficult to detect.

Druker describes how amazingly successful the biotech lobby has been – and the extent to which the general public and government decision-makers have been hoodwinked by the clever and methodical twisting of the facts and the propagation of many myths. Moreover, it appears that a number of respected scientific institutions, as well as many eminent scientists, were complicit in this relentless spreading of disinformation.

Chapter 5 shows how the key step in the commercialization of GE foods occurred through the unbelievably poor judgment – if not downright corruption – of the US Food and Drug Administration (the FDA). This regulatory body is supposed to ensure that new additives to foods are safe before they come to market, and it had a responsibility to require that GE foods were proven safe through standard scientific testing. But the information that Druker pried from the agency's files through a lawsuit revealed that it apparently ignored (and covered up) the concerns of its own scientists and then violated a federal statute and its own regulations by permitting GE foods to be marketed without any testing whatsoever. The evidence further shows how the agency assured consumers that GE foods are just as safe as naturally produced ones – and that their safety has been confirmed by solid scientific evidence – despite the fact it knew that no such evidence existed.

Druker makes the case that it was this fraud that truly enabled the GE food venture to take off. And he asserts that the fraud continues to deceive the public and Congress, despite the fact that the lawsuit he initiated thoroughly exposed it. His description of the proceedings surrounding this lawsuit was, to me, one of the most astounding and chilling parts of the book.

And what of the role of the media? How have the American public been so largely kept in the dark about the realities of GE foods – to the extent that until quite recently, a vast majority of the populace did not even know they were regularly consuming them? Druker describes, in Chapter 8, how the mainstream media have been highly selective in what they report – and have consistently failed to convey information that would cause concern about these engineered products. Moreover, Druker demonstrates that the policies imposed by the media magnates have been, in his words, "not merely selective, but suppressive." And he relates several dramatic incidents in which journalists who tried to bring unsettling facts to light had their stories altered or totally quashed by higher level executives. So it is not surprising that the American public, and a good many key decision-makers, believe that there are no legitimate concerns regarding GE foods.

I am personally grateful to Steven Druker for writing this book. It has been a monumental task and reflects the passionate desire of a man with a true scientific spirit to reveal, as precisely as possible, the truth behind the misrepresentations of the truth. Nonetheless, despite its integrity, *Altered Genes, Twisted Truth* can be expected to meet fierce criticism from those who promote the GE food venture; and, like all who attempt to disclose the venture's underside, its author will probably be attacked and branded as anti-science and anti-progress. BUT it seems to me that it is not those who point to the problems of the venture who are anti-science: it is quite the other way around. Nevertheless, Druker will almost surely be subjected to the same sort of criticisms as those leveled against Rachel Carson when she published *Silent Spring* in 1962.

I think it is important that you read this book carefully, assessing for yourself how firmly it is grounded in fact and logic. You may well come to the same conclusion as I have: that Steven Druker is upholding the tradition of good science. Then read some of the books and articles written by pro-GE scientists – especially some of those by prominent biologists – and you may well decide that their standards often fall significantly short of his.

In fact, he points out several instances in which it appears that such publications are downright deceptive, not only portraying genetic engineering in a misleading manner, but even misrepresenting some basic features of biology. Further, although these scientists may genuinely believe that GE foods are the solution for world hunger, it appears that many of them have vastly overestimated the benefits of these foods – and that even *if* these products did *not* entail higher risks, it's doubtful they could significantly reduce malnutrition or solve any major problems of agriculture.

Although this book tells a story that's in many ways distressing, it's important that it has finally been told because so much confusion has been spread and so many important decision-makers have apparently been deluded. Fortunately, the final chapter shows how the story can have a happy ending, and it clearly points the way toward realistic and sustainable solutions that do not involve genetic engineering. Thus, just as my own books aim to instill hope, this book is ultimately a hope-inspiring one too. For it describes not only some of the mistakes that we have made but how they can be rectified in creative and life-supporting ways.

Druker has, without doubt, written one of the most important books of the last 50 years; and I shall urge everyone I know, who cares about life on earth, and the future of their children, and children's children, to read it. It will go a long way toward dispelling the confusion and delusion that has been created regarding the genetic engineering process and the foods it produces.

To me, Steven Druker is a hero. He deserves at least a Nobel Prize.

– Jane Goodall, PhD, DBE and UN Messenger of Peace

INTRODUCTION

How I Reluctantly Became an Activist

– And Uncovered the Crime that Enabled the Commercialization of Genetically Engineered Foods

Most people would be surprised to learn that Bill Clinton, Bill Gates, and Barack Obama (along with a host of other astute and influential individuals) were all taken in by the same elaborate fraud.

They'd be even more surprised to learn that it was not perpetrated by a foreign intelligence agency, an international crime syndicate, or a cabal of cunning financiers but by a network of distinguished scientists – and that it did not involve change in the climate but changes to our food.

And, if they're Americans, they would be shocked to discover that the US Food and Drug Administration has been a major accomplice – and that because of its deceptions, for more than fifteen years they and their children have been regularly ingesting a group of novel products that the agency's scientific staff had previously determined to be unduly hazardous to human health.

This book tells the fascinating and frequently astounding story of how such a remarkable state of affairs has come to be; and I'm uniquely positioned to tell it, because I uncovered one of its key components.

In early 1996, I did something few Americans were then doing: I decided to learn the facts about the massive venture to restructure the genetic core of the world's food supply. And the more I learned, the more I became concerned. It grew increasingly clear that the claims made in support of genetically engineered foods were substantially at odds with the truth – and that there were strong scientific grounds for viewing such products with a cautious eye.

Of special concern was the behavior of the Food and Drug Administration (FDA), which has refused to regulate genetically engineered foods and instead has energetically promoted them.[1] I found it problematic this agency had adopted a presumption that genetically engineered (GE) foods are as safe as natural ones and was allowing them to be marketed not only without testing but even without labels to inform consumers about the genetic reconfiguration that had occurred. I believed this was unscientific, irresponsible, and fundamentally wrong.

I also had a hunch it was illegal – a hunch my research eventually confirmed.

As my knowledge grew, there also grew a conviction that a lawsuit should be brought against the FDA to overturn its policy on GE foods and compel it to require the safety testing and labeling that consumers were being wrongfully denied. At that point, I didn't envision playing an active role in the legal proceedings or even getting extensively involved in the developmental phase of the suit. My intention was to present the idea to others who had greater expertise and resources and inspire them to carry it out. Although I have a law degree from the University of California at Berkeley, practicing law has not been the central focus of my professional life, and I had scant experience in litigation. Further, I was immersed in a project that was dear to my heart and didn't want to get sidetracked.

Yet, in the process of trying to inspire others to do the lawsuit, I gradually became the main person organizing it and driving it forward. The executives of public interest organizations with whom I spoke all thought the suit was a great idea, but none felt ready to take it on. After some weeks of attempting to find an organization that would shoulder the suit, I discussed the situation with a molecular biologist who was concerned that in the push for rapid commercialization of GE foods, the risks were being unduly discounted and testing irresponsibly neglected. As I explained how my ideas for the lawsuit had been uniformly greeted with enthusiasm but that none of the groups was prepared to turn them into reality, he said: "Steve, don't you realize this is your baby? If you don't do it, it's not going to happen." Much as I desired to have someone else do the suit so I could get back to my other project, and much as I wanted to reject his assessment, deep down I had an inescapable feeling he was right.

So I set my project aside, founded the Alliance for Bio-Integrity (a nonprofit public interest organization), and as its executive director, devoted myself full-time to organizing the lawsuit. In a few months, I gained the collaboration of the International Center for Technology Assessment, a respected public interest organization in Washington, D.C. with a skilled team of lawyers. They had substantial experience in litigation with federal administrative agencies, and they agreed to be the attorneys of record, on the condition that I would continue to coordinate the various elements of the project and to raise the necessary finances. In time, I also became actively involved as an attorney, undertaking key research and contributing to the briefs and other documents filed with the court.

During the preparation phase, a primary goal was to attain an impressive set of plaintiffs. Over the following months, through numerous phone calls, emails, and journeys to personal meetings, I assembled an unprecedented coalition to join the suit and sign the complaint against the FDA that was submitted to the court. For the first time in US history, a group of scientific experts became involved in a lawsuit challenging the policy of a federal administrative agency, not as advisers or expert witnesses, but as plaintiffs – plaintiffs who formally objected to the policy on scientific grounds. In a bold move highlighting the unsoundness of that policy, nine well-credentialed life scientists (including tenured professors at UC Berkeley, Rutgers, the University of Minnesota, and the NYU School of Medicine) stepped up to sue the FDA and formally assert that its presumption about the safety of GE foods is scientifically flawed because they pose abnormal risks that must be screened by rigorous testing.

Equally unparalleled, they were co-plaintiffs with a distinguished group of spiritual leaders from diverse faiths who objected to the FDA's policy on religious grounds. Within this group were the President of the North American Coalition on Religion and Ecology, the chaplain at Northeastern University, and a lecturer in theology at Georgetown University. In all, there were seven ordained priests and ministers from a broad range of Christian denominations (including Episcopalian, Lutheran, Baptist, and Roman Catholic); three rabbis (Orthodox, Conservative, and Reform); the chancellor of the Americas Dharma Realm Buddhist University; and a thousand-member Hindu organization from Chicago. These plaintiffs stated that in their view, the manner in which biotechnicians are reconfiguring the genomes of food-yielding organisms is a radical and irreverent disruption of the integrity of God's creation – and that they felt obliged to avoid consuming the products of such interventions as a matter of religious principle. They alleged that by failing to require proper labeling, the FDA was unavoidably exposing them to these foods and preventing them from the free exercise of their religious beliefs. (Some of the religious-based reasons for rejecting GE foods are more fully described in Chapter 14.)

Although proponents of GE foods attempt to portray any religiously motivated opposition as due to ignorance about the facts of genetic engineering and a resultant failure to appreciate its similarity to traditional breeding, these plaintiffs *were* well-informed; and they therefore understood how deeply it does differ from natural processes. (These differences are thoroughly discussed in Chapter 4).

Alliance for Bio-Integrity, et al. v. Shalala, et al. was filed in US District Court in Washington, D.C. in May 1998. The first named defendant was Donna Shalala because, as the Secretary of the Department of Health and Human Services at that time, she oversaw the FDA, which is one of the agencies within that department. The acting commissioner of the FDA was the other defendant.

The suit quickly achieved a major effect because, as part of the discovery process, it forced the FDA to hand over copies of all its internal files on GE foods. Eager to delve beneath the agency's public pronouncements and see if they jibed with what it really knew and how it had actually operated, I assumed responsibility for analyzing this trove of documents. As I combed through the more than 44,000 pages of reports, messages, and memoranda, I made several startling discoveries. By the time my investigation was finished, I had compiled extensive evidence of an enormous ongoing fraud. It revealed that the FDA had ushered these controversial products onto the market by evading the standards of science, deliberately breaking the law, and seriously misrepresenting the facts – and that the American people were being regularly (and unknowingly) subjected to novel foods that were abnormally risky in the eyes of the agency's own scientists.

This fraud has been the pivotal event in the commercialization of genetically engineered foods. Not only did it enable their marketing and acceptance in the United States, it set the stage for their sale in numerous other nations as well. If the FDA had not evaded the food safety laws, every GE food would have been required to undergo rigorous long-term testing; and if it had not covered up the concerns of its scientists and falsely reported the facts, the public would have been alerted to the risks. Consequently, the introduction of GE foods would at minimum have been delayed many years – and most likely would never have happened.

So it's vital that the story of the FDA's crime be fully told; and this book does so in a comprehensive and vivid manner, disclosing how a government agency with the duty to safeguard the nation's food supply was induced to perpetrate such a fraud, how the fraud was carried off, and how, even after being exposed and conclusively documented, it has maintained its strength and continued to deceive the public.

Moreover, in fully telling this story, the book relates a much bigger one, a story in which the FDA's behavior does not stand as an isolated aberration but forms an integral part of a broader pattern of misconduct. It presents a graphic account of how the genetic engineering venture arose, the stages through which it has advanced, and how, at every stage, the advancement relied upon the sustained dissemination of falsehoods. In line with its title, it demonstrates that the broad-scale altering of genes has been chronically and crucially dependent on the wholesale twisting of truth – and shows how for more than thirty years, hundreds (if not thousands) of biotech advocates within scientific institutions, government bureaus, and corporate offices throughout the world have systematically compromised science and contorted the facts in order to foster the growth of genetic engineering, and get the foods it produces onto our dinner plates.

Thus, the narrative that unfolds in the following pages is fundamentally a story about the corruption of science and its concomitant corruption of government, not through the machinations of a scientific fringe group in league with a pack of powerful political ideologues, but through the workings of the mainstream scientific establishment in concert with large multi-national corporations – and their co-optation of government officials across the political spectrum, and across the globe. Further, by the time the story ends, it will be clear that the degradation of science it depicts has not only been unsavory but unprecedented: that in no other instance have so many scientists so seriously subverted the standards they were trained to uphold, misled so many people, and imposed such magnitude of risk on both human health and the health of the environment.

A variety of documents (including transcripts of scientific conferences, statements by government agencies, newspaper reports, journal articles, and books by historians of science) collectively chronicle the bioengineering venture. Together, they amply illumine its underside, revealing how the integrity of science and the integrity of government have both been routinely sacrificed so the enterprise could advance. I have drawn deeply from these resources, often crystallizing key facts that were not widely known. Additionally, because I was engaged in the campaign to properly regulate GE foods for many years on several continents (meeting a broad range of government officials, interacting with scientists and journalists, and participating in conferences and debates), I have repeatedly witnessed the corrosive processes firsthand; and the narrative has been enhanced by a number of these experiences.

Further, many striking accounts of the corrosion were imparted by scientists who have striven to stop it. One of the foremost is the eminent biologist Philip Regal, who for twenty years spear-headed the endeavor to get the genetic engineering enterprise aligned with solid science and tempered by responsible regulation. His story, which forms part of several subsequent chapters, illustrates the diverse and often shocking ways in which the scientific establishment and the government consistently frustrated this endeavor – to the extent he became convinced that when dealing with GE foods, the US executive branch would not honor science and the law unless compelled by a court, and so decided to become a plaintiff in the lawsuit I organized. By sharing his insights and experiences with me over the course of many personal meetings, phone conversations, and emails, and by giving me the extensive set of recollections he had recorded, he has enabled me to expose the infirmities and delinquencies of the bioengineering venture in a much richer way than would otherwise have been possible.

Like Dr. Regal, a growing number of experts have recognized that this enormous venture rests on shaky assumptions and relies on questionable claims – and that increased creativity is required to chart the best way forward. Among them is Evelyn Fox Keller, a distinguished professor of the history and philosophy of science at the Massachusetts Institute of Technology. In her book, *The Century of the Gene*, she notes that the apparent efficacy of genetic engineering provides no assurance that it's free from unintended harmful effects.[2] She further points out that with the rise of this technology, an "unprecedented" bond has grown between science and commerce – and that as this bond has tightened, scientists have become increasingly invested in the rhetorical power of a persuasive mode of "gene talk" that imputes a precision and predictability to bioengineering that it does not possess.[3] Keller emphasizes that the "shortcomings" of such gene talk necessitate its transformation.[4] Her book concludes with the hope ". . . that new concepts can open innovative ground where scientists and lay persons can think and act together to develop policy that is both politically and scientifically realistic."[5]

The following chapters aim to help clear the way to such innovative ground by revealing that the most scientifically realistic policy can easily coincide with the most politically realistic one – and that it's only because the politics of genetic engineering became detached from the scientific realities that the current problems we face were allowed to arise. It's my hope that the information they contain and the insights they convey will end the confusion that has caused the split and speed the implementation of needed reforms, the reinstatement of scientific standards, and the growth of an agricultural system that yields abundant wholesome food in a safe and sustainable manner.

Ways to Enhance Your Enjoyment of this Book: Utilizing the Executive Summary and Easily Accessing the Endnotes

I've endeavored to make this book a good story and have employed a narrative style as much as feasible. But because the story is about science – and the corruption of science by many of its practitioners – it was necessary to explain many technical facts and examine some rather complex scientific issues. And because I've aimed to produce a book that's not only accessible and enjoyable for the general reader but also serves as a reliable and comprehensive resource for experts, some chapters discuss a substantial amount of information. Many readers will find these discussions stimulating and will appreciate their depth; but others may, at some stage in one of the longer chapters, develop a desire to simply get the gist of the remainder and move on to the next chapter.

In the event such a feeling arises, you can skip to the Executive Summary and read that chapter's main points. (It can be downloaded at: http://alteredgenestwistedtruth.com/executive-summary/) You can also look at a chapter's summary after you've completed it in order to crystallize the basic facts. And even if you read the entire book without glancing at the summary, you may then wish to read it to gain a holistic overview and solidify your understanding.

Of course, some individuals with limited time may prefer to read the Executive Summary first and later read the entire book (or selected chapters) to gain more detailed knowledge.

However, I don't encourage this, because if you read it first, it might spoil the experience that can be gained by allowing the story to unfold chapter by chapter. Several of those who reviewed the book have remarked that it's engaging and often imbued with drama, and some have described it as a "page-turner." But the drama could be dampened by reading a summary of each chapter ahead of time.

So, if you intend to read the entire book, I advise that you initially ignore the Executive Summary. Further, if you want to examine the issues even *more* thoroughly than is done in the main text, you will find that many significant points are discussed in greater depth in the appendices and the endnotes – which leads to an important note about these notes.

For those of you reading the e-book version, hopping to an endnote and returning to the text is simple. But if you're reading the printed book, it would ordinarily be a lot more complicated and time consuming. So to make the endnotes more readily accessible in this situation, they're located not only at the end of the physical book but also online at http://alteredgenestwistedtruth.com/endnotes/. That way, you can download the endnote section and either print it or store it on your computer, tablet, or e-reader. Then, as you read a chapter, you can have a copy of its endnotes nearby and easily transition between the two.

Further, so you won't need to travel back and forth between the notes and a bibliography that contains the full references for the sources that are cited, when a source is cited in a chapter's note section for the first time, it will be fully referenced (even if it's already been fully referenced in the notes for an earlier chapter). Then, subsequent citations of that source will indicate at what preceding note within that section the full reference can be found.

A Note Regarding Terminology

The term "biotechnology" is sometimes broadly employed to refer to all techniques that utilize (or modify) biological processes, including ancient practices that rely on fermentation such as making wine, brewing beer, and leavening bread. But the term can also be used in a narrower sense, to refer exclusively to modern techniques, such as genetic engineering, that depend on highly artificial interventions and that have no established history of safe use. In this book, I employ the terms "biotechnology" and "biotech" in their restricted sense to denote only this latter group of techniques that have not stood the test of time.

Further, because instances of "misrepresentation," "misstatement," "misinformation," "inaccuracy," and "falsehood" can occur through ignorance of the truth, and none of the terms necessarily denotes an intent to deceive, I do not use them to imply that one existed – even though it may have. Instead, I reserve the words "fraud," "lie," "deception," and "disinformation" to denote deceit. Moreover, when I refer to a fraud, deception, or disinformation campaign that was propagated by many individuals, I do not imply that every person who in some way abetted it has been guilty of deception – merely that some have. Furthermore, due to the difficulty of discerning who spoke from ignorance and who did not, unless I specifically assign guilt, it should not be assumed that anyone in particular has been accused.

CHAPTER ONE

THE POLITICIZATION OF SCIENCE

– And the Institutionalization of Illusion

As he returned the phone to its cradle, Philip Regal knew that his scientific career was about to enter an important and distinctly challenging phase. Ernst Mayr had just urged him to assume a crucial role in connection with the most profound technological revolution since the splitting of the atom.

Mayr was a towering figure in the life sciences. Numerous colleagues, including several of his fellow Harvard professors, considered him to be the greatest biologist of the 20th century, and he was widely regarded as the most influential theorist in the field since Darwin.[1]

For several weeks during that year of 1983, he and Regal had been engaged in a series of discussions via phone and mail about the unprecedented power of genetic engineering and the pressing need to manage it wisely. But this conversation had taken a new turn. Besides endorsing Regal's concerns about the deficiencies in the way the venture was being conducted and the damage that might result from pushing ahead absent adequate knowledge, Mayr asked him to do something about it. He encouraged him to take the lead in organizing a concerted endeavor to induce change and ensure that genetic engineering would be deployed in accord with sound scientific principles – and that the novel organisms it produces would not be released into the environment without sufficient forethought. He counseled him to continue his risk analyses, to stimulate similar assessments by others, and to foster a dialogue within the scientific community that would engender fuller understanding of this technology and a more responsible manner of employing it. Mayr believed that unless there was such deliberation and dialogue, life scientists, the biotechnology industry, and government regulators would not be prepared to intelligently handle the new potencies that had been brought within human grasp.

Yet, even as Mayr urged Regal ahead, he warned him to proceed with caution. He reminded him that the biotech industry and its allies in the molecular biology establishment wielded great economic, academic, and political power – and noted that any attempts to subject their projects to thorough scientific scrutiny would be regarded not only as unnecessary impediments to progress but as major provocations. Then, his voice growing more solemn, Mayr spoke words that still resonate in Regal's memory: "They will try to crush you." Accordingly, he advised Regal that although his credentials were excellent and he was well-respected, he should not go it alone and should get other respected biologists to join him.

Mayr's words were compelling, and despite the difficulties that would be entailed, Regal resolved to undertake the task. But what he didn't realize at that time was just how formidable a task it would turn out to be – and how massive would be the resistance, not only within the confines of the biotechnology industry, but within the corridors of government and the halls of academia as well. Nor did he foresee that over the next three decades, the resistance would in large part prevail.

Regal's concerns about genetic engineering were first aroused in the early 1980's when word spread among life scientists that all its practices and products were soon to be fully deregulated. Because for several years the proponents of this revolutionary technology had been promising that it *would* be carefully regulated, he was surprised at this news – and equally surprised at how many biologists were elated by it. At the University of Minnesota, where Regal was a professor in the College of Biological Sciences, the college's dean enthusiastically announced that the molecular biologists in the National Institutes of Health and the National Academy of Sciences, along with key officials in government, had decided that genetic engineering was safe and were going to give unconditional approval to all its applications.

But Regal did not share the enthusiasm – nor, as he was to learn, did numerous other scientists. For one thing, he found it strange that genetic engineering was being treated as a process that could be considered safe in itself irrespective of the diverse uses to which it was put – and that its proponents assumed this inherent quality of safety would then automatically adhere to all its various products. This approach struck him as fundamentally flawed, because these products could be enormously different from one another in many biologically important ways.

Genetic engineering (technically termed "recombinant DNA technology" and also referred to as "bioengineering" and "gene-splicing"[2]) comprises a set of novel and powerful procedures that restructure the genomes of living organisms by moving, splicing, and otherwise re-arranging pieces of DNA in ways that were formerly impossible. Through it, a wide range of outcomes can arise. It can endow an organism with extra copies of some of its own genes, reconfigure the sequences of some of its genes, and re-program the ways in which its genes are turned on or off, or transplant genes from a distinct and distant species into its genetic program. Further, it can transform any kind of organism, whether a bacterium, a plant, or an animal; and each transformation could give rise to a unique set of effects (both intended and unintended) depending on the organism involved, the genetic alterations performed, their location on the DNA molecule, and the environment in which the organism is placed. Therefore, Regal regarded the claim that genetic engineering would always be safe to be just as bizarre as the claim that all art would be non-offensive.

Yet, molecular biologists promoted this claim as scientifically sound; and most were so sure of it that they shunned discussing the issue with any scientists who disagreed, even if those scientists possessed greater expertise in some relevant areas of knowledge. Nor were they prepared to consider whether their own expertise was broad enough to adequately manage all the facets of genetic engineering.

Regal had first encountered this insular attitude while serving on a committee at the University of Minnesota that reviewed graduate degree programs. To keep the university apace with the latest developments in biotechnology, a new graduate curriculum in microbial engineering had been proposed. As was typical of such programs at other universities, the course work largely consisted of chemistry, biochemistry, molecular genetics, and some physiology. During the committee's discussion of the proposal, Regal expressed the opinion that the students should also study ecology, biological adaptation, and population genetics (fields in which he had expertise) so they could better comprehend the full dynamics of genetically engineered organisms. He emphasized that without such expansion of the curriculum, the graduates would only know how some of the microscopic components of these new organisms functioned in isolated biochemical pathways but would not be able to understand how they functioned as wholes, especially in relation to other organisms. He pointed out that because biotechnicians were planning to release their creations into the environment, it was important that they be able to assess how these living entities would interact within ecosystems.

But his input provoked an indignant response from the promoters of genetic engineering, who flatly asserted that broader training was not necessary because gene-splicing would be invariably safe. They further maintained that genetic engineering was an intensely competitive field, that no universities required budding practitioners to "waste time" in studying the topics Regal had suggested, and that if the University of Minnesota did impose such an extraneous burden it could not keep up with the other schools.

Regal was both stunned and stirred by these statements. As he later wrote:

> I went away from that meeting walking slowly across campus, eyes on the pavement, pondering the flock of serious questions that had been roused in my thinking. How could people whose expertise was limited to chemistry be so sure that radical modifications of complex biological organisms living on farms or within broader populations in nature would necessarily be safe and effective? How could they be so certain? The promoters of genetic engineering at that committee hearing had not the slightest scientific credentials for estimating ecological adaptations and disturbances. It was not simply that they did not have degrees or had not taken courses. One can certainly be self-taught. But they had no credible knowledge. Yet, they were claiming they did not need to acquire any additional understanding or seek advice from experts beyond the bounds of their narrow training – and that no other molecular biologists recognized such a need either. This was an astonishing prospect for me to contemplate at the time, but it turned out this was indeed the prevailing attitude among molecular biologists the world over.[3]

Regal believed it was highly misleading for scientists whose expertise was restricted to molecular biology to present themselves as fully qualified to estimate the ecological effects of genetically engineered organisms. In his mind, it was like someone who knows the details involved in the printing of dollar bills purporting to be an expert forecaster on how the dollar will be valued against the euro and the yen, despite the fact his technical knowledge of dollars was limited to the realm of engraving plates, inks, and printing presses and he had no training or meaningful experience in economics and the intricacies of international currency markets. Nonetheless, the categorical claims of the molecular biologists would increasingly be accepted as authoritative, and would powerfully shape government policy.

Given the boldness of their pronouncements, someone hearing the molecular biologists in 1983 for the first time would have been surprised to learn that they had not always exuded such unqualified confidence in the safety of genetic engineering – and had even called for major precautions. But that was a decade earlier, when the technology was a startling new phenomenon and they openly acknowledged their limited ability to predict and control its effects. The story of how their initial message mutated, and their influence concurrently expanded, provides a striking example of the politicization of science – and the minimization of the role of evidence in setting public policy that's supposed to be science-based.

The Advent of an Astonishing Technology

In 1969, the attention of people throughout the world was riveted on a novel pathogenic microorganism that threatened global devastation of human life. Nothing like it had ever been encountered and the most sophisticated control strategies were being foiled by its awesome destructive capacity. What's more, this malevolent microbe was the product of human invention.

But the invention was purely literary, and the ominous entity came to life only within the pages of a book – Michael Crichton's best-selling science-fiction thriller, *The Andromeda Strain*. And although in this story the deadly organism makes its appearance through the efforts of the US Army to obtain biological weapons, it has not been created by scientists. That's because Crichton aimed for realism; and at that time, it would have been fanciful to portray this novel creature as the product of human engineering. Since DNA was still largely unmanageable, a technology that could precisely copy genes and then splice them into living organisms was well beyond the realm of what was practically achievable. Consequently, it seemed more plausible that ultra-lethal (and completely novel) microbes would be found beyond earth's atmosphere than formed within its laboratories; and Crichton crafted a plot in which the army sends satellite probes into space to collect pathogens for the bioweapons program. In this scenario, the new microbial menace arrives in a satellite that crashes to earth instead of emerging from a terrestrially-bound test tube.

After genetic engineering had become a reality, Crichton seized on it and made it an essential feature of *Jurassic Park*, the best-seller he published in 1990. But when he began to write *The Andromeda Strain*, even though scientists had detailed knowledge about the structure of DNA and the nature of the genetic code, they were far from the stage of controlled gene-splicing; and while there was a buzz about the possibility of "genetic engineering" among biologists who believed that the means for such a radical technology would eventually be developed, it appeared that no one was anywhere close to doing so.[4]

Yet, as improbable as it might have seemed when *The Andromeda Strain* first hit the bookstores in 1969, earth-based laboratories would soon supplant plummeting space probes as the most likely point of entry for perilous new microbes. The next year, scientists finally discovered the means by which the DNA molecule could be cut with precision; and within four more, a team of researchers succeeded in copying a gene from one organism and splicing it into the DNA of another, creating the first genetically engineered bacterium.[5] (The steps of this process are described in Chapter 4.)

Soon, dozens of other new microbial strains had been similarly produced. And, although these novel organisms were created on earth, in the minds of many people, they were almost as alien as if they'd come from outer space. Not only did they contain unprecedented combinations of genetic material, it was highly unlikely that most of these conglomerates could have arisen under natural conditions. Instead, they owed their existence to extensive human contrivance. Further, regardless of the degree to which people considered them alien, a large part of the public feared that some of these creatures might prove to be nearly as dangerous as the unearthly terror portrayed in Crichton's book. Moreover, they were not alone in their apprehension. It was to a significant extent shared by the life science community. In fact, the concerns of the public were sparked by warnings that had issued from the mouths and pens of molecular biologists.

Scientists Sound the Alarm

In the early phase of the recombinant DNA revolution, several molecular biologists became struck by the enormity of the new powers with which they'd suddenly been endowed – and deeply concerned about their capacity to cause widespread harm. It seemed that unless this technology was managed very carefully, even the best-intentioned researchers could produce a high degree of accidental damage.

One of the first scientists to apprehend the danger, and voice concern, was Robert Pollack, who was running a laboratory at Cold Spring Harbor, Long Island that was directed by the Nobel laureate James Watson, a co-discoverer of the structure of DNA. In the summer of 1971, he learned that the Stanford biochemist, Paul Berg, was planning to construct a piece of recombinant DNA that contained a gene from a virus that can induce malignant tumors in monkeys, rodents, and humans.[6] And the gene Berg was going to employ was the gene that causes the tumors. What's more, he intended to insert that recombinant segment into the DNA of another virus that infects a bacterial species (named *E. coli*) that abundantly inhabits the intestines of humans and many other animals. Although Berg hoped to gain valuable knowledge from such an experiment, and had not intended to put the virus into the bacteria, Pollack was concerned that such an incursion could inadvertently happen, transforming an ordinarily friendly occupant of our gut into an agent of disease. This would in turn create a risk that such radically transformed microbes might escape the lab, widely infect the intestines of people and livestock, and cause a lot of cancer.[7]

So he called Berg, explained his concerns, and asked if he had also been troubled by such considerations. Berg said that he hadn't; but Pollack's call got him thinking. As Berg later recounted, "I began to ask myself if there was a small possibility of risk. And if there is, do I want to do the experiment?"[8] He then consulted another scientist, who told him there was potential for harm and that he would have to accept responsibility for any mishaps. According to Berg, "At that point I stepped back and asked, 'Do I want to go ahead and do experiments which could have catastrophic consequences, no matter how slim the likelihood?'"[9] He decided that he didn't; and the experiment was placed on hold.

He also decided it was important to initiate a dialogue within the scientific community so that the potential problems would be appreciated and adequate safeguards employed. And so did a number of other biologists.

One of these discussions occurred at the Gordon Research Conference on Nucleic Acids in June 1973. It resulted in a letter that appeared in the September 21, 1973 issue of the influential journal *Science* cautioning that the new ability to transfer genetic sequences between organisms was "a matter of deep concern" – and that "[c]ertain . . . hybrid molecules may prove hazardous to laboratory workers and to the public."[10] Airing this concern in such a prominent forum was a bold step, and one of the editors of *Science* reportedly questioned the wisdom of doing so.[11] Many conference participants also had reservations about going public, and the resolution to publish the letter only passed by a six-vote margin (48 to 42).[12]

Soon thereafter, the National Academy of Sciences established a committee on recombinant DNA (rDNA), which issued a letter that went much farther than its forerunner by urging scientists to refrain from specific types of genetic engineering "... until the potential hazards of such recombinant DNA molecules have been better evaluated or until adequate methods are developed to prevent their spread. . . ."[13] This letter became known as "the Berg letter" because its lead signatory was Paul Berg, who was the driving force behind the committee's creation and the letter's production. Like its predecessor, the letter was published in *Science*; and it spurred even greater repercussions. It was unprecedented for a group of scientists to voluntarily restrict their research and call on their colleagues to do the same. Not only did it show an admirable level of social responsibility, it revealed the formidable uncertainties that surrounded genetic engineering – and legitimized concerns about them.

The Berg letter asked the National Institutes of Health (NIH) to establish research guidelines and oversee experimentation. It also sought involvement of the broader rDNA research community, so it proposed an international meeting to "discuss appropriate ways to deal with potential biohazards of recombinant DNA molecules."[14]

Restricting the Release of Engineered Organisms

Both recommendations soon bore fruit. On October 7, 1974 the NIH established an advisory panel (eventually named the Recombinant DNA Advisory Committee [RAC]) which played a significant role in policy formation over many years. And the following February an international meeting of over a hundred researchers convened at the Asilomar Conference Center in Monterey, California. Its main focus was on formulating guidelines that were sufficiently rigorous to prevent catastrophes yet liberal enough so biologists could end their broad moratorium and get on with research. As an article in *Science* described the outcome: "After much haggling, the group settled on a set of safety guidelines that involved working with disabled bacteria that could not survive outside the lab. The guidelines not only allowed the research to resume but also helped persuade Congress that legislative restrictions were not needed – that scientists could govern themselves."[15]

In reaching their decisions, the molecular biologists did not seek input from other perspectives, and no avenues were provided for public interest groups to participate. Further, it's clear this was not an oversight but an essential aspect of policy – a policy to restrict those outside the molecular biologists' fold from influencing the ways in which rDNA research was conducted and applied. James Watson unabashedly acknowledged that he and his colleagues at Asilomar embraced such an exclusionary policy: "Although some fringe groups . . . thought this was a matter to be debated by all and sundry, it was never the intention of those who might be called the molecular biology establishment to take the issue to the general public to decide. We did not want our experiments to be blocked by over-confident lawyers, much less by self-appointed bioethicists with no inherent knowledge of, or interest in, our work. Their decisions could only be arbitrary."[16] In the words of Susan Wright, a historian of science at the University of Michigan who is an authority on bioengineering's first decade: "[P]olicy-making decisions were claimed to be the right and responsibility of scientists alone."[17]

Accordingly, most of the molecular biologists expected the self-imposed research restrictions to assuage public concerns and allow them to maintain exclusive control over the ways in which the genetic engineering enterprise would develop. Watson has written that as they departed Asilomar, they were "as exhilarated as they were exhausted" because "[h]aving demonstrated their integrity, they naively believed that they would now be free of outside intervention, supervision, and bureaucracy."[18]

However, contrary to the expectations of its practitioners, rDNA research did not stay free from government supervision. The day after the Asilomar conference ended, planning began for NIH research guidelines; and the initial set was issued on June 23, 1976. Despite the absence of legal penalties for violating them, there were constraints, because they applied to any organization receiving NIH funds – and they were eventually extended through presidential order to encompass all federally funded research. So funding could be curtailed if a project ignored them. Further, the NIH guidelines went beyond those agreed upon at Asilomar and banned the deliberate release into the environment of *any* organism containing recombinant DNA.

Uneasy Equilibrium

Because the open airing of concerns had stirred widespread anxiety, the ban on releasing gene-spliced organisms was necessary to calm the public enough so that laboratory research with rDNA technology could move ahead. But many scientists grew dissatisfied with the restrictions and regretted the readiness with which early apprehensions were publicized. It had become clear that bioengineering was a highly volatile issue and that any misgivings expressed by its practitioners would be seized upon by the media. Already, headlines had appeared proclaiming: "Genetic Scientists Seek Ban – World Health Peril Feared" (*Philadelphia Bulletin*), "Scientists Fear Release of Bacteria" (*Los Angeles Times*), and "A New Fear: Building

Vicious Germs" (*Washington Star News*).[19] Even the staid *Atlantic Monthly* published an article entitled "Science that Frightens Scientists."[20] Such reports significantly unsettled the citizenry.

Not only were a large number of molecular biologists disappointed by this outcome, as one observer notes, most "felt betrayed."[21] Although they had hoped their self-imposed ban would convince the public that they could be trusted to manage this new technology without government supervision, it instead had fanned public fears and induced the imposition of such supervision. Further, during 1976 more than a dozen bills were introduced in Congress to regulate rDNA research.[22] And one, initiated by Senator Edward Kennedy, called for regulation by a presidential commission.[23]

As the effort to impose restrictions gained momentum, American molecular biologists worried they would fall behind scientists in countries where research was unregulated – and that the US would lose its lead in the field.[24] Accordingly, many publicly disavowed their former precautionary stance. In one of the more dramatic turnabouts, James Watson, a signatory of the Berg letter, declared that the danger initially imputed to bioengineering was "an imaginary monster,"[25] and he registered regret that he'd signed the letter.[26]

In retreating from their previously-voiced concerns so they could assert the safety of bioengineered organisms, these scientists were falling back on the foundational faith of their field. Molecular biology was developed as a distinct discipline during the 1930's largely through the efforts of the Rockefeller Foundation, under the leadership of Max Mason and Warren Weaver.[27] These two mathematician/scientists were uncomfortable with quantum mechanics, which during the first third of the 20th century had ascended to prominence in physics. This new theory was much more complicated than the classical theory it supplanted, and, as Weaver acknowledged, he and Mason disliked what they regarded to be its "essentially unpleasant 'messiness.'"[28] Further, they thought it would eventually be replaced by something that would be simpler and "more elegant" – and consequently "much more satisfying."[29]

And, having realized that they themselves could not reshape physics along the lines they desired, they enthusiastically embraced the opportunity to do so for biology. In fact, they wanted to ground biology in physics; and they believed that by turning it into an extension of the latter, they could develop a science of life that would be essentially simple, precise, and predictable. Phil Regal has observed that their approach was fully reductionist:

"The social sciences and humanities will ultimately be reduced . . . to biology with no residue. . . . Biology will in turn be reduced to chemistry, which will reduce to physics, which will reduce to a simple deterministic unity that will allow precise predictions at all levels of life."[30] This precision would enable comprehensive control. As Weaver has written, it was "reasonable" to expect that a well-founded biology could furnish "a similar degree of control over many of the aspects of living matter" as the physical sciences exert over nonliving matter.[31]

Mason and Weaver instilled their faith in the ultimate simplicity, predictability, and controllability of life processes in the physicists and chemists they recruited to become the pioneers of molecular biology.[32] In their vision, this new science would solve most of humanity's major problems through precise genetic and chemical manipulations that would be comprehensively controlled by human intelligence – with scant space for unintended consequences. Thus, as Regal has remarked, "The agenda for molecular biology and the engineering of life . . . was infused with complete optimism from the start, and there was only a positive view of the promise of the new science and the biotechnologies it was supposed to produce. Risks and other negative developments were not considered or planned for."

Moreover, when confronted by the possibility of adverse outcomes, the bioengineers displayed unrealistic confidence in their ability to manage them. For instance, at a conference Regal attended in 1984, a molecular biologist gave a talk describing all the hoped-for benefits of rDNA technology as if they were virtually certain outcomes. When someone asked, "What if you accidentally create a new disease?" she seemed offended, but unhesitatingly declared, "We'll develop a cure for it." Regal then queried, "Don't you think it would be a good idea for genetic engineers to first develop cures for AIDS and the common cold before making such bold promises?" She appeared stunned and was unable to muster a response.

Regal notes that over time, the evidence has increasingly countered the molecular biologists' convictions in the precision and predictive power of their discipline. "Abundant data has exposed a big discrepancy between the world they initially envisioned and the world as it really is – and shown that nature is more frustratingly subtle than they'd assumed both at the microscopic level and on the level of ecosystems."

Among U.S molecular biologists, the denial of the risks of gene-splicing was so deeply seated that many maintained it could not cause harm even if purposely employed to do so. Ken Alibek, who played an important role in the Soviet Union's bio-weaponry program before emigrating to the US, says he encountered "an alarming level of ignorance" about biological weapons within the expert community of his adopted country. He reports: "Some of the best scientists I've met in the West say it isn't possible to alter viruses genetically to make reliable weapons. . . . My knowledge and experience tell me that they are wrong."[33]

Regal confirms Alibek's observation. "I had long heard the same naive opinions from leading American biotech advocates. . . . My sense is that many of them had talked themselves into sincerely believing that rDNA had no weapons potential because they felt constantly on the defense and experienced a need to protect the image of biotechnology – and to sustain their own faith in the fully benign nature of their manipulations. These arguments spread and took hold as 'common wisdom' among American biotechnologists, despite their dissonance with reality."

Yet, not all molecular biologists were averse to acknowledging risk; and several spoke forcefully about the problems they perceived. An especially strong warning was released by one of the field's major pioneers, Erwin Chargaff. In an essay in *Science* titled "On the Dangers of Genetic Meddling" he called bioengineering "warfare against nature" and emphasized its irrevocable consequences. He declared: "You can stop splitting the atom; you can stop visiting the moon; you can stop using aerosols . . . But you cannot recall a new form of life. . . . It will survive you and your children and your children's children. . . . Have we the right to counteract irreversibly the evolutionary wisdom of millions of years in order to satisfy the ambition and the curiosity of a few scientists?"[34] In contrast to the molecular biologists who argued for less regulation, Chargaff urged *greater* government intervention. Further, he expressed doubt that the RAC could handle the various problems, and he deplored that almost all its members were proponents of genetic engineering.[35]

Another eminent molecular biologist who advocated precaution was Jonathan King, a professor at the Massachusetts Institute of Technology. Moreover, like Chargaff, he critiqued what he perceived to be the RAC's promotional proclivities and alleged that it functioned "to protect geneticists, not the public."[36] And Harvard biology professor George Wald, a Nobel laureate, warned that rDNA technology entails "problems unprecedented not only in the history of science, but of life on the Earth."[37] He emphasized that the radical type of intervention it performs "must not be confused with previous intrusions upon the natural order of living organisms"[38] – and branded it "the biggest break in nature that has occurred in human history."[39] He cautioned that "going ahead in this direction may be not only unwise, but dangerous."[40]

There were also individuals in the biotech industry with misgivings. As Phil Regal sought perspective from its members, he encountered several of them, including a friend from graduate school who had advanced from corporate researcher to administrator. Not only was his friend pleased to hear from him, like Mayr, he urged him to take on the safety issue. As he explained:

> Phil, we badly need input from ecologists and organismic biologists like you. We molecular biologists are out here by ourselves on this, and we've got no way of evaluating the safety of our own work, or of even knowing if our hype about social benefits makes sense. We never studied the sorts of things you guys studied. There was never the time or the interest. This is a very competitive business. A lot of people are trying anything they can think of when a new technique comes along or a new gene is available. "You've isolated a new gene? Lend it to me and let me see what I can get it into. Let's see what happens."
>
> Competitive gene jocks are a dime a dozen. The way to outshine the next guy, to get an offer from another company, the way to get a raise, is to do something sensational. There's a competition to do sensational things. Nobody has time to think deeply about safety or really how much good will come from this.

To some extent, the conflicting pressures exerted by the various factions in the genetic engineering controversy sustained an equilibrium over a few years which, though not deeply satisfying to any one group, did not tilt very far in any direction. The overall level of concern remained high enough so that some federal oversight was maintained, but not so high as to trigger the imposition of additional rules.

Then, in 1977, the equilibrium decisively shifted in favor of the biotechnicians. Public concern waned; and the initiative for regulation on Capitol Hill lost its momentum.[41] So substantial was the shift that, as Susan Wright puts it, "by 1979 the hazard question was almost a non-issue."[42] The main factor behind this transformation also underlay the genetic engineers' display of new-found certitude about the safety of their creations. It was the alleged emergence of important new evidence.

The Rise of 'Molecular Politics' – and the Force of Phantom Evidence

The pivotal news about new evidence arose as the result of three meetings held to evaluate the safety of engineered organisms. The first occurred in 1976 in Bethesda, Maryland, the second during the following year in Falmouth, Massachusetts, and the third in 1978 in Ascot, England. Collectively, they conveyed the impression that sufficient evidence had amassed to demonstrate that genetically engineered organisms are safe – and that there were no longer any concerns among experts. However, this impression was misleading.

For one thing, although the meetings purported to be scientific, they differed in significant ways from standard scientific gatherings. In contrast to conventional norms, the organizers carefully controlled who attended, how issues were discussed, and what information got disseminated. The conferences were not announced by normal procedures, participation was by invitation only, and the invitees predominantly favored minimal controls on rDNA research.[43] Jonathan King of MIT, one of only two scientists at the Falmouth conference who advocated stronger precaution, noted that many like-minded experts who ordinarily would have attended "were rather upset . . . to find out that a risk-assessment conference was taking place and they didn't even know about it until after the fact." [44] The Bethesda meeting had gone even farther than Falmouth in maintaining privacy, to the extent that a decade after it occurred, even the identities of the participants (other than the two chairmen) had not been officially revealed. And the organizers of the Ascot meeting did not invite any members of the British Genetic Manipulation Advisory Group (GMAG), an omission that seemed highly irregular and prompted one member of that group to state: "It might be thought a discourtesy to run an international conference on an important policy question without involving the corresponding organization in the host country. . . ." He surmised that the GMAG was snubbed because it featured "strong representation . . . of the public interest" and "would have supplied a critical presence." [45]

Susan Wright has observed, based on thorough study of the transcripts and her interviews with participants, that the meetings did not merely engage in the technical assessment of risk but were at least as concerned with how public perceptions of risk could be managed.[46] This concern was especially salient at Bethesda. Wright notes that a "strong informal theme" of the conference "was a shared sense of a pressing need, beyond containing possible hazards of recombinant DNA work, to contain the spread of the controversy as well."[47] She reports that the discussions reveal "a siege-like feeling . . . , a shared sense of threat, of polarization, of scientists versus society"; and she notes a tendency to employ "polarized categories" and speak in terms of scientists versus "the sky-is-falling people" and "the prophets of doom."[48]

This polarized mood and the meeting's political as well as scientific aims were manifest in the chairman's opening remarks: "Part of the agenda today is to get you guys involved and get your voices heard . . . If I could say to the prophets of doom: 'Look, these guys have come out and said that there is nothing to worry about here, so let's . . . get on with serious business.' That's what I hope we can accomplish."[49]

This aim for consensus played out in the way issues were handled. Although the participants recognized that rDNA technology could entail several hazards, the focus was systematically narrowed to research employing one particular type of bacteria called *E. coli* K-12, because it appeared to pose virtually no threat.

As previously noted, *E. coli* is a bacterial species that inhabits the intestines of humans and several other animals; and *E. coli* K-12 is a distinct strain that was developed in laboratories for research purposes. Because K-12 has been used for so many years in labs, it has become quite weak in comparison to other bacteria (including other strains of *E. coli*) and would have great difficulty surviving outside the protected lab environment. As one microbiologist puts it: "K-12 . . . wouldn't stand a chance in the hugely competitive environment that is your gut where bacteria are constantly evolving to keep their 'cutting edge' and not be pushed out by other microbes. Getting K-12 to establish itself in the gut would be like trying to qualify for a Formula 1 race with a car from 1922 (which is when K-12 was taken from somebody's gut)! It was competitive at the time, but is now way off the pace."[50]

Consequently, experts could feel confident that no matter what foreign genes got implanted within *E. coli* K-12, there was scant likelihood such feeble bacteria could cause an epidemic if they escaped the lab (which accounted for their frequent utilization in rDNA research). However, many of the conference participants did have other concerns. For one thing, NIH guidelines didn't bar research with microorganisms better equipped to survive outside the lab than K-12.[51] Further, even if research remained confined to K-12, there was recognized potential for problematic genes to transfer from it to other organisms which could then become agents for novel diseases. One participant pointed out a few potential scenarios and remarked: "To me, those are frightening."[52]

Yet, as Wright observes, these and other outstanding safety issues "tended to be factored out of consideration rather than confronted."[53] She says that instead, "the sense . . . that biomedical research was threatened came increasingly into focus," accompanied by warnings that science was under "very serious attack."[54] She reports that the transcript reveals a meeting "dominated" by "visions of laboratories swathed in red tape," and that in this context, the argument that K-12 could not become an epidemic-causing pathogen was seen as the best means for "defusing" controversy.[55]

According to Wright, most participants appear to have accepted this "political strategy."[56] As one biologist stated: ". . . in terms of PR, you have to hit epidemics, because that is what people are afraid of and if we can make a *strong* argument about epidemics and make it stick, then a lot of the public thing will go away."[57] She notes that at the end of the morning session, one participant "summarized the sense of the group" by stating that the primary task was to convince the public. He then declared: "[T]hat is very easy to do. It's molecular politics, not molecular biology. . . ."[58]

In reporting the results of the Bethesda meeting to the RAC, the chairman stated there was consensus that the possibility of epidemics is "extremely remote" – and a shared opinion that this concept "should be discussed in a public forum."[59] Accordingly, an organizing committee was formed, and in June 1977 the Falmouth conference convened. However, the facts indicate that the call for a public forum was merely public relations – and that the only thing the organizers wanted to make public was an advantageous outcome, not the process through which it would be produced. Otherwise, they would not have kept the conference a private affair to which the media were not invited (and about which they were uninformed) – as had also been the case at Bethesda, and would continue to be at Ascot.[60]

The conference managers likewise followed the Bethesda strategy in keeping the focus on *E. coli* K-12. Even so, participants raised controversial issues; and they debated whether foreign genes inserted in K-12 could then transfer to robust organisms – or instead, while remaining within it, could propagate dangerous toxins or hormones to the surroundings.

According to Susan Wright, the published proceedings reveal that these "troublesome questions" were not resolved.[61] The inconclusiveness of the discussions is evident from a list of proposals for further research, introduced by a statement that ". . . from the cauldron of vigorous scientific debate will finally emerge critical experiments to assess the potential hazards in recombinant DNA technology."[62]

Thus, even in an event where participation was almost exclusively limited to scientists who wanted minimal restrictions on rDNA research, and where the format was so tightly controlled that one attendee characterized it as "choreographed" and another as "a real set-up,"[63] potential hazards were acknowledged – along with the fact that "critical" experiments to accurately assess them had yet to be done. However, neither the public nor the wider scientific community was given the impression that the participants recognized the need for hard scientific evidence and "vigorous scientific debate" to stimulate its production. That's because, with the press excluded and the official conference report left unpublished until eleven months had elapsed, there was leeway for selective communication.

The main information released in a timely manner was in a letter sent immediately after the conference ended by the chairman of the organizing committee, Sherwood Gorbach of Tufts University, to the NIH Director. This letter, which was widely circulated in the summer of 1977, primarily shaped public perceptions of the results. Susan Wright says that it centered on the epidemic pathogen question "to the virtual exclusion of other issues" and presented "an essentially soothing view . . . in which uncertainties and unresolved issues were obscured by the emphasis on the remoteness of possible hazards."[64]

However, some of the participants tried to offset what they considered to be a misleading report of what had happened, including Richard Goldstein, one of the conference organizers. He sent a letter to the NIH Director pointing out that "though there was general consensus that the conversion of *E. coli* K12 itself to an epidemic strain is unlikely (though not impossible) . . . there was *not* consensus that transfer to wild strains is unlikely." He then stated: "On the contrary, the evidence presented indicated that this is a serious concern."[65] Several other participants wrote concurring letters.[66]

But, as a researcher with the Stanford School of Medicine observed, it was Gorbach's summary that "drew attention on Capitol Hill and in the media."[67] And the media, which assumed it was accurate, relayed its message without qualification. The *Washington Post* declared the scientists had "unanimously concluded that the danger of runaway epidemics [was] virtually nonexistent;" and a headline in the *New York Times* announced "No Sci-Fi Nightmare After All."[68] Further, as Susan Wright notes, this version of the results was not only accepted by the press and public but "quickly achieved scientific respectability" and was advanced by distinguished biologists.[69] Moreover, many of their statements (including an editorial in *Science*) exceeded the claim that *E. coli* K-12 could not become pathogenic and asserted there was consensus that *all* research employing it was safe.[70] The National Academy of Sciences (the NAS) even extended the distortion, declaring the evidence showed that the risks of genetic engineering *in general* were insignificant.[71]

Most important for the biotech proponents, and congruent with the aims of the conference, the Gorbach report became a powerful political tool. Armed with its purportedly evidence-based assurances, both the industrial and academic components of the molecular biology establishment mounted a massive lobbying campaign, described by Susan Wright as "one of the largest" ever related to a technical issue.[72] Participants included leading investigators at the American Society for Microbiology and also the NAS; and universities weighed in through a lobbying group called "Friends of DNA," whose members included presidents of "the most prestigious American academic institutions."[73] Harvard even hired two professional lobbyists to help out.[74] So extraordinary was the campaign in both membership and magnitude that some Congressional staffers remarked "they had never seen anything like it."[75]

The goal of these scientist/lobbyists was to thwart regulation, and a key target was the proposed legislation championed by Senator Kennedy, the bill that had achieved the most formidable momentum. Susan Wright reports that it had "sailed through" the relevant Senate committees when introduced and seemed "assured of approval" at the time the biotech proponents initiated their campaign.[76]

So they swiftly set out to scuttle it. Less than a week after the Falmouth conference, a group of eminent biologists met with Kennedy and argued that in light of the "new information," his proposed legislation was unnecessary and should be dropped.[77] But he held his ground and reasserted the need for a regulatory commission.

However, many legislators were more readily won over, and less then three months after the proponents of unfettered rDNA research were rebuffed by Kennedy, their persistent campaign had effected a decisive shift in the legislative mood. Senator Adlai Stevenson III expressed this new attitude in a speech to his colleagues on September 22nd asserting that "recent evidence" about the decreased risks of such research required them to "carefully" reassess whether the benefits of regulation would outweigh its adverse impacts on scientific research.[78]

With so many legislators now aligned against regulation, Kennedy was finally compelled to capitulate. On September 27th, in a speech to the Association of Medical Writers, he announced that he would no longer support his own bill, stating that "the information before us today differs significantly from the data available when our committee recommended the . . . legislation."[79]

According to Susan Wright, this reversal was a major event in the history of genetic engineering, ". . . demonstrating the power of the biomedical research community to retain control over regulating the field and to dictate the terms of technical discourse on the hazards."[80] It also demonstrated that this power could be gained and maintained through promotional claims that were unsubstantiated and seriously dubious, so long as they were professed to be science-based.

Moreover, the fabrications from Falmouth were not the only deceptive data employed to quash the Kennedy bill. A report on research conducted by Stanley N. Cohen of Stanford University also played a key role. Cohen, a co-inventor of recombinant DNA technology, was among the scientists who were not content merely to argue for the safety of research with *E. coli* K-12. Instead, he maintained that the technology he helped develop is *in general* safe – and even averred that it *could not* entail special hazards.[81] In 1977, he performed a study to support his stance. He wanted to demonstrate that the kinds of genetic recombinations achieved in test tubes also occur naturally in living organisms – and thus, that the splicing of genes between unrelated species is not a radically new and artificial development but something that's been innocently occurring in nature for eons. When the results were in, he declared success, because he (and his collaborator, Shing Chang) had been able to create a situation in which fragments of mouse DNA were taken up by *E. coli* K-12 and then integrated with some of the DNA that they carried.[82]

Cohen claimed broad implications for his research, arguing it showed that "scientists can only duplicate what nature can already do."[83] He sounded this theme even more boldly in a letter to the NIH Director on September 6, 1977 in which he asserted that the outcome was "compelling evidence" that recombinant DNA molecules constructed in the laboratory "simply represent selected instances of a process that occurs by natural means."[84]

Further, it appears that Cohen timed the release of his news to aid the lobbying campaign. Not only did he take what he admitted to be the "unusual step" of issuing the announcement about his findings well in advance of their publication in a scientific journal, he said he did so due to their "importance with regard to the regulation of recombinant DNA."[85]

The campaigners seized on his premature pronouncement, and because it maintained that bioengineering *as a whole* is essentially natural (and therefore safe), it strongly augmented declarations from the Falmouth conference. Accordingly, it helped convince legislators that their prior concerns were unfounded; and due to its breadth and its apparently evidential basis, Senator Kennedy relied on it as the main justification for his momentous reversal.[86]

However, as in the case of the claims from Falmouth, the impression that Cohen's claims derived from sound evidence was illusory. Although he avowed that the experiment was conducted under natural conditions, the reality was otherwise; because in order to induce the bacteria to accept the foreign DNA, not only did he and Chang have to treat them with a calcium salt, they also had to subject them to a major heat shock (by rapidly raising the temperature by 42 degrees Centigrade, which equals a boost of 107.6 degrees Fahrenheit).

These conditions were far from natural; and most scientists knew they were. Moreover, the NIH had special reason to be aware of it. Only six months before Cohen's letter declaring the naturalness of the conditions under which he'd induced the inter-species exchange reached the Director's desk, the prominent microbiologist Roy Curtiss had sent one with a starkly contrasting view. Ironically, though Curtiss's was also instrumental in the campaign against regulation (it was an open letter that was widely distributed), it undermined the claim that Cohen would later make because its argument for the safety of rDNA research was in part based on the fact that the conditions Cohen imposed were highly unusual. In contending that the insertion of foreign DNA into *E. Coli* K-12 "offers no danger whatsoever," Curtiss asserted that even if such DNA were later released, there was scant chance that other bacteria would take it up, unless they were treated with a salt and also subjected to a rapid 42-degree Centigrade rise in temperature – conditions which, he pointed out, "were unlikely to be encountered in nature."[87]

Despite the fact that the letters contradicted one another, the NIH used both as supporting evidence for its policy statements befriending biotechnology, while never noting the glaring discrepancy between them.[88] The agency was finally forced to confront the illegitimacy of Cohen's claim during a meeting the Director held with his advisory committee in December 1977, when the artificiality of the research setup was emphatically driven home by the distinguished biologist Robert Sinsheimer.[89] Although this potent dis-creditation deterred the NIH from citing the research in subsequent publications, its response remained minimal, and it apparently did nothing to correct the false impressions that had been instilled within the minds of Congress and the public. Thus, legislators were never properly informed that the purportedly evidence-backed proclamation on which they'd so strongly relied was bogus; nor was Senator Kennedy made aware that, half a year prior to his capitulation based on that pronouncement, the NIH possessed information undercutting it in advance – and that less than three months after his reversal, its infirmity was again revealed to the NIH, this time so directly and before so many experts that the agency didn't dare refer to it again.[90]

Ascot Compounds the Confusion

Despite the anti-regulatory victories of 1977, restrictions remained on some forms of rDNA research, and many virologists were dissatisfied that the NIH guidelines continued to classify the cloning of animal virus DNA in *E. coli* as "high risk."[91] Encouraged by the way the Falmouth conference altered perceptions, they hoped that a similar conference focused on their area of research could achieve like results. And so the Ascot meeting was held in January 1978. As was the case at Falmouth, discussion was limited to scenarios involving *E. coli* K-12; and there was likewise a meager store of evidence on which to form definitive conclusions. Based on her review of the proceedings, Wright notes: "The tenor of these discussions . . . shows that at many points, predictions were speculative. Too little was known about the mechanisms of viral infections and transformation to be able to predict the effects of cloning these genes."[92] As one participant remarked: "You see, the whole discussion has [the feeling of] a sort of Aristotelian academy because we are really just discussing extremely theoretical things and we're deriving models which are based on no experiments whatsoever. . . . that's why we're talking so much."[93] Nonetheless, the conference's final "consensus" statement confidently asserted that hazards to the public from cloning viral DNA were "so small as to be of no practical consequence."[94] As Wright observes: "The overwhelming impression produced by the report was one of reassurance. Almost all hazard scenarios were considered 'remote,' 'most unlikely,' or 'impossible.'"[95] She further notes that because the sole experiment to assess the risks of cloning viral DNA was a year away from yielding results, such conclusions "were surprisingly emphatic."[96] Moreover, it's evident that the consensus was not as broad as the document implied and that several participants had concerns that were never adequately addressed. Instead, when apprehensions were expressed about one or another perceived risk, they were rebuffed by assertions that the Falmouth conference had determined such a problem could not occur. In the words of one participant: "The trouble with the Ascot meeting was that the moment one raised a scenario, one would be shouted down by [those] saying that the Falmouth meeting had said that the clones were not mobilizable, that they would never get out of *E. coli* K12 . . . and could not become an epidemic strain."[97]

If the actual conference report from Falmouth had been available, it would have been clear that the participants had *not* reached such conclusions and that the possibility of foreign DNA transferring from K-12 to robust organisms had *not* been ruled out – and was a lively concern in the minds of many. But that report remained unpublished for another five months, and the only seemingly official account then at hand was the overly assuring (and in some ways misrepresentative) Gorbach letter. Thus, those who opposed a precautionary approach to genetic engineering prevailed over colleagues who raised legitimate safety issues by citing the authority of an illusory scientific consensus in order to claim that those issues had been definitively resolved – a practice that would become routine over succeeding years.

In all, any Ascot participant could justifiably have felt manipulated; and some clearly did. As one remarked: "It was very obviously a political meeting . . . We were being used in the name of being a disinterested group of virologists but it was fairly clear by the end of the meeting that [the organizers] wanted to go back with a result that could be exploited for deregulation."[98]

"Political" Science Prevails

The lopsided report from the Ascot meeting complemented the Gorbach summary from Falmouth, and their combined effect was substantial. Not only did proponents of biotechnology proclaim that employing *E. coli* K-12 in recombinant research is safe, several went much further (as had Stanley Cohen the year before) and claimed there was new evidence demonstrating that rDNA technology *as a whole* poses negligible risk.[99] This misleading version of the facts quickly spread. In March 1978, a few months after the Ascot meeting, it was vigorously advanced by members of both the academic and industrial sectors at a conference co-sponsored by the World Health Organization in Milan.[100] The same month, the Senate Subcommittee on Science and Technology prepared a report stating that rDNA research presented no unusual risks;[101] and the next month the NIH Director declared that the burden of proof should shift from the technology's promoters to those who wanted to regulate it – a shift that did occur, along with revision and substantial weakening of NIH guidelines.[102] This transfer of burden was historic, because, as will be described in the next chapter, it would carry over to all subsequent government policy on genetically modified organisms (GMOs).

Further, the influence of the inflated pronouncements extended well beyond America. Susan Wright notes that they impacted regulatory systems in many nations because "[o]nce the discourse of . . . 'negligible hazard' became established in the United States, the powerful geopolitical position of that country virtually assured the diffusion of the discourse elsewhere."[103]

And so was born molecular politics, through which overgeneralizations and unsubstantiated opinions have been passed off as sound scientific conclusions based on hard evidence. Because of the credentials of those making the assertions, neither the media nor the populace doubted the existence or solidity of the purported evidence; and even individuals as astute as Senator Kennedy were led to believe in it despite the fact it was just as chimerical as the expert consensus that was claimed to be based upon it. Further, due to the boldness and persistence with which these assertions were advanced, the bulk of the life science community came to accept them as well, including many biologists who should have realized how exaggerated they were. So powerfully did these false impressions take hold that they were essentially impervious to contrary input, no matter how well founded. Even a debunking by the eminent journal *Nature* had little effect. Although its report on the Milan conference stated that "the new evidence . . . does not seem substantial" and that the attendees "witnessed some unseemly clutching at straws," there was no retardation of the biotech juggernaut.[104] Thus, although the Falmouth and Ascot meetings had little data to go on and only reached a consensus about the improbability of *E. coli* K-12 being transformed into an epidemic pathogen, an illusion was inculcated within the minds of nonscientists and scientists alike that new evidence had been presented which uniformly convinced the participants that rDNA technology *in general* is essentially safe.

Moreover, when genuine evidence *was* garnered (as increasingly occurred after the Ascot meeting), it often clashed with the standard promotional claims. According to Susan Wright: "In many respects, this new evidence posed more problems than it resolved . . . [and] many in the scientific community . . . saw some of the results as surprising and therefore as raising new questions about hazards."[105] Yet, Congress and the public had virtually no idea that such surprising evidence was emerging, because the molecular biology establishment impeded communication of the facts. Time after time, when faced with research results they didn't like, the biotech proponents would routinely fail to acknowledge them – or else substantially mischaracterize them.

A prime example is the Rowe-Martin experiment, one of the most influential ever conducted on bioengineering, which was supposed to provide definitive answers to persistent questions about the safety of rDNA research.[106] Susan Wright reports that during 1975 and 1976, there were still "serious differences" among experts about whether some aspects of the research might be unreasonably risky – and insufficient evidence to rule out the possibility that a seriously harmful organism could in some circumstances be created.[107] She relates that such concerns surfaced at the NIH Recombinant Advisory Committee meeting held in December 1975 and that because there was no evidence demonstrating that gene-splicing was thoroughly safe, one molecular biologist proposed that a "dangerous" experiment should be performed that would attempt to make *E. coli* K-12 hazardous.[108] If it failed to do so, it would strengthen the case that the extensive rDNA research employing these bacteria is safe.

The committee liked the proposal, and one of its members, Wallace Rowe, assumed responsibility to implement it in conjunction with Malcolm Martin, a colleague at the NIH research lab he directed. As part of their planning, they organized the Bethesda conference, which they co-chaired, to furnish advice on how the experiment should be designed.

As the preceding examination of the conference indicates, Rowe and Martin intended it to do more than merely advise them on their research, and they initiated broader discussions that they hoped would convince legislators and the public that gene-splicing is safe. They led the discussions about their prospective research in the same spirit, focusing on how it could best be fashioned to calm public fears. In this vein, one participant argued they should demonstrate that *E. coli* "can't kill a mouse" no matter what's done to it. This idea was well-received, and someone suggested it could be effected by splicing DNA from a virus that can induce cancerous tumors in rodents into *E. coli* K-12 and then implanting the altered bacteria within mice. However, some of the scientists protested that such an experiment would, at best, only relate to manipulations of K-12 and would have little bearing on the safety of rDNA research in general. Further, they emphasized that because the K-12 strain was so debilitated, there was little chance it could do any damage. They argued that the experiment would therefore be of slight scientific value – and that the researchers should "take the opportunity to do a good experiment" by employing an organism with a greater capacity for harm.[109]

However, Rowe and Martin, along with most of those present, appear to have been less interested in securing the experiment's scientific value than in maximizing its political clout.[110] So the discussion stayed focused on public relations, exemplified by a scientist who advocated the use of *E. coli* K-12 by noting that because there was scant chance it could be made harmful, the study would be a "'slick New York Times kind of an experiment'" that would gain lots of positive publicity.[111] Accordingly, the majority eschewed the type of experiment that could have revealed embarrassing risks in favor of one that was almost sure to be image-enhancing – opting for less than optimal scientific worth in exchange for the apparent certainty of a soothing outcome.

Therefore, when Rowe and Martin adopted this PR-driven approach, they, along with the community of pro-GE scientists, expected their study to yield fully favorable results. So when it concluded, no one was surprised that such results were claimed for it. And the claims were by no means modest. At a 1979 press conference, the two scientists unequivocally declared they had demonstrated that the recombinant research they investigated was "perfectly safe."[112]

However, when one probes beneath their rosy representations and examines the actual data, it's clear that the term "perfectly safe" was imperfectly applied.[113] The investigation encompassed several aspects of the *E. coli*-based research system, and (contrary to the expectations of the researchers – and the gist of their public pronouncements) not all of them were found to be problem-free. For instance, cleaving the DNA of the cancer-causing virus (which must be done in order to work with its discrete genes) substantially increased its capacity to induce tumors.[114] There were other troubling results as well, and some eminent biologists warned they showed that splicing viral genes into the bacteria could enable the virus to expand its infective range.[115] But none of the adverse findings were mentioned at the press conference or in the other references to the research that were employed for promotional purposes. Accordingly, Congress and the American people were led to believe that the results wholly supported reduction of regulation, remaining unaware that significant problems had been discovered – and that several experts viewed them as signaling the need for stronger safeguards.

Nor were they informed that Rowe and Martin had not even employed the strain of *E. coli* routinely used in rDNA research but a strain that had been purposely rendered much weaker, to the extent it had become (in the words of one biologist) "severely disabled."[116] This occurred because, despite *E. coli* K-12's infirmities, NIH guidelines barred the transfer of tumor-causing genes into it without an exception from the Director; and he refused to grant one. So the researchers had to use the more enfeebled strain instead. Consequently, although the experiment's problematic findings *were* applicable to the hardier strain of *E. coli* actually used in most research, the favorable results were *not*; and, as Susan Wright points out, it was "not justifiable" to treat them as if they were.[117] But most people were unaware of this fact; and the biotech proponents felt no need to acknowledge it, or be restrained by it. Nor were they prepared to acknowledge, or to inform the public, that even if the Rowe-Martin results *had* been fully applicable to the strain of *E. coli* that researchers actually used, and even if they *had* all been fully favorable, they would still have been irrelevant to gene-splicing with other organisms, which was to become a prevalent practice.[118]

Thus, the key experiment designed to reassure the public primarily did so by not being fully publicized; and, with its deficiencies undisclosed, the promoters of bioengineering were able to milk it for far more than its scientific worth. Besides employing it to calm qualms and preserve the hands-off attitude on Capitol Hill, they used it to substantially reduce NIH research restrictions and significantly expand gene-splicing's permissible range. In the process, just as they had portrayed the limited discussions at Falmouth and Ascot as pertaining to bioengineering in general, they frequently stretched the relevance of the Rowe-Martin experiment well beyond legitimate bounds – not only averring it had demonstrated the safety of all forms of recombinant research, but sometimes even claiming it had done so for genetic engineering as a whole.

And these false claims continued for more than three decades. One of them was present on the website of The National Institute of Allergy and Infectious Diseases until at least November 2010. That institute is part of the NIH, and thus part of the United States Government. The falsehood appeared on the page that described the credentials and accomplishments of one of the institute's long-serving laboratory chiefs: Dr. Malcolm Martin. Thus, it's reasonable to assume not only that he was familiar with the content of the statement, but that he wrote it. And, due to the authoritative context, anyone who didn't know the details of the experiment that he and Wallace Rowe conducted would have also been led to assume that the statement was accurate – a statement which, without a trace of qualification, declared that the experiment "established the safety of recombinant DNA."[119]

On balance, not only were the claims that abetted the rapid – and largely unregulated – advance of the bioengineering venture during its first seven years more political than scientific, the scientists making them displayed the parochial attitude of a typical special interest group more predominantly than the public-spiritedness traditionally associated with the scientific endeavor. As Susan Wright puts it:

> [T]he refusal of the scientific establishment in the United States to call for hard experimental evidence . . . and the alacrity with which biomedical researchers in general rallied round to promote the public results of brainstorming sessions as 'new evidence', both suggest that the most immediate concern . . . was neither public safety nor scientific rigor. In fact, the history of the controversy indicates something entirely different: the insistence of research scientists that their freedom of investigation take precedence over the competing needs of the public.[120]

In the following years, as the molecular biologists consolidated their political power, their agenda would expand and increasingly prevail; and the needs of the public would continue to be compromised.

CHAPTER TWO

THE EXPANSION OF THE BIOTECH AGENDA

– And the Intensification of the Politicization

A New Phase: The Push for Environmental Release

Despite their success in whittling down the regulatory framework, the biotechnicians eventually grew restless with what remained and sought a radical change. This occurred because biotechnology broadened.

During the early years of genetic engineering, the main focus was on medical research and applications, and the altered entities were microorganisms that could be fully utilized within laboratory settings. Accordingly, environmental release was viewed not as a goal but an unwanted accident. Therefore, so long as rDNA technology was limited to the medical arena, though its practitioners could resent some of the restrictions they faced, they had no cause to fight the ban on releases; and the bioengineering venture could co-exist with it.

But the situation drastically changed as the enterprise expanded to agriculture. Genetically modified organisms (GMOs) designed to serve as agricultural crops must be grown in open fields, and much of the research on their efficacy has to be performed in outside settings as well. Further, microorganisms tailored for agricultural applications likewise must leave the labs. Thus, in order for this major phase of biotechnology to advance, the blanket ban on releases had to be lifted.

The stakes seemed high, because it appeared to many within both the private and public sectors that the greatest benefits of genetic engineering would accrue from applying it to agriculture. There were grand expectations that engineered crops would boost yields, increase nutrient levels, and reduce dependence on synthetic fertilizers and pesticides. But none of the high hopes could be realized unless GMOs were allowed to be deployed beyond laboratory walls.

However, lifting the ban would not be easy, because for years it had been the key factor in calming peoples' fears. Further, the claims about the inability of engineered organisms to cause widespread damage were related to biomedical laboratory research. But now the issue was not whether laboratory bacteria enfeebled by decades of confinement in artificial conditions could survive in the external environment and cause an epidemic, but whether genetically altered plants and microbes flourishing in farmers' fields could cause ecological harm.[1] And there was no evidence to show they would not. Attesting to the uncertainty, the then Deputy Director of Biotechnology Permits at the USDA has remarked: "In the 1970s, we were all trying to keep the genie in the bottle. Then in the 1980s, there was a switch to wanting to let the genie out. And everybody was wondering, 'Will it be an evil genie?'"[2]

Expanding the Arguments for Safety

To counter the renewed anxieties about bioengineering, molecular biologists insisted that GMOs designed for agricultural applications are just as safe as bacteria that cannot survive outside the laboratory. They argued that the risks are no greater because genetic engineering, like chronic lab confinement, crimps an organism's capacity to survive beyond its intended location.

However, although this claim was presented as science-based, there was no evidence to support it; and it was backed by a set of assumptions which, due to their proponents' lack of training in the relevant fields of biology, clashed with scientific knowledge. But this infirmity was not exposed for many years (a development described later in this chapter). Consequently, the claim passed as a scientific fact for a substantial span of time.

Moreover, even when some biotech advocates did argue along different lines, their contentions were still un-buttressed by either evidence or sound logic. For instance, in 1978 a group of plant specialists met to discuss the containment conditions that would be needed when organisms that had been engineered for agricultural purposes were field tested. One of the main issues involved the use of microbes that are naturally pathogenic to plants and plant pests. Although the scientists did not attempt to establish an analogy between these altered organisms and crippled laboratory bacteria, they nonetheless anchored their argument in flimsy thinking. They claimed that because the natural parental forms of these pathogens had apparently caused no harm in agriculture, the gene-spliced versions would not create problems either – even though they'd been endowed with traits that the parents did not possess.[3] While this conclusion delighted the biotech community, several experts were astonished that credentialed scientists would so readily presume practical equivalence between the bioengineered and natural versions of various viruses and bacteria in the absence of supporting evidence. The former director of the National Biological Impact Assessment Program said that he reacted with "disbelief."[4]

The Impact of Illusion

Despite the dearth of data and the misgivings of many experts, the illusion that *all* GMOs (even robust plants and microbes destined for farmers' fields) had been scientifically determined to be safe was so artfully instilled that it largely took hold; and it became big capital on Capitol Hill. Most legislators now believed they had an even stronger basis for relaxing regulation of the biotechnology industry and instead focusing on how to foster its growth and attain US dominance of the field, thereby boosting the nation's economy.[5]

This dovetailed with the agenda of the Reagan administration, which came to power in January 1981. Under President Reagan, the federal executive branch became committed to promoting industry and substantially freeing it from regulatory restrictions – and grew inherently hostile toward regulation. This attitude became more prevalent within Congress as well, since the Republicans had won control of the Senate in the election that brought Reagan to the White House, increasing the number of legislators who were ideologically opposed to regulation in general.[6] Accordingly, within both the executive and legislative branches, attempts to regulate GMOs were increasingly regarded as unnecessary and unwelcome.

The Campaign to Deregulate Bioengineering

Coinciding with the ascendance of an anti-regulatory agenda within government, some prominent molecular biologists launched an initiative in April 1981 to remove all mandatory restrictions from genetic engineering. Two members of the Recombinant DNA Advisory Committee (RAC) for the National Institutes of Health submitted a proposal calling for that agency's guidelines to be transformed into a completely voluntary code of conduct, which would have permitted experiments that were previously banned, including environmental release of GMOs, to proceed without oversight.[7] Other molecular biologists promptly sounded the theme that because GMOs posed no extraordinary risks, the NIH restrictions imposed an unnecessary burden. Ironically, several scientists who supported excision of the guidelines' regulatory muscle had seven years earlier signed the Berg letter – and thus been instrumental in establishing those guidelines in the first place.[8] Even Paul Berg called for rescinding the mandatory restrictions that had ensued from the Berg letter.[9]

It's noteworthy that when they signed that letter in 1974, Berg and his colleagues had advocated restrictions on research "until the potential hazards of . . . recombinant DNA molecules have been better evaluated . . ." And they specified that an appropriate level of evaluation would at minimum entail that "some resolution of the outstanding questions has been achieved."[10] Accordingly, members of the public would have been justified in assuming that their turnabout in 1981 signaled that extensive risk assessment *had* been conducted – and justifiably dismayed to learn that the only risk-related research was still limited to one enfeebled strain of bacteria unfit to survive outside laboratories, that this research was not entirely conclusive (and raised some reasonable doubts), and that this limited data served as the sole evidentiary basis for the claim that it was safe to proceed with unbridled research on, and release of, GMOs in general.[11]

Yet, despite the significant uncertainties, the deregulation bandwagon steadily gained members and momentum. However, many biologists refused to get on board. Moreover, they pointed out that several who rode it had financial ties to biotech enterprises. One of the most glaring cases of alleged conflict of interest was Nobel laureate David Baltimore, who had co-authored the proposal to eliminate the mandatory guidelines.[12] Phil Regal notes that in many instances, the stakes were huge, and the incentives went far beyond receipt of consulting fees: "Molecular biologists had become entrepreneurs and not merely consultants to industry. Many had bet their personal finances as well as their careers on the financial success of biotech."

Letting the Genie Loose

As the pressures to reduce restrictions mounted, the RAC substantially loosened them. And even before the campaign to eradicate all controls had begun, it removed the main precautionary measure. It lifted the blanket ban on releases. It took this momentous step in June 1980 by approving a request to field test a type of bioengineered corn, despite the fact it had not received complete information about how the transformation of the corn would be achieved – or even specific information about where the test field was located.[13] Nor had there been a thorough description of how the pollen was to be prevented from spreading beyond that particular field.[14]

As it turned out, not only was the approval of the request premature, so was the request itself. As Chapter 4 describes, it took several more years before any GE plants could actually be created, and even longer before corn could be transformed. Thus, because the applicants' aspirations outstripped their technical capacities, they were never able to implement their plan.

Nonetheless, this incident revealed that although the RAC was willing to let the genie loose, it was unprepared to do so with reasonable care – and was in over its head. The committee had been primarily established to deal with closely contained rDNA biomedical research, but now it was also responsible for passing judgment on proposed releases of bioengineered plants (and microorganisms involved in their cultivation). As the RAC eventually went on to approve additional releases, there was renewed concern within both the public and Congress, because it became obvious that such releases should be regulated by an agency with broader expertise.

However, no other agency seemed up to the task either. According to the 1983 report of a congressional subcommittee, ". . . no single agency or entity [possessed] both the expertise and authority to properly evaluate the environmental implications of releases from all sources."[15] In particular, the report referred to the capacity of the Environmental Protection Agency (EPA) as "unknown," the experience of the Department of Agriculture (USDA) as "limited," and the expertise of the RAC as "inadequate." It also noted that the USDA displayed "a disinclination toward oversight in this area."[16]

Moreover, the USDA was not the only federal agency with a disinclination to regulate GMOs. Consistent with the Reagan Administration's policy, the Food and Drug Administration (FDA) and the National Science Foundation also favored minimal oversight; and although the NIH was providing some monitoring, its sentiments were in harmony with those of the White House, and it did not want to implement a rigorous review system either.[17]

Countering the Generic Safety Arguments with Solid Science

Because the government was ideologically inclined toward deregulation, and because the biotech advocates pressed their campaign for it so vigorously, they achieved steady progress toward their goal. It was in 1983, when they were on the brink of success, that Ernst Mayr and Philip Regal resolved to press back with genuine science.

Regal reports: "Over the next several months, I talked with as many molecular biologists and biotech promoters as I could so I'd be able to make a systematic list of the arguments that were being used to support deregulation and deal with them point by point."[18] He refers to these arguments as "generic" because they extended to virtually all GMOs, based on the simplistic assumption that when it came to assessing their safety, they could be treated as a uniform class.[19] He soon recognized that the various ideas being expounded boiled down to a few basic arguments that were not only "disturbingly superficial" but based on outdated notions about both ecology and biological adaptation.

According to one of these notions, the biosphere is so tightly integrated it affords no niches for GMOs. Regal says this idea stems from the belief that evolution has so finely tuned the biosphere that any unnatural alteration will drop off unless sustained by human intercession – that "nature will cleanse itself of anything artificial."[20] Those who held this belief argued that the engineered entities would inevitably be so impaired by the modifications to which they'd been subjected that they could not compete with other organisms outside controlled agricultural settings – and could therefore not spread through the environment and cause damage. An article in *Genetic Engineering News* in 1984 expressed this idea by asserting that each species has become adapted for a particular ecological niche and that ". . . any genetic modification introduced in the laboratory is infinitely more likely to impair rather than to improve the adaptation, unless the environment is also changed."[21]

Regal states that although the belief that species are optimally adapted had seemed scientific in the 19th century and the first decades of the 20th, research by ecologists and experts in biological adaptation eventually revealed it to be unsound. As he notes: "There is abundant evidence that organisms are only *adequately* adapted for survival and are not *optimally* or *perfectly* adapted. Careful biomechanical analysis and comparative studies show that there is usually room for improvement."[22] He emphasizes that gene-splicing would not always cripple an organism's survival capacity – and that some alterations might bestow a competitive edge that would enable it to flourish in the wild and become a major pest.

Of course, even if this "no available niche" argument had been scientifically sound, it still would have been psychologically inconvenient because it portrayed GMOs as unnatural – and unnaturally impaired. In contrast, many biotech proponents took an approach in which they could depict gene-splicing as essentially natural instead of entirely artificial. Nonetheless, though they alleged the technology's naturalness, they still maintained that it would inevitably reduce the fitness of its progeny. They did so through a two-step argument. They first asserted that bioengineering is akin to traditional breeding via sexual reproduction because each is merely a process of combining genes. They then claimed that just as creating domesticated lines of plants and animals through the traditional process renders them unfit to survive without human support, producing organisms through the more modern mode likewise curtails their capacity to compete in the wild. They could thus declare the naturalness of genetic engineering while yet insisting that its creations would not become environmental pests.

However, though it may have sounded scientific to many ears, Regal deemed this argument to be just as flawed as the other because, as he points out, rDNA technology combines genes in "a radically different way" than does traditional breeding. This difference is obvious when one considers the facts that the promotional argument ignored.

There are alternative versions of every gene, which are referred to as its *alleles*. Each gene has multiple alleles, and some possess many. Different alleles give rise to different traits. For instance, the gene that determines the shape of garden peas has one allele that makes them smoothly round and another one that instead gives them wrinkles.[23]

In the process of domestication (which usually entails multiple cycles of selective breeding), several alleles possessed by the forms of the species that exist in the wild (referred to as *wild-type* alleles) are gradually replaced by other alleles, giving rise to new characteristics. Thus, it's a process of trade-offs; and, as Regal explains, the trade-offs are not easy to manage.

> In practice, the breeder cannot normally swap single alleles at only one site at a time, and thus flanking chunks of unwanted alleles may 'hitch-hike' along, and the alleles that originally occupied all of those sites can get lost in the process. Thus, in traditional breeding there are typically trade-offs that sometimes require the breeder to swap the genetic features that contribute to survival in nature for those that the breeder wants for commercial reasons. Consequently, many traditionally bred organisms have lost some of their natural competitiveness and are quite unlikely to become ecological pests under normal circumstances. For example, corn has been so highly domesticated that it cannot compete in nature. The seed coats have become thin, which makes for easy eating, but which provides little protection. Moreover, the seeds stay on the cob, a boon for harvesting but a handicap in the wild, because a competitive plant should have seeds that fall off and scatter.

In marked contrast, biotechnicians splice in new genes while maintaining all the others, adding new traits without sacrificing any the organism already possesses. Regal emphasizes that in this novel process, one does not have to "trade away" the natural vigor of an organism and can thereby increase the competitiveness of an already competitive wild-type organism – something "nearly impossible" to do with traditional breeding.[24]

Yet, despite their dissonance with reality, the generic safety arguments usually went unchallenged; and, backed by the prestige and influence of their proponents, they were accepted as authoritative by those in government and the media. Nor had these arguments, or the environmental safety issue itself, been seriously analyzed by the scientific community. Ernst Mayr informed Regal that within the National Academy of Sciences (NAS), discussions had been limited to the escape of disabled laboratory microbes – and that the issue of whether it was reasonable to presume that all other GMOs would be equally impaired had not been properly addressed. He explained that although the Academy should have conducted such an examination, the internal politics had prevented it. The molecular biologists were too wary they would lose control of the issue.

So, as Regal began to systematically analyze the environmental hazards of engineered organisms, and the arguments of those who sought their unregulated release, he was breaking new ground. Mayr collaborated, as did Peter Raven, who, like Mayr, was a highly influential biologist, a member of the National Academy of Sciences, and an authority whose expertise extended well beyond the level of molecules and cells. He too was troubled by the rate at which bioengineering was being commercialized in the face of ongoing ignorance about risks; and he agreed that it was imperative to initiate genuine dialogue within the bioscience community – especially because the biologists who *were* familiar with the recent advances in ecological genetics and the study of adaptation had never met to achieve a general integration of all this new information, let alone to examine how it applied to the release of GMOs.[25]

Accordingly, Mayr, Raven, and Regal believed it was essential to present the preliminary analysis Regal had prepared to other experts in these fields for their assessment. Regal notes:

> Peter, Ernst, and I were not sure what the outcome might be. Would these other experts endorse my analysis, or would they find holes in it that none of the three of us could find? I truly hoped that they would find holes in it, because if I was right, the implications were profoundly disturbing. Society would be moving into the era of deliberate releases with each project a veritable crap shoot. And in the worst case, there would be a rapidly growing number of floating

crap games, with the stakes rising over time as the rDNA techniques became each year more powerful.

Besides involving the leaders in ecology and related fields, it was necessary to confront leading molecular biologists with the current state of the evidence – and to make sure they understood its implications. So Mayr and Raven encouraged Regal to organize a workshop at which such interactions could be facilitated.

As the first step, he flew to Washington, D.C. to explore funding possibilities. He also wanted to get first-hand knowledge of how well the government was dealing with the various issues surrounding GMOs. As he visited division and program directors at the National Science Foundation, and key individuals in other agencies, the disclosures he received were eye-opening – and often prefaced with words he was to hear repeatedly as he interfaced with government officials: "If you quote me, I'll deny it, but it is important for you to know that . . ." Many were deeply concerned and were glad he was getting involved and willing to "stick his neck out." They expressed hope that he'd be better at tackling the tough issues than the Washington insiders, who, due to the political climate in the capital, were afraid to stick their own necks out.

These officials explained that it was difficult for them to take appropriate actions because eminent molecular biologists and leaders from the biotech industry had captured the ear of the Reagan Administration and convinced it that biotechnology was crucial for reviving the economy and should therefore be given special treatment. As Regal relates: "I was informed that the Administration had taken the position that most of the nation's economic problems could be solved by on the one hand reducing government regulations on credit, trade, and environmental pollution, and on the other, promoting a colossal shift to high-tech industry. Further, because it had proved difficult for US corporations to maintain a monopoly over the computer industry, the Administration was determined that the nation would preserve its supremacy in biotechnology.[26] So it had charged federal officials to foster the rapid development of biotech – and (according to those with whom I spoke) it was giving the molecular biologists virtually anything they wanted."

Not only were the program directors at the NSF under pressure to promote biotech, Regal learned they were "in something of a panic" because the Reagan Administration was about to cut nearly all funding for any biological research that was not directly contributing to the national effort in biotechnology. Thus, out of ninety-three research proposals in ecology suggested by the NSF staff, the only three regarded favorably by the Administration involved the study of bacteria that inhabit hot springs, because besides their potential to yield important thermal stable enzymes, they might facilitate the development of engineered microorganisms that could be incubated at higher temperatures and thus produce chemicals at faster rates. Regal reports that the directors were "bitterly lamenting" this abandonment of basic science.

In such a climate, Regal's requests could not be fulfilled. Although several NSF officials wished to assist, they acknowledged it would be too politically risky for them to fund a workshop that would raise safety issues regarding GMOs. And they explained that the difficulty was compounded by the fact the Reagan Administration was not even willing to talk with ecologists, let alone fund a forum that would parade their concerns. Its members held the misconception then common among politicians that the term "ecologist" was synonymous with "environmentalist," and they believed that individuals to whom the label applied were merely advocates of a set of values and policies that were inimical to economic development. They were unaware that, unlike environmentalism, ecology is not a set of policy preferences but an established science that investigates the complex interactions between groups of organisms and between organisms and their non-living environment. Nor did they realize that ecologists vary in their political leanings and value systems – and that some strongly disagree with many environmentalists on particular policy issues.

To make matters worse, many molecular biologists likewise confused ecology with environmentalism; and even those who recognized that it's a branch of biology and not merely a policy agenda still viewed it with disdain. Because the majority of them had been trained in physics or chemistry, they had a limited understanding of what traditional biology entails and tended to regard its main activity as the mere collection and categorization of the various life forms – which led several of the most influential to dismiss it as "stamp collecting." [27] These scientists believed that the traditional biologists were not studying life in a way that reveals its basic laws; and they were convinced that only *they* had the correct approach. As Regal explains: "In their minds, the truly legitimate way to study the living world is from the molecules up. Because they were *the* experts on molecules, and because molecules are the basic building blocks of life, they believed they were in prime position to deduce how everything up the chain of complexity should behave – far better positioned than even the organismal biologists, ecologists, and other scientists who studied higher levels of complexity directly. So they saw little value in communicating with these scientists and were predisposed to reject whatever input they might offer that countered their own assumptions. Thus, those with the worm's eye view scorned the perspective of those with the bird's eye view."

Having come up long on inside information but short on funding, Regal had to return to Washington for another try at securing the resources that would enable the overdue dialogue between ecologists and molecular biologists to commence. One of the main destinations was the stately building that houses the National Academy of Sciences, where he hoped to convince administrators of the National Research Council (NRC), the arm of the NAS that conducts studies and issues reports, to sponsor the workshop he envisioned. But he quickly learned that his hopes were misplaced.

While meeting with a group of NRC senior administrative staff, he was informed that the "power players" in the Academy feared that any ecological analysis of GMOs that was not tightly managed could give ammunition to those who opposed them. Further, it soon became obvious that NAS policy was colored not only by fear of ecological analysis but by disrespect for ecologists. He relates: "I was told that the most influential factions in the Academy (the *de facto* 'bosses' of the people with whom I was speaking) defined true science as based squarely on physics and chemistry – and insisted that only such science could be sufficiently predictive. They therefore were averse to allowing ecologists any role in evaluating GMOs."

Struck by the incongruity between the internal workings of the nation's premier scientific institution and its august public image (which it cultivates by, among other things, describing its headquarters as "a Temple of Science"), Regal called the administrators' attention to the dangers of allowing national policy to be dictated by the proponents of one narrow viewpoint about science that was rejected by numerous scientists and could not justly claim superiority over alternative, and broader, perspectives.[28] Although they agreed it would be valuable to broaden the outlook of those at the helm of the NAS/NRC, and to get them to foster a serious dialogue between molecular biologists and ecologists (and other "traditional" biologists), they said such developments were highly unlikely in the foreseeable future.

Regal notes that as he was departing, one of the administrators said, "Let me show you something before you leave." He then led him outside to a serene elm and holly grove in which stood a grand memorial to one of the Academy's greatest members. Regal reports:

> I marveled at a magnificent bronze statue of Albert Einstein, sprawled out like a fascinated child, with the universe as the floor of

> his playpen. "Look," my host said, pointing to an inscription around the base of the sculpture conveying Einstein's admonition to those engaged in the scientific endeavor: *The right to search for truth implies also a duty; one must not conceal any part of what one has recognized to be true.* After I'd read it, he looked me in the eye and smiled knowingly until he was sure that he'd gotten me thinking deeply about its potent message.[29]

With a resolve that would have earned Einstein's applause, Regal persisted in his effort to get the evidence from ecology into the biotech arena, and he finally obtained funding from the one federal agency which at that time recognized both the need to carefully regulate the release of GMOs and the value of ecological science: the EPA.

Not only did the EPA agree to fund the workshop, it desired to be actively involved in the planning; and an EPA scientist (Jack Fowle) became Regal's co-organizer. Mayr and Raven were also active in the planning; and Regal spent extensive time on the phone with each, discussing who should be invited and what should be included in the workshop's agenda.

Another important consideration was what the workshop's character would be. Increasingly aware of how political preferences and pressures had been degrading the discourse of scientists in regard to genetic engineering, Regal attempted to mitigate such influences. "I wanted all the scientists to speak frankly, as scientists, about a subject charged with enormous political sensitivity," he says. "I insisted that all the participants from government and industry should be scientists, even if they had gone into administration after getting their PhDs. I made it clear that this was to be a scientific discussion – not a policy discussion. And to minimize the temptation for grandstanding, I did not invite the press." This decision was prompted not only by his awareness that media presence would induce excess from some attendees, but that it would have an opposite, inhibiting effect on government scientists, several of whom had expressed fear that they'd lose their jobs if their candid comments were published. Thus, in contrast to the organizers of the Bethesda, Falmouth, and Ascot meetings, who excluded the media so that policy-driven presumptions could be passed off as scientific conclusions, Regal excluded them to preserve the scientific integrity of the proceedings.

Although the EPA was pleased with how plans were progressing, the NAS was not. From the standpoint of its leadership, the prospect of a workshop on the environmental risks of GMOs that was led by ecologists – and at which they would have equal representation with the molecular biologists and ample opportunity to critique their ideas – was worrisome. So it attempted to intervene. Regal reports: "When word spread that I was going to get EPA funding for the workshop, the NAS tried to take it over. I got a call from Jack Fowle informing me that the NAS had contacted the EPA and offered to conduct their own study of the risks if EPA would drop its plans. But as the NAS envisioned it, only one ecologist would be allowed to participate: me. Fowle asked if I found the idea appealing. It was such a ridiculously transparent attempt to keep molecular biologists in tight control of the issue that I laughed out loud. Fowle told me that the folks at EPA did not like the idea either and had assumed I would reject it."

In August 1984, the workshop finally got down to work – at the Banbury Center of the renowned Cold Spring Harbor Laboratories in New York State. Regal explains that the EPA selected this site because the laboratories were then directed by James Watson, one of the most influential and outspoken proponents of genetic engineering. The agency wanted to literally bring the issue of environmental risk home to him, not only so he'd be faced with the latest evidence from ecology, but to find out if he could discern any flaws in Regal's analysis. An added benefit of holding the event there was that Barbara McClintock lived at the facility and could participate. Like Watson, she was a giant of genetics and had won a Nobel Prize for a groundbreaking discovery; but unlike him, she had broader knowledge of the organismic level of biology.

Not only was the workshop a revelation for all who attended, the surprises it delivered were unsettling to both sides. The ecologists and other organismic biologists quickly concurred with Regal's analysis of the flaws in the generic safety arguments – and affirmed that the assumptions on which they rested were outdated. Moreover, Regal reports they were "shocked" to learn that so many influential scientists had been claiming that all GMOs would be safe on the basis of notions that they regarded as "scientific nonsense." "They were further shocked," he says, "at some of the projects that people from government and industry told us were going on. They had no inkling that such patently hazardous organisms were even being developed, let alone slated for near-term release."

In turn, the molecular biologists were "shocked and incredulous that their safety arguments were so completely disreputable in the eyes of those who held expertise in such matters." However, their discomfort was mitigated by the fact that although the ecologists rejected the arguments that *all* GMOs would be harmless, they believed that most would not cause problems. As Regal relates: "All the ecologists agreed that the great majority of even ecologically competent GMOs would not be threats to the environment. However, we concluded that a small fraction could well cause vast and irreversible problems. We made it clear to the molecular biologists and government scientists that although the probability of producing harm was low, in the small fraction of releases that did become harmful the damage could be enormous. This should have convinced them that it's necessary to carefully evaluate each GMO – and foolish to presume that such precaution can be dispensed with."

The ecologists went to some lengths to explain why the assumptions that underlay the generic safety arguments were flawed; and at one point, Barbara McClintock entered the discussion and avidly assisted in making the case. Regal remarks: "Many of the non-ecologists received yet another shock as one of the icons of modern genetics dismissed as simple-minded and misleading the views of biological adaptation that they had until then assumed were scientifically solid and even self-evident."

To the dismay of many molecular biologists, the subjection of their presumptions to open, science-based critique would not cease with the close of the Banbury conference but would recur nine months later on an even larger scale – at a conference inspired by Banbury that was not only bigger, but open to the media. Further, this conference (which convened in Philadelphia in June 1985) had much broader backing than the Banbury meeting and was sponsored by the American Society for Microbiology along with sixteen scientific groups and government agencies.

Yet, despite the broad sponsorship, there was considerable discord. Regal, who was invited to deliver the opening address, recounts: "Overall, the meeting was far from congenial, and many molecular biologists were utterly furious that ecologists had been invited to comment on 'their' science." Their anger was intensified because, with the science press covering the meeting, these comments might gain wide publicity. They worried that such an unbridled discussion of risks would not only stoke public fears, but sour investors and policy-makers.

Further, so minimally had the revelations at Banbury penetrated the molecular biology community that many of its members who came to Philadelphia were completely unaware of what had happened there. Consequently, Regal's presentation was their first exposure to some sobering scientific realities. In it, he described the advances that had occurred in the study of ecology and ecological genetics in the preceding decades and how they invalidated the arguments that all GMOs would be safe. Over the course of the meeting, the other ecologists more fully explained the evidence – and how it exposed the flaws in the generic safety arguments. Regal says, "They did a splendid job of demonstrating that their concerns were based on systematic science and not, as some critics were accusing, on an emotional objection to progress."

The impact was substantial, and, as the science press reported, the net effect of the conference was to clearly demonstrate the need for input from ecologists in setting biotech guidelines and in risk assessments of GMOs. This translated into changes in governmental policy. Regal was informed by Washington insiders that due to the influence of the Banbury and Philadelphia meetings, plans to deregulate biotech were reevaluated, scheduled cuts in funding for basic research in ecology were canceled, and the EPA began to more actively seek advice from ecologists regarding GMOs.

Some Ill-Conceived Projects Collide with Reality

The recognition of risks came none too soon. Regal says it's estimated that at the time of the Philadelphia conference, hundreds of the novel microbes being cultivated in various laboratories were on the fast track for release. And several posed risks that were both obvious and ominous.

Creating new forms of microbial life designed to thrive outside of laboratory conditions is a risky endeavor because the intent that the microbes will behave in a particular manner does not reside within the microbes but only in the minds of the people who create them; and those minds cannot fully fathom what the microbes will ultimately do. Regal says that although it's fairly straightforward to create a new type of bacteria through gene-splicing, adequately predicting how it will behave in nature is almost impossible. Science knows too little about which species will be interacting with the new bacteria let alone what the dynamics will be.

Thus, even if the gene inserted into a new bacterial strain codes for a trait that's presumed to be harmless, there's still a lot of uncertainty; and if it instead codes for a trait *known* to be harmful, the situation is especially risky. Nonetheless, such high-risk projects were underway.

One of the more alarming was spawned by a biotech corporation in search of a new way to destroy garden pests. Pursuant to this goal, its technicians created a novel strain of bacteria into which they'd spliced the gene that renders a particular toadstool toxic. The hope was that when spread in the garden, the poison-packed bacteria would eradicate snails and other pests, making garden owners happy and the corporation wealthy. But amid all the calculations of the steps of the bioengineering, the costs they would incur, and the profits that could accrue, scant attention had been given to the issue of how the bacteria might affect the myriad other species that inhabit gardens – and the pets and children that play in them. There had only been some poorly designed studies on potential impacts on honeybees and earthworms. Nor had there been analysis of whether the bacteria could migrate beyond the gardens – and what would happen if they did. As Regal remarks, "It was simply assumed that the bacteria would kill only the pests while cooperatively staying put."

Fortunately, due to the growing realization of risks, this project and several equally ill-conceived others were curtailed. Yet, industry shortsightedness was far from cured. Misbegotten projects continued to arise, even though many entailed risks that were not only major, but manifest. For instance, one biotech corporation developed food-yielding plants endowed with the venom-producing genes from scorpions. The aim was to make the crops invulnerable to predators, since any hapless insect that bit into them would at once suffer the same result as having been stung by a scorpion. However, in engineering these plants that could in effect bite back, the inventors had been surprisingly naïve. As with so many other biotechnicians, they operated under the assumption that they were dealing with a biologically simple situation and that no unexpected risks would be generated by their cross-species manipulation.

After their venomous plants were flourishing in greenhouses, the executives wanted to conduct field trials on farms. Luckily, they decided it would be prudent to first get a fuller assessment of the risks; and they called in Regal to perform a review. As he conversed with the staff scientists, it was apparent that up until then their concerns had been limited to the potential toxic effects on the humans and other mammals that would eat the altered organisms. Consequently, they were unprepared for many of the questions he posed.

Had they studied the effects of the toxin on bees that would gather pollen and nectar from the plants? No, they hadn't – but perhaps someone else would deal with that down the line. *What about studies to determine whether beneficial insects or birds that ordinarily feed on the plant would be poisoned – or whether those that feed on the plant's predators would be harmed by consuming carcasses laced with scorpion toxin?* They hadn't done any of those studies either. As they explained, their job had been to build a new food crop that would fend off pests; and their focus had centered on how the plant's known pests, and mammalian consumers, would be affected. *What about assessments of effects on the soil and its indwelling microorganisms? Could wide-scale cultivation of plants whose roots continually exuded this potent toxin, and whose decaying stalks and leaves were laden with it, cause long-term disruptions to the ecology of the surrounding soil?* That was another one they hadn't considered. *What about the plants' potential for pollinating weedy relatives with the scorpion gene and intensifying their adverse affects?* Blank stares all around.

Their greatest embarrassment came when Regal pointed out a potential threat to human health they had overlooked. Previously, they felt confident there would be no risk to human consumers because the venoms of the particular species of scorpions from which they had taken the genes were considered non-toxic to mammals. However, their confidence was shaken when Regal explained that toxicity is quite distinct from allergenicity, and that a substance could be non-toxic yet highly allergenic. He informed them that many people are severely allergic to insect venoms and that thousands either die or are badly sickened each year from adverse reactions when stung by bees and spiders. *What would be the implications,* he asked, *of putting these foods onto supermarket shelves? Additionally, how might people involved in the production and processing of the crops be affected? Would the dust from the crops cause problems for farm workers and grain handlers? How might it impact the air around farming communities?* The scientists were visibly stunned that they had taken the project so far without considering these issues – and due to the dose of reality Regal injected, it did not advance farther.

Yet, even as bioengineers developed better understanding of the risks, they were loath to discuss them openly. As Regal notes: "The industry's debts had increased enormously, and the pressures to rush on, show profits, and pay off investors remained intense. There were fears that any forthright, properly scientific discussion of risks would discourage investors, entail legal vulnerabilities, and call down government regulation with real teeth in it."

Through the practice of molecular politics, such teeth never formed – although, as will be seen, the public was led to believe that a strong set of them was in place.

Politics Continue to Preempt Science

The airing of scientific knowledge at the Philadelphia conference made it difficult to overtly deregulate genetic engineering; and the Reagan Administration felt obliged to respond to lingering public concerns.[30] However, its aim was to manipulate public perceptions rather than implement new safeguards, thereby placating demands for greater environmental and consumer protection without impeding its agenda to spur the growth of the US biotechnology industry.

To provide the industry an open road, the Reagan team determined there should be no new laws on biotechnology, and it had sufficient clout on Capitol Hill to block any new legislation. Additionally, because the federal administrative agencies could yet have issued some new regulations specifically tailored to GMOs under the existent laws, the White House ordered them to stay within the regulations then on the books and to refrain from making new ones.[31]

Moreover, it revised lines of authority in order to restrict the role of the EPA. This seemed necessary because that agency wanted to supervise GMOs and was arguing that because they contained new chemical substances, it could regulate them under the Toxic Substances Control Act, which empowered it to require testing of industrial chemicals that may pose an environmental or human-health hazard.[32] This ambitious attitude perturbed those at the White House. So to the fullest extent possible, they vested responsibility for the environmental safety of GMOs, not with the agency possessing the broadest level of environmental expertise, but with the USDA, because of its friendlier attitude toward biotechnology and reluctance to subject it to regulation.[33]

However, the EPA could not be totally stripped of authority over GMOs, since it had the statutorily granted power to regulate pesticides and was therefore entitled to oversee organisms that express pesticidal proteins. Nonetheless, through appointments and other means, the White House (under Reagan and then George H.W. Bush) was increasingly successful at bringing the agency's outlook more into line with its own.

To formalize its promotional policy, the Reagan Administration wanted a publication that would demarcate the partition of authority it favored, set principles to guide the various agencies, and convince the public that science-based oversight was being provided. The result was the Coordinated Framework for Regulation of Biotechnology, signed by the president June 18, 1986.

A chief feature was the incorporation of a White House directive "to regulate the product, not the process." In this approach, GE organisms were to be dealt with based on their specific characteristics rather than their method of production – preventing them from being subjected to special requirements merely because they'd been generated through rDNA technology.

However, applying this principle presented additional issues, because it's hard to determine what all the effects of a GMO's specific characteristics will be without conducting tests. So the crucial question was how much testing would be considered necessary. In line with the White House aim to keep regulation minimal, the USDA adopted a key presumption: the products of bioengineering were to be regarded as environmentally safe unless proven otherwise.[34] Eventually, the EPA also adopted this presumption; and the FDA then extended it to food safety (as will be discussed in Chapter 5). In consequence, the kinds of tests needed to detect potential dangers were not required and instead left to the manufacturers' discretion – effectively foreclosing meaningful regulation. As Phil Regal notes, "Notwithstanding the litany about regulating the product and not the process, the main thrust of the policy was to avoid examination of the product."

Thus, the notion that it's scientifically justified to regard GMOs in general as safe, initially promulgated on the basis of the alleged "new evidence" presented at the Bethesda, Falmouth, and Ascot conferences, was instituted as a foundational principle of US regulatory policy; and the shift in burden of proof that the NIH implemented in 1978 in the wake of those conferences became standard throughout the regulatory agencies. This occurred despite the reality that the "new evidence" was essentially a set of conjectures; that the limited data on which the conjectures were based, as well as the logic they employed, were largely irrelevant to genetically engineered plants and animals (or even to microorganisms designed to survive outside the laboratory); and that the evidence that *was* available did not support a general presumption of safety – and instead indicated there were good reasons to exercise caution.

Although the White House presented this outcome as science-driven, it's clear that the dominant forces were political. As Mary Ellen Jones' extensive study led her to conclude in a doctoral dissertation in Science and Technology Studies at Virginia Polytechnic Institute: ". . . the U.S. Coordinated Framework for Biotechnology Regulation is based principally in political criteria, not solidly based in science as its proponents claimed."[35] So salient was the role of politics that she titled her dissertation "Politically Corrected Science." However, reality notwithstanding, the impression that the Framework was science-based largely took hold; and, as had the RAC Guidelines a decade earlier, it substantially calmed the public.[36]

The NAS Adds Its Assurances

Yet, the Framework could not entirely close the controversy, and several scientists and public interest groups criticized it for failing to provide adequate safety. As the need to field test more GMOs increased, and concerns again began to mount, the National Academy of Sciences endeavored to furnish greater assurance by producing a short position paper, which it issued in August 1987. While the paper mentioned some points raised by the ecologists at Banbury and Philadelphia, its main thrust was to downplay problems. It asserted there are no "unique hazards" associated with GMOs and that the risks of releasing them into the environment are the same as in the case of unaltered organisms.[37] It further stated that many of the prospective projects "are either virtually risk-free or have risk-to-benefit ratios well within acceptable bounds,"[38] and it concluded that "strict and rigid controls" for all releases of bioengineered organisms "are not justified."[39]

Overall, the paper served to promote the rapid deployment of GMOs. As an article in the *Harvard Journal of Law and Technology* notes: "The NAS report, which was widely publicized, appeared to vindicate the view that the risks from deliberate release are overstated, and that the real danger is that excessive regulation could stifle the young biotechnology industry."[40] Accordingly, two seasoned observers scored it "a major victory for the biotechnology sector."[41]

But many experts did not view the victory as fairly won. Ecologist David Pimental of Cornell University charged that the composition of the five-member panel that wrote the report "was heavily weighted toward genetic engineering,"with only one ecologist included;[42] and Sheldon Krimsky, of Tufts University, argued that the assertion about *no unique hazards* "has less to do with good science than it does about political correctness within the scientific fraternity."[43]

Phil Regal also regarded the report as driven more strongly by political than scientific considerations. He soon received stunning confirmation that he was right. During a break at an academic conference, he approached one of the report's authors (with whom he was already acquainted) and attempted to engage him in a discussion of the biology he thought had been mishandled. But the man cut him off, exclaiming, "Phil, you keep insisting on treating this as a scientific issue; but I can't discuss it with you on that basis. It's a political issue." Regal realized it was futile to persist. "So we went off and had a beer with some other biologists," he reports, "and talked about other matters."

Because the 1987 paper had been criticized as too short and superficial, the NAS endeavored to produce a more complete and authoritative statement; and it issued a much longer and more extensively referenced report in 1989 (through the National Research Council, one of its divisions). According to Regal, the authors were not only obliged to acknowledge key concerns raised at Banbury and Philadelphia, they were "forced to admit that there *were* risks and that ecological input would be needed in the design and evaluation of GMOs to be released into nature." These admissions occurred in the report's mid-section, which contained scientifically meaningful analysis – along with disclosure that no conclusions could be reached on some issues of serious concern, especially the release of bioengineered microorganisms, which the authors considered to be a big question mark. However, as Regal points out, this "scientific meat" in the middle was sandwiched between opening and closing chapters of a distinctly different character. He says: "These chapters were largely written by the NRC staff, employees directly obligated to the power structure in the Academy, which had a substantial stake in advancing biotechnology. Accordingly, their text contained several broad generalities expressing great optimism about the safety of genetic engineering – and provided rich material for its advocates to quote."

Habituation of Exaggeration

The advocates liberally exploited the opportunity that the NRC functionaries had furnished them. In doing so, not only did they misrepresent the report's content by ignoring its cautions and citing only the positive pronouncements appended to it, they claimed that these pronouncements applied to GMOs in general, thereby misrepresenting its scope. The report dealt *solely* with the issue of field trials of GE crop plants and microorganisms in the continental US. Accordingly, it did not even pertain to field tests in Hawaii and Puerto Rico, let alone to large-scale commercial release within the US or other countries. Moreover, it was focused exclusively on environmental effects and did not touch on the question of food safety, which is a distinct and unrelated issue, since a plant can be environmentally benign and yet devastating to human health. Regal emphasizes that it contained "no scientific data or theory whatsoever for any extrapolation beyond its narrow confines." Therefore, even if it *had* been legitimate to extract the report's soothing generalities from the context that qualified them and to accept them at face value, it was illegitimate to apply them beyond the discrete set of issues the report officially addressed.

Nonetheless, just as they'd exaggerated the relevance of the limited discussions of the Bethesda, Falmouth, and Ascot conferences (and also the limited findings of the Rowe-Martin research), most biotech promoters presented these general pronouncements as authoritative conclusions regarding *all* the various facets of bioengineering, including food safety; and they frequently cited them as scientific backing for the permissive US regulatory policy in these areas. They likewise stretched the application of the previous NAS report well beyond its rightful range, purporting that it dealt with GMOs in general even though, as a technical matter, it was just as narrowly focused as the report that followed it. Further, just as the American media had uncritically circulated the exaggerations about the Bethesda, Falmouth, and Ascot conferences and the Rowe-Martin research, so they tended to accept and disseminate the false claims about the NAS reports without reservation, even though a quick skim of the actual documents could reveal their limited scope – especially since the limitations were explicitly stated early on.

Regal recalls several instances in which scientists who were misrepresenting one of the reports employed dramatic flourishes (such as waving the document while speaking) to be more persuasive. One of the most amusing occurred at a meeting of a task force that was deliberating whether the state of Minnesota needed to enact regulations. He recounts: "One of the molecular biologists brought in a stack of the 1989 reports, ceremoniously passed them around, and adamantly asserted that the experts had concluded that everything was safe. When I asked if he had actually read the report, he got angry and said 'no' but that he and his fellow molecular biologists knew very well what was in it. In this case, his authority games and theatrics didn't fly because enough people on our committee were familiar with the report's contents, but how many Americans were going to read that report and avoid being taken in?"[44] As it turned out, not very many.

Due to the scientific credentials of those who advanced the exaggerations, their persistence in doing so, and the unquestioning repetition of the claims by prominent institutions and the media, the intended illusions broadly took hold within the United States. Moreover, they have strongly endured, to the extent that even the Environmental Media Service, which favored a precautionary approach and endeavored to cut through promotional hype, was so chronically misled that the media guide it published in 2000 portrayed the 1989 report as having concluded there is "no reason" to treat GMOs differently than conventional organisms in any respect – even when it comes to food safety.[45]

However, spreading the illusions abroad was more difficult. The deferential attitude of the American media toward biotech proponents did not catch on in Europe, and the attempts to pass off the NAS reports as scientific conclusions about the general safety of GMOs failed because journalists and policy analysts on that continent routinely checked the promotional claims against the actual documents. Regal remembers how refreshing it was to be approached by journalists who had actually read the reports, in contrast to his experience with their American counterparts. And he notes that for these astute Europeans, the gross disparity between the concrete text and the inflated claims engendered "anger and mistrust."

The fact that European journalists exercised their critical faculties by reading the original documents (despite the fact they were written in a foreign language) while most American journalists did not (even though it would have been easy to do so) is indicative of basic differences between their approaches to biotechnology – which may go a long way toward accounting for the substantial differences in public awareness and attitudes in the two regions.

Perpetuation of the Politicization

During the presidency of George H.W. Bush, the Coordinated Framework was maintained (as it has been by all subsequent administrations); and the pro-biotech, anti-regulatory policy of the Reagan Administration continued, with scientific issues routinely resolved according to economic and political priorities. In 1990, the President's Council on Competitiveness, chaired by Vice-President Dan Quayle, assumed oversight of federal policy on biotechnology and made it clear that boosting the development of the biotech industry would remain a major objective – and should not be constrained by concerns about safety.

This occurred even though Phil Regal had published three peer-reviewed scientific articles that collectively went farther than had the discussions at Banbury and Philadelphia in refuting the various generic safety arguments – and the Ecological Society of America had published a widely praised article (co-authored by Regal and six other scientists) that explained the environmental risks of GMOs and the kinds of regulatory oversight they necessitated.[46] Yet, according to Regal, most of the molecular biologists promoting genetic engineering ignored these articles and continued to use the old arguments and analogies that had been thoroughly discredited. As he explains: "They were able to proceed in this manner because they had political muscle on their side. The Bush Administration shared their desire for effective deregulation and was similarly ready to dismiss any evidence and genuine scientific analysis that ran contrary to this goal. So it accepted their assertions as 'sound science' because this allowed the Council on Competitiveness to claim scientific backing for what they felt was sound economic policy."

Eisenhower as Prophet: The Ascendancy of a Scientific-Technological Elite

When he delivered his presidential farewell address in 1796, George Washington issued a strong warning against entangling alliances with foreign nations. One hundred and sixty-five years and thirty-three presidents later, such alliances were deemed essential for national security. So in 1961, no one was surprised that the farewell address of the thirty-fourth president, Dwight D. Eisenhower (who had been the first Supreme Commander of the North Atlantic Treaty Organization), contained no cautions about them. But most people *were* surprised that this former general launched a strong warning about another type of entangling alliance – not between the government and foreign powers but between the government and a colossal domestic power he referred to as "the military-industrial complex." [47]

Further, even most of those who *are* familiar with this particular caveat are surprised to learn it was part of a larger warning that was not limited to the military-industrial complex but extended to a broader phenomenon. Eisenhower noted how scientific research had been dramatically transformed into an endeavor practiced on a large scale by "task forces of scientists" with massive funding; and he noted the substantial interconnections that had developed between scientists and the government. He then cautioned, ". . . in holding scientific research and discovery in respect, as we should, we must also be alert to the equal and opposite danger that public policy could itself become the captive of a scientific-technological elite." [48]

In light of the subsequent history of the biotechnology venture, those words appear prophetic.

During its first two decades, the genetic engineering establishment gained so much influence over public policy that virtually its entire agenda was adopted and ardently promoted by two consecutive national administrations – and would continue to be by the next three as well. Its individual members received (and would continue to receive) lavish government grants to pursue their research; and its corporate constituents were given license to develop and deploy a slew of novel products with minimal oversight, even though numerous experts had concluded they might entail enormous risks. Moreover, it exerted broad control over the dissemination and interpretation of information and could deftly manipulate the impressions of government officials, the media, and the public – passing off conjectures as hard evidence and limited conclusions as general truths, while suppressing facts that threatened its interests.

So great was its power that it was even able to avoid any inhibiting consequences from the biggest documented catastrophe caused by a product of genetic engineering – a food-borne epidemic that dealt extensive death and disability throughout America – by instilling the illusion that no such thing had ever happened.

CHAPTER THREE

Disappearing a Disaster

How the Facts About a Deadly Epidemic Caused by a GE Food Have Been Consistently Clouded

A Set of Unsettling Facts

In September 1989, as the soothing generalizations in the National Research Council's report first circulated, other reports were emerging that induced a distinctly opposite effect. These reports did not come from institutions engaged in theoretical discussions of risk but from offices of medical doctors and public health officials; and instead of upbeat pronouncements geared to calm concerns, they contained startling accounts of an unusual new disease.

During that year, thousands of people throughout the United States experienced the onset of severe muscle and joint pain accompanied by swelling of the legs and arms, extensive skin rashes, and significant breathing difficulties. Some also developed congestive heart failure, while others succumbed to complete paralysis, with a respirator required in order to breathe. But even if they avoided these latter two outcomes, most of those with the basic set of symptoms suffered greatly.

One woman from California reported: "I was in so much pain – joints, bones, skin, everything – that I could barely stand to be touched. I lost about 60 percent of my hair, had no energy, and was usually asleep. At various times, I . . . had mouth ulcers, nausea, shortness of breath, severe muscle spasms, itching and painful rashes all over, edema (swelling of extremities), concentration and memory difficulties, handwriting problems, balance problems, irritable bowel syndrome, weight gain, visual perception problems, just to name a few symptoms!" [1]

An ordained Catholic deacon in Cincinnati recalls: "The pain was so intense in my body that if I were to lie on the mattress at nighttime when I went to bed, it would hurt too bad. I would sit up on the side of the bed and try to sleep sitting up because of the intensity of the pain. My legs became – you wouldn't believe it unless you saw it – they became as big as a telephone pole. They split and water oozed from them. No amount of medicine they gave me . . . calmed the pain."[2] After six years of such agony, during which time he couldn't work, his stamina finally started to improve; but he continues to endure constant muscle pain and physical disabilities.

A woman from Skokie, Illinois who had always enjoyed excellent health and abundant energy initially developed a rash all over her body, then a horrible cough, and eventually a degree of muscle weakness and pain so extreme that "it was hard to walk, hard to do anything." At times, her hand or jaw would suddenly clamp shut or another muscle would abruptly lock down. As things deteriorated, she had to leave work. Eventually, she became bedridden for six months, with pain so strong that the mere act of rolling over was an almost unbearable ordeal that took two full minutes.[3]

As these accounts indicate, the pain associated with this strange set of symptoms was unusually intense. The chief of the Division of Rheumatology at The Graduate Hospital in Philadelphia, who treated several afflicted individuals, described it as "the severest" he had seen in his entire practice. Further, besides the high level of pain, the level of the white blood cells called *eosinophils* also ran high. These cells fight infections and also control mechanisms associated with allergies. Their normal count is about 100 to 200 per microliter of blood. For someone with an allergic reaction or asthma, the count can rise to 600 or 800 – sometimes even to 1,000 or more. But people with this novel malady had average counts of 4,000; and many had counts that ran much higher.[4] When the level of these cells goes too high, the arsenal of molecules with which they're armed to battle invaders start attacking the body's normal tissue instead, resulting in massive systemic damage and intense pain.

Doctors were baffled by this extraordinary disease, and the treatments they attempted were largely ineffective. Moreover, although many people were stricken during the summer of 1989, because the symptoms often varied and the outbreak was dispersed (with most practitioners observing but a single case), it took several months before the medical community even recognized that a new disease had arisen, let alone that it was surging as a nation-wide epidemic. It took even longer to learn what was causing the disease. Finally, by early November, there was enough data to determine the critical commonality between the victims: they had all been ingesting L-tryptophan supplements.

L-tryptophan (LT) is one of the amino acids, a class of chemicals that form the building blocks of proteins. It's essential for human life, and, among other things, it participates in the production of the neurotransmitter serotonin, which promotes relaxation and sleep. Some of the best natural sources are dairy products, soybeans, fish, poultry and meat. In the 1980's, LT was also available as an over-the-counter supplement. For many years, doctors had recommended it in cases of insomnia, premenstrual tension, stress, and depression. Further, although at one point approximately 2% of the US population was taking it, there had never been documented problems when it was properly employed.[5]

But now it had become associated with a novel and nasty disease. Because this ailment was characterized by an elevated eosinophil count (a condition called *eosinophilia*) along with severe muscle pain (*myalgia*), it was named *eosinophilia-myalgia syndrome* (EMS). By early December, there were 707 reported cases in 48 states, with at least one death. By April 1990, 1,411 tryptophan-linked EMS cases had been reported, along with 19 deaths. According to the final estimate of the Center for Disease Control (CDC), between 5,000 and 10,000 people were stricken.[6] Of these, at least 80 died and around 1,500 have been permanently disabled.[7]

Identifying the Source of the Disease-Linked Tryptophan

It was important to know if all tryptophan supplements were truly dangerous or if there was something unique about the supplements the EMS victims had consumed. Many different retail brands of LT were involved with the disease, and investigators wanted to learn if, beneath the differences in brand names, the various batches of EMS-associated LT had any features in common. Because there were far more retail brands than actual manufacturers, and because the connections between brands and manufacturers were unclear, the first thing to ascertain was the production facility at which each case-related batch had originated.

Only six manufacturers, all Japanese, supplied L-tryptophan to the US market. In the early months of 1990, CDC researchers diligently traced the batches of LT that were associated with EMS back through the complex network of wholesalers, distributors, tablet makers, encapsulators and importers to their point of origin. In late April, they announced an important discovery. Their investigation revealed that every batch of EMS-associated LT that could be definitively traced back to a manufacturer (accounting for 95% of all such batches) came from a single source: Showa Denko KK, Japan's fourth largest chemical company and the biggest supplier of LT to the United States.[8]

Trying to Determine the Deadly Difference

The next step was to determine if there was something distinctly different about Showa Denko's tryptophan – something that set it apart and made it uniquely harmful.

For many years, all manufacturers had used a method in which bacteria are induced to synthesize LT through fermentation. Because additional (and unwanted) substances get generated as well, the contents of the fermentation tank are then put through a multi-staged process of purification, culminating with carbon filtration. Rigorous analytical tests are then conducted to assure that the end product is pure.

Investigators soon discovered that Showa Denko's LT did differ from the products made by other manufacturers. For one thing, it was uniquely contaminated. Not that the others were contaminant-free. It's virtually impossible to remove every bit of unwanted substances, and analytical testing revealed that every manufacturer's LT contained trace contaminants. But Showa Denko LT contained more than 60, a much greater number than did the others.[9]

Further, it was clear that one or more of these contaminants packed an abnormally potent punch. That's because Showa Denko routinely tested its LT to make sure that it met the United States Pharmacopoeia standards for purity (at least 98.5 % pure). In fact, the levels of each contaminant were extremely low: 10 or fewer parts per million.[10] Consequently, even though SD's product contained a greater number of contaminants than usual, none was present at a level high enough to pose problems in the usual case. So the fact that one (or more) of them made thousands of people very sick meant that it was (or they were) extraordinarily toxic.

Moreover, another important difference had come to light between Showa Denko's LT and the products of competitors: it had been manufactured in a different manner. In order to get the bacteria to yield substantially more LT, Showa Denko had broken new ground and altered their genomes via recombinant DNA technology.

The news that Showa Denko's deadly LT had been produced by genetically engineered bacteria was first announced in July 1990 in the *Journal of the American Medical Association.*[11] It soon spread to the popular press. *Newsday* led the way with an article titled "Genetic Engineering Flaw Blamed for Toxic Deaths."[12] In it, Michael Osterholm, an epidemiologist with the Minnesota Health Department who had been researching the epidemic, asserted that Showa Denko had "cranked up" its bacteria to increase LT production and that "something had gone wrong." He then remarked, "This obviously leads to that whole debate about genetic engineering."

Biotech proponents watched in dismay as numerous other newspapers followed with stories that linked the EMS catastrophe to genetic engineering, and they hoped for an authoritative rejoinder that would blunt the force of Osterholm's allegations. The FDA promptly rose to the occasion. When a reporter from *Science* interviewed an agency official, he "blasted" Osterholm for "propagating hysteria;" and he declared that it was "premature" to suggest that the epidemic was related to genetic engineering – "especially given the impact on the industry." [13] But Osterholm stood his ground and countered: "Anyone who looks at the data comes to the same conclusion. . . ." [14]

The *Science* article went on to disclose that the FDA's concern for protecting the image of biotechnology was so strong that, although the agency had known about Showa Denko's use of genetic engineering for months, it had withheld this information from the public "apparently hoping to keep the recombinant link quiet until they could determine whether it in fact did play a role in the outbreak." [15] However, as will be seen, it was unduly charitable to have presumed that the FDA was earnestly seeking the truth – or would have voluntarily divulged any findings adverse to the interests of the biotech industry.

The Quest for Clarification

Fortunately, several investigators *were* dedicated to discovering the relevant facts and ascertaining whether bioengineering played a key role in the calamity. To do so, it was necessary to determine which contaminant (or combination of contaminants) caused the EMS and how it had come into existence.

The first major step was reported in August 1990 by *The New England Journal of Medicine*. Researchers determined that one of the contaminants was not only associated with EMS cases but that it was a novel chemical substance formed by the fusion of two LT molecules, something never seen before.[16] They dubbed this new substance "EBT." [17] However, although they knew its chemical structure, they didn't have enough evidence to know if it was the cause of the epidemic.

For almost two years, EBT was the only contaminant known to be associated with EMS. Then, in June 1992, researchers determined there was at least one more, a compound called 3-phenyl-amino-alanine (3-PAA). While EBT had never been seen before its appearance in SD's

L-tryptophan, 3-PAA had; but it had *never* been found in food grade LT produced through conventional means. However, as in the case of EBT, there was insufficient evidence to conclude that it had caused the epidemic.

Eventually, four other contaminants were determined to be case-associated as well. But none could be proclaimed the cause of EMS either. The evidence was still too scanty to establish that any of the six case-associated contaminants was the culprit – or even a minor accomplice. That's because mere association does not equal causation, and far more data is required to prove that a substance caused an epidemic than to show that it's merely associated with it. Chemicals can qualify as associated even if they are found in only a small portion of the batches that cause illness.

The uncertainty about what had caused the contamination was not merely puzzling for researchers, it was deeply disturbing for biotech proponents. By the time the epidemic hit, insulin synthesized through genetic engineering was in wide use and an enzyme to substitute for animal rennet in cheese production was being primed for sale. As in the case of Showa Denko's LT, these substances were churned out in large quantities by microorganisms that had been artificially endowed with new genetic material. If employing such altered organisms had induced deadly side effects in the production of LT, their use in producing these other substances might likewise be risky. Further, if bioengineering had caused ordinarily trustworthy bacteria to generate unexpected toxins, it might do the same when used in producing more complex organisms such as fruits, grains, and vegetables. Consequently, the future of genetic engineering was to a large extent riding on whether or not the technology would be implicated as a cause of the EMS, because if it were, the projects employing it might lose their commercial viability. Moreover, even absent conclusive proof of the technology's guilt, lingering suspicions about its involvement could hamper its continued development. So its advocates strove mightily to exonerate it.

A Key Question: At What Stage of the Process Did the Contamination Occur?

Fallacies Regarding Filtration

One of their main defenses was to pin the blame on something else. They pointed out that just before the epidemic-related batches of LT were produced, Showa Denko had cut costs by reducing the amount of charcoal used during the final phase of the filtration process; and they argued that this change, not the bioengineering, was the key factor in the contamination.

According to the thrust of this argument, it was no longer important to discern the role played by the gene-splicing, because once it was clear that the contaminants had not been properly contained, it little mattered how they had arisen.

But it was illogical to deny the relevance of the contaminants' source. Placing sole condemnation on the change in filtration was like asserting that a soldier's death was caused by a defective helmet while ignoring the bullet that pierced the helmet, and the gun from which it was fired. As an article in *The New England Journal of Medicine* pointed out, although the reduction in carbon may have been a contributing factor, it did not explain how the lethal agent entered the product in the first place.[18] Thus, contrary to the impression induced by the biotech proponents, it was still critical to assess the role of the gene-splicing – and to learn whether the killer contaminant was its side effect.

The irrationality of centering blame on the reduction of carbon was underscored by the fact that due to the potency of the lethal contaminant(s) at extremely low concentrations, cases of EMS still arose (albeit at a lower rate) from bioengineered LT produced during periods when the carbon was restored to adequate levels. Accordingly, the reduction in carbon was not the key event that caused the EMS; it merely allowed the disease to strike more people.

Further, the issue of carbon levels is irrelevant to GE fruits and vegetables because, in contrast to isolated substances, whole foods don't pass through filters prior to sale. Therefore, any toxins formed during production would be fully present when consumed – which highlights the importance of knowing whether those in Showa Denko's LT emerged via bioengineering.

Nonetheless, despite the illogic of fixating on the reduction of carbon, many biotech proponents continued to do so because it fed confusion and deflected attention from the bioengineering. And the confusion was widely spread. For instance, *Lords of the Harvest*, an influential book about genetic engineering by a science reporter for National Public Radio, indicated that "inadequate filtration" might be fully to blame for the epidemic.[19]

Did a Different Part of the Purification Process Generate the Toxin?

Yet, even scientists who recognized the fallacy in this fixation couldn't jump to the conclusion that, because the reduction in carbon had not caused the contamination, the bioengineering must have. There was another possibility: that the toxins had instead been generated during the part of the purification process that *preceded* the charcoal filtering.

While it may seem odd that a toxic substance would be generated during the very process that's designed to remove toxins, it can happen; and at that point of the investigation, there was insufficient evidence to rule it out. So researchers sought to determine at what stage of production the lethal contaminant had been generated. If it was already present in the fermentation broth *before* filtration began, that would imply it was an effect of the bioengineering; but if it only appeared *after* the broth had entered the filters, that would imply it formed during the latter process – and that the bioengineering was innocent.

If the altered bacteria that Showa Denko used in producing the contaminated LT had been available, researchers could have resolved the issue by using them to make new batches under the same conditions that SD had used and analyzing the contents before and after filtration. But the bacteria were not at hand, and an article in *Science* reported that Showa Denko had destroyed them when the problems first arose.[20]

So researchers had to proceed in less direct ways. Although the bacteria SD used could not be employed, other features of its production process could be; and researchers found that the system it used might have generated some of the case-associated contaminants *after* the altered bacteria had done their work. One team found that when tryptophan was purified using the company's procedures, it could generate EBT.[21] Another team then discovered that PAA could also be formed from chemicals present during that particular process of purification.[22]

Advocates of bioengineering were quick to declare that the technology had been exonerated by this evidence, and this claim has been persistently repeated and widely spread.

But it's false. It's based on the assumption that either EBT or PAA was critical to the epidemic, and this assumption ignores a substantial body of evidence that indicates they were not. For instance, a study published in the *New England Journal of Medicine* examined twelve case lots that were all linked to the epidemic and did not detect *any* EBT in three of them (25% of the total).[23] So this study alone shows that EBT was not a necessary factor in causing the disease. Further, besides being unnecessary for the causation of harm, EBT was not even significantly related to a lot's harmful status, a fact revealed through subsequent statistical analysis that compared lots manufactured within a short time of one another (only some of which were disease-associated).[24] Other research found that when case-associated LT was administered to rats, it caused more immune cell activation than did EBT administered alone, even though that dose of EBT was over 100 times higher than the amount the rats received from the LT.[25] In light of these (and additional findings), it's evident that EBT itself was not the key cause of the EMS and that the crucial role must have been played by something else.[26]

And that something else was *not* PAA. The evidence showed that its relationship to the epidemic was even weaker than in the case of EBT.[27] Thus, the fact that these two contaminants could have been formed during purification instead of during fermentation was irrelevant to the question of whether the genetic alterations were to blame for the epidemic.

Moreover, the rest of the available evidence did not resolve this important question either. While researchers had identified the chemical structures of three other case-associated contaminants, comprehensive analysis revealed that they were no more strongly linked to the epidemic than were EBT and PAA. But it did reveal that one other case-associated contaminant *was* significantly related.[28] However, little was known about this substance (referred to as AAA), and its chemical structure had not been identified. So there was no clue as to whether it had been synthesized within the bacteria or within the purification system. The issue was further complicated by the fact that, in the eyes of several scientists, the evidence as a whole suggested that the EMS had been caused by multiple factors acting together – and the composition of the crucial combination was not known.[29]

Why Bioengineering Could Have Been the Key Cause

However, although there was no basis for making any final judgment about whether genetic engineering had been instrumental in causing the fatal contamination, there were sound reasons to think that it *could* have been. Bioengineering has inherent potential to disrupt the normal processes within a living cell and create unintended and unusual side effects that can give rise to deleterious substances. When this technology is employed to accelerate bacterial synthesis of LT, such unintended effects could readily occur.

Professor Charles Yanofsky of Stanford University, a leading authority on tryptophan biosynthesis, has stated: "Genetic engineering results in the formation of higher than normal concentrations of certain enzymes and products; these could provide the basis for the synthesis of higher levels of toxic substances."[30] And he noted that merely increasing the rate of tryptophan synthesis (the goal of SD's gene-splicing) can lead to such ill effects: "The more tryptophan is produced in the cell, the greater the chance that some side reaction will occur at a greater rate, producing more of some contaminant."[31] As he has further explained: "Overall this would mean that the bacterium is producing large amounts of about 10-15 metabolites that are not normally produced in excess. The accumulation of these metabolites would, in some cases, lead to modification by other enzymes, to give products that normally are never produced by the bacterium. One or more of these unnatural products could be a compound toxic to man. Similarly, the overproduction of enzymes of the aromatic and tryptophan biosynthetic pathways could lead to the synthesis of unnatural products by side reactions that normally do not occur. Again, toxic products could be produced."[32]

Further, in the case of the bacteria used by Showa Denko, some unusual side reactions might have occurred as acts of self-defense rather than as undirected accidents, since an overabundance of LT is toxic to them. So they may have activated uncommon mechanisms as a means of self-protection.

Moreover, not only was the presence of novel contaminants consistent with what could be expected from genetic engineering, so were the fluctuations in their levels. For instance, while the lots of LT that Showa Denko produced in March, April, and May of 1989 contained high amounts of overall contamination, the level of one particular contaminant dropped substantially toward the end of April, and many others were markedly diminished within a year.[33] Such variation suggests erratic biological activity rather than changes in the manufacturing process; and there are several ways in which genetic engineering could have induced it.

A Major Issue: When Were GE Bacteria First Employed to Produce LT?

However, defenders of biotech had an ostensibly powerful argument to parry the thrust of the foregoing evidence. And the FDA wielded it artfully. In July 1996, freelance journalist William Crist phoned the agency's biotechnology manager, James Maryanski, in an attempt to gain clarification about the cause of the EMS and was told that the evidence pointed away from genetic engineering. As Maryanski put it: ". . . we are aware of close to two dozen cases of L-tryptophan linked EMS that occurred before Showa Denko began using their engineered strain. So, there would have to be a cause other than just the mere engineering of the strains." While he conceded that genetic engineering could not be conclusively ruled out, he maintained that "the more likely cause" was L-tryptophan itself, or LT "in combination with something that was the result of the purification process."[34]

Although this information seemed to absolve genetic engineering, Crist's research had already made him wary of the FDA's reliability when the reputation of biotechnology was on the line. So he decided to do more digging. What he unearthed was startling. Yes, there were cases of EMS that pre-dated the epidemic; in fact, there were far more than two dozen. However, rather than exonerating genetic engineering, the existence of these earlier cases instead implicated it. But the implication was only visible in the light of other evidence that had gone largely unnoticed – and that the FDA was averse to disclose.

Crist compiled this evidence in stages. He first sought to learn if LT from a manufacturer other than Showa Denko had been linked to any of the early EMS incidents. He searched the scientific literature and found three studies by the CDC that pegged pre-epidemic EMS to Showa Denko's LT but no studies involving the product of any other company. He also contacted about a dozen law firms that had represented EMS victims and learned that all the lawsuits (including those based on pre-epidemic cases) had been brought against Showa Denko – and that none of the firms knew of an EMS incident connected with a different manufacturer.

This evidence, in conjunction with the extensive data relating to the epidemic, clearly refutes the contention that LT itself could have caused EMS. As CDC epidemiologist Edwin Kilbourne has pointed out, if LT were the cause, then all products of equal dose from different companies should have had the same effect – a scenario unsupported by the evidence.[35] Gerald Gleich (a medical doctor who studied the epidemic thoroughly while at the Mayo clinic) has sounded a similar note: "Tryptophan itself clearly is not the cause of EMS in that individuals who consumed products from companies other than Showa Denko did not develop EMS. The evidence points to Showa Denko product as the culprit and to the contaminants as the cause."[36] Moreover, Crist eventually discovered that the evidence not only pointed to SD's product, it revealed that during the four and a half years preceding the epidemic, *all* of the company's tryptophan had been produced with genetically engineered bacteria.

The engineered strain that caused the epidemic was introduced in December 1988. It was named Strain V, which implies there were at least four earlier strains. Crist learned that such strains had in fact existed – and that all but one were developed via genetic engineering. That lone non-engineered line was called Strain I. All the others had been created from it through successively more powerful forms of gene alteration, yielding progressive increases in the output of LT.

Although this information was apparently not well-known, it had appeared in a scientific journal in September 1994 – almost two years before

Maryanski told Crist that the pre-epidemic cases were linked with non-engineered strains.[37] Given the critical bearing of such information on an issue about which the FDA had displayed keen interest, it's reasonable to presume that the agency would have been aware of it. But even if this article had somehow escaped the FDA's attention, it didn't really matter, because the agency had already learned the facts through a different channel.

Crist discovered this when he obtained a copy of a fax the FDA had sent a journalist listing the various strains of engineered bacteria SD had used and describing the genetic manipulations through which each had been created. Almost as surprising, the fax was dated September 17, 1990. Moreover, the FDA had acquired the information much earlier. According to an attorney who sued Showa Denko on behalf of an EMS victim, the company sent it to the agency the preceding February.[38] So, shortly after the epidemic was first detected, the FDA had learned about these other engineered strains; yet, for years thereafter, it professed that no such strains were ever used.

In fact, SD started producing LT with genetically engineered bacteria in October 1984 and continued using the technology from then on. And as each successive strain was manipulated to produce more LT than its predecessor, it appears to have also produced more disease, with the incidence of EMS steadily rising until the upsurge induced by Strain V.[39] Further, the total number of pre-epidemic cases, while far smaller than the number caused by Strain V, was substantial. Employing data from CDC researchers, Crist estimated that between 350 and 700 people were stricken.[40]

Moreover, although it took many years (and an epidemic) before those early cases could be linked to SD's bioengineered tryptophan, during that earlier period it became clear to the company that the product had problems. For instance, SD internal documents show that in the summer of 1988 (months before Strain V was used), a German firm found a suspicious impurity in a shipment of its LT – and that its scientists were unable to determine whether or not the substance was toxic because they couldn't figure out what it was.[41]

But that was not the only conundrum confronting Showa Denko scientists during that year. For an extended time, they had to grapple with a more baffling one. According to SD's documents, problems "broke out" with one engineered strain due to an onslaught of viruses.[42] Moreover, the viruses were not invading from the outside. They had inhabited the preceding strains of bacteria all the way back to the initial non-engineered version, but they had existed in a quiescent state and were therefore unnoticed.

What sparked their transformation from peaceful lodgers to hostile aggressors? When I posed this question to a renowned virologist, Adrian Gibbs, he said such changes are triggered by stress to the bacteria "that stirs up their metabolism."[43] And forcing them to churn out a lot more LT clearly would have stirred up their metabolism. Accordingly, he remarked that the critical stress could have resulted from "mucking about with the bacteria" on the genetic level. The likelihood that such "mucking about" did activate the viruses increases in light of documents indicating that the eruption entailed major difficulties, and that substantial time elapsed before a virus-free strain could be isolated. When I informed Dr. Gibbs of these facts, he noted that if the problem had resulted from a localized stress like heat shock, there would probably have been a reserve of undisturbed bacteria – and that because the entire stock of the bacterial strain SD was then employing seems to have been affected in a sustained manner, "it suggests that the problem was genetic in origin."

SD was beset by other serious problems as well. The records reveal that when Strain IV was first used in commercial production, SD pulled it after only two weeks and reverted to Strain III. Further, SD stayed with that earlier strain for eight months before attempting to use IV again (in early September 1988), which suggests there was an unexpected difficulty that took a long time to clear up.[44] Then, after only a few days with IV, production was apparently shut down for more than three weeks.[45] Since SD had employed bioengineering to *increase* LT production, this long lapse implies that another major dilemma had arisen. Further, because SD documents state that the virus problem had by then been solved, the difficulty must have involved something else. And, although the virus outbreak had come as a surprise, the emergence of additional problems did not, as evidenced by a scientist's memo written after virus-free bacteria were again in use predicting there would be "other troubles."[46] Moreover, it appears that even after production finally resumed, SD still had qualms about Strain IV, because it once more reverted to Strain III and didn't employ IV again until mid-November – and then for only a five-week run, whereupon it was supplanted by Strain V (which was created through further manipulations to it).

Thus, the evidence strongly suggests that as the bacteria were altered to output increasingly higher levels of LT, there was concomitant increase in stress, creating disturbances that made trouble for the technicians. Further, it's plausible that besides causing headaches for SD's staff, the metabolic imbalances in the bacteria induced toxins that caused chronic aches for thousands of consumers that were far more excruciating. And it's *indisputable* that any account of the epidemic which ignores the earlier strains of engineered bacteria is itself seriously imbalanced.

The FDA Sustains Its Distortions

Despite the importance of the evidence about the pre-epidemic GE strains, the FDA stubbornly refrained from mentioning it – and consistently evaded confronting it. On October 9, 2001 Crist sent letters by certified mail to both Maryanski and Joseph Levitt, Director of the FDA's Center for Food Safety and Applied Nutrition, pointing out how the FDA's 1990 fax proves that it knew of this crucial evidence but nonetheless insisted that no engineered strains pre-dated the epidemic. He then asserted, "It appears that FDA has tried to defuse and downplay the issue of genetic engineering by shifting the blame to tryptophan itself, using pre-epidemic EMS . . . cases as justification . . ."[47] Finally, after a thorough exposition of the other discrepancies between the FDA's pronouncements and the facts (as reported in standard scientific journals), he stated: "I am left with the perplexing question: Did FDA have <u>any</u> solid evidence at all supporting its position on L-tryptophan?" While for Crist this question was perplexing, Maryanski and Levitt seem to have found it vexing. They never replied.

The FDA was equally unresponsive when Crist confronted it with other uncomfortable questions. One sought to clarify how diligent the agency had been in trying to obtain SD's bacteria. As previously noted, the best way to determine whether the fatal toxin(s) had been produced by the gene-altered bacteria or the purification process would have been to obtain Strain V and run tests. Although it was generally believed that Showa Denko destroyed the bacteria before investigators could apprehend them, when Crist contacted Don Morgan, an attorney who represented SD, he was informed that it had instead tried to cooperate. According to Morgan, although the FDA inspected SD's plant in May 1990, it didn't request samples of the bacteria then and only asked for them subsequently. But the company was "reluctant" to mail the bacteria overseas because that might induce changes that would impair the accuracy of the tests. Morgan explained that SD wanted to turn the bacteria over to FDA representatives and show them how to properly handle the cultures but that the agency never followed through on this offer. He further revealed that although the company eventually destroyed the bacteria, they waited until 1996 to do so, providing the FDA ample opportunity to send someone to get them.[48]

To learn the FDA's side of the story, Crist sent a letter to Sam Page, a scientific director at the agency, recounting Morgan's allegations and asking that he respond to them. But he never did. Nor was the agency responsive to the various Freedom of Information (FOI) requests he sent. Although he did receive some perfunctory replies, he was not given the information asked for. Instead, he was told that the information "was lost" or "could not be found" and that the individuals involved had all left the FDA. However, Crist verified that these people were still at the FDA, but when he raised this point to an FDA staff member, he was again told that they had left. His FOI requests to the Centers for Disease Control were likewise rebuffed.[49]

Crist has noted that the evasion of these important questions suggests something sinister: "For more than a decade the question of whether SD's genetically engineered bacteria were a causal factor in EMS has been downplayed or denied outright by these agencies. Now, it appears that they both may have known all along that the GE strains did play a crucial role in EMS and that they concealed this information to protect the U.S. biotech industry."[50]

Culturing the Clouds of Confusion

In all, the efforts of these government agencies, along with those of other biotech proponents, have indeed protected the industry. The sustained suppression of facts, conjoined with the steady spread of falsehoods, has created so much confusion that, although genetic engineering cannot be ruled out as the main cause of the catastrophe, and although there are good reasons to think that it played the key role, neither the public nor most of the journalists and scientists who have sought the truth are aware of this.

A prime force for delusion has been the claim that the filtration process was proven guilty – and that the gene-splicing has thus been acquitted. Despite the strength of the evidence arrayed against it, this assertion has been so staunchly maintained that it has even misled experts who endeavored to stay informed. More than a decade after the epidemic, I met with a distinguished biologist at a leading university who, despite having followed the story of the toxic tryptophan more closely than most scientists, had become convinced that the lethal contamination emerged during the steps of purification. He was quite surprised to learn that there was no evidence to compel this conclusion – and that it was quite plausible the epidemic stemmed from the genetic alteration. This news would have also surprised the biologist who wrote a book on biotech published by Oxford University Press in 1993 asserting that "the problem was eventually traced to a chemical generated during the (perfectly conventional) purification procedure, and had nothing to do with recombinant DNA."[51]

Due to the volume and persistence of the misinformation, the confusion has compounded over time and spawned extreme outcomes. Some of the most striking appeared in the 2001 report of New Zealand's Royal Commission on Genetic Modification. This blue-ribbon panel was supposed to help the government set policy by assessing the main issues regarding genetic engineering. After conducting months of hearings, it issued its report. One section of the fourth chapter dealt with the toxic tryptophan. In a methodical manner and an authoritative tone, it fully absolved genetic engineering of responsibility for the lethal contamination. But it did so by radically reshaping reality.

Besides distorting several facts, it seemed to pull some out of thin air, making assertions that were not only utterly unfounded but completely novel. For instance, it falsely (and uniquely) stated that all the disease-linked bottles came from a single batch, which implied the problem was a rare quirk that was not associated with the GE production process in any ongoing way.[52] It additionally declared that other manufacturers besides Showa Denko had marketed LT derived through bioengineering – and that because none of these products harmed anyone, it's unlikely the gene-splicing had caused EMS.[53] Yet, there was no evidence to support this contention, it contradicted common understanding, and there was no indication it had previously appeared in print.

The report then moved on to ostensibly close the case. In its most astounding pronouncement, it claimed that US courts had actually resolved the issue and had determined that the epidemic was not caused by the genetic modification but by another aspect of the manufacturing process.[54] However, notwithstanding the boldness of the assertion and the august aura of the document in which it's contained, it is flat-out false. And it's hard to comprehend how the commission even came up with it.

When I first saw the report, I was especially struck by this statement because, although I had extensively researched the epidemic, not only had I never read nor heard such an allegation, I had strong grounds to doubt it. So I contacted Crist because he had studied the topic far more thoroughly. He, too, was amazed, because he'd never encountered such an assertion either. At his advice, I phoned Don Morgan, whose law firm defended Showa Denko in all of the more than 2,000 lawsuits brought against it in the US He told me that almost all the suits were settled out of court and that only three went to trial. Further, he said that due to the nature of product liability law, the basic issue was whether SD's product had caused harm – and that consequently, it had not been necessary to determine the role of genetic engineering. Therefore, the issue was never raised, and none of the verdicts in any way touched on it.

Not only is it surprising that the commission advanced so many demonstrably false assertions, it's hard to know who fabricated them and how they got included. That's because there was a serious deficiency in the way the references for the LT section were provided. Not only did this make it practically impossible to ascertain the source of any particular statement, it allowed the possibility that some of the false ones were not even based on the public hearings but instead derived from input that slipped in through irregular channels.[55]

However, although it's unclear how the falsehoods entered the document, it *is* clear they could never have done so if the facts about the toxic LT had not become thickly clouded. It's only because so much confusion had been sown for so long that a major report by a royal commission could have harbored so many bogus assertions – and that these fabrications could have been accepted by the media and the New Zealand government.

The Enduring Effacement of Facts

Although it was mistake-riddled, at least the commission's report acknowledged that GE bacteria were associated with the EMS. But over time, this key fact has gradually faded from general awareness; and the lapse is not wholly attributable to mass forgetfulness. To a substantial degree, it's the result of a sustained endeavor to befog that fact.

And one of the chief befoggers was the FDA. Whenever the agency could control the flow of communication, it stayed silent about genetic engineering while liberally impugning tryptophan itself. A striking example occurred on July 18, 1991 when the deputy director of the Center for Food Safety and Applied Nutrition, Douglas Archer, appeared before a congressional committee to present the FDA's official position on the epidemic. Although by that date it was well-known within the agency that the disease-linked bacteria had been genetically engineered, Archer did not mention that fact – nor did he even refer to the technology. Instead, he targeted tryptophan in general, using the epidemic as a means to advance the agency's protracted campaign against dietary supplements. He asserted that it confirmed FDA's warnings about the hazards of such products and that the deaths and injuries "demonstrate the dangers inherent in the various health fraud schemes that are being perpetrated on segments of the American Public."[56]

For a long time, the FDA had argued that to protect the public from such schemes, all vitamins, minerals, and amino acids in dietary supplements should be brought under its supervision; and Archer acknowledged there was an agency desire to "closely regulate" them. However, most Americans wanted free access to natural health supplements, and Congress sided with them, amending the Food, Drug and Cosmetic Act in 1976 so as to limit the FDA's authority to regulate vitamins and minerals. The agency's reach over supplements was further restricted by several court decisions, including two that blocked attempts it made to remove over-the-counter LT from the market.

But the FDA's desire to restrict supplements still simmered. And it was substantially fueled by dubious motives. This is clear from a report by its Dietary Supplement Task Force stating that deliberations had included ". . . what steps are necessary to ensure that the existence of dietary supplements on the market does not act as a disincentive to drug development." [57] Reflecting on this "particularly disturbing" statement, an article in the Rutgers Law Journal noted that the agency's policy in this area "has far more to do with eliminating competition in the pharmaceutical industry than preserving the public health." [58]

In the wake of the EMS epidemic, the agency saw an opportunity to advance this anti-competitive aim and achieve what it had twice failed to do in court; and it banned all LT supplements. Of course, to do so, it had to pretend that LT could have caused the EMS all by itself (and ignore the compelling evidence to the contrary) while remaining mute about the bioengineering employed in its production. But its insincerity was revealed by its inconsistency. Despite its professed concern about the hazards of LT, the agency displayed significant selectivity in restricting it, forbidding its sale as a nutritional supplement while allowing pharmaceutical companies to vend it as a prescription medicine (at around five times the price it had borne as an over-the-counter supplement) – thereby eliminating a popular and relatively inexpensive competitor to Prozac® and other antidepressant prescription drugs which, like LT, enhanced the level of serotonin.[59] So disingenuous was the agency's anxiety about L-tryptophan that it even permitted the chemical's continued use in baby food.

Thus, by misrepresenting the details of the EMS incident, the FDA was able to simultaneously advance three of its cherished aims. It created ostensibly solid grounds for taking LT supplements off the market, it strengthened the case for strictly regulating *all* supplements, and it shielded genetic engineering. Moreover, the shielding was effective. Even though the facts warranted a precautionary policy on bioengineering, the unequivocal nature of Archer's pronouncements, conjoined with the confusion that had already been created, diverted attention from it. Accordingly, Congress did not investigate further, and the media routinely failed to mention the technology in regard to the epidemic, while frequently parroting the FDA's indictment of unregulated supplements.

By 1994, the FDA's effort to efface the facts of the toxic tryptophan incident had grown so brazen that the agency not only ignored the role of bioengineering in the supplement's production, it even pretended that the supplement had never been produced. Thus, an FDA publication about biotech foods released in that year contained not a word about Showa Denko's GE-derived food supplement – while declaring that the enzyme for cheese production had been "the first biotechnology food product," despite the fact it was not introduced until six years *after* the tryptophan supplement was initially marketed.[60]

Moreover, even when interacting with experts who could not be fooled about the fact the toxic supplement had once existed, the FDA still tried to pin sole blame for the toxicity on LT itself. An especially egregious attempt occurred at a scientific conference in 2004. According to Stephen Naylor, who investigated the epidemic thoroughly while he was a professor of biochemistry, molecular biology, and pharmacology at the Mayo Clinic, an FDA representative made claims about the role of LT in the causation of EMS that were so "bizarre" they "defied belief."[61]

So, although when queried about the relation between the epidemic and genetic engineering, the FDA has sometimes conceded that it cannot be ruled out as the cause, the net effect of the agency's EMS-related statements has been not merely to rule it out, but to blot it out. And the FDA has not been alone in giving the technology the silent treatment. Whenever they could, other individuals and organizations that promote bioengineering have also avoided its mention in regard to EMS. Further, they've increasingly proclaimed that no food produced by it has ever caused any disease. For example, a brochure by the Australia/New Zealand Food Authority touting the safety of genetically modified foods declared: ". . . there has been no case reported worldwide of a GM food causing an adverse effect on human health. . . ."[62] In the exceptional cases where a propounder of such claims has been challenged by someone with knowledge of the facts, he or she would contend that because bioengineering had not been proven to be the epidemic's cause, it was valid to assert that none of its products had caused any ailment – and unwarranted to state that one had. But this argument is clearly false. It's accurate to assert that a food produced through bioengineering caused a disease if it's obvious that one did. This does *not* entail that the process was the cause, but it does imply the possibility. Only by acknowledging that a GE-derived food has caused a problem can one then address the question of whether the technology was a significant factor. But by denying that any GE food has caused disease, one distorts reality and implies there's nothing to investigate.

Despite their illegitimacy, these denials (and related deceptions) have continued; and as the resultant illusions took hold, scientists felt free to ignore the EMS disaster when writing books that promote GE foods. For instance, although they purport to be balanced, neither *Mendel in the Kitchen* (by a molecular biologist who's a member of the National Academy of Science) nor *Tomorrow's Table* (by a plant scientist at the University of California) mentions anything related to the epidemic – which enables them to present a more appealing picture.[63] Nonetheless, these influential books have been praised for their scientific approach, even though they also omit (or distort) many other unfriendly facts, as the next chapter reveals.

The ongoing inaccuracies and omissions have caused widespread delusion, even among people who would ordinarily have maintained clarity. Thus, although several journalists did keep sight of the fact that GE was used in producing the toxic product, they were misled by other misrepresentations. For instance, the science reporter who authored *Lords of the Harvest* stated that because cases of LT poisoning occurred before the introduction of bioengineering, it was unlikely to have caused the epidemic. He then declared: "Indeed, if the tryptophan case showed anything, it was the dangers residing in food supplements that often are sold in health food stores, not genetically engineered foods."[64] Worse, due to the extent of the distortions, many commentators didn't even realize that genetic engineering was part of the picture. Among them was the British scientist, Susan Aldridge. In her book, *The Thread of Life: The Story of Genes and Genetic Engineering,* not only did she fail to note the technology's involvement, she made an assertion about the epidemic's cause that was devoid of evidentiary support. She indicated the problem was inherent in the bacterial strain itself – and remarked that SD's technicians "were unlucky" because they chose one that produced a toxic contaminant. [65] That a seasoned science writer committed such a blunder, and that the editors at Cambridge University Press let it slide by, indicate how muddled the facts had become.

Equally indicative is a report on L-tryptophan issued by a prominent natural health center. In discussing the contamination that caused the EMS, it states: "The manufacturing error was identified and corrected relatively quickly."[66] This report was written in December 2009 by a medical doctor who is a past-president of the American Holistic Medical Association, and it has circulated widely for several years, appearing within a number of health-related magazines and websites. Further, it's noteworthy that the health center sponsoring the article appears to be opposed to GE foods (another article on its site cautions against eating them because they're "not natural").[67] Nonetheless, despite the preference for natural approaches shared by the center and the author, and despite the latter's expertise, not only was he unaware that bioengineering had been used to produce the disease-linked LT, and that it may well have been the epidemic's cause, he absorbed the false impression that a simple manufacturing error was conclusively identified as the causative factor. Further, the mass of misinformation has been so confounding that many commentators even lost sight of the epidemic, including two journalists who covered GE foods for years but whose popular books on the topic failed to mention the calamity at all.[68]

Not surprisingly, the confusion is more widespread within the general public than among experts. The vast majority of people with whom I've spoken over the years had no inkling that a food supplement produced through bioengineering was associated with a major catastrophe; and it's likely that most readers of this book are learning about it for the first time.

The Thalidomide of Genetic Engineering

While the obfuscation of the epidemic has been significantly nefarious, its initial detection was largely fortuitous. The discovery was due to anomaly; and if the symptoms of EMS had not been so unusual, the epidemic would probably have gone unnoticed. Crist, along with a biochemist and a medical doctor, emphasized this point in an article comparing the GE tryptophan to thalidomide, a drug used between 1957 and 1961 that eased morning sickness in pregnant women while unexpectedly inducing severe deformities in their fetuses. They stated that "if thalidomide had happened to cause a type of birth defect that was already common, e.g., cleft palate or severe mental retardation, we would still not know about the harm, and pregnant women would have kept on taking it" because "the fractional addition to figures that were already relatively large would not have been *statistically* significant."[69] They noted that the adverse effects were detected only because they were extraordinary (major malformations of the arms and legs) and that, similarly, the disease caused by SD's tryptophan "stood out" because it was novel. They observed that if instead it had caused the same quantity of a common illness, "we would still not know about it." Likewise, "if it had caused delayed harm, such as cancer 20 - 30 years later, or senile dementia in some whose mothers had taken it early in pregnancy, there would have been no way to attribute the harm to the cause."

It's sobering that, despite the novelty of EMS, many years had to pass (and an epidemic had to erupt) before it was finally detected; and even after that dramatic outbreak, several months elapsed before the detection could be accomplished. Further, several more months passed before the disease could be linked to SD's tryptophan. This provides grounds for questioning the safety of the many supplements and additives derived from GE bacteria that are currently in use. The mere fact that they satisfy normal standards of purity does not rule out the presence of contaminants that are highly toxic at extremely low concentrations (as was the case with SD's product). Nor does the fact that there's been no observed link to disease, since a toxin might be causing a common malady that's going undetected. The uncertainty is underscored when one realizes that if SD's lethal tryptophan was first appearing today, it could enter the market just as freely as it did twenty-five years ago, in Europe as well as the US.[70] Accordingly, many experts have warned that GE-derived additives should undergo thorough safety testing before they're approved for sale – warnings ignored by those with the authority to implement the reform they call for.

The Evidence Implicates Bioengineering as the Most Likely Cause of the Calamity

As we've seen, the EMS story is replete with anomaly, surprise, and paradox. Although genetic engineering cannot be ruled out as the calamity's cause, due to an exceptional degree of misinformation, not only do most experts believe it's been absolved, many don't know it was even involved. Moreover, only a few of the people who comprehend that the engineering *could* have been the critical cause realize how strongly the evidence implies that it actually *was*. So strong is the case against the technology that, although the Royal Commission falsely asserted that US courts had found it blameless (despite the fact its involvement was not at issue), if its role actually *had* been the decisive factor, the verdicts would most likely have deemed it culpable. That's because in a civil trial, where only monetary damages (not the defendant's life or liberty) are at stake, the plaintiff does not need to prove his case beyond a reasonable doubt. It's sufficient to demonstrate that the *preponderance* of evidence is on his side. That means the EMS victims could have won simply by showing it was *more likely than not* that GE caused the critical toxicity. And the evidence clearly tilts toward such an outcome:

- Even people consuming high doses of conventionally produced LT did not contract EMS, which indicates that LT alone was not the cause.
- All the LT that was definitely linked to EMS was produced by Showa Denko.
- SD was the only manufacturer that used genetically engineered bacteria.
- Not only were all the epidemic cases of EMS that could be traced linked to GE bacteria, every pre-epidemic case appears to have been linked to strains of GE bacteria as well.
- As SD's bacteria were increasingly altered to output greater levels of LT, it appears there was concomitant increase in stress, resulting in metabolic imbalances. It also appears that as the genetic manipulations became more powerful, the LT became proportionately more harmful.
- (a) The presence of unusual contaminants, (b) the lethal toxicity of at least one of them at an extremely low concentration, and (c) the odd way in which their concentrations fluctuated over time are phenomena more readily explicable as effects of genetic engineering than as outcomes of another aspect of the production process – especially since there is no evidence indicating that the critical contaminant was generated during the purification phase.

Important New Evidence Increases the Likelihood that Bioengineering Was the Cause

Further, the preceding summary only reflects facts that were available when the suits were decided. Today, the case against genetic engineering is even stronger. As the neurobiologist David Schubert points out, although for many years it was hard to explain how an extremely minute contaminant (well below 0.01% by weight) could have caused fatal dysregulation of the immune system, we now have a better perspective. He notes that several studies have revealed that metabolites (derivatives) of LT control important steps of the immune response, which presents the possibility that unusual (but analogous) metabolites induced by the overproduction of LT could have displaced the ordinary versions and disrupted people's immune function in disastrous ways.[71]

Moreover, the structure of the contaminant most closely correlated with EMS has finally been ascertained; and it's a novel, metabolically derived compound of tryptophan. As previously noted, that contaminant was dubbed AAA, and its structure had remained unknown, even though analysis by CDC scientists showed it to be the only contaminant linked with EMS to a statistically significant degree – prompting them to urge that "high priority" be placed on the quest to identify it.[72] Nonetheless, despite this plea, it took another six years before the identification was achieved; and, as I write, only a few people are aware it has happened.

I learned of it because I heard that Stephen Naylor and Gerald Gleich, who had identified the structures of the five other case-associated contaminants while they were at the Mayo Clinic, had also investigated AAA during their tenure there. So I contacted Dr. Naylor, and a series of communications ensued during which he conveyed the details of the research.

Before the investigation began, Naylor had hypothesized that in order for a toxic contaminant ingested in extremely minute quantities to induce EMS, it would have to avoid immediate excretion and remain in the body for a prolonged time so that the effects of successive doses could accumulate. This would require it to be fat soluble rather than water soluble, which would enable it to lodge within fatty tissues and then slowly seep into the surroundings. But the structures of the other five case-associated contaminants were water soluble – and hence fat aversive. Would the structure of AAA, the contaminant linked with EMS to an exceptionally high degree, prove to be an exception in this regard as well?

In the latter part of 1998, Naylor and Gleich had a chance to find out. They obtained some tablets from a batch of SD tryptophan that had caused an extraordinarily large number of EMS cases. Then, employing sophisticated analytical separation techniques, they determined that, while the overall concentration of AAA was minute, it was nonetheless quite high in relation to its levels in tablets from less toxic batches. Further, it was the *only* contaminant of the six case-implicated compounds that was markedly elevated.

The next step was to ascertain its structure. Through mass spectrometry, and great perseverance, they ultimately determined that AAA had been formed by the fusion of two compounds. One was the LT molecule (minus a single hydrogen atom), and the other was a long chain hydrocarbon derived from a fatty acid that's found in all bacteria.[73] Based on this structure, the researchers could draw some important conclusions.

According to Dr. Naylor, it's a biological certainty that such a compound could not have arisen during purification. When I spoke with him, he made this point at least twice, without a trace of qualification. He further noted that chains of this kind are formed within bacteria – and that they would have to be synthesized via biologically produced enzymes. Consequently, he stated there's a "high probability" that the AAA molecules had an intra-bacterial birth. Moreover, although he acknowledged the possibility they were produced outside the bacteria (within the fermentation broth *before* it entered the purification phase), he said this possibility had "low probability." He added that even in that case, the synthesis would have relied on enzymes from the bacteria acting upon LT; and he emphasized that in either instance, the synthesis would almost surely be attributable to the alteration of the bacterial DNA and the massive overproduction of LT it induced, which most likely destabilized the organisms' metabolism and caused unusual side reactions resulting in the formation of AAA.

Thus, it's been demonstrated that the contaminant most significantly associated with EMS is a novel compound formed by bacterially produced enzymes acting on L-tryptophan well before the purification process could have exerted an effect – which strongly implies that the engineering of the bacteria was the root cause of its creation. Further, according to Naylor, "Not only is it virtually certain that this contaminant was formed through the action of bacterial enzymes, its chemical structure renders it fat soluble. In contrast to the other case-associated contaminants, these unique chemical properties of AAA facilitate uptake by fatty tissue, allowing accumulation and concentration by the body – potentially resulting in the stimulation of eosinophils and, ultimately, the onset of EMS."[74] Therefore, although we cannot say for sure that AAA triggered the illness, we *do* know that it's quite plausible; and we also know it's highly probable that this novel and potentially toxic compound emerged through abnormal metabolic activity caused by the hyper-production of L-tryptophan.

However, despite the importance of this discovery, the results have yet to be published. When I asked Dr. Naylor why, he explained that although he and Gleich had definitely determined that AAA consists of an LT-like skeleton conjoined with a nine carbon linear chain, there is still some uncertainty about two minor points. First, while it's clear that the chain is attached to the skeleton, it's not clear at which of two adjacent places the attachment is made. Further, although they know that the chain contains eight single bonds and one double bond, it's not evident at which of two neighboring positions the double bond occurs.

But is the existence of these two small uncertainties relevant to the issue of how AAA emerged? When I posed this question, Naylor replied that the ambiguities had absolutely no bearing on that issue. However, they did affect the chances of getting the research published. He explained that scientific journals would require the determination of AAA's structure to be complete – and that after he left Mayo in 1999 to take another position, the research could not be sustained with the same intensity. Further, a few years thereafter Dr. Gleich also moved on to assume new responsibilities; so it's not clear when either of them will have the time or resources required to gain the final bits of knowledge. Yet, because the evidence already acquired has profound implications, Dr. Naylor has agreed for it to be presented in this chapter as "unpublished work." [75]

With the unveiling of the evidence about AAA, the EMS story comes to a close – at least for now. It's a story that begins with agony and ends in irony. Even without the revelations regarding AAA, the facts plainly point toward genetic engineering as the underlying cause of the EMS; but they've been so befogged that many scientists don't even know the toxic tablets were produced through it, while most of those who do know believe that it's been proven innocent. Moreover, not only has there been systematic obfuscation of the epidemic's cause, there's been substantial obfuscation of the epidemic itself – to such an extent that, although the GE-linked disaster was detected only through its uncommon symptoms, most people are as oblivious of it as they would be if the symptoms had instead been commonplace.

Thus, just as the smooth advance of the bioengineering venture had, in the years preceding 1989, depended upon the clouding of unfavorable facts, so its continued progress was enabled by obscuring the facts about the toxic tryptophan, the first ingestible product of recombinant DNA technology. Yet, in the latter case, the facts were not merely clouded but essentially shrouded – and ultimately buried. Absent such interment, the development of bioengineered foods would almost certainly have been delayed, and probably derailed.

Moreover, in order to keep GE foods on the fast track to commercialization, it was insufficient merely to obfuscate the disaster one had caused. In the following years, as the campaign to bring them to market accelerated and its attendant controversies flared, their proponents would find it increasingly necessary to distort and even suppress key facts about the very process by which such foods are produced.

CHAPTER FOUR

Genes, Ingenuity, and Disingenuousness

Reprogramming the Software of Life while Refashioning the Facts

> "... there is a seamless continuum between conventional and 'new' GM [genetic modification]."[1] Henry I. Miller, Founding Director, FDA Office of Biotechnology (on the relation between conventional breeding and recombinant DNA technology)

> "Recombinant DNA technology faces our society with problems unprecedented not only in the history of science, but of life on the Earth. It places in human hands the capacity to redesign living organisms. . . . Such intervention must not be confused with previous intrusions upon the natural order of living organisms. . . .[2] [It is]the biggest break in nature that has occurred in human history."[3] George Wald, Nobel Laureate; Professor of Biology Emeritus, Harvard University

Pressures to Repress the Facts

In 1993, Oxford University Press published *Biotechnology from A to Z*, a guide to the terminology and techniques of what had become one of the most important and controversial fields of applied science. It was written by a professional biologist, it presented a positive picture of this remarkable new phase of human enterprise, and it was praised by several scientific journals. The introduction was written by the president of a corporation at the forefront of biotechnology and expressed what for many years has been a standard theme in the statements of its proponents: that it is crucial to educate the public about this innovative endeavor and to ensure that the information they receive is accurate.[4]

But in a subsequent section of the book, its author, William Baines, indicated that such an educational initiative would likely backfire. He noted that research has revealed an inverse relation between the public's knowledge about biotechnology and their acceptance of it, with people less receptive the more they learn the details. He observed that in light of this phenomenon, biotech advocates might have greater success by providing the populace fewer facts and more "mythic images."[5] And he underscored the significant role of myth-making in the biotech venture by giving the topic its own distinct heading: *Mythogenesis*.

From such a perspective, it appears there's no field of biotechnology in which the need for myth creation has been greater than that of genetically engineered food. Because food safety is such a vital and visceral issue, people tend to be especially wary about what they perceive as artificial tinkering with the DNA of plants and animals that are basic to their diets, and most display significant reservations when they first hear about the agricultural biotech agenda. Further, because this initial resistance generally intensifies as peoples' knowledge of the facts increases (a trend that continues to be confirmed by research in a variety of nations),[6] biotech advocates have frequently found it expedient to follow the course suggested by Dr. Baines and opt for creativity over candor – fashioning a group of mythic images to aid their cause. For instance, a memorandum from the world's largest public relations firm, Burson-Marsteller, to the European biotech industry (which was leaked to a public interest group) counseled it to eschew "logic" and instead employ "symbols," particularly those "eliciting hope, satisfaction, caring and self-esteem."[7]

However, at least one aspect of the public's wariness could not be easily assuaged with evocative symbols: the perception that producing new varieties of crops through genetic engineering is a radical and unnatural departure from traditional breeding. So proponents of genetic engineering tried to overwhelm this perception by inducing a compelling counter-impression: that the process is merely a minor extension of traditional breeding practices. Moreover, many prominent advocates have insisted that the connection is quite close, with Henry I. Miller, the founding director of the FDA Office of Biotechnology, proclaiming (in a widely circulated statement) that there's "a seamless continuum" between genetic engineering and what came before.[8]

And to further blur the distinction between the novel and traditional processes, biotech proponents transformed their terminology. Although they had initially referred to the use of recombinant DNA technology as "genetic engineering" because of the positive associations they expected it to convey, they eventually learned that in most peoples' minds, the term did not primarily connote control and precision but artificial – and potentially detrimental – intervention. So they decided to recast the process as mere "genetic modification," which seemed to strike the public as less threatening. Moreover, whereas "genetic engineering" had been exclusively applied to rDNA technology, the new term of choice was not so restricted and was employed in reference to *all* forms of breeding (even simple sexual reproduction), with gene-splicing presented as the "new" or "modern" phase of genetic modification.

Additionally, while downplaying differences that could cause concern, the proponents advanced the idea that this new technology *does* differ from conventional practices in one key respect. They claimed that it's more precise; and they contended that by virtue of this precision, it is more predictable than conventional techniques and is consequently a *safer* way to generate new varieties of food. And these claims could be markedly immodest. During a BBC interview in 2000, the president of Britain's Royal Society (who for five years had served as the government's chief scientist) declared that genetic engineering is "vastly safer" and "vast, vastly more controlled" than conventional breeding.[9]

However, notwithstanding its prevalence, and the prominence of many who advanced it, the portrayal of genetic engineering as a minor, precision-enhancing extension of natural breeding was starkly at odds with reality. It was yet another instance of myth-making: biotech's own "creation myth," depicting the genesis of GEOs in an attractive but essentially fictitious manner.

The extent of the fiction becomes evident when one examines the various manipulations that are necessary to produce a new type of food-yielding organism via genetic engineering and discovers how imprecisely they function, how many natural barriers they had to surmount, and how unpredictable have been the results. As we shall see, not only is there a deep disparity between the elegant procedures through which new organisms are generated and sustained under natural conditions and the crude contrivances of genetic engineering, in several respects, the two methods induce opposing outcomes.

How Nature Functions: The Essential Dynamics of Living Systems

Living organisms are comprised of living cells.[10] The simplest organisms, such as bacteria, consist of a single cell, while plants and animals contain millions or trillions of diverse and specialized cells that give rise to a variety of tissues and organs. But whether an organism consists of one cell or a myriad, it stands in stark contrast to its nonliving surroundings. Every living entity displays a high degree of organized complexity, and each integrates a multitude of diverse parts into a harmoniously functioning whole. Such orderliness is absent within the inanimate realm of nature, and none of its structures comes anywhere close to the degree of organization exhibited by even the simplest bacteria. Further, due to the distinct way in which organisms are structured, they can alter the ordinary course in which energy flows. In the nonliving world, energy tends to dissipate, and it diffuses in a fairly uniform and undirected manner. But energy flows otherwise within an organism. It is systematically absorbed, stored, and then efficiently utilized to power a host of precise manufacturing processes, yielding an enormous range of products that sustain the organism's existence.[11]

These processes, and the energy transformations that drive them, occur through a vast variety of chemical reactions. And, as with the other features of organisms, most of these reactions are in a significant way unique to the animate world. The vast majority never happen in nonliving nature, and of those that do, most occur far too infrequently to fulfill the needs of life. If organisms had to wait for these reactions to occur at their normal pace, they could not survive – and could never have come to exist. Fortunately, and marvelously, all organisms possess the ability to induce reactions that never occur in the nonliving world and to profoundly increase the rate of those that do. They accomplish these feats by producing a special set of tools in the form of proteins. While many proteins serve as components of a cell's structure, those that serve as its reaction-enhancing tools are catalysts – agents that facilitate the interaction and transformation of other chemicals while not being changed themselves.

Although catalysts exist in nonliving nature, they are far more abundant (and more varied) within living cells. Such cell-dwelling catalysts are called *enzymes*, and they are the biggest class of proteins within an organism. The average mammalian cell contains about 3,000 of them.[12] Even the simplest bacterium requires hundreds of enzymes to function, and without them, there would be no life on earth.

Moreover, cells must not merely be able to create catalysts, they must do so selectively. It's necessary that they stimulate the production of specific reactions only as needed – and just as necessary that they keep the myriad production processes coordinated. Otherwise, in attempting to make enough of the materials on which they depend, they could be overwhelmed by too much of a particular product even when it is needed, or disrupted by its appearance (even in a minute amount) when it isn't.

The remarkable processes by which living organisms create the kinds of enzymes they need, at the right times, and in the proper amounts and places, rely on a source of order that is highly stable while enabling great adaptability. Since information theory recognizes a close relationship between information and order and defines information in terms of orderliness, it's not surprising that the basis of the order underlying the stability and flexibility of life processes is an exquisite information system.

DNA as a Repository of Foundational Biological Information

This information system is within every cell; and each cell, whatever its type, functions as a powerful information processing machine.[13] Although portions of the information are dispersed throughout several cellular domains, a large and essential part of it is encoded within an extraordinary molecule referred to as *deoxyribonucleic acid*, or DNA, that resides at the heart of the cell. Bacterial cells usually contain one main DNA molecule while higher organisms, which possess larger and more complex information systems, are endowed with a substantial number.[14] But whether contained within one molecule or spread among many, the information encoded by DNA is foundational for the coordinated growth and function of the organism. It is therefore immense – and must be highly condensed. In fact, it is more densely and efficiently stored than the information in any man-made system.[15]

DNA's profound information-bearing capacity is due to its structure. The basic constituents are called nucleotides, and they're composed of a phosphate molecule, a five-sided sugar molecule, and a nitrogen-bearing molecule called a base. While the phosphates and sugars are the same from nucleotide to nucleotide, the bases vary. There are four different ones, each with a distinct chemical structure: adenine, thymine, cytosine, and guanine (commonly designated by their first letters A, T, C and G).

Nucleotides naturally pair up because there's a chemical attraction between adenine and thymine and between cytosine and guanine, which causes them to bond together. This results in segments that have phosphate and sugar molecules at each end and base pairs of either adenine and thymine or cytosine and guanine in the center. (*See Figure 4.1*) In a DNA molecule, numerous segments are aligned in a ladder-like structure, with the phosphate and sugar complexes forming the outer rails and the bonded base pairs the rungs. Further, this ladder is not essentially a flat two-dimensional structure but is twisted into a helix, so that in three-dimensional space, it is more like a spiral staircase. (*See Figure 4.2*)

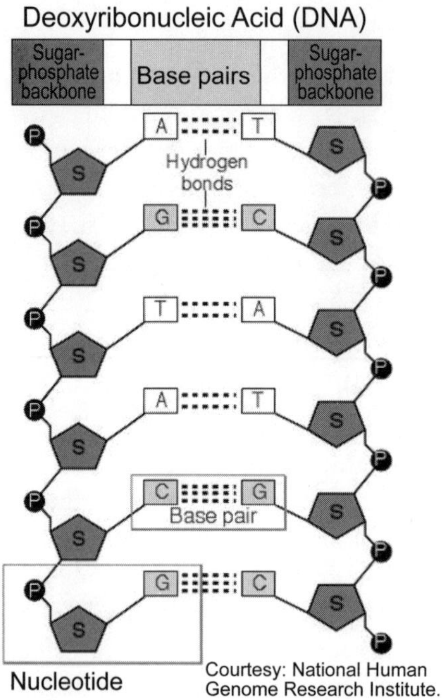
Deoxyribonucleic Acid (DNA)
Sugar-phosphate backbone
Base pairs
Sugar-phosphate backbone
A T
Hydrogen bonds
G C
T A
A T
C G
Base pair
G C
Nucleotide
Courtesy: National Human Genome Research Institute.

Figure 4.1

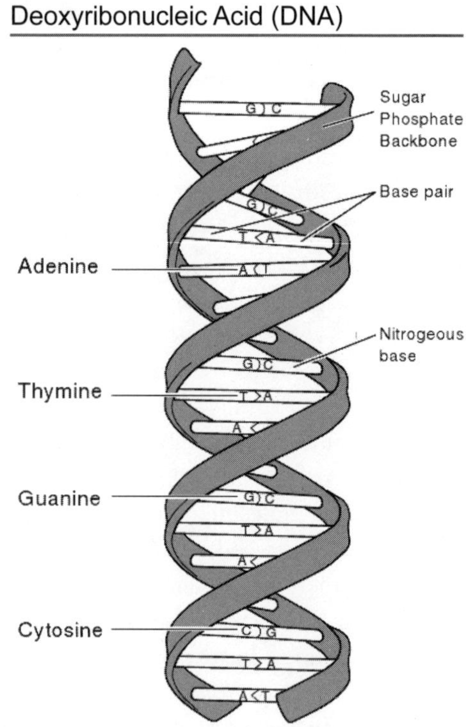
Deoxyribonucleic Acid (DNA)
Sugar Phosphate Backbone
Base pair
Adenine
Nitrogeous base
Thymine
Guanine
Cytosine
Courtesy: National Human Genome Research Institute.

Figure 4.2

Although its spiral structure endows DNA with important properties, the key factor underlying its profound information-bearing capacity is the variability of the bases embedded within the spiral – and the selective way they bond to one another. These bases convey information through the sequence in which they appear, because the sequence serves as a code. The code consists of equal-sized units of meaning comprised of three bases. These three-base units are referred to as *codons*, and what they specifically code for are amino acids, the building blocks of proteins. Because the meaning is in the sequence, three contiguous thymines (TTT) bear a different significance than do two thymines followed by adenine (TTA), and each of these units codes for a different amino acid. However, because there are twenty basic amino acids, and because the four bases can be arranged to form sixty-four codons, most amino acids are signified by more than one. For instance, both TTT and TTC code for phenylalanine, while TTA, CTA, and four other codons denote leucine.

Proteins are formed from chains of linked amino acids, and each type of protein has a distinct sequence of them. These protein-specifying sequences of amino acids are derived from corresponding sequences of codons within special regions of DNA. These information-rich coding regions are called *genes*. Humans have over 20,000 of them, and even some bacteria contain 5,000.

Cells have several finely tuned tools to convert the sequences of codons within genes into proteins, and the process has two basic stages. In the first, a specialized enzyme travels the length of the gene and transcribes its information into a strand of another (but similar) type of nucleic acid called *ribonucleic acid*, or RNA. In the second stage, this RNA strand becomes a messenger and carries the information to an intricate structure that can translate it into a chain of amino acids that will then fold into a protein.[16] (*See Figure 4.3 on the next page.*)

Propagation and Progress: Continuity Enriched by Diversity

Besides DNA's essential role in the development and survival of the individual organism, it enables the survival of the organism's species. By virtue of its unique attributes, organisms can propagate new organisms endowed with the essential stock of genetic information that they themselves possess, preserving the species' fundamental characteristics.

Although this process of preservation through propagation occurs in all species, its mechanics vary. For single-celled entities like bacteria, propagation occurs through cell division, which yields two organisms in

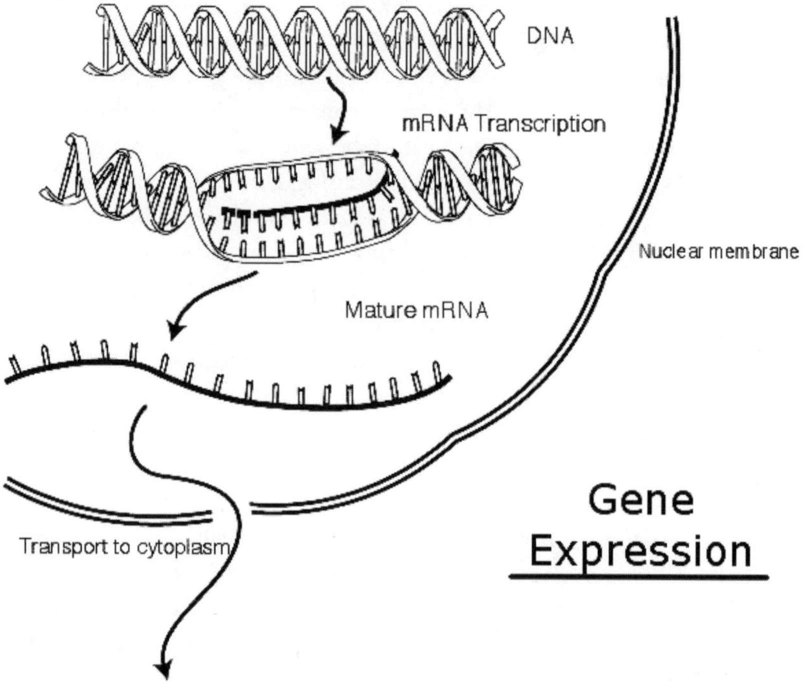

Courtesy: National Human Genome Research Institute.

Figure 4.3

Author's Note: *This illustration depicts transcription of DNA into messenger RNA (mRNA), and it shows that the mRNA travels outside the nuclear membrane into the surrounding cytoplasm for translation into an amino acid chain. But it does not depict that translation process.*

place of one. This is possible because the DNA molecule can be replicated, furnishing an identical copy around which another cell can coalesce. (*See Figure 4.4*) However, while this process assures continuity, it does not foster diversity – and lack of diversity can lead to problems, because if a species' genome remains uniform from generation to generation, it has difficulty adapting to environmental change. Of course, genomes can change via spontaneous mutations; and although most are maladaptive, some are beneficial and can be conserved in the species over time. But bacteria have additional ways to increase their genetic diversity – ways through which they acquire genes from other bacterial species.

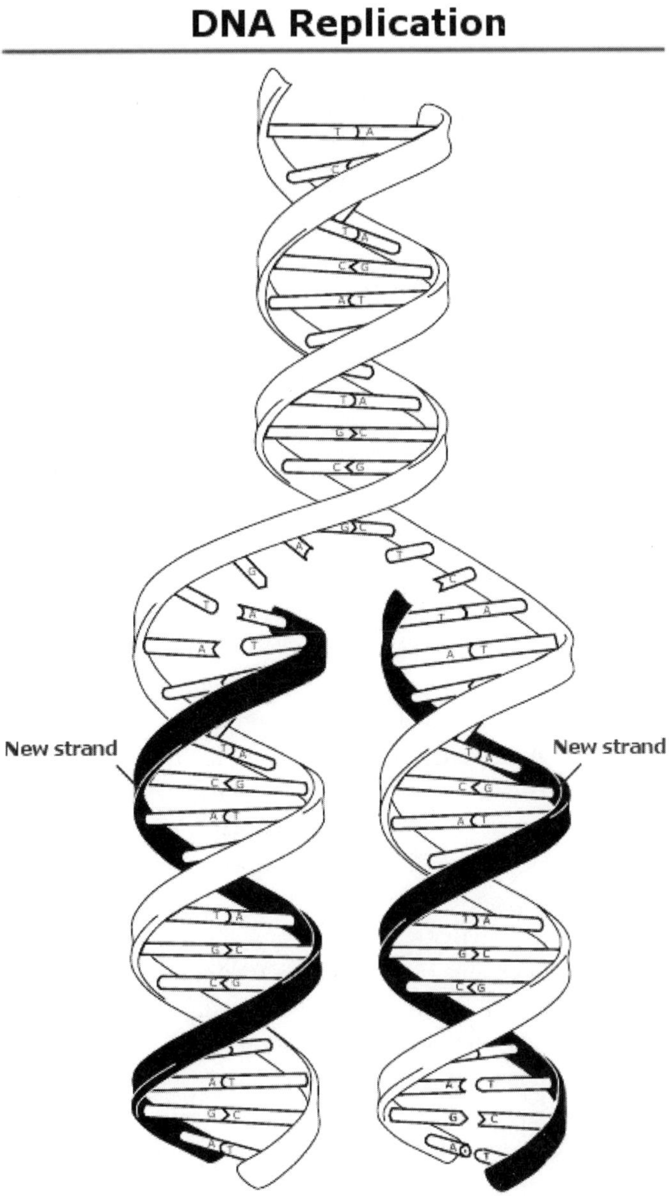

Courtesy: National Human Genome Research Institute.

Figure 4.4

One way, called *conjugation*, relies on direct contact between two bacteria, one of which is the donor and the other the recipient. A tubule extends from the donor, and copies of some of the genes are transmitted to the recipient's interior. Although the entire genome is rarely transferred, a substantial amount of genetic material can be conveyed. There are two other ways in which some bacteria can gain foreign genes, and while neither requires contact with another bacterium, one does depend on the agency of another entity – but in this case, it's a virus. Sometimes, when infecting a bacterium, a virus pulls one (or a few) genes from its DNA, moves on to infect a bacterium of a different species, and transfers the foreign genetic material to it. Further, in some instances, decomposing bacteria release a bit of DNA that's directly absorbed through the outer wall of another bacterium.

In contrast to bacteria, most plants and animals enhance their genetic diversity in a more comprehensive manner: sexual reproduction. In this process, DNA from two organisms is combined to form a new one; and even when the parents are from the same species (the usual situation), there's a significant increase in genetic diversity. That's because of the way DNA is arranged within higher organisms – and is deployed during their reproduction.

As previously noted, while most bacteria have but one main DNA molecule (which is generally circular), plants and animals have several. These molecules are usually linear and, in combination with specialized proteins, each forms an organized structure referred to as a *chromosome*. Moreover, each chromosome has a partner, which contains the same genes in the same sequence.[17] However, although the corresponding genes are the same, they can still differ from one another. That's because (as was discussed in Chapter 2) there are alternative versions of a gene, called its *alleles*, just as there are different versions of a particular model of a car. In preparation for sexual reproduction, an organism forms special cells (called *gametes*) to which it contributes only one set of chromosomes. But before the partner chromosomes are separated and encased in separate gametes, they exchange some complementary sections of DNA. In this way, each ends up with a different set of alleles than it previously possessed, enhancing diversity.

Gametes come in two basic types: male and female. And most animals come in distinct male and female types too. A male animal produces only male gametes (sperm), females produce only female gametes (eggs), and the sperm from the males combine with the eggs of a female. In contrast to animals, most flowering plants are bi-sexual, and a single organism commonly creates both male and female gametes. A plant's male gametes are usually encased in pollen grains that travel via either the wind or insects to fertilize the female gametes of other plants, while its own female gametes receive pollen flowing in from others.[18]

Whatever the species, when the male and female gametes unite, every chromosome in the sperm is partnered with the corresponding chromosome in the egg. The resultant cell is thus endowed with a full complement of chromosomes and can develop into a mature organism. Further, because that organism's genome is a blend of chromosomes from each parent, and because several combinations of alleles within those chromosomes were re-arranged prior to gamete formation, the organism will not only be genetically distinct from each parent but will possess some features found in neither.

The Modes of Conventional Breeding

For millennia, farmers added direction to the reproductive process of cultivated plants by selecting the most desirable specimens from each year's harvest and replanting their seeds. Then, in the modern era, breeders learned they could more closely guide the process by selecting which plants would mate. By taking pollen from a plant with one set of valuable traits and placing it on the pollen receptor of a plant with other desirable features, offspring exhibiting both sets of qualities could result.

Through the natural modes of breeding, tremendous diversity has arisen. For instance, over 100,000 varieties of rice have been developed.[19] However, while nature promotes abundant genetic variety *within* the various species, it restricts the exchange of genes *between* them. Within nature's system of boundaries, not only is it impossible to interbreed distant and unrelated organisms, many species cannot even be crossed with their cousins. Thus, there are no avenues for mating tomatoes with fish, and, although peaches and cherries are closely related, placing the pollen of one on the receptors of the other will not be productive.

During the 20th century, agronomists sought ways around the natural barriers. One of the techniques they developed, called *embryo rescue*, enables the maturation of some types of seeds that would otherwise be infertile. Such enfeebled seeds can result when plants of related species are interbred. In many cases, breeders are able to revive them by placing them in a nutrient medium conducive to their growth. Moreover, besides developing ways to widen the range of interspecies gene commingling, breeders also created new intraspecies alterations by mutating an organism's DNA through radiation or chemicals.

However, although the various techniques significantly expanded the range of genomic change, there were still major restrictions on what could be accomplished through them. Embryo rescue is not an option unless two species are sufficiently similar to produce some form of rescuable seed; and the vast majority of combinations are incompatible. Further, when employing radiation and chemicals, breeders cannot select a specific gene and mutate it in a particular way. Instead, they have to irradiate (or inundate) thousands of separate cells and hope that some beneficial new trait will be created in at least one of them by a fortuitous alteration of one or more of its genes.

Species barriers also limited the extent to which scientists could induce genetic transfer between bacteria. Generally, only closely related species conjugate; and because viruses usually infect a limited range of bacteria, they don't provide avenues for unrestricted gene transfer either. Further, it appears that only a very small percentage of bacterial species can ordinarily absorb DNA fragments from the environment.[20]

Genetic Engineering: Breaking New Ground by Breaking Ancient Boundaries

As molecular biology advanced, several of its practitioners dreamed of overcoming the constraints of nature by developing the powers to isolate and precisely manipulate individual genes – and to selectively move them between distant and disparate species. But, as we saw in Chapter 1, even after discovering the structure of DNA and the nature of the genetic code, they were still so far from this goal that it was beyond the realm of serious science fiction.

The seemingly insurmountable obstacle was inherent in the nature of DNA. In order to examine a gene and then to copy it, biologists needed to isolate it from the surrounding DNA. But this was no small task, because DNA is no small, or easily divided, molecule. Although scientists could isolate DNA from living tissue, they could not take the isolation process further by differentiating any of its components. A standard genetics textbook notes that within the test tube, the molecule is a "tangled mass of DNA threads" that "looks like a glob of mucus" – and that it therefore "seemed impossible" to isolate individual genes from it.[21] There were no mechanical means for neatly separating DNA into manageable segments, nor could it be done with any of the chemicals then available. Consequently, up until the early 1970's, all knowledge about genes had come from indirect inferences.[22]

And biologists would have remained at the stage of indirect knowledge had it not been for a lucky break – that enabled a breakthrough. In 1970, researchers began discovering a class of chemicals that can cleave DNA into discrete, manageable packets. These chemicals are enzymes that exist within several species of bacteria, and scientists eventually found hundreds of them. These chemicals can defend against viruses by restricting their activity, which is why they came to be called *restriction enzymes*.

When unrestricted, viruses are prolific parasites that usurp the resources of living cells. They do so because they are *not* cells, and they lack the capacity to reproduce and sustain themselves. In order to replicate its genes or to transform its genetic information into proteins, a virus must commandeer the resources within a living cell and compel them to generate more of its own components – which, compared to even the simplest bacterium, are minimal. They consist of the viral DNA (usually with fewer than thirty genes) and a set of proteins that act as a surrounding coat. Besides providing protection for the genes, the coat enables the virus to attach to a target cell.

Numerous species of virus are specialized to target bacterial cells. After one of these viruses binds with such a cell, it injects its DNA into the interior, leaving the coat on the outside. This viral DNA then re-directs the metabolic machinery to make copies of itself and to synthesize the various proteins that it codes for. These freshly-formed genes and proteins then combine to make new, coat-encased viruses. As the viruses accumulate, they stress the invaded cell and eventually burst it. However, when a bacterium harbors restriction enzymes, they can cut up the naked viral DNA before it starts to reproduce.

What makes restriction enzymes so important in genetic engineering is not merely their capacity to cut DNA, but their ability to do it in a precise manner. Each particular restriction enzyme recognizes a specific sequence of bases and cuts the DNA only at locations along the strand with such a sequence. Since every DNA molecule (regardless of species) contains restriction enzyme target sites purely by chance, bioengineers can utilize the various restriction enzymes to cut any DNA into short enough segments to work with.

There's yet another feature of restriction enzymes that has greatly aided the practice of bioengineering. Besides making it possible to consistently cleave DNA, they also enable the selective fusion of diverse fragments produced by the cleaving. That's because many of them make staggered cuts in DNA in such a way that any segments created by the same restriction enzyme will possess protruding ends that are complementary to one another and can readily bind together – which is why they're called *sticky ends*.[23] The staggered cutting performed by the enzymes and the complementary ends that are generated made it possible to neatly splice a segment cleaved from one DNA strand into a strand of a different species.

Thus, without the discovery of restriction enzymes, genetic engineering would have remained an unrealized dream; and they continue to be indispensable for its practice. Consequently, the GE venture has, from its outset, projected a paradox, because the role these enzymes have been made to assume in the laboratory starkly contrasts with the essential role they play in nature. While their main natural function is to *prevent* foreign genes from entering a cell and altering its operation, biotechnicians have employed them to promote that very thing, effecting the forced entry of alien genes into creatures that have never known them. So deep is the dichotomy between natural function and human application, one could reasonably argue that the latter does not merely contort nature, but stands it on its head.[24]

Of course, biotech proponents would probably argue that there's no perversion of function because, while the DNA that the enzymes attack within their home cells is pathogenic, the DNA that's transferred to other organisms in the laboratory is not harmful to the recipients. Yet, as we shall see, such alien inserts tend to induce stress, and in other key respects as well, they function more like viruses than cooperative constituents of the organism.

Creating the First Transgenic Organisms: Ingenious Incursions into Bacteria

Biotech advocates often impart the impression that trans-species bioengineering is merely a matter of taking a gene from one organism and popping it into the DNA of another, where it's gracefully received and fully ready to function. In reality, many steps are required, entailing extensive manipulation and modification.

Before a gene can be utilized, it must be isolated; and isolating a particular gene is a big job. Biotechnicians can't just go in and cut it out with restriction enzymes unless they know where within the glob of DNA that gene resides, and they can only gain that knowledge by doing extensive analysis – which itself entails a lot of cutting (usually employing several restriction enzymes).

Once the gene has been isolated, it must be copied. And one or two copies won't suffice. For reasons that will soon become apparent, a vast number are needed; and several steps are required to achieve the massive multiplication.

Then comes the task of getting the genes into the target organisms. During the first phase of the bioengineering venture, the targets were limited to the simplest organisms: bacteria – with *E. coli* the usual microbe of choice. But a vehicle was needed to convey the foreign DNA into the bacteria, and the most obvious option was to employ the entities the bacteria use when transferring genes among themselves. These entities are called *plasmids*, and they're small, usually circular DNA molecules that reside within bacteria but are not part of the bacterial chromosome. (*See Figure 4.5*) If a gene is spliced into a plasmid and the plasmid is transferred to the bacterium, that gene can be transcribed into RNA and translated into protein; and because the plasmid will replicate, the gene will continue to appear in successive generations of bacterial cells.

To prepare the plasmids for the insertion of the genes, they're cut open with the same restriction enzymes that were used to cut the genes from their surrounding strand of DNA. In this way, the open ends in the plasmids will be complementary to the ends of the segment carrying the gene, enabling the ends to fit together. Further, because the complementary ends are "sticky" in relation to one another, the attractive force between the complementary bases facilitates bonding.

However, there's a sticking point: the sticky ends are not sticky enough. Although bonds form between the complementary bases on the inner surfaces of the adjoining segments, gaps remain between the sugar-phosphate

Figure 4.5

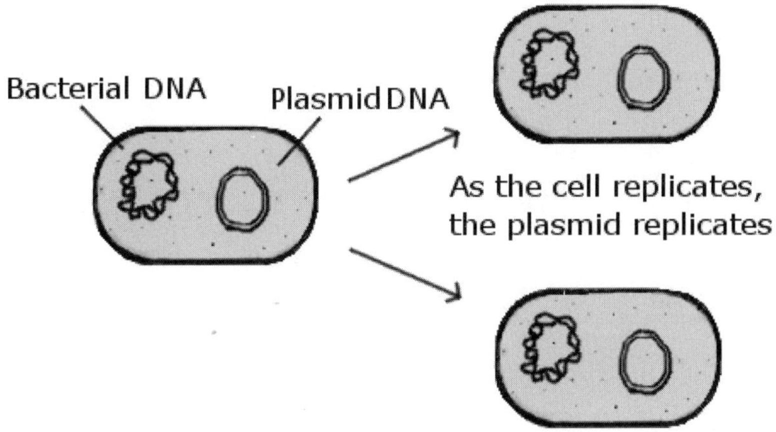

backbones on the outer surfaces because there's no attractive force between neighboring units. And unless these units are fused, the recombinant molecule can readily come apart, since the bonds between the bases are not, on their own, strong enough to keep the inserted segment in place. So biotechnicians have to apply an enzyme (called a *ligase*) that cells ordinarily use to repair breaks in their own DNA. In this way, a stable bond is formed.

The next step is to get the plasmids into the *E. coli*. However, although these bacteria can receive plasmids conveyed via direct contact with members of their own species, they won't ordinarily take up isolated DNA from their surroundings – and it appears that the vast majority of the other bacterial species will not either. So to render the *E. coli* receptive, biotechnicians resort to artificial manipulation.[25] As we saw in Chapter 1, a common approach is to subject the bacteria to calcium salt and a major heat shock. However, even this approach fails in the case of large plasmids (which are needed for carrying big DNA inserts). In such situations, biotechnicians must employ a more drastic method. Instead of heat shock, they apply electrical shock, opening pores in the cell wall with pulses of high-voltage current.

Moreover, even prior to inserting the genes in the plasmids, other artificial interventions are necessary – directed not at the plasmids or the bacteria, but at the DNA that's going to be transferred to them. First, because large numbers of engineered plasmids are going to be mixed with large numbers of bacteria, and because only a small fraction of the bacteria will end up containing one of the plasmids, biotechnicians need some way to identify the ones in that latter group. So they add another gene to the gene they want to transfer, one that will produce a distinct and easily observed effect. In most cases, the genes employed as markers confer resistance to a particular antibiotic, enabling biotechnicians to isolate the plasmid-endowed bacteria by dousing the entire batch with that anti-biotic – which kills all the cells except those with resistance born of the engineered plasmid.

Further, in most cases, the initial alteration of the DNA is not limited to the addition of marker genes. That's because the majority of the chemicals that bioengineers want bacteria to produce in great volume come from the genes of organisms far more complex than the bacteria, such as plants, animals, and people. And, due to the vast biological gulf between bacteria and these species, the microbes are not up to the task of expressing their genes.

For one thing, the genes of higher organisms contain elements that the genes of most bacteria don't possess.[26] In bacteria, all the DNA within a gene is expressed into protein, but the genes of plants and animals contain many segments of nucleotides that do not get expressed, and these non-expressed segments are interspersed between the segments that are expressed. The expressed regions are called *exons*, and the non-expressed ones are called *introns*. But the introns were not discovered until 1977, after the genetic engineering venture was well under way. Not only were they a major surprise, for a long time, they were a mystery. Although they're transcribed into RNA along with the exons, they are then edited out before the RNA travels to the cell's protein assembly sites. So biologists couldn't figure out why they were there in the first place. Eventually, it became clear that introns do have an important function; but for several years, from the perspective of the biotechnicians they were more than nonfunctional, they were *dysfunctional*, because the gene expression machinery of the bacteria into which they were inserted couldn't deal with them, preventing expression of the exons that neighbored them.

So, in order to overcome this natural barrier, bioengineers had to devise a way to get the introns out of plant and animal genes. They eventually succeeded; and the first method they employed was to construct the desired gene nucleotide by nucleotide, synthesizing the exons while excluding the introns.

Later, they developed another method; and it has become predominant. Rather than directly constructing the desired gene, they take the messenger RNA transcripts that arise from it *after* the introns have been excised and then reverse the transcription process, producing a DNA segment from the RNA that contains the desired sequence of exons yet is *devoid* of intervening introns. But they can only accomplish this feat by using an enzyme derived from a *retrovirus*. Retroviruses (such as the AIDS virus) have an exceptional feature: they don't contain any DNA, and their genetic information solely resides within RNA. Nonetheless, they achieve replication in an *indirect* manner. They make an enzyme that transcribes their RNA into DNA, this DNA is then inserted into the target organism's genome, and from there it is transcribed into new viral RNA.

It's the enzyme in the first stage of this process that's employed by bioengineers to induce reverse transcription. Yet, because it doesn't work as well with the RNA transcripts of higher organisms as with the RNA of viruses, they need to do some tinkering. They must add chemical primers to the transcripts so the enzyme can transcribe them with greater efficiency.[27]

Nonetheless, regardless of the way it's accomplished, shedding the introns still cannot shed all the difficulties. Even though the genetic code is utilized by both bacteria and higher organisms, in significant respects, there's a language barrier between them, because they use the code in different ways. This occurs because, as previously discussed, most amino acids are designated by several distinct codons; and bacterial DNA tends to play favorites, routinely employing specific codons while excluding their counterparts. On the other hand, plants and animals often favor the codons that bacteria shun. So even after a gene from a higher organism has been shorn of its introns, bacteria still won't effectively express it due to the presence of incompatible codons. Accordingly, biotechnicians need to substantially reconfigure such genes, replacing the codons that bacteria dislike with those they prefer.

Yet, even when devoid of introns and endowed with compatible codons, genes from higher organisms require additional modification before bacteria can express them. That's because, besides the regions that code for protein, a gene is flanked by elements that regulate its expression; and there's significant disparity between the regulatory elements of higher organisms and those of bacteria. One of these elements is called a *promoter*, because it promotes the process of expression. However, its regulatory role is broader than this name implies. It does not merely promote expression, it prevents the process from starting when inappropriate, and it shuts the process down when expression of the gene is no longer necessary. In other words, it's the gene's basic on/off switch; and it deploys in one or the other mode depending on the organism's requirements. Accordingly, it's finely attuned to specific biochemical signals so that the expression of the gene harmonizes with the organism's needs. Therefore, when a gene is taken from one species and spliced into an unrelated one, the promoter will rarely (if ever) receive signals to which it's sensitive – and will keep the gene inactive. Hence, before transferring a plant or animal gene to a bacterium, biotechnicians have to remove the native promoter and affix one that will function within the foreign surroundings.

Further, in cases where the bioengineers need the foreign genes to be expressed at extremely high levels, they've sometimes fused them to promoters that operate fully outside the plant's regulatory system and can boost their expression to abnormal heights. These promoters come from viruses that infect the bacteria; and they enable the process by compelling incessant transcription of the virus's genes regardless of bacterial welfare. For instance, when Showa Denko wanted its bacteria to crank out excessive tryptophan, it not only endowed them with additional copies of some of their own genes, it placed a viral promoter in front of one of them, which kicked it into aberrant overdrive.

But replacing the promoter is not enough. There's another element that's essential for expression of the gene – and it too must be removed and replaced. While the promoter resides at the start of the gene, this entity comes at the end; and its role is to demarcate this terminus, and halt the transcription process. Such demarcation is crucial, because the enzyme that transcribes the gene into RNA needs to recognize when transcription should stop or else it will extend the process beyond the bounds of the gene, adding extraneous information to the RNA that will thwart the proper production of protein. Therefore, because this regulatory element halts transcription, it's called a *terminator sequence.* And because terminators from plants and animals are not adequately recognized by the transcription enzymes of bacteria (and therefore do not induce termination when they're supposed to), they need to be replaced by sequences that the bacterial enzymes do recognize.

So, contrary to the simplistic impression that biotech proponents often impart, genetic engineering is not merely a process of transferring a gene from one organism to another. It's a process of transferring a gene along with the regulatory elements that will enable it to function in the alien organism – conjoined with a marker gene that will enable bioengineers to detect which cells it has entered. Thus, genes are not transplanted alone but only as part of a conglomerate that includes a promoter, terminator, and marker. Such conglomerates are referred to as *cassettes*. And the cassettes are typically designed to be hyper-active.

However, as we'll soon discuss, this hyper-activity entails risks. Moreover, regardless of the health risks it may pose, the massive output of foreign protein has caused big technical problems. That's because, as these chemicals accumulate in a non-native environment, a substantial number can condense and become denatured or otherwise inactivated. These clumps of nonfunctional proteins are called *inclusion bodies*, and, in the words of one scientist, "they were the bane of early recombinant DNA production methods."[28] Prior to rDNA technology, inclusion bodies only appeared as a result of viral infection, since that was the sole way in which hyper-production of foreign protein occurred. So, having induced their formation through unnatural means, bioengineers had to struggle with them for several years before discovering methods that reduced their occurrence.

Fathoming the Depth of the Dominant Deception

As the preceding discussion makes obvious, getting a gene from a higher organism into a bacterium – and inducing it to function there – can hardly be likened to a natural process. Nevertheless, despite the disparity between bacteria forced to bear the genes of higher organisms and those functioning under natural conditions, bioengineers have routinely asserted their essential congruence; and, as Chapter 1 demonstrated, such assertions were instrumental in forestalling government oversight. Now we're in a better position to discern the degree of deception behind the claim that carried the most clout: Stanley Cohen's declaration (discussed in Chapter 1) that genetic engineering merely "duplicates" processes that occur within nature. Cohen professed he had proven this point by demonstrating that *E. coli* will assimilate genes derived from a mouse. However, as we saw, in order to induce the bacteria to accept the foreign DNA, he and his collaborator (Shing Chang) had to subject them to calcium salt and a major heat shock – a fact that undercuts his claim of naturalness.[29] As we also saw, because that fact was not publicized, the boldness of Cohen's contentions convinced a majority of those in Congress to drop plans for regulating GMOs. We'll now see how his unequivocal claims not only evoked false impressions about the naturalness of his experimental design, but about what the experiment actually achieved.

Because Cohn's assertions pertained to the bioengineering process in general, and because the central goal of that process is not merely to transfer genes into the DNA of an alien organism but to also achieve production of the proteins that they code for, one could readily assume that he had accomplished the second step as well as the first.[30] I certainly did; and it appears that most others did as well. Yet, though I assumed he had succeeded in getting the mouse genes to express, I wondered how he had been able to do so.

As we've seen, bacteria cannot accurately express genes taken from the nuclei of animals because they're unable to deal with the introns, the different types of promoters and terminators, and some of the animal-preferred codons. But all the genes within the nucleus of a mouse cell contain such components. So Cohen could not have induced such genes to express within bacteria unless he had substantially modified them.

Yet, there are genes *outside* the nucleus that don't possess all the inhibiting features of their nuclear-bound neighbors. These genes inhabit the small structures, called *mitochondria*, that serve as the cell's power stations. Mitochondria produce the energy source that drives cellular processes; and their genes are more like the genes of bacteria than are those in the nearby nucleus. Not only do they lack introns, they resemble bacterial genes in other respects; and by the 1970's, many biologists had accepted the theory that they originally derived from bacteria.

Therefore, as Cohen and Chang noted, if any mammalian DNA could be expressed in bacteria, mitochondrial DNA was the "probable candidate."[31] So they bypassed the mouse's main genome, which resides in the cellular nucleus, and employed genes from the mitochondria instead. However, even though these genes bear significant resemblance to bacterial genes, they still differ from them in several ways; and due to these differences, the *E. coli* weren't able to accurately express them. The evidence indicates that only some of these genes were translated into protein; and that even then, none of the proteins was complete. Thus, instead of producing the full-sized proteins that are formed within a mouse's mitochondria, the bacterial expression machinery could only produce versions that were significantly truncated.[32]

Accordingly, Cohen's categorical claim about having duplicated the feats of genetic engineering through natural means entailed a three-fold deception. First, it instilled the false impression that the uptake of the alien genes occurred naturally. Second, it induced belief that the proteins encoded by the foreign genes were adequately expressed, when in fact, they were not – even though he had employed the one class of mammalian genes that might have been expected to properly function within bacteria in their natural state. Third, it implied that his experiment pertained to *all* plant and animal genes, when in reality, it was only relevant to mitochondrial genes. Therefore, even if its conditions *had* been fully natural and its aims *had* been fully achieved, it would have been essentially irrelevant to the general practice of genetic engineering, in which the genes employed come exclusively from the cellular nuclei. And these genes cannot be expressed in bacteria unless they're first subjected to extensive modifications that *never* occur under natural conditions.

However, one could not have known that Cohen's claim about his research severely distorted reality without reading the report that was published in a scientific journal; and, as was pointed out in Chapter 1, Cohen issued his claim long before that report was published. His well-publicized letter to the NIH director was sent on September 7, 1977, but his scientific paper didn't appear until November – more than a month after his pronouncement had derailed a bill by Senator Kennedy that would have established better safeguards on the applications of genetic engineering. Moreover, because that paper is dense with technical language (I had to consult a molecular biologist to fully comprehend the facts), even if any legislators had subsequently seen it, they would almost surely have failed to realize that they'd been hoodwinked.

Thus, despite its deep discord with the truth, Cohen's declaration played a key role in restricting the regulation of genetic engineering by thwarting the most promising legislative effort and effecting a change of mood on Capitol Hill that set the stage for the lax approach the US government has taken ever since.

Ironically, when proponents of genetic engineering were not proclaiming its safety to legislators but were instead pitching its commercial potential to investors, they were eager to contrast it with natural processes to highlight how it would yield valuable products that could not otherwise be obtained. Accordingly, they emphasized the virtual impossibility that any useful gene from a higher organism would ever be fully expressed within bacteria through solely natural means. For instance, in a 1975 report to prospective funders, the Cetus corporation pointed out that "no process of mutation or evolution" would ever enable a bacterium to manufacture a valuable animal protein because "[t]he changes in DNA necessary to do that are so complicated that it is statistically valid to say that they will never happen randomly." It then declared, "Gene splicing can and will make these things possible."[33]

And when biotechnicians did finally induce bacteria to produce a mammalian protein, they had to employ substantial artifice. Not only did they eliminate the impeding introns by building a synthetic version of the necessary gene in the manner previously discussed, they avoided codons that bacteria don't express well by substituting synonymous codons that the microbes can work with. Moreover, they didn't include the mammalian promoter and terminator sequences, which wouldn't have functioned properly. Instead, they put their synthetic gene under the control of a bacterial promoter and terminator.[34]

This breakthrough was reported in *Science* on December 9, 1977, only ten weeks after Senator Kennedy's capitulation. And, although Kennedy and most other legislators no doubt heard of this broadly-trumpeted triumph, they probably never learned the details of how it had been accomplished – or realized how these facts revealed that they'd been so recently, and so deeply, deceived.

Yet, despite the degree to which Cohen and other bioengineers had deceptively described their doings in order to thwart regulation and induce public acceptance, in the years beyond the 1970's this subterfuge could not suffice. And as bioengineers extended the range of their genetic restructurings from bacteria to edible plants, they found it necessary to *expand* the scale of the deception.

Creating Genetically Engineered Plants: Broadening the Breach of Natural Barriers

As difficult as it had been to endow bacteria with genes from disparate species, even greater obstacles were encountered in the case of plants; and the techniques employed to slip alien DNA through the membranes of the former did not work against the defenses of the latter. So daunting were the difficulties, and so numerous the failed attempts, most scientists concluded that the only way plants would accept genes is through pollination.[35] Although biologists finally devised the means to surmount the barriers against the piecemeal reprogramming of plant life, the task was so arduous that close to nine years elapsed between the appearance of the first genetically engineered bacterium and the creation of the first functional engineered plant. And, in order to achieve such a transformation, it was again essential to enlist the powers of a pathogen.

However, this time the disease-dealing agent was not a virus but a bacterium – one with a skill that's virtually unique among members of the bacterial kingdom. Like a virus, it can trick plants into expressing some of its genes – for its benefit but to their detriment. The bacterium's technical name is *Agrobacterium tumefaciens*, and the genes it deploys to infect plant cells are part of a large plasmid that it carries.[36]

The marauding starts when a horde of bacteria detect chemicals that are released when a plant is wounded. They then advance to the injury, surge through the break in the surface, and form mating tubes that connect with adjacent plant cells (as would happen if they were conjugating with other bacteria). This allows them to send large segments of DNA from their plasmids through the walls of the targeted cells and into the nuclei, where the segments are integrated within the native DNA. The alien genes then start a process that transforms their hapless hosts into tumor cells producing substances that serve as bacterial food. And, as the afflicted cells multiply, they form a prominent bulge, referred to as a *gall*.

As the feats of this bacterium came to light, several scientists realized it could serve as a vehicle for transferring genes of their choosing – if they could excise the tumor-inducing (Ti) genes and replace them with the ones they wanted. However, reconfiguring the plasmid in a workable way was far from easy. In the words of one molecular biologist, it was "a laborious process, involving several painstaking steps."[37] Eventually, the pain paid off, and biotechnicians were able to transfer genes they selected into the DNA of several species of plants using *Agrobacteria* as vehicles.

Yet, they still faced other difficulties. As had been the case when altering bacteria, achieving integration of a gene into the DNA of an organism to which it was an utter stranger could not in itself induce the expression of that gene. For one thing, just as genes from plants are, in their natural state, incompatible with the gene expression machinery of bacteria, so genes from bacteria clash with the expression system of plants. But several of the bioengineers' prized projects involved such inter-kingdom transfers. Therefore, the codons of a bacterial gene had to be revised so they would mesh with the predilections of plants. Further, the promoters and terminators that adjoined the bacterial genes destined for GE foods had to be removed, because they couldn't coordinate with the plants' processes either. However, there were some exceptional promoter/terminator sets in the bacterial realm that *could* function in plants, and they were the ones enabling the Ti genes of *Agrobacteria* to work their mischief. Similar to plant viruses, these bacteria had evolved promoters and terminators that were sufficiently plant-like to commune with (and commandeer) a plant's transcription machinery. So although the bioengineers discarded the Ti genes, they used the promoter/terminator sets of those genes to achieve the expression of the "designer" genes that replaced them.

Nevertheless, even though the inserted DNA was now getting expressed, there was still a problem. The *Agrobacterium's* promoters didn't reliably generate the level of protein that was needed in most commercial applications. So they, too, had to be replaced.

Fortunately for the biotechnicians, a much stronger surrogate was available, one that could boost expression far beyond the limit of any promoter yet tried. But to acquire it, they again had to draw on the resources of the viral realm – in particular, those of a virus adept at victimizing vegetables such as cauliflower, cabbage, and broccoli. Its infective prowess was in large part due to the potent promoter that forced the invaded plant to express its genes copiously and constantly. This pathogen was named the *cauliflower mosaic virus*, and, for technical reasons, its muscular gene activator was called the *35S promoter*. Not only was this promoter extremely powerful, it was versatile. It could subjugate the transcribing enzymes of almost any plant into which it was inserted.

As it turned out, not only did biotechnicians resort to the 35S promoter when inserting bacterial genes in plants, they even had to recruit it when transferring genes from one plant to another. That's because promoters in higher organisms tend to be species specific, which means that when a gene from one plant is placed within a plant that's not closely related, the promoter attached to it will seldom be activated. Consequently, the 35S has been affixed to the foreign genes in virtually all the GE foods currently on the market.[38] In fact, because the bioengineers usually seek a much higher level of expression than most genes will deliver even within their home environment, the 35S usually replaces the native promoter when a plant is given an extra copy of one of its own genes. Further, because there are some species in which even the 35S is insufficiently forceful, biotechnicians had to alter its structure to make it hyper-active so it could adequately function within these plants as well.

Yet, although the souped-up 35S could spur prodigious transcription in virtually any plant species, *Agrobacteria* could not deliver it (and its affixed foreign gene) into all of them. Despite their broad infective range, several plants would not succumb to their transformational powers; and, to the dismay of the bioengineers, among them were the two most valuable crops: corn and soy.

Soy was a curious case. Although the bacteria could infect many of the cells the bioengineers presented to them, those that had been infected (and were thus endowed with a foreign gene and an accompanying marker gene) could not be separated from those that hadn't. When the lethal chemical was applied to kill the vulnerable cells that lacked the marker, their neighbors that did have resistance to it nevertheless died along with them – in what's been termed a "cooperative collapse."[39]

Corn was even more recalcitrant. As was the case with several other types of grain, *Agrobacteria* could not infect any of its cells in the first place. Moreover, attempts to transform corn through electrical shock didn't work either.[40]

So another technique was needed, but the only one being attempted seemed so outlandish (and so destined for failure) that most biologists scoffed at it. Yet, despite the doubts and derision, after several years of effort, this bizarre technique finally succeeded. It was referred to as *particle bombardment* (or *bioballistics*), and it was implemented by a device called a *gene gun*. Initially, the gun that was employed fired a .22-caliber bullet coated with metallic particles which in turn had been coated with DNA. When the bullet slammed into a barrier, the particles would fly into a mass of corn cells in a petri dish. Numerous cells would be destroyed, but the foreign DNA would work its way into the genomes of a tiny fraction of the survivors (only "one in a million," according to a Monsanto scientist).[41] As the gun evolved, macroscopic bullets were no longer used, and the microscopic particles were propelled by a blast of air.

After their success with corn, biotechnicians next trained their sites on soy, targeting a type of cell less susceptible to cooperative collapse. Eventually, by blasting DNA-dusted particles into clumps of these cells, they were finally able to produce genetically transformed soy plants.

Unnatural, Uncontrollable, and Unpredictable

Thus, the two chief GE crops were born of micro-ballistic mayhem, and the remainder of those on the market were wrought with the weapons of an aggressive pathogen. Further, besides being coarse, these contrived modes of gene transfer are highly imprecise. The fragments of foreign DNA enter the target genome in an essentially random manner; and research indicates that they usually disrupt the DNA of the regions into which they wedge.[42] According to scientists who conducted a review: "It is apparent that small and large-scale deletions, rearrangements of plant DNA, and insertion of superfluous DNA are each common occurrences. . . ."[43] Moreover, the insertions also cause disruptions throughout the genome. According to molecular biologist Michael Antoniou of King's College London School of Medicine, ". . . the gene transfer process in general is known to introduce . . . hundreds or even thousands of additional mutational defects in the DNA, with potentially devastating consequences on global host gene function."[44] In one striking instance, scientists used microarray technology to study how gene expression in human cells was effected by insertion of a single copy of a human gene. They discovered that 5% of the assessed genes underwent significant changes in their expression levels (either upward or downward).[45] According to molecular biologist David Schubert, a professor at the Salk Institute, the complexity of insertional effects is compounded by the fact that each cell type of the organism tends to respond differently.[46] He further notes that when organisms are altered by the insertion of a gene that is foreign instead of native (as is almost always the case with engineered food crops), the magnitude of the changes could be far greater.[47]

The disruptive potential is amplified by the presence of the 35S promoter in each haphazardly placed fragment. Because this viral-derived booster is so powerful, it can induce erratic expression of some native genes – or activate biochemical pathways that are ordinarily inactive.[48] Each of these outcomes could spur the production of unintended toxins or induce damaging imbalances.

Moreover, due to their *always-on* promoters, the transplanted genes act independently of the host organism's intricate control system, as do the genes of an invading virus, in contrast to the harmonious coordination that exists among the native genes. Consequently, not only is every cell of the organism forced to produce substances that have never been in that species, it's forced to produce them in an unregulated manner – which can disrupt complex biochemical feedback loops (and induce unintended toxins). As we saw in the chapter on the EMS epidemic, the forced over-expression of a few genes apparently caused a disruption that resulted in the formation of one or more toxic by-products. And in that incident, the hyper-expressed elements were not foreign but were merely extra copies of some of the organisms' own genes. When foreign genes are involved, the risks could be greater.

This unrestrained activity is highly unnatural, and it violates a basic principle of living systems: energy efficiency.[49] Organisms ordinarily adhere to this principle strictly. For instance, to power their activities, *E. coli* can draw on two types of sugar: glucose and lactose. But to utilize lactose, they must synthesize one more enzyme than is required when using glucose. Therefore, as long as glucose is present, the promoters that govern the production of the lactose-processing enzymes remain idle – even if there's lots of lactose around. They only rev up if all the glucose has been consumed, which renders the more energy intensive process they trigger essential rather than optional.

In contrast, the viral promoters push organisms off the path of frugality and force them to expend considerable energy to produce substances they do not need. This incessant energy drain may be the reason that GE crops sometimes underproduce. For instance, Monsanto's GE Roundup Ready® soy bean was determined to have a 5% decrease in yield that's directly attributable to the genetic alteration.[50]

So, not only have the bioengineers been adept at keeping government regulation of their enterprise minimal, they've been able to deregulate the foreign genes they insert in the target crops. But in the latter case, the deregulation has been complete – and the regulatory system that's been evaded belongs to nature.

There's yet another unnatural facet of GE crops. Because transferring DNA to a cell by infecting it with *Agrobacteria* or blasting it with a gene gun does not produce a fertile seed (as does the union of gametes), biotechnicians cannot grow it into a mature plant by putting it directly into the soil. Instead, they must first develop it with an artificial process called *tissue culture*, in which it is coaxed to mature via applications of hormones and antibiotics in ways that would not naturally occur. In fact, the phrase "coaxed to mature" may be a bit mild. Some scientists have described the process as one in which the cell is "forced to undergo abnormal developmental changes."[51] And some species require substantial forcing. For instance, as challenging as it had been to insert foreign genes into the cells of corn, it was even *more* difficult to get the transformed cells to develop into viable corn plants, and considerable effort was needed to make it happen.[52]

Further, whatever the species, tissue culture imparts a broad jolt known as "genomic shock."[53] This shock induces extensive genetic perturbations – and is another way in which unintended harmful substances can be formed.[54]

Moreover, not only are native genes frequently destabilized due to the unnaturalness of the developmental process, the inserted foreign genes are frequently destabilized due to the unnaturalness of their presence. Organisms are geared to defend against the invasion of foreign DNA, and they have mechanisms to inactivate it. Accordingly, the alien genes are frequently incapacitated by these defenses and prevented from expressing in subsequent generations. By 1994, this phenomenon was already a significant problem. Reflecting its seriousness, a review was published that year in the journal *Biotechnology* titled: "Transgene inactivation: plants fight back!" In it, the authors stated, "While there are some examples of plants which show stable expression of a transgene these may prove to be the exceptions to the rule. In an informal survey of over 30 companies involved in the commercialization of transgenic crop plants . . . almost all of the respondents indicated that they had observed some level of transgene inactivation." They also noted that the problem is likely to be bigger than generally perceived because, according to many of the respondents, "most cases of transgene inactivation never reach the literature."[55] And as the cultivation of GE crops expanded, there have been several conspicuous cases of failed performance that may have been caused by such genetic shut down.[56]

Thus, not only do GE plants get created through multiple breaches of nature's boundaries, the foreign genes within them can only continue to function by escaping the plant's natural defenses. Contrary to the contentions about the naturalness of such organisms, both their birth and their ongoing utility are critically dependent on the foiling of natural systems that preserve genetic integrity.

Consequently, due to their unnaturalness, and to the disruptions caused by their creation, few GE plants thrive; and a sizable percentage don't even survive. This is in marked contrast to natural propagation. Pollination is rarely a lethal event, and the vast majority of seeds produced through it grow into normal adults. But a substantial portion of the cells altered through bioengineering are killed outright; and a large fraction of those that aren't either develop with gross deformities or fail to adequately express the intended trait.[57]

Reaping Bizarre Results

Besides being numerous, the unintended outcomes of genetic engineering are frequently surprising, and often bizarre. Some of the strangest arose when pigs were engineered with a human gene that codes for a growth hormone. Although the aim was to produce fast-growing "superpigs," the novel creatures did not grow faster than their parental stock; but they did grow a lot weirder. One female was born absent an anus – and also without a vagina. Several of the others were too lethargic to stand up.[58] Additional afflictions included arthritis, enlarged hearts, ulcers, dermatitis, kidney disease, and impaired vision.[59] And because their immune systems were dysfunctional, they were prone to pneumonia.[60] But most of the engineered embryos didn't do much growing in the first place. Only one in 200 reached the state of malformed maturity.[61] And when researchers inserted a foreign gene into cow DNA to alter the composition of the milk, most attempts to generate live calves from the engineered cells failed. Moreover, the one calf that did result was "unexpectedly" born without a tail.[62]

Plants have proven as unpredictable as animals. For instance, when tobacco was engineered to produce a particular acid, it also generated a toxic compound that's not a natural component of tobacco.[63] And when bioengineers altered yeast to increase its fermentation, they were surprised to find that a toxin that naturally occurs at low levels appeared at levels 40 to 200 times higher. The shock was intensified by the fact that (as in the case of Showa Denko's tryptophan-producing bacteria) no foreign genes had been inserted, just extra copies of some of the yeast's own genes. In discussing the incident, the scientists acknowledged that it "may raise some questions regarding the safety and acceptability of genetically engineered food. . . ."[64] Another shock arose when scientists tried to suppress an enzyme in a potato and inadvertently raised its starch content. One of them admitted, "We were as surprised as anyone." He added: "Nothing in our current understanding of the metabolic pathways of plants would have suggested that our enzyme would have such a profound influence on starch production."[65]

Even the *Wall Street Journal*, which usually presents biotech in a favorable light, couldn't resist featuring its freakish effects; and an article titled "Genetic Vegomatics Splice and Dice With Weird Results" described several.[66] For instance, when biotechnicians employed gene-splicing to improve the shelf life of tomatoes, some of the fruit had regions that ripened very fast along with those that didn't ripen at all. According to one of the scientists, the "green islands" of unripe tissue interspersed among patches of red made the tomatoes look like "Christmas tree bulbs." When other scientists tried to create a small, sweet red pepper they got some surprising specimens, including "squat-shaped peppers [that] grew straight up, perched on stems." One of the researchers noted, "You won't see that anywhere else in the world." And when another team tried to create bruise-resistant potatoes by endowing them with synthetic genes built with bits of chicken and moth DNA, many of the spuds were shockers. Some had "alligator-hide-type skin," some had "thumb-like protrusions and eyebrows," some had "arms," and others had "noses that look like Pinocchio's." Reflecting on the grotesque results, one scientist remarked, "It's not nice to fool nature. Sometimes you get slapped. And some people get slapped around a lot."

A Need to Refashion the Facts

However, despite repeated slaps from nature, the biotechnicians pressed on with the venture to reprogram our food – all the while claiming that their creations conform to what's natural. But to sustain this claim, not only did they have to distort the story of what they were doing, they had to recast some basic truths of biology to boot.

And some highly influential scientists have created considerable confusion. A prime example is the molecular biologist Nina Fedoroff, a member of the National Academy of Sciences and a recipient of the National Medal of Science, who served as the Science and Technology Adviser to the US Secretary of State from 2007 to 2010 and then became President of the American Association for the Advancement of Science. In her book, *Mendel in the Kitchen: A Scientist's View of Genetically Modified Foods*, she blurs the distinction between traditional plant breeding and genetic engineering by asserting that farmers have been modifying plants' genes for more than 10,000 years and that the transformation of a wild plant into a food plant, as well as the transformation of a low-yielding plant into a higher yielding one, entails a change in the genes.[67] She has also claimed: "There's almost no food that isn't genetically modified."[68]

But she consistently fails to note that when genes move between organisms via natural pathways, they do not undergo the alterations that occur through bioengineering. New genes do not abruptly appear, nor are the internal structures of those already present routinely modified in radical ways. The primary change is in the way the genes are combined. There are new arrangements of the various alleles of particular genes that have been in the species' genome for millennia. Further, the genes continue to be governed by the same regulatory elements, and the coordination between them is preserved. Even in the rare instances when a spontaneous mutation arises and is maintained in the species' gene pool, the change in the gene's structure ordinarily occurs at just one point (with only a single base pair affected).[69] Moreover, the regulatory elements are usually conserved.[70]

Genetic engineering stands in stark contrast. Genes are inserted into species that have never contained them; and, prior to insertion, they're subjected to significant restructuring. In several cases, introns are excised and codons reconfigured, or terminator sequences are replaced. And in the case of virtually every GE plant currently on the market, the native promoter has been removed and replaced with a powerful one derived from viruses – a promoter that operates outside the finely tuned regulatory system of the target cells, perturbing their patterns of energy consumption and potentially causing other imbalances as well. Additionally, the insertions are random, and they tend to cause extensive disruption to the functioning of native genes.

Nonetheless, despite the fact that such disruptive effects have been documented in numerous scientific journals, not only does Federoff fail to acknowledge them, she flatly asserts that no such problems occur. In a 2008 *New York Times* interview, she declared that GE techniques "introduce just one gene without disturbing the rest."[71]

She further distorts her depiction of bioengineering by failing to note the necessity for excising introns and re-writing codons when transferring genes from plants and animals to bacteria – and for also revising codons when moving genes from bacteria to the higher kingdoms.[72] Acknowledging such realities would make bioengineering look a lot more unnatural than in the picture she paints.

Moreover, to enhance the impression that the technology is closely aligned with traditional practices, she claims it's essentially equivalent to the ancient art of grafting. However, she can only do so by significantly mischaracterizing the nature of that time-honored technique.

Grafting is widely used in horticulture, and it plays an important role in fruit production. In the latter, a bud or branch from one tree is spliced onto the rootstock of another; and the composite then grows into a mature tree. In many cases, grafting is done to increase hardiness, as when the upper part of the tree is joined to the rootstock of another type of tree better suited to particular variety of soil than is its own stock. It serves other ends as well, a chief one being the preservation of genomic integrity from generation to generation.

Grafting can preserve a distinct genome because it's a form of cloning. The branch that's joined to the rootstock will produce fruit genetically identical to the fruit on the tree from which it was taken. In contrast, if the fruit on the donor tree pollinates with neighboring trees, genetic variation will occur and the precise characteristics of the fruit will not be fully maintained in the progeny, even when pollination is limited to the same species.

While grafting generally involves members of the same species, the parts are sometimes drawn from different ones. However, although such trees contain components from distinct species, as do molecules constructed through rDNA technology, the similarity stops there. While the latter are formed from fusions of diversely-sourced DNA fragments, there's no such commingling of DNA in grafting. The distinct parts of the tree cooperate in the movement of water and nutrients between them, but the genes in their nuclei stay put.[73] Each gene remains within the nucleus of its own cell, under the control of the cell's regulatory system. Further, whereas rDNA technology can move DNA between biological kingdoms, the range of grafting is quite limited. Only similar species can be grafted, and some trees (like the cherry) can't even be fused to their closest cousins. Moreover, there's no way to merge elms or oaks with fruit trees.

Of course, there's another great gap between grafted trees and engineered organisms. Whereas all the genes in the former are in their natural state, those in the latter are not. As previously described, every cell of almost every GMO contains one or more genes that have been artificially restructured.[74]

Nonetheless, although the discrepancies between grafted trees and genetically restructured organisms are glaring, Fedoroff glibly glosses over them – declaring that the latter are "no different" than the former. And in a final flourish, she proclaims that a GMO is just as natural, and no more artificial, than an apple tree.[75]

But her only support for this astonishing claim is the simplistic assertion that genetic engineering and grafting both require human intervention – along with the false assertion that "each technique combines genes from different species."[76] In fact, to the best of our knowledge, there's no greater fusion of foreign genes within a grafted tree than in a bowl of mixed fruit.[77]

The inanity of her assertions is underscored by remembering that the two techniques serve opposite aims. Whereas bioengineering *alters* genomes in radical ways, grafting *prevents* genetic change and *preserves* the status quo. Yet, she insists that the two processes yield the same kind of results.

Additionally, in her ardor to portray bioengineering as more natural than it is, Fedoroff creates further distortion by equating the manipulations of biotechnicians with the behavior of a bacterium. As do most biotech boosters, she points to *Agrobacterium tumefaciens* as a "natural genetic engineer" and argues that because it has moved its genes into plants for ages, genetic engineers are hardly doing anything new when they employ it for the same purpose.[78] Rather, she says that the novelty lies in the expansion of possibility: "What's new is that the pool of possible genes from which plant breeders can now choose has grown very much larger."[79]

However, this claim disregards the fact that when the *Agrobacteria* function naturally, they create tumors in plants, not new varieties of plants. They infect pre-formed plants, not single cells that will be grown into them. The alien genes are only injected into a portion of the plant's cells, and it's extremely unlikely they'd be transferred to subsequent generations.[80] Nor do the foreign genes enter the fruit of infected trees. In contrast, when the bacteria are employed in bioengineering, they infect isolated cells that will become whole plants. Therefore, the alien genes that are transferred end up in every cell of the mature plant (including the fruit of engineered trees); and they pass to the plant's progeny. This is a radical re-working of a natural process.

Of course, it's a lot harder to pass the gene gun off as something natural, so Fedoroff says little about it. She only mentions it in a few sentences, and she gives no indication that it's the technique through which most of the GE food that's been eaten was generated.[81]

Unfortunately, Fedoroff's proclivity to recast reality in regard to GE foods is shared by many (if not most) of their other proponents. And a phenomenon they're particularly prone to refashion is the critical role of viral promoters. After all, it's especially tough to make the case that genetic engineering is substantially similar to natural breeding when virtually all the foreign genes it implants in food crops cannot even get expressed without the aid of a potent booster that not only enables their expression, but propels it nonstop, independent of the organism's regulatory system and contrary to its needs. So biotech advocates rarely give a fair account of what such promoters are doing. For instance, while Fedoroff does discuss the use of the 35S promoter and acknowledges that it's powerful, she does not explicitly disclose that it deregulates the expression of the foreign genes and compels their transcription 24/7; and most readers who didn't know this already would probably not draw such an inference from what's written. Moreover, she downplays the possibility that the use of that promoter poses risks, despite the fact many experts think that it does.

In order to obscure the promoter's true effects, some proponents even misrepresent how it operates. An especially egregious example occurred in a guide for consumers published in 2000 by the food standards agency for Australia and New Zealand (which, like the FDA, endeavors to foster biotechnology). In describing how a foreign gene is expressed in a GE organism, the document stated that the 35S promoter "tells the plant to turn the gene on" and that the terminator sequence "tells the plant to turn the gene off."[82] This is flat-out false. In reality, the promoter is both the "on" and the "off" switch. Although the terminators stop the transcribing enzymes from proceeding beyond them, they do not terminate the transcription process itself. As long as the promoter remains in the "on" mode, one transcribing enzyme after another will affix to the DNA strand and travel along it until reaching the terminator. Thus, the terminator sequence acts like an instruction that tells a computer's printer to stop after the twelfth page of a very long document, while the viral promoter acts like an instruction to keep churning out copies of that twelve-page section.

By misrepresenting the role of the 35S promoter, the consumer guide actually guided consumers into the mistaken notion that the foreign gene is being regulated and that its transcription is shut off as needed. Accordingly, they were given no inkling that the promoter's innate purpose is to overwhelm a plant with the products of its own genes, in disregard for the health of that plant, and that any gene attached to it will therefore be transcribed incessantly.

Moreover, just as the viral promoter relentlessly expresses its adjoined gene with no regard for the plant, the agency continued to disseminate the falsehood with no regard for the truth; and despite being formally apprised of the error (as well as several others), it continued to dispense the misleading guide to the public for a long time.[83] Thus, regulatory integrity was breached at the governmental level to becloud its breach at the level of the engineered genome.

So awkward a topic is the dependence on viral promoters that most biotech advocates prefer to completely avoid it. This is the course taken by Susan Aldridge in *The Thread of Life: The Story of Genes and Genetic Engineering*, published by Cambridge University Press in 1996, and also by William Baines in *Biotechnology from A to Z*.[84] Avoidance is also the rule in *Tomorrow's Table: Organic Farming, Genetics, and the Future of Food*, co-authored by Pamela Ronald, a professor of plant pathology at the University of California, Davis, and her husband, Raoul Adamchak, an organic gardener. Published in 2008 by Oxford University Press, it has been highly influential and highly touted. In an opening section, the authors proclaim that it is "accurate, thorough, and balanced;" and several authorities have hailed it as such. For instance, the journal *Nature Biotechnology* called it "One of the best, most balanced accounts of transgenic agriculture," and the similarly prestigious journal *Science* praised the authors' "clear, rational approach" while noting that "the balance they present is sorely needed."[85] Yet, accolades for its balance notwithstanding, none of the book's chapters even mentions promoters.[86] Apparently, to have acknowledged the technology's need for viral promoters, and to have described their effects, would have clashed with the book's contention that GE foods are so similar to naturally produced ones that they ought to be allowed in organic agriculture.

Moreover, the use of viral promoters is not the only important aspect of bioengineering that this purportedly balanced book ignores. It's also silent about the use of particle bombardment – and the fact that the two main GE crops were created by it.[87]

Thus, in their attempt to impute closeness between bioengineering and conventional breeding, several supposedly reliable publications have stretched the truth in truly unseemly ways. But the most extreme assertion of similitude appears in the 1989 report of the National Research Council (the NRC, a division of the National Academy of Sciences). It boldly declares that "no conceptual distinction exists between genetic modification of plants and microorganisms by classical methods or by molecular methods that modify DNA and transfer genes."[88]

This is an astounding statement. It does not merely assert that genetic engineering is substantially equivalent to all other forms of plant breeding (including natural sexual reproduction), it proclaims that the technology is essentially *identical* to them. After all, if one cannot make a conceptual distinction between GE and sexual reproduction, they must necessarily be the same thing. But that's obviously absurd. The rest of the report draws numerous conceptual distinctions between them. Furthermore, merely by categorizing traditional methods as "classical" and recombinant techniques as "molecular," the statement itself makes such a distinction – and is thus self-contradictory.

It also contradicts the understanding of scientists who employ genetic engineering but are not prepared to distort its realities. For instance, the molecular geneticist Michael Antoniou, who uses genetic engineering in human gene therapy, states that it "technically and conceptually bears no resemblance to natural breeding."[89] Thus, the assertion is absurd not only as a matter of logic, but as a matter of fact.

How could such an absurdity have made its way into a report by the United States' premier scientific institution? Although (as we saw in Chapter 2) the National Academy of Sciences had an agenda to promote biotechnology, it's still surprising that such an outlandish contention even entered the first draft, let alone survived subsequent layers of editorial review.[90] Apparently, the personnel involved were so intent on establishing the congruence of genetic engineering and other methods that they lost sight of how irrational their words had become. And the fact that throughout succeeding decades, their absurd statement has been perceived as authoritative, repeatedly cited, and unwittingly relied on, attests to the staying power of even the most preposterous claims advanced on the technology's behalf.

Public Confusion: The GE Venture's Most Consistently Produced Effect

While the effects of genetic engineering have been quite variable (and often surprising), the distortions dispensed to advance it have functioned far more reliably than the technology itself. Unlike the latter, they have consistently achieved their intended result: widespread confusion. And there's been particular success in obscuring the technology's most unappealing aspects – especially its dependence on viral promoters. I traveled and lectured extensively about GE foods between 1998 and 2004, and I rarely met a member of the general public on any continent who knew about them. Further, most people were astonished when I provided the facts, because eminent individuals and institutions had repeatedly told them that the transplanted genes behave just like the native genes.

Moreover, confusion about viral promoters also exists among life scientists who haven't specialized in molecular biology – even when they've studied its applications extensively. A striking example is the biologist Eric S. Grace, who was invited by a publisher to write a book about biotechnology. The result was *Biotechnology Unzipped: Promises and Realities*, which appeared in 1997.

Because Dr. Grace (who received his PhD in zoology) did not know the details of genetic engineering, he needed to do substantial research. As he reports in the preface: "My goal was simply to find out what was going on, using as many different sources as possible, and to make my own interpretations as I went along."[91] He then states: "It was a reasonably straightforward matter to find information about the techniques of biotechnology – such as how to cut and splice DNA . . . or make transgenic organisms (animals, plants, or microbes carrying the active genes of other species). . . . The difficult part was evaluating the final outcomes, separating facts from speculation, and science from politics."[92]

However, discovering the details about the creation of transgenic plants and animals was not as "straightforward" a process as he was led to believe, because, despite his extensive efforts, he came up short on some key facts. For instance, in the chapter, "Tools in the Genetic Engineering Workshop," although he correctly notes that promoters are like switches that regulate the expression of genes, he then states: "When genetic engineers transplant genes for making products, they must include the switches that control gene expression as well as the genes themselves."[93]

This statement is doubly erroneous. It implies that a gene's own promoter is transferred along with it, while also implying that the promoter can induce adequate expression of that gene within a foreign cell. Accordingly, there's no mention of the need for viral promoters; and the book never sets things straight.

But it does twist other important facts. In one of its biggest distortions, the book declares: "Outside of its cell, there is no distinction between a human gene, a cat gene, a wheat gene, or a bacterial gene."[94] Yet, as we've seen, there are *major* differences between bacterial genes and those of plants and animals. And because of them, genes from bacteria must be extensively modified before they can function within the cells of higher organisms, just as genes from those organisms must be reconfigured before they can operate within cells of bacteria. Somehow, Dr. Grace did not discover this essential information. This lack of knowledge (coupled with his confusion about the ways of promoters), caused him to convey the impression that such cross-kingdom transfers are rather simple matters. Further, there are other respects in which his presentation is not as solid or balanced as he apparently thought it was.

Thus, as Dr. Grace attempted to unzip biotechnology, the zipper got stuck – which, in an important respect, reveals more about the GE food venture than all the information that he *did* accurately convey. When he began research for the book, he already had a good grasp of the fundamentals of biological science. Further, he was an accomplished science writer. Yet, some of the most basic facts of genetic engineering were so obscured that even someone with his knowledge and abilities did not discern them. Moreover, although he was sincerely endeavoring to communicate the facts, it seems that others associated with his book were quite content for unattractive ones to remain hidden – and for falsehoods to stand in their stead.

Accordingly, it should not be not surprising that the book was published by the National Academies Press, an organ of the National Academy of Sciences.

As we've seen, the Academy has been a long-time proponent of bioengineering; and seven years after releasing Grace's book, it published Nina Fedoroff's *Mendel in the Kitchen*, which presented a highly partisan case for GE foods. Of course, Grace endeavored to be accurate and nonpartisan, but those responsible for editing and vetting his book apparently did not. It was thoroughly analyzed by a team of reviewers that included several scientists, and its back cover contains a strong endorsement from Fedoroff expressing her intention to use it in her university courses.[95] Somehow, its erroneous discussion of promoters did not trouble her or the other experts who reviewed it, or anyone on the publishing staff. Nor did its false attribution of similarity between bacterial genes and those of higher organisms. Otherwise, there would have been demands for revision – demands that almost surely would have been made if the mistakes had worsened the image of genetic engineering instead of making it look far less radical, and much more natural, than it really is.

In light of the evidence discussed above, it's not an exaggeration to say that any citizens who endeavored to get an accurate picture of how GE foods are produced by reading the accounts provided by scientists and scientific organizations that endorse the process have had the deck stacked against them. Even reading books on the topic published by prestigious academic presses would have left them ignorant of several key facts and with false impressions of many others. A person could have read two books published by Oxford University Press, one by Cambridge University Press, and one by the National Academies Press from cover to cover and never known about the need to add viral promoters to the transferred genes. And even if he or she studied another book from the National Academies Press that did note the use of such promoters (Fedoroff's), there would still have been no explicit indication that they cause the genes to behave in an uncontrolled – and potentially disruptive – way. Moreover, after finishing all five books, he or she would have acquired a picture that was not only incomplete, but severely distorted by extensive misstatements. And this would have been the result of consulting books that met the standards of premier presses. The sources from which most people gain their information (newspapers, magazines, radio, and television) have tended to present statements from biotech advocates that are even more misleading.

The degree of difficulty in gaining basic knowledge is conveyed by the following fact: sixteen years after Eric Grace began his quest "to simply find out what was going on," he still did not know about the need for viral promoters in the production of almost all GE foods – or the abnormal effects they induce. His reply to an email query made this clear.[96]

If a biologist who extensively studied genetic engineering in order to write a book about it for the National Academies Press, and who evidently remained keenly interested in the topic over the subsequent sixteen years, had not yet learned one of the crucial facts that render GE foods different from their conventional counterparts, how much more confused must the average citizen be?

Sadly, the confusion has been thickened because the biologists who foster it claim to speak on behalf of the scientific community, portraying their opinions as the official stance of science and any opposing views as scientifically illegitimate. Accordingly, many of them impart the impression that only nonscientists have concerns. For instance, Pamela Ronald declared on a nationally broadcast television program that "the scientific community is perfectly comfortable with genetically engineered crops."[97] And while Nina Fedoroff at least acknowledges that some scientists oppose the use of genetic engineering in food production, she says there are only a "few" – and that "they are rarely those who know this new science well."[98]

However, as we've already seen (and as succeeding chapters will make more obvious), there are numerous well-credentialed and well-informed experts who regard genetically engineered foods as deeply different from, and more dangerous than, those produced by *any* prior means. These scientists have tried to counteract the confusion spread by Fedoroff and so many other biotech boosters – and to correct the false but widely held impression that genetic engineering is substantially similar to conventional forms of breeding. This chapter began with a statement from one of them: an assertion by a Nobel laureate biologist that genetic engineering is "the biggest break in nature that has occurred in human history." And at this closing stage, it's fitting to feature words from another: a molecular biologist who was selected to represent the position of the scientific community at a forum on GE foods in Washington D.C. sponsored by the American Association for the Advancement of Science. On the morning of May 8, 1998, in the stately edifice that housed the organization over which Nina Fedoroff would later preside as president, Dr. Liebe Cavalieri delivered the day's first substantive address, systematically refuting the kinds of arguments for the similarity between genetic engineering and conventional methods that were by then quite common – and that Fedoroff would subsequently advance in her book.[99]

However, although Dr. Cavalieri's concerns were obvious, no one in the audience (except me) knew that they had motivated him to sign on as a plaintiff in the lawsuit that the Alliance for Bio-Integrity would file against the FDA later that month – and that in a few weeks he would be back in Washington speaking at a press conference the day the suit was filed. On that occasion, he elaborated on his prior assertion that it is "simplistic, if not downright simple-minded" to claim that genetic engineering is substantially the same as traditional breeding; and he emphasized the irresponsibility of doing so. His concluding comment was forthright. Looking into the row of cameras, he declared it's "disgraceful" that eminent scientists have engaged in this practice – and he denounced it as a "sham." [100]

Thus, through the ingenious circumvention of several natural barriers, the bioengineering venture managed to create foods with functional foreign genes; and through the disingenuous depiction of what the process entailed, it maintained government support and mitigated public concern. Yet, it still faced a formidable barrier of a distinctly different kind, imposed not by the laws of nature but by the laws of the United States.

It was confronted by a venerable consumer protection statute that for more than thirty years had been regulating the introduction of foods with new additives. This statute mandated that such foods be proven safe, thus blocking the fast track to commercialization that biotech proponents desired. To circumvent this obstacle while conveying the impression that no laws had been broken (or even bent) would require another round of ingenuity – and an additional layer of deception. It would also require the collusion of the US Food and Drug Administration.

CHAPTER FIVE

ILLEGAL ENTRY

The Governmental Fraud that Put GE Foods on the US Market

Trying to Bring Food Safety to the Fore

During the mid-1980's, as Philip Regal contemplated the deep differences between traditionally produced foods and those being developed via genetic engineering, he realized that the issue of whether these novel entities might harm the environment had tended to overshadow the issue of whether they were safe to eat – and that the risks they posed to consumer health were in some respects more serious.

Of special concern was the potential for unintended toxins. As he recounts: "I thought it was important to recognize that plants are little biochemical factories that produce scores of bioactive compounds in order to mess up the metabolism of a great variety of predators and pathogens. That's why all the first drugs were discovered in plants. Even in small amounts some plant chemicals are good for us and some are dangerous, so ethnobotanists seek knowledge from local tribal people about which are which."

Because he knew that a plant's biochemical pathways can "switch tracks," and that a small shift can make a big difference in the suitability of a plant for food, Regal was convinced that edible plants produced through bioengineering must be carefully tested for food safety. He notes: "We have extensive evidence that genetic engineering can disrupt biochemical traffic in ways that cause bizarre side effects. And if the pathways involved are ones that produce bioactive compounds, and if they are induced to reroute into new paths or to open up auxiliary ones, that can spell trouble, because it can lead to the synthesis of novel substances that are harmful. That's why I kept stressing to government regulators and industry scientists that even if we determined that a particular crop bioengineered in a specific way would be ecologically safe, this was no assurance it was safe to eat. To determine whether it was going to be a wholesome food would require other sorts of studies. My associates and I were trying to get the molecular biologists, the government people, and the industry to think about areas of scientific knowledge and levels of biological complexity that they had previously ignored."

Token Gestures, Broken Promises

In light of the various facts, Regal thought it was mistaken to presume that GE foods would be invariably as safe as their conventionally produced counterparts – and more realistic to view them as entailing higher risk. Further, he learned that many other biologists shared this view. He therefore sought to alert the people with whom he interacted not only that there were significant health risks, but that they could in some ways be more problematic than the environmental ones.

In several cases, industry scientists readily understood and agreed that there were grounds for concern. They confessed they were in a race simply to get new DNA into plant cells (and then to function within the foreign surroundings) and that this race took all their attention. But they assured him that they and their colleagues would give more heed to the food safety issues once they had finally produced viable transgenic plants. Further, virtually all the biotech proponents he met, whether or not they acknowledged that food safety had been slighted up till then, guaranteed that ample precautions would be implemented. As Regal relates:

> During a string of workshops and conferences between 1984 and 1988, I had numerous informal conversations during meals, breaks, and walks with biotech advocates who insisted I could banish all concerns about the possibility of some genetically engineered foods causing public health disasters. They asserted that no bioengineered product would ever get into grocery stores unless it had been thoroughly tested and confirmed safe; and they assured me that any problems of the sort that I had raised would be dealt with by what is called "good agricultural practice," and "good laboratory practice," and "good industrial practice." Industry and government people emphasized repeatedly that while they needed my advice on the potential ecological problems that GMOs might cause, issues of food safety would be diligently addressed – and the integrity of the American food supply carefully guarded. Time and again I was urged not to think about food safety issues because the FDA and USDA had very strict regulations that would protect the public.

> I knew little then about the biotech food industry and was naive enough to largely believe the assurances of its various proponents. And I suspect that most of them believed their own rhetoric. Even the term "genetic engineering" implies faith in a degree of precision that has always been far beyond the capacity of recombinant DNA technology; and many biotech enthusiasts firmly (albeit unrealistically) expected this technology would routinely match the exactness and reliability of electrical and mechanical engineering.

Moreover, it seems that most biotech advocates could no longer even discern the difference between hard fact and promotional fluff. As one industry insider confided to Regal: "We've hyped ourselves so much about genetic engineering that we don't know what's hype and what isn't anymore. Financing this whole business has absolutely depended on hype and on promising the world to venture capitalists and even to each other."

Not only did Regal accept the assurances he'd been given, he regularly passed them on to representatives from public interest groups who questioned him about the direction in which food safety policy was heading. But he stopped doing so after he was invited to participate in a 1988 interagency biotech policy meeting in Annapolis, Maryland attended by numerous government officials from nearby Washington, D.C. Although the conference started on a reassuring note, with spokespersons for USDA, FDA, and EPA asserting there would be science-based procedures to guarantee the safety of genetically engineered foods, ominous overtones soon arose. For instance, after making such a pledge, the Acting Director of the FDA's Center for Food Safety and Applied Nutrition confessed: "I wish I could tell you that we know how we will approach transgenic plants. In fact, we organized this conference with USDA and EPA because there are important scientific questions that we have not resolved." But he then went on to say, "We should set aside questions for the moment that cannot be answered."[1]

Regal found this far from comforting, and he considered it fanciful to promise that safety would be scientifically certified while important unresolved scientific issues were ducked. More troubling, over the course of the conference he learned that many, if not most, of the experts in attendance believed that the question of safety could never be fully resolved in favor of GE foods because significant uncertainty is inherent in the technology and cannot be eliminated. He recalls: "In informal discussions at meals and on walks, government scientist after scientist insisted that there was no way to be sure about the safety of genetically engineered foods."[2] In their view, a rigorous system of testing could be helpful but not fully reliable. It could weed out many dangerous GE foods, but it might not catch them all. While it could identify a product as harmful, it could never certify safety.

Though some scientists were more optimistic about the potential of testing, virtually all concurred that no matter how great the degree of scientific uncertainty or how long it might persist, the commercialization of GE foods would not be delayed on account of it. Nor did they believe that a meaningful system of testing would be instituted, because they knew it would entail a cost that the biotech corporations were unwilling to bear – and amply able to avoid. These corporations had already invested far more in the development of GE foods than is normal for conventional ones. While it takes 10 years to develop a GE seed and only between 5 and 8 years to develop a new non-engineered one, the cost differential is far greater.[3] Major Goodman, a plant breeder at North Carolina State University, has pegged it at fifty times higher.[4] And that doesn't include any price tag for tests related to food safety. Experts have estimated that just a combination of short and medium-term toxicological feeding tests would cost at least an additional $25 million – which, according to Goodman's figures, would boost the overall expense by more than 40%.[5] Full long-term testing would be substantially more expensive. Further, any meaningful safety testing would significantly extend the time required before marketing could commence.

Such extra cost (in both time and money) for each product would have been a burden that the manufacturers could probably not have borne. They were under pressure to recoup their vast investments and project a healthy image on Wall Street, and this image would have been sullied by news that each insertion event entailed risks that necessitated extensive, and expensive, tests.

The industry's priority was to get the new products marketed as quickly as possible, not to minimize the attendant risks; and it had become so powerful that it could largely bend the institutions of government to its will. So pervasively had industry preference merged with government policy that several scientists employed by regulatory agencies confided to Regal that although they were distressed by the way things were going, they feared reprisals if they raised concerns that could retard the introduction of GE foods.

In such circumstances, it was obvious that a sound system of testing was not going to be a feature of the GE food enterprise, and the biotech proponents were no longer even attempting to persuade Regal that it would be. He also learned that the other hopes they'd instilled in him during the previous four years were equally baseless. As he says: "The 1988 conference was sobering, and significantly shocking, because although many of us had been led to expect that genetically altered foods would be thoroughly tested and that the agencies were on top of things, it was now being openly admitted by every agency or industry person with whom I had a cup of coffee or took a walk that the foods would not be well-tested – and that instead of being on top of the technical problems, the agencies had no definite plans for dealing with them. In fact, our break-out work groups revealed that industry and government scientists were only just starting to think seriously about the problems."[6] Moreover, an attitude he repeatedly heard expressed, not only from industry representatives but from government officials, was: "If the American people want progress, they are going to have to be the guinea pigs."

As it became increasingly clear that the pressures of economics and politics would trump the basic requirements of science, Regal felt dismayed and more than a little betrayed. "After years of being fed assurances that GE foods would be thoroughly tested – and then passing on these assurances myself – I was now hearing that meaningful testing would not be implemented because it was financially impractical, regardless of the degree to which it was technically feasible and scientifically important. Not only had my efforts to get the enterprise in line with science been largely rejected, I had served as an unwitting accomplice in spreading misinformation to many others."

Retreat from Reality

When Regal expressed his opinion that the public needed some genuine protection and asked if any was to be provided, he was frequently told that sufficient protective mechanisms would be provided by the free market system. The threat of lawsuits would motivate the manufacturers to keep their products safe, and if any did cause harm, consumers could get compensated by winning judgments in court, which would ultimately force harmful products to be withdrawn. Further, the industry had promised that to facilitate the just imposition of liability, GE foods would be labeled, making it easier to establish causal links.

However, while these arguments seemed appealing to those who made them, they were largely detached from reality. As Regal and other experts well knew, unless a particular GE food regularly caused acute illness after only one or a few exposures, it would be extremely difficult for anyone to prove in court that he or she had been damaged by it, even if it had induced widespread harm. Most food-borne illness arises over a long time through repeated exposure, and the maladies are often not unique but are ones (like cancer) that can arise from diverse causes. So in most cases, plaintiffs would be unable to meet the burden of proof. And because the weakness of the consumers' position was obvious, there would be little incentive for industry to incur the costs of meaningful safety tests – as was already quite evident.

When Regal pointed this out, he received no satisfactory responses. In most cases, he was met with shrugs and looks of resignation. One undaunted advocate suggested the food would first be sold abroad and not introduced in the US until it was clear that no catastrophes had occurred in foreign populations. Among the many individuals at the conference with positions of responsibility, none seemed willing to do anything to re-direct the course on which the GE food venture was proceeding or reduce the rate at which it was accelerating. As Regal recounts:

> The more I interacted with biotech developers over the years, the more evident it became that they were not creating a science-based system for assessing and managing risks. And as momentum built and pressures to be on the band wagon mounted, people in industry and government who were alerted to potential problems were increasingly reluctant to pass the information on to superiors or to deal with it themselves. Virtually no one wanted to appear as a spoiler or an obstruction to the development of biotechnology. By the end of the Annapolis event, it was clearer than ever that the careers of too many thousands of bright, respected, and well-connected people were at stake – and that too much investment needed to be recovered – for industry or government to turn back. The commercialization of GE foods would be allowed to advance without regard to the demands of science; and the supporting rhetoric would stay stretched well beyond the limits of fact.

A Disregarded Dimension: US Law

Ironically, one of the many realities to which the biotech proponents and most of their government allies had become oblivious was that the nation's legal system formed a bulwark against what they were planning to do – and that to push GE foods onto the market absent meaningful testing, they would not only have to break their promises to Regal and his collaborators, they would have to break the law as well.

What the proponents failed to grasp was that the federal food safety laws do not intend for market forces to be the basic mode of protection – and that they were designed to reduce the chances that a novel product will cause mass misery and only thereafter be driven from the market by liability lawsuits. Accordingly, in the case of new products like genetically engineered foods, the laws require the kind of science-based testing that the biotech boosters were resolved to shirk.

Of course, not all the government officials were ignorant of what the food safety laws required. Those in the FDA were well aware of the mandates, and the agency had even created some of them. But too many of these officials were prepared to circumvent the requirements in order to advance government policy and speed the entry of GE foods into the nation's supermarkets. However, due to the rigor of the law's provisions, significant cunning would be required to do so.

Confronting a Classic Consumer Protection Statute

The precautionary features of US food safety law are neither new nor nuanced. They're explicit, and they've been in place since 1958, when Congress significantly upgraded the Food, Drug and Cosmetic Act (the FDCA). That statute, initially signed into law in 1938, established standards for the safety and purity of the three categories of products named in its title; and over the years, it became one of the nation's most venerated consumer protection laws. From its inception, the statute took a strong precautionary stance in the case of drugs, mandating that each be proven safe before it could be marketed. But in 1938, it appeared to the legislators that the safety of foods could be maintained without such strictures because, in contrast to drugs, most foods had been safely consumed for centuries. Thus, while it had been found necessary to regulate the conditions under which many were produced in order to prevent their adulteration, it seemed unreasonable to demand that the safety of the foods themselves be demonstrated. In their pure forms, they were presumed to be safe and were only expected to pose risk in the event they became tainted. Consequently, foods could be purveyed without prior testing, and protective measures only came into play if particular batches clearly produced harm, in which case the FDA could recall or seize them. Moreover, although many substances were added to processed foods, a large proportion (like salt, vinegar, and spices) had been safely used throughout the ages, and many of the others did not appear threatening. So (except for substances such as formaldehyde, which were specifically banned) additives were presumed safe unless proven otherwise – and the burden of proof fell on the FDA.[7]

Yet, though this approach seemed sensible in 1938, by 1958 it was significantly outdated. During those twenty years, the volume of processed foods dramatically burgeoned, as did the number of substances added during processing, resulting in what's been termed a "chemogastric revolution."[8] Not only had the majority of these additives never appeared in food, many had never even existed and were recent concoctions of chemistry labs. Accordingly, there was no evidence to show that they were safe, and good reason to suspect that many might not be.

Therefore, Congress decided to strengthen the FDCA so it could better deal with the influx of novel additives. The result was the Food Additive Amendment of 1958. Its firm intent was to prioritize precaution, even if that moderated the pace of technical innovation – a fact made clear by an official Senate report which stated: "While Congress did not want to unnecessarily stifle technological advances, it nevertheless intended that additives created through new technologies be proven safe before they go to market."[9] Under the new provisions, additives to food were no longer presumed safe but instead were presumed to be harmful; and the industry had the burden of proving they were not. Absent such proof, a new additive was prohibited from entering the food supply.

Further, because it would be unreasonable to demand that things like salt and spices undergo such safety testing, the amendment established a category of additives exempt from this requirement: those that were "generally recognized as safe" (commonly referred to as "GRAS"). Substances with a history of safe use prior to 1958 were placed in this category, and they could continue to be employed without any testing. Moreover, Congress recognized that even if a substance's safety had not been established through common experience before 1958, there would still be cases in which it had been subsequently demonstrated, obviating the need for further testing. Consequently, the law stated that a substance could also be deemed GRAS if there was a consensus among experts that its safety had already been firmly established through "scientific procedures."[10] And, as we shall see, the standard of proof was rigorous.

These provisions of the Food Safety Amendment have been widely regarded by both liberals and conservatives as necessary; and when the Republican Richard Nixon was president, he directed the FDA to tighten its regulation of the GRAS list. Moreover, there has been no serious attempt to deregulate food additives and allow them to be governed solely by market forces.[11]

Therefore, if the FDA was going to let GE foods onto the market without prior proof of safety, it would have to do more than merely bend a well-respected law. It would have to break it. Moreover, not only would it have to disregard the law, it would have to disregard the opinions of its own scientists.

FDA Scientists Voice Their Concerns

In the early 1990's, the FDA established a scientific task force, with experts from all the relevant disciplines, to aid in developing its policy on GE foods. And as their input accumulated, it expressed a common theme: the need for caution. In memo after memo, agency experts explained how GE foods pose problems that their conventionally produced counterparts do not – and asserted that their safety cannot be presumed but must be established via testing.

For example, microbiologist Dr. Louis Pribyl criticized the administrators' efforts to equate the unintended effects of bioengineering with those of conventional breeding. He wrote: "The unintended effects cannot be written off so easily by just implying that they too occur in traditional breeding. There is a profound difference between the types of unexpected effects from traditional breeding and genetic engineering. . . ." He branded the assertion of similarity as "the industry's pet idea" and pointed out "that there is no data to back up their contention. . . ."[12] He also noted that because the added genes cannot be inserted precisely but instead enter the DNA strand randomly, ". . . it seems apparent that many pleiotropic [unintended] effects will occur." He further explained that several aspects of gene-splicing "may be more hazardous," including the risk that the promoters fused to the foreign genes will activate dormant ("cryptic") metabolic pathways. As he explained: ". . . breeders have not had to face the issue of new, powerful regulatory elements being randomly inserted into the genome. So there is no certainty that they will be able to pick up effects that might not be obvious, such as cryptic pathway activation. This situation IS different than that experienced by traditional breeding techniques."[13] (*emphasis in original*)

Dr. Pribyl also pointed to the risk that the enzyme produced by the inserted gene ". . . while acting on one specific, intended substrate to produce a desired effect, will also affect other cellular molecules, either as substrates, or by swamping the plant's regulatory/metabolic system and depriving the plant of resources needed for other things." He added, "It is not prudent to rely on plant breeders always finding these types of changes (especially when they are under pressure to get a product out)."[14] In a subsequent memo, he again warned about the risk associated with the hyper-activity of the foreign gene, citing "the potential for the newly introduced gene (or gene product) to swamp the plant's resources." He noted that this could result in "shutting down other genes. . . ." [15]

Dr. E. J. Matthews of the FDA's Toxicology Group likewise emphasized the risk of unintended effects. He asserted that "... genetically modified plants could . . . contain unexpected high concentrations of plant toxicants . . . ," and he cautioned that some of these toxicants could be unexpected and could ". . . be uniquely different chemicals that are usually expressed in unrelated plants." [16]

The Division of Food Chemistry and Technology similarly cited the risk of novel toxins, while calling attention to a few others as well; and it noted the need for testing. Its memo stated: ". . . some undesirable effects such as increased levels of known naturally occurring toxins, appearance of new, not previously identified toxicants, increased capability of concentrating toxic substances from the environment (e.g., pesticides or heavy metals), and undesirable alterations in the levels of nutrients may escape breeders' attention unless genetically engineered plants are evaluated specifically for these changes. Such evaluations should be performed on a case-by-case basis, i.e., every transformant should be evaluated before it enters the marketplace." [17] The memo further advised that the evaluations should include toxicological tests. Moreover, it stated that tests should extend to the plants' edible extracts, noting that "it would . . . be necessary to demonstrate that edible seed and oils produced from genetically engineered plants do not contain unintended potentially harmful substances at levels that would cause concern." [18] In another memo, the Division again warned about unintended outcomes, stating that "DNA insertion may affect the expression of many genes." [19]

Another expert who discussed the risks of unintended effects (and the related need for toxicological testing) was Dr. Carl B. Johnson of the Additives Evaluation Branch. He noted that the "inability of analytical or molecular methods to detect the presence of an unknown toxin produced by activation of a previously cryptic gene" provides "justification" for toxicological testing. And he criticized the proposed policy statement's implication that if gene insertion occurred at only one site, the risk of such unintended effects would be negligible. He stated that limiting insertion to a single site would "reduce, but not eliminate" the likelihood of unintended effects; and he twice pointed out that the document provided no evidence to show that the risk would be negligible.[20]

The Director of FDA's Center for Veterinary Medicine (CVM) also underscored the need for scrutiny – and the absence of evidence. He wrote: "I and other scientists at CVM have concluded that there is ample scientific justification to support a pre-market review of these products. . . . It has always been our position that the sponsor needs to generate the appropriate scientific information to demonstrate product safety. . . ." Further, he noted that GE foods pose exceptional problems, stating: ". . . CVM believes that animal feeds derived from genetically modified plants present *unique* animal and food safety concerns."[21] [*emphasis added*] Among the unique concerns he listed was the risk that residues of unexpected toxicants in engineered animal feed could appear in meat and milk products, making them unsafe for humans. And he pointed out that a slight shift in chemical composition could have a major impact. "Unlike the human diet, a single plant product may constitute a significant portion of the animal diet.. . . Therefore, a change in nutrient or toxicant composition that is considered insignificant for human consumption may be a very significant change in the animal diet."[22] This warning was especially pertinent because GE crops were expected to comprise a large part of the diet of US livestock.

Further, besides stressing the need for testing, Dr. Guest decried the agency's attempt to exploit its absence. In critiquing a draft of the proposed policy he wrote: "I would urge you to eliminate statements that suggest that the lack of information can be used as evidence for no regulatory concern."

Like Dr. Guest, the head of the Biological and Organic Chemistry Section emphasized that lack of proof that a GE food is dangerous does not confirm its safety, noting that "in this instance ignorance is not bliss."[23] He also faulted the proposed policy statement because it "turns the conventional connotation of *food additive* on its head" – and "conveys the impression that the public need not know when it is being exposed to '*new food additives*'. . . ." [*emphasis in original*] He additionally remarked that a particular section of the statement "seems very arbitrary."

The pervasiveness of the opinion about the uniqueness of bioengineering and the distinctness of its risks is attested by a memo of January 8, 1992 to the Biotechnology Coordinator written by Dr. Linda Kahl, an FDA compliance officer responsible for monitoring the expert input. In it, she protested that the agency was ". . . trying to fit a square peg into a round hole . . . [by] trying to force an ultimate conclusion that there is no difference between foods modified by genetic engineering and foods modified by traditional breeding practices." She then declared: "The processes of genetic engineering and traditional breeding are different, and according to the technical experts in the agency, they lead to different risks."[24]

Industry Agenda Trumps Expert Assessment

Although the consensus among FDA scientists was clear, their collective opinion would have little bearing in shaping a policy that was supposed to be science-based. That's because agency administrators were apparently much more willing to ignore the advice of their experts than to ignore the wishes of the biotech industry.

As we've seen, for many years those wishes had been effectively translated into priorities of the U.S government; and the industry's influence was still on the rise. Further, the rising power within that industry was Monsanto. This St. Louis-based corporation had become the largest developer of genetically engineered plants, and its influence was enormous. As a January 2001 article in the *New York Times* reported, through three consecutive administrations (Reagan, Bush, and Clinton), "What Monsanto wished for from Washington, Monsanto – and, by extension, the biotechnology industry – got."[25] (This phenomenon would continue through the following two administrations as well.)

Moreover, the article revealed that the industry had essentially *dictated* policy. In the words of Henry I. Miller, who dealt with biotechnology issues at the FDA between 1974 and 1994 and directed the Office of Biotechnology for five of those years, "the U.S. government agencies have done *exactly* what big agribusiness has asked them to do and *told* them to do." [*emphasis added*]

What Monsanto wanted (and demanded) from the FDA was a policy that projected the illusion that its foods were being responsibly regulated but that in reality imposed no regulatory requirements at all. However, although the administration of George H.W. Bush embraced this plan and apparently induced the FDA decision-makers to go along with it, the agency's technical experts were not persuaded that science was an enterprise to be freely shaped by political and economic agendas. As their memos attest, they refused to serve as spin doctors for Monsanto, and they instead endeavored to fashion a policy based on the best available evidence and supported by sound scientific reasoning. Consequently, early drafts of the policy significantly clashed with the goals of the politicians and the politically appointed administrators.

The tensions between the approach of the scientists and the agenda of the politicians to a large extent reflected the disparities between the food safety laws and the Coordinated Framework for Regulation of Biotechnology that the Reagan Administration had fashioned in 1986. As we saw in Chapter Two, the Framework mandated the administrative agencies "to regulate the product, not the process." But this precept entailed the presumption that genetic engineering is essentially as safe as traditional breeding – and that it poses no risks of unintended effects that warrant safety testing.[26] As FDA compliance officer Kahl noted, it was due to this mandate that the agency's administrators tried "to force an ultimate conclusion" that GE foods are essentially the same as their conventional counterparts. [27]

But although the Framework induced FDA officials to force the adoption of an unscientific precept, it lacked the force of law – and it had no power to modify laws. Therefore, it could not technically override the statutory requirement that safety be demonstrated through scientific procedures – and that decisions be based on evidence rather than speculation. So in calling for tests to screen for bioengineering's inherent risk of unintended consequences, and in stressing that the presumption of safety lacked evidentiary back up, the FDA experts were adhering to the law. And in trying to conform to the Coordinated Framework, the administrators were skirting it.

Because the dissonance between the scientific staff and the administrators was impeding development of a policy statement in line with White House preference, the FDA decided to bring in someone to get the process moving more decisively toward the desired outcome. So in July 1991, a new position was created, Deputy Commissioner for Policy; and the individual appointed to fill it was not only wise in the ways of the agency, but adroit in advancing the aims of the industry. He was Michael Taylor, a man who had worked at the FDA from 1976 to 1981 as a staff lawyer and Executive Assistant to the Commissioner. He then became a partner in the law firm of King & Spalding, where the clients whose interests he represented included Monsanto and the International Food Biotechnology Council. Thus, his selection as the official to oversee the development of FDA policy on GE foods was yet another indication of Monsanto's clout.

With Taylor in charge, as successive drafts of the policy statement were written, sections describing differences between GE foods and their conventional counterparts were progressively purged. However, scientists were displeased by the excisions, and microbiologist Pribyl was moved to protest. "What has happened to the scientific elements of this document?" he demanded. "If the FDA wants to have a document based upon scientific principles, these principles must be included, otherwise it will look like and probably be just a political document. . . . It reads very pro-industry, especially in the area of unintended effects. . . ." [28]

But the fact that the document had a "very pro-industry" slant was an essential part of the plan; and a memo from the office of the FDA

Commissioner, David Kessler, to the Secretary of Health and Human Services dated March 20, 1992 touted its conformity with the White House's economic aims. As the memo pointed out, "The approach and provisions of the policy statement are consistent with the general biotechnology policy established by the Office of the President. . . . It also responds to White House interest in assuring the safe, speedy development of the U.S. biotechnology industry."[29] The document additionally emphasized that the policy statement would play a "critical" role in helping the industry "win public acceptance of these new products" by "assuring" consumers about their safety.

Yet, from the standpoint of the upper echelons, even this version of the policy statement, which Pribyl had derided as short on science and long on pro-industry pap, was not sufficiently sanitized. For instance, an administrator in the office of the Assistant Secretary for Health at the Department of Health and Human Services was disconcerted by the extensive airing of potential problems that GE crops could pose for the environment. So he fired off an objection which declared: "The extensive twelve page discussion seems to be . . . dangerously detailed and drawn-out. . . . In contrast to the sections on food safety, which properly imply that biotechnology is a fundamentally innocuous tool of food production and that the fruits of biotechnology will be substantially equivalent to those with which we are already familiar, the [environmental] section gives an incorrect impression that biotechnology raises significant new agricultural and environmental concerns."[30] In response to his complaint, the environmental section was drastically cut and revised so as to be cleansed of "dangerously detailed" discussion – and the impartation of any impression deemed "incorrect" by non-scientist superiors.

But despite its successive surgeries, in the eyes of the White House team, the document was still in need of cosmetic adjustments. And James B. MacRae, Jr., an administrator in the Office of Management and Budget, sent a memo to President Bush's White House counsel that called for several corrections.[31] He termed the "tone" of one paragraph "inappropriate" because it implied there might be "obligatory FDA review and oversight," and he proposed an alternative paragraph to "stress the role of decentralized safety reviews by producers." He also faulted the document for insufficiently stressing that the method by which a food is produced is essentially "irrelevant" to the issue of its safety, and he suggested that two offending sentences be replaced with text that he set forth. And to more fully minimize the importance of method, he even recommended removing the reference to "plants developed by recombinant DNA techniques" from the document's title and changing it to read: "Statement of Policy: Foods Derived from New Plant Varieties." Moreover, he said that the document "should state" that rDNA techniques "actually may produce safer foods." Accordingly, he requested the inclusion of the following sentence: "Since these techniques are more precise, they increase the potential for safe, better characterized, and more predictable foods."

Because MacRae's requests came from the executive office of the president, they were extremely effective.[32] Whereas the FDA's experts labored for many months to fashion a science-based footing for the policy statement, only to have most of it deleted or deformed, a mere eight days after MacRae wrote his memo, the final document was published in the Federal Register with all his proffered text included – and the altered title he proposed at its head.

Slighting the Law through Sleight of Hand

Just as the Bush Administration had accelerated the policy's completion, it likewise hastened public introduction. On May 26, 1992, three days before the policy was published, Vice-President Dan Quayle stood before a crowd of executives and reporters and presented it as a major achievement in the White House's campaign for what was termed *regulatory relief.* "The reforms we announce today will speed up and simplify the process of bringing better agricultural products, developed through biotech, to consumers, food processors and farmers," he declared. "We will ensure that biotech products will receive the same oversight as other products, instead of being hampered by unnecessary regulation." And, to soothe concerns that in revving up the process, hazards could go undetected, he avowed: "We will not compromise safety one bit."[33]

Moreover, the policy statement itself went Mr. Quayle one better by alleging that there really wasn't an issue about the safety of GE foods in the first place. It described rDNA techniques as mere "extensions at the molecular level of traditional methods," and it consistently depicted their products as no riskier than other foods.[34] Then, to cap things off, it proclaimed: "The agency is not aware of any information showing that foods derived by these new methods differ from other foods in any meaningful or uniform way, or that, as a class, foods developed by the new techniques present any different or greater safety concern than foods developed by traditional plant breeding."[35]

But this outright denial was an outright lie. As we've seen, the FDA had received abundant information from its own experts alerting it to several differences that were indeed meaningful – and that created a need for safety testing. However, the memos the experts had written were lodged in the agency's files, and most of their input to the policy statement had been excised by the time the final draft was produced. So citizens were not aware of the fraud.

Nor were they aware that the Vice-President's assertion that GE foods would be given the standard degree of oversight was also false. In his rendition, while the law would be fully applied, it would not be bloated by "unnecessary regulation" that would hamper the development of an important technology. But in reality, GE foods were not being freed from extraneous burdens; they were being illegally exempted from the central provisions of one of the nation's most important consumer protection statutes.

As previously noted, the Food, Drug and Cosmetic Act stipulates that new food additives without a history of safe use prior to 1958 must be proven safe before they are marketed. And it defines a food additive as "any substance the intended use of which results or may reasonably be expected to result, directly or indirectly, *in its becoming a component or otherwise affecting the characteristics of any food. . . .*"[36] [*emphasis added*]. This is a broad definition, and it clearly covers the transferred genetic material in bioengineered foods as well as their expression products. However, the statute modifies the definition by stating that a substance will not be classified as an additive if it is "generally recognized" by experts to be safe.[37] This is the GRAS exemption; and the FDA's policy statement claimed that in virtually every instance, the DNA inserted into GE foods and the substances it produces could be reasonably presumed to qualify.[38]

But it's difficult to accord this claim respect, because the FDA clearly knew that GE foods were not generally recognized as safe by experts. First, the overwhelming consensus of its own experts was (a) that these products pose potential problems, (b) that their safety cannot be presumed, and (c) that each must be demonstrated safe via testing. Second, FDA officials knew there was not a consensus about the safety of GE foods among scientists outside the agency either – a fact acknowledged by the FDA's Biotechnology Coordinator in a letter to a Canadian health official dated October 23, 1991. In commenting on a document that discussed GE foods, he stated: "As I know you are aware, there are a number of specific issues addressed in the document for which a scientific consensus does not exist currently, especially the need for specific toxicology tests."[39]

Moreover, even if there had been a genuine consensus about safety, the law also requires that it be based on solid evidence. These requirements are quite strict. They're contained in the statute itself and elaborated in its associated regulations. Such regulations result when Congress authorizes the agency that's going to administer a statute to make additional rules that will facilitate its implementation. These regulations then become part of the law.

In the case of the GRAS exemption, Congress had already mandated that a substance without a prior history of safe use cannot qualify unless evidence for its safety has been generated through "scientific procedures;" and in 1971 the FDA issued a regulation that strengthened the mandate by stipulating that this evidence has to be of the "same quantity and quality" as is required to gain approval as a food additive – and that it must demonstrate a "reasonable certainty" that the substance is not harmful under its intended conditions of use.[40] The regulation further stated that because the evidence must be widely known by experts, it should "ordinarily" be reported in "published studies, which may be corroborated with unpublished studies and other data and information."[41]

Thus, the GRAS exemption does not relax the degree of required evidence but rather relieves a producer from performing new tests for substances already known to be safe on the basis of previous ones. And the FDA was well aware that there was no reliable evidence demonstrating the safety of GE foods. As we've seen, memos by several of its experts had noted the paucity of such evidence – and rebuked the administrators for pretending that a lack of data demonstrating harm can count as concrete proof of safety. Further, microbiologist Pribyl had emphasized that there was no evidence to support the claim that the unexpected effects of genetic engineering are similar to those of traditional breeding. And Dr. Johnson of the Additives Evaluation Branch had emphasized there was no backing for the notion that the risks of such effects could be rendered negligible. Perhaps the strongest acknowledgement of the evidentiary void appeared in a memo from compliance officer Kahl, who demanded: "[A]re we asking the scientific experts to generate the basis for this policy statement in the absence of any data?"[42]

However, although FDA officials were willfully breaking the law by treating GE foods as if they were GRAS, they were able to create the impression that nothing illicit was happening. They had grappled with the issue of how to pull this off for some time, and a document written during the summer of 1991 provides insight into their calculations.[43] In discussing the basic options open to the agency, the document noted that "some regulatory middle ground is needed" between complete lack of pre-market regulation and "routine imposition of the food additive/GRAS regime, with its requirement of petitions as the only basis for obtaining any FDA involvement in the task of safety assurance."[44] However, it observed that such a route would have drawbacks, because by allowing GE foods to be marketed without food additive petitions and the safety testing that must accompany them, the FDA would give ". . . the appearance of loosening requirements of an existing regulatory category to fit biotech foods. . . ."[45] Therefore, it said that procedures would have to be implemented to avoid such an appearance – and also "[t]o avoid the appearance of complete industry self-regulation . . ." In elaborating on procedures that might do the trick, the document suggested that the FDA could promulgate guidelines through which the manufacturers would make their own safety assessments, determine that their products are GRAS, and then inform the agency of their determinations.

This is the basic approach that the 1992 policy statement adopted. And, although the whole process was completely voluntary, because it projected the appearance of regulation, and because the pronouncements of government and industry routinely maintained that responsible regulation was in place, most people were fooled. Thus, the vast majority of Americans (including those in Congress) were led to believe that GE foods were being carefully regulated and rigorously tested when in reality the biotech industry had, in effect, been granted self-regulation and the voluntary assessment process was incompetent to assure the safety of any product that passed through it.

Contrary to the government's pretensions, this process is not only "informal" but remarkably superficial. As the FDA's Biotechnology Strategic Manager has described it, when in a candid mode: "The FDA requests that firms submit a summary of their assessment to the agency. The FDA does not request the original data and, therefore, does not conduct a scientific review of the firm's decision."[46] Moreover, the agency does not even make a determination that the firm's decision is valid – and thus has never officially determined that any GE food currently on the market is actually safe.

Taking the FDA to Court

Although the FDA had deluded the public and Congress, it seemed highly improbable that the charade would survive the scrutiny of a federal court. And when the Alliance for Bio-Integrity and its co-plaintiffs filed their lawsuit in May 1998, there was ample basis for optimism. For one thing, numerous scientists and scientific organizations (such as the British Medical Association) had indicated that they were not convinced that GE foods are as safe as conventional ones, so it was clear that there was not a general recognition of safety within the scientific community. Further, the mere filing of the suit established that point, because nine of the plaintiffs were scientific experts who alleged that genetic engineering poses higher risks than traditional breeding – and that its products cannot be presumed safe. This group comprised seven professors from institutions such as UC, Berkeley, Rutgers, and the NYU School of Medicine along with the Associate Director of Targeted Mutagenics at Northwestern University Medical School and a computational biologist with the Human Genome Project. Accordingly, it contained more than enough experts to refute the FDA's claim that there's a general recognition of safety – especially considering that in a prior judicial proceeding, the FDA had defeated a manufacturer's GRAS claim by producing just two experts who contested it.

To drive home the invalidity of the GRAS presumption, three experts submitted declarations detailing the differences between genetic engineering and traditional breeding, explaining the kinds of tests that are required to assess the safety of GE foods, and pointing out that none of these foods had yet to be demonstrated safe on the basis of such testing. Two of the declarations came from plaintiffs, and one was submitted by Richard Lacey, M.D., Ph.D., a Professor of Medical Microbiology at the University of Leeds in the U.K., who couldn't be a plaintiff because he is not a US citizen. Dr. Lacey, a leading authority on food safety, had written five books on the topic and published over 200 articles in scientific journals. He had also distinguished himself by predicting a major food-borne peril, which not only demonstrated his prowess, but ultimately demonstrated how crucial it is for governments to heed the warnings of experts – and how disastrous it can be when they forgo precaution. As his declaration explained:

> In 1989, I anticipated that there could be serious health risks to the British cattle and human populations from the practice of feeding cattle rendered meat from sheep and other animals. I published my warnings in *Food Microbiology,* 1990. In this article, I explained the nature of the malady that could result. This was the first prediction of what eventually became the "mad cow" epidemic in the United Kingdom. Unfortunately, the governmental authorities were slow to respond to my warning. Had they properly assessed and acted upon the information I presented, much hardship would have been

> avoided, and the citizens would not have been subjected to as high a degree of risk. (Because of the long latency period between exposure to the infectious agent and development of symptoms, there is a potential for widespread incidence of infection within the British public over the next forty years.)[47]

In the case of GE foods, Dr. Lacey had again issued warnings for the sake of averting widespread harm; and he hoped that his declaration to the court would have a more immediate effect on US governmental policy regarding those products than his warnings about mad cow disease had on the policy of the U.K. government.

His declaration in part asserted:

> It is my considered judgment that employing the process of recombinant DNA technology (genetic engineering) in producing new plant varieties entails a set of risks to the health of the consumer that are not ordinarily presented by traditional breeding techniques. It is also my considered judgment that food products derived from such genetically engineered organisms are not generally recognized as safe on the basis of scientific procedures within the community of experts qualified to assess their safety. . . .
>
> Recombinant DNA technology is an inherently risky method for producing new foods. Its risks are in large part due to the complexity and interdependency of the parts of a living system, including its DNA.
>
> . . . whereas we can generally predict that food produced through conventional breeding will be safe, we cannot make a similar prediction in the case of any genetically engineered food. Therefore, the only way even to begin to assure ourselves about the safety of a genetically engineered food-yielding organism is through carefully designed long-term feeding studies employing the whole food. . . . Even if the most rigorous types of testing were performed on each genetically engineered food, it might not be possible to establish that any is safe to a reasonable degree of certainty, as is possible in the case of most ordinary chemical additives. However, we at least would be in a far better position than now to have greater confidence in these new foods.

One of the two plaintiffs who submitted declarations was John Fagan, a molecular biologist who had led a research group at the National Institutes of Health and had received a Research Career Development Award from the National Cancer Institute. The other was Philip Regal. He had become a plaintiff because suing the FDA appeared to be the only remaining path that held promise. During the previous fifteen years, he had interacted with the scientific community, the corporate establishment, and the government agencies through standard procedures for educating and persuading. He had endeavored to generate awareness of the risks and the implementation of practices that would make genetic engineering safer. He had diligently responded to the repeated requests of government officials (including President Reagan's "Biotechnology Czar" and administrators from the National Science Foundation and the regulatory agencies) for input that would help shape the government's GE food policy – and had otherwise cooperated with them as they asked. He had met with dozens of corporate executives and representatives. He had attended forty-four scientific conferences and workshops, had presented papers at thirty-two, and had organized four of them (including a meeting of the American Association for the Advancement of Science). He had published nine articles on bioengineered organisms in scientific journals and contributed six chapters on the topic to academic books. Yet, little had come of it; and regardless of what he'd been promised, hardly any concrete reform had occurred. Further, the FDA's actions had convinced him that its administrators were determined to promote GE foods regardless of scientific data or legal duty – and would not desist unless coerced by a court.

Accordingly, his declaration denounced the agency's misbehavior. After a detailed explanation of why every GE food should be presumed to entail higher risk than its conventional counterpart, and after asserting that he was unaware of any reliable scientific study demonstrating that even one such food was safe, he stated:

> To ignore the fact that the living systems involved have complex biochemical and developmental dynamics and that unusual high-technology genetic interventions have the real potential for unpredicted deleterious side effects is, at best, biologically naive.
>
> However, while it might be no more than naive for a layperson to make such simplistic assumptions, it is otherwise in the case of a government regulatory agency that has been repeatedly informed of the facts. When such an agency persists in ignoring these facts, even though the safety of the food supply is at stake, its behavior is not merely scientifically unsound but morally irresponsible.

> It is because I view the FDA's policy and practices regarding genetically engineered food to be irresponsible – and because I regard the consequent risk posed for public health to be substantial – that I have taken the step of joining the above-named lawsuit as a plaintiff. By standing as a plaintiff rather than merely participating as an expert witness, I hope to make clear to the public and the Court not only the extent to which I disagree with the agency's assumptions as a purely intellectual matter, but the degree to which I deplore its behavior on ethical grounds. Ultimately, I was compelled by my conscience to become a plaintiff, and I am proud to stand with so many other scientific experts who have similarly acted on the basis of ethical as well as strictly scientific principles.[48]

After these declarations were submitted, the plaintiffs' position was strengthened even further by numerous documents of equal potency: documents from the FDA's own files. The contents of these files were unknown to us when we initiated the suit and only became accessible because the FDA was required to hand over copies of all its records relating to GE foods as part of the discovery process. This mass of information contained extensive evidence demonstrating that GE foods are not GRAS, and every agency document quoted in the previous pages was part of it. Considering the expected combined impact of these revelations, the involvement of our nine scientist-plaintiffs, and the power of the submitted declarations, I and the other attorneys on the plaintiffs' team had good reason to expect victory.

(As the introduction explains, the attorneys of record were with the International Center for Technology Assessment in Washington, D.C. They had extensive experience in litigation with federal agencies, and they managed the bulk of the lawsuit. I spent considerable time in the ICTA offices assisting with several aspects of the suit, and I contributed significantly to the sections of the briefs dealing with the GRAS and labeling issues. When I use the terms "we" and "our," I'm referring to myself and the ICTA lawyers collectively.)

Despite the strength of the evidence against its claims, the government was not about to concede; and it committed substantial resources to defeating our suit. During the ensuing months, many government attorneys engaged in the fight; and they did not always fight cleanly. In one of their submissions, they impugned the plaintiffs' motives and alleged that the suit was filed for a number of unsavory purposes, including "fear mongering."[49] This aspersion was especially egregious, considering that several of the plaintiffs were eminent scientists and ten were ordained members of the clergy. And there were several other respects in which it seemed that we were confronting cagey industry lawyers instead of public servants dedicated to upholding law and justice.

How the Suit Proceeded: Arguments, Counterarguments, and a Counterfactual Ruling

Over the years, numerous people have asked me to recount the high points of the trial, and they're usually surprised when I inform them that a trial was not held. But lawyers aren't surprised, because trials are only necessary to resolve disputes about the facts, and in our case, the relevant facts were contained within the documents that we filed and the 44,000 pages of the FDA's administrative record. Since these facts were not disputed, the parties agreed that the case should be determined via *summary judgment*, in which each side submits arguments attempting to show how, in light of the facts, the law supports its position.

In all, both sides submitted three written arguments: a motion for summary judgment, a critique of the other's motion, and a reply to the other's critique. Although we expected that the judge (Coleen Kollar-Kotelly) would then request an oral argument so she could probe more deeply into points raised in the written submissions, she did not call for one. On September 29, 2000, more than a year after the last round of arguments was submitted, she released her opinion, addressing each issue in the order it had been raised in the written arguments.

FDA Prevails on Procedural Issues Due to the Rebuttable Nature of the GRAS Presumption

The first issue was procedural. It involved the mandate of the Administrative Procedure Act that before making a substantive rule, a federal agency must implement formal notice and comment proceedings so that interested parties can submit their points of view. We had argued that because the statement of policy is in effect a substantive rule, the FDA had violated this law by not going through such procedures. However, the judge held that the statement is not a substantive rule because it isn't binding on either the FDA or the industry. She stated that the FDA is not bound because its GRAS presumption is *rebuttable* rather than final – and that industry is not bound because the policy does not "impose any new . . . obligations."[50]

The next issue was also procedural, and it pertained to the requirement of the National Environmental Protection Act (NEPA) that before implementing any action that could significantly affect the environment, an agency must perform an Environmental Assessment or prepare an Environmental Impact Statement. We argued that because the FDA had done neither, its policy had been issued in violation of that particular law. But the judge again disagreed due to the insubstantial nature of what the agency had done. She concurred with the FDA's contention that its policy is not subject to these requirements because it is not a significant federal action. She once more emphasized the rebuttable nature of the GRAS presumption; and she held that the policy represents inaction, stating that it is "not properly an 'agency action.'" She agreed with the FDA that the policy does not regulate GE foods any differently than was the case prior to its issuance.[51]

Although we technically lost on these two procedural issues, the net effect was helpful, since the court certified that the FDA is not exercising one iota of pre-market regulation over GE foods.[52]

Was There General Recognition of Safety? The GRAS Presumption Becomes Non-Rebuttable

Judge Kollar-Kotelly next addressed the central issue: *Is it legal for the FDA to presume that all GE foods are generally recognized as safe?* As she was no doubt aware, if the answer was "no," there would be far-ranging consequences, because it would entail that legitimate safety concerns existed – and that every product on supermarket shelves with an ingredient derived from a genetically engineered plant was being sold in violation of the law. And, from all appearances, she was reluctant to instigate such a system-shaking outcome.[53]

She began by acknowledging that to qualify as GRAS, a substance ". . . must meet two criteria: (1) it must have technical evidence of safety, usually in published scientific studies, and (2) this technical evidence must be generally known and accepted in the scientific community."[54] She also noted that "'A severe conflict among experts . . . precludes a finding of general recognition.'"[55] Regarding the second criteria, our submissions had asserted that there was clearly a severe conflict among experts. The judge agreed, stating: "Plaintiffs have produced several documents showing significant disagreements among scientific experts."[56]

Most people would think that, with such a finding, the case was closed. After all, the judge had in effect acknowledged that as of the time we filed our suit, there was not a general consensus about safety in the scientific community. However, she stated that the critical time for assessing the existence of consensus was not May 1998 (when we filed the suit) but May 1992 (when the FDA issued its policy statement). She said that as a matter of administrative law, she could only consider the information the FDA administrators had before them at that time. Consequently, she ruled that all the new evidence we had introduced was legally irrelevant.

Although the approach the judge took may seem unfair or illogical, in the general case it's sound. A judge is not entitled to usurp the role of the administrators and fashion the policy that he or she thinks best. The judge's job is to make sure that the administrators have followed proper procedures in setting the policy; and it's an established principle that unless the administrators have acted arbitrarily or capriciously, a reviewing court should grant them great discretion – and defer to their decision. As long as there was a reasonable basis for the decision, a court cannot legally classify it as arbitrary or capricious. And in order to fairly assess whether such a reasonable basis existed, the judge is only supposed to consider the information the administrators had before them at the time their decision was made.

As noted, this approach makes sense in the usual case. But our case was not usual. The judge emphasized that the FDA's GRAS presumption does not amount to ordinary agency action because it is fully rebuttable and in no way binding. This entails that evidence beyond May 1992 *has* to be relevant, since it is only through such evidence that the presumption could be rebutted.

However, although the judge in effect conceded that the evidence we introduced rebutted the presumption, she nonetheless held that it and all other post-1992 evidence must be ignored, thus converting a rebuttable presumption into a non-rebuttable one. The paradox is glaring, since on the one hand she excused the FDA from following the requirements of the Administrative Procedure Act and NEPA because its presumption is rebuttable by future evidence, while on the other, she fully insulated the presumption from all such evidence.

Yet, even deprived of the post-1992 evidence, our case was still sufficiently strong because the FDA's own files contained extensive evidence demonstrating that GE foods do not meet the GRAS criteria – and this evidence was known to the administrators prior to their 1992 decision. For one thing, the numerous memos from FDA experts about the unique hazards of GE foods clearly indicated that general recognition of safety did not even exist within the FDA's scientific staff. We pointed out that this should have barred the agency from making a GRAS presumption, especially since it had previously convinced a federal court that the opinions of even two experts are sufficient to deny an additive GRAS status.[57] Moreover, in that case, the experts merely stated that they were not aware of any studies in the standard literature demonstrating the substance was safe. In the case of GE foods, the FDA experts pointed to several hazards that are posed.

Surprisingly, the judge ruled that the evidence contained in the experts' memos was also irrelevant. She argued that because these scientists were "lower-level FDA officials," the agency's administrators did not have to pay attention to their opinions when setting policy.

But just as her transformation of a rebuttable presumption into a non-rebuttable one was illogical, so was her argument stripping the scientists' statements of legal import. For one thing, the written opinions of the agency's scientists represented far more than mere policy preferences. They constituted solid evidence that a significant number of experts did not recognize GE foods as safe. In effect, the judge said that the administrators were entitled to presume that there is an overwhelming consensus among scientists that GE foods are safe despite the obvious fact that most of their own experts did not regard them to be. Moreover, not only did the judge allow the administrators to disregard their experts' warnings, she herself ignored the fact that they covered them up and issued a false statement implying that no such warnings existed. As previously noted, their official policy statement declared: "The agency is not aware of any information showing that foods derived by these new methods differ from other foods in any meaningful or uniform way. . . ."[58] Although our briefs fully documented this fraudulent misrepresentation, the judge neglected to mention it, notwithstanding the fact it demonstrates that the administrators were acting in a manner that was not only arbitrary and capricious, but immoral.

Biased Citation from the Parties' Submissions

Remarkably, although the judge failed to note many significant facts and pertinent cases that we had brought to her attention (and even ignored the glaring falsehood in the policy statement claiming that the FDA was "not aware" of information showing that GE foods are different), she uncritically repeated one of the spurious claims in the government's submissions to her – even though we had demonstrated that it was false and misleading. She said: "Moreover, pointing to a 44,000 page record, the FDA notes that Plaintiffs have chosen to highlight a selected few comments of FDA employees, which were ultimately addressed in the agency's final Policy Statement." In fact, as our briefs had shown, the cautionary comments made by FDA staff were not "few" but numerous; and their authors were not mere employees but scientific experts, many of whom held positions of significant responsibility, including the head of the Biological and Organic

Chemistry Section, the Director of the Center for Veterinary Medicine, and even the Biotechnology Coordinator.

Further, we had demonstrated that the safety concerns reflected the dominant opinion of the agency's scientists – and the FDA didn't produce a single memo from an agency scientist asserting that GE foods can be regarded as safe. Additionally, we had pointed out that the final policy statement did not "address" the experts' concerns (as the FDA alleged) but instead disregarded them (sometimes over their authors' protests), suppressed them, and ultimately made misrepresentative statements about them.

However, although the judge included the FDA's false assertion in her opinion, that did not constitute a formal determination about the degree of disagreement within the FDA; and her reference to the claim that we had selected only a few comments was not a definitive finding that there were in fact only a few. In light of the evidence that the comments were extensive, she could not have made such a finding – although many people have been misled into believing that she did.

Ignoring that General Recognition of Safety Didn't Exist Outside the FDA Either

Of course, even if the input of the FDA's scientists is disregarded, the administrative record contains ample evidence showing that scientists outside the FDA had similar concerns. As we pointed out, there were cautionary statements by experts from the Department of Molecular Biology of the Centre for Plant Breeding and Reproduction Research in The Netherlands[59] and the United Kingdom's Ministry of Agriculture, Fisheries and Food.[60] And we also emphasized the fact that, as previously described, there was a letter in the files written by the FDA's biotechnology coordinator acknowledging that there was *not* a consensus about safety within the scientific community.[61]

The administrators were definitely not entitled to disregard such evidence, and there was no legitimate way the judge could argue that they had been. So she didn't try to make that argument. Instead, she ignored this evidence and never mentioned it, despite the fact it unequivocally established (a) that there was not general recognition of safety and (b) that this was known by the FDA in May of 1992.

Ducking the Issue of Whether There Was Technical Evidence of Safety

But the issue of whether a genuine consensus existed was not the sole issue. As we've seen, the law also requires that a consensus must be based on "technical evidence" of safety. And in regard to this crucial issue, our position was likewise strong. In fact, it appeared to be invincible. As was discussed earlier in this chapter (and as we pointed out to the judge), the administrative record unequivocally attests to the utter absence of the requisite evidence – with an FDA compliance officer complaining: "[A]re we asking the scientific experts to generate the basis for this policy statement in the absence of any data? . . . it is an exercise in hypotheses forced on individuals whose jobs and training ordinarily deal with facts. . . ."[62]

Her assertion that FDA's policy relied on hypotheses is especially significant because the agency had previously maintained that GRAS status cannot be based on hypotheses and inferences but must be grounded in solid evidence; and the courts have concurred.

The FDA strongly advanced its argument about the insufficiency of hypotheses in a proceeding against a supplement for swine feed (called Ferro-Lac) that it alleged contained unsafe food additives.[63] In that case, the manufacturer presented an affidavit from a scientific expert alleging that because the three contested constituents of the compound were GRAS when used alone, "'it is a reasonable scientific certainty'" that their use in combination would also be safe. The affidavit further claimed that such a conclusion is based upon "principles of chemistry" and that "any chemist would 'necessarily recognize' the result stated." In opposition, the FDA, submitted two expert affidavits asserting that the use of the three ingredients in combination was a new use – and that their safe use in isolation did not support an inference that they could be safely used together. These affidavits stated that the only way to determine whether the compound is safe is through "actual testing . . . to demonstrate that long term ingestion of potential residues of the chemical in edible tissues will not be harmful to humans."[64] Both experts also stated that they were not aware of any reports of tests of this particular compound in the pharmacological-toxicological literature.

The court ruled in favor of the FDA, stating that the affidavits it submitted established that the evidentiary underpinning for a general recognition of safety was lacking. It dismissed the affidavit submitted by the manufacturer because it was solely based on "theoretical evaluation" and contained "at best, an inference that safety might be shown by scientific testing and procedures."[65]

In our submissions, we explained that there were even stronger reasons to regard GE foods as containing non-GRAS additives than had been the case with Ferro-Lac. First, while the contested constituents of Ferro-Lac were each recognized to be safe in separation, most of the intended expression products of the genes inserted via bioengineering are not themselves recognized as safe. Rather, the FDA administrators infer them to be safe. Second, FDA scientists had pointed out that the bioengineering process could yield a wide range of unintended and unexpected deleterious substances. Third, even though there was testimony that the safety of the concerted action of the components of Ferro-Lac could be inferred with a reasonable scientific certainty, the court held this was insufficient to establish the supplement's safety. GE foods are in an even weaker position, because in their case, such an inference cannot justifiably be made. Scientists both within and without the FDA stated that the dynamics of DNA and living systems are so complex, and the disruptive potential of rDNA technology so great, that it is not possible to infer the safety of any GE food with reasonable certainty.

Despite its significance, Judge Kollar-Kotelly took no account of the Ferro-Lac case, and it is not mentioned in her opinion. Nor did she note the established principle that GRAS status cannot be based on inference and hypotheses. In fact, she disregarded the entire issue of whether the requisite technical evidence existed. Although in her initial statement of the law, she had acknowledged the necessity of such evidence, she then *completely* sidestepped the topic and avoided further discussion of it. Had she confronted it, there's no rational way in which she could have upheld the presumption that GE foods are GRAS.[66]

As With GRAS, So With Labeling: Decreeing an Intimate Linkage

The opinion then moved on to another important issue. The law requires that "material" facts about food be disclosed through labeling.[67] Contrary to our arguments, the FDA contended that the fact a food was produced via rDNA technology is *not* material, even if there's widespread consumer interest in knowing that fact. The agency asserted that unless a process entails unique risk to human health or causes a uniform change in the food, its use is legally immaterial, and there's no duty to inform consumers about it.

The judge upheld this interpretation of the law, and she also deferred to the FDA's determination that GE foods do not (in the agency's words) "present any different or greater safety concern than foods developed by traditional plant breeding."[68] Accordingly, she ruled that the FDA had no obligation to label them. Thus, she linked the labeling issue to the GRAS issue, in effect holding that because the FDA was entitled to presume that GE foods are safe, it was also entitled to reject consumer demands for labeling.

Summing Up: What the Judge Actually Said – and What She Ignored in Order to Say It

In one of its submissions to the court, the FDA claimed that it had been delegated "completely unfettered discretion" to implement the Food, Drug and Cosmetic Act in whatever way it wants.[69] When I and the other attorneys read this pronouncement, we were astonished by its arrogance; and because our submissions demonstrated that courts routinely restrict agency discretion when they detect the kinds of derelictions that the FDA had displayed, there were ample grounds to expect that Judge Kollar-Kotelly would rebuff its argument. However, to our amazement, she essentially granted the FDA the unfettered discretion to which it laid claim. In effect, she gave the administrators the right to ignore legal precedent, their own prior policies, the reasoned opinions of their own experts, and any other facts they found inconvenient – and then to lie about the whole affair by denying they had received any information contrary to the presumption they were pushing.

And in issuing her ruling, she herself had to ignore a substantial amount of critical information. She declared that in May 1992, FDA administrators had a rational basis for presuming that GE foods are generally recognized as safe – even though it's clear from the FDA's own files that: (a) such general recognition has never existed and (b) the technical evidence of safety upon which such recognition is required to rest has never existed either.

Further, in reaching her decision, not only did she ignore the above two facts, she ignored the established principle that GRAS status cannot rely on hypotheses. She also disregarded the fact that the FDA's 1992 policy sharply reversed its previous position on GRAS – while she likewise disregarded the line of judicial decisions (repeatedly called to her attention) asserting that such shifts deprive an agency's decision of the deference it would ordinarily deserve.

Unfortunately, most people have no idea how much the judge had to overlook in order to render her ruling; nor do they understand how limited it really was. She did not rule that GE foods have actually been shown to be safe; nor did she determine that there ever was a general recognition of safety among the FDA scientists or within the scientific community. She did not even say that the FDA could justifiably continue to presume that GE foods are safe. Her decision was strictly limited to the particular exercise of discretion made by the FDA in May of 1992. She ruled that at that specific point in time, the FDA had been entitled to presume that there was a general recognition of safety among scientific experts; but she indicated that we presented evidence showing there was not a general recognition of safety at the time we filed our suit. And she emphasized that the FDA's presumption is supposed to be *rebuttable* by evidence it receives to the contrary.

Imagine the reaction of America's mothers on learning that the genetically engineered foods they've been routinely feeding their children have never been generally recognized as safe among experts, have not been proven safe as required by law, and were determined by the FDA's own scientists to entail unusual risks. And consider whether they'd be comforted by the knowledge that a federal judge allowed these foods to stay on the market, not because she concluded that they met the standards of the law, but because she ignored the evidence that demonstrated they did not – and ruled that FDA executives had discretion to do the same.

(Appendix A provides additional analysis of the errors in the judge's opinion; and it more fully demonstrates how her arguments are undercut by prior decisions of federal courts.)

Enhanced Anomaly: How the FDA Policy Was Saved by Withdrawal of a Plan to Reform It

In light of the opinion's serious flaws, you may be wondering why it was not reversed on appeal. The explanation is not only remarkable, like so many other facets of the lawsuit's story, it contains anomalous twists.

In January 2001, after we had filed an appeal but before our arguments were submitted, the FDA proposed a new rule on GE foods. This rule changed very little. It maintained the GRAS presumption, and it refrained from requiring safety testing or labeling. It merely added a mandate that manufacturers must give the FDA notice at least 120 days before they market a GE food. Nonetheless, despite the minimal degree of proposed reform, had the rule been implemented, it would have had a major impact on our lawsuit. By replacing the informal policy decision against which the suit had been brought, it would have made the suit irrelevant and rendered the appeal a waste of time, because we would have needed to proceed against the new rule by filing a new action. Further, starting a new lawsuit would have been advantageous, because all the evidence that the judge excluded in our initial suit would have been admissible, since it was known to the FDA at the time the new rule was proposed. Moreover, during the notice and comment period on the proposed rule, the FDA had been openly informed about other recent evidence that GE foods were not GRAS, including a January 2001 report by the Royal Society of Canada that declared it is "scientifically unjustified" to presume they're safe. With all this evidence in play, it would have been virtually impossible for a court to uphold the FDA's GRAS presumption.

Therefore, the Alliance for Bio-Integrity and the other plaintiffs dropped our appeal, intending to bring a new suit when the rule took effect. However, after we did, the FDA delayed final action on the proposed rule for more than two years and then announced that it was being withdrawn. But by then, our appeal could not be revived.

Thus, if the FDA had improved the 1992 policy by requiring notification before a new GE food hits the market, the policy would have been struck down by a court, because its GRAS presumption would no longer have been insulated from the abundant evidence that refutes it. But by abandoning the proposed reform, the agency continued to shield the policy from such evidence, thereby saving it.

Few people realize how vulnerable the proposed rule was. It would have been quashed not only through the force of opposing evidence, but through self-contradiction. When published in the Federal Register, it was accompanied by extensive supplementary information describing, among other things, its background and why it was needed; and this 32-page document was patently at odds with itself. On the one hand, it asserted that the 1992 GRAS presumption was still valid, while on the other, it acknowledged that many impending GE foods might pose safety issues that would bar them from GRAS classification. For instance, it noted ". . . that because breeders utilizing rDNA technology can introduce genetic material from a much wider range of sources than previously possible, there is a greater likelihood that the modified food will contain substances that are significantly different from . . . counterpart substances historically consumed in food. In such circumstances, the new substances may not be GRAS. . . ."[70] The text further acknowledged that the inserted genes "may disrupt or inactivate an important gene or a regulatory sequence that effects the expression of one or several genes, thereby potentially affecting adversely the safety of the food. . . ."[71] And it observed that as biotechnicians increasingly insert multiple genes into the target organisms, such unintended effects "may become more common."[72] Elaborating on the potential for these unintended effects to cause harm, the document continued: "FDA believes that the use of rDNA techniques in plant breeding may lead to unintended changes in foods that raise adulteration or misbranding questions. These unintended changes may cause a food to be adulterated because the food may be rendered injurious to health. . . ."[73]

Accordingly, the document stressed the need for pre-market notification, explaining: "Because of its role in ensuring the safety of the U.S. food supply, FDA needs to be aware of the modifications to food source plants from the application of rDNA technology and any unintended effects in food that result so that the agency can evaluate whether the foods from such plants are adulterated or misbranded."[74] Additionally, it underscored the uniqueness of the problems posed by rDNA technology, explaining that although notification is not needed for new plant varieties produced by other methods, ". . . rDNA techniques have a greater potential, relative to conventional methods of breeding, to result in the development of foods that present legal status questions."[75]

These acknowledgements regarding the unintended effects of genetic engineering and their uniquely problematic nature were a dramatic departure from the 1992 policy statement. In fact, they were the same type of science-based assertions that had been methodically excised from the drafts of that document due to political pressures from the Bush White House. However, although in 1992 such passages were purged in order to preserve the plausibility of the GRAS presumption, in 2001, the FDA attempted integration. It tried to merge a measure of scientific reality with the core of the 1992 statement. But the attempt failed. Although the agency pretended that the two components were harmonized, as the two preceding paragraphs reveal, they were inherently incompatible. The assertions about the problematic nature of bioengineering were either directly contradicted by assertions held over from 1992, or else their import was mitigated. For instance, the statement stressing the agency's need to be informed about unintended effects of the genetic manipulations seemed insincere in light of the claim that the voluntary assessment program was still adequate to assure food safety, even though it did not even require superficial tests, let alone the rigorous toxicological testing that's necessary for detecting unintended effects.

Although I can't prove it, I suspect that the FDA decided to withdraw the proposed rule when awareness dawned that, due to its internal infirmities and the evidence arrayed against it, it could not survive a lawsuit. I assume that upper level officials realized that the only way to preserve the GRAS presumption (and keep the biotech industry free to forgo meaningful testing) was to retreat to the confines of the 1992 policy, because it had been upheld in court by a ruling that was no longer subject to appeal. I further assume that in retracting the proposed rule, the administrators hoped that its embarrassing admissions would quickly fade from public memory – which, by all appearances, has happened.

A Decision that Has Stayed Unscrutinized

Even though Judge Kollar-Kotelly's decision had, through a strange turn of events, escaped the scrutiny of an appellate court, one could have reasonably expected that its defects would be apparent to astute observers and exposed in the media and professional journals. But this never occurred, despite the fact that the participation of nine expert plaintiffs should have made it obvious that GE foods were not GRAS – and raised questions about how the judge was able to discount their significance. Instead, the newspaper reports on the suit's outcome uncritically accepted the judge's rulings, and none of the many I've seen noted the significance of the scientist-plaintiffs.

More surprising, although there have been several articles in legal journals discussing the case, they have not discerned the opinion's key errors either. For one thing, it seems that none of the authors obtained copies of the briefs we filed, so they weren't aware of what the judge ignored in order to reach her ruling. It further appears that some authors even neglected to read the entire ruling (or at least neglected to read it carefully). Such a lapse seems to have affected the first article, which appeared in the *Temple Environmental Law and Technology Journal*.[76] This article stated that the plaintiffs "failed" to show a sufficient conflict among experts – and that due to this failure, the court rejected their contention that the GRAS presumption was invalid.[77] In fact, the judge stated: "Plaintiffs have produced several documents showing significant disagreements among scientific experts." Somehow, the author missed this statement. Consequently, she also failed to realize that the judge upheld the GRAS presumption only because, as a procedural matter, she ruled that this evidence of conflict could not be taken into account. Nor did the author recognize that the FDA had the burden of demonstrating there was technical evidence of safety – and that the judge had avoided confronting that issue.

Subsequent articles have yet to set things straight. Like the initial one, several also indicated that we failed to show sufficient disagreement among experts, missing the fact that we had – and the fact that the judge upheld the GRAS presumption in the face of evidence that GE foods were clearly not GRAS.[78] Further, of the many articles I've read, even those that avoided making such an erroneous assertion nevertheless failed to note that we had shown substantial disagreement among experts. And none seemed to realize that the judge treated the GRAS presumption as rebuttable for some purposes and non-rebuttable for others.

Nor did any detect that she had dodged the question about technical evidence; and several authors seemed unaware that such evidence was legally required. For instance, one article stated that under FDA policy, the party challenging a GE food has the burden of "presenting physical evidence of a safety hazard" – never even noting that the law explicitly places the burden on the party defending the food.[79] Nonetheless, a faculty panel at the University of California, Berkeley School of Law awarded the article a prize for government law writing, which indicates that the members shared the author's confusion.

Unfortunately, such confusion is widespread within the legal community, and even jurists who have striven for proficiency in the legal aspects of biotechnology have not stayed clear of it. For instance, another article that similarly misrepresented the law was co-authored by the then Chief Justice of the Supreme Court of Ohio, who was a driving force behind the founding of a resource center for preparing judges to deal with cases involving biotechnology.[80]

How a Reign of Confusion Preserves Industry's Free Rein

The FDA's policy on GE foods has survived only through widespread confusion. Most people (including most anti-GE activists) believe that the FDA has followed the law, when in reality, the agency has been willfully violating it for two decades. So comprehensive is the confusion that even seasoned journalists have been taken in – including Bill Lambrecht, who reported on genetic engineering for the *St. Louis Post-Dispatch* for fifteen years before publishing a book on the topic. In that book, he stated that the FDA had "religiously" applied the same regulatory standards to GE foods that apply to other foods, when in fact, the agency had scorned those standards by illicitly exempting GE foods from their mandates.[81]

Further, most critics of GE foods believe that they've suffused the American market because the law is too weak to properly deal with them when, in fact, the law is so strong that had it been obeyed, there's little likelihood that any GE food would yet have entered the nation's kitchens. And that's not because the law is unreasonably demanding. It's because the law's sensible requirement for demonstration of a *reasonable certainty of no harm* could not have been met by any of these novel products. As this chapter has already shown, and as Chapters 10 and 11 establish in greater detail, the tests that would be even minimally adequate are much longer, more rigorous, and more costly than the manufacturers were prepared to employ. Moreover, the mere announcement that GE foods could not be presumed safe and must be subjected to rigorous testing would most likely have induced a wave of concern that would have doomed the enterprise that was producing and promoting them (a point that's dramatically elucidated in Chapter 6).

Thus, in regard to the safety assessment of bioengineered foods, the US government has reversed the statutorily imposed burden of proof with hardly anyone catching on – not even commentators in law reviews. This transformation occurred in stages. As we saw in Chapter 1, it began in 1978 when the NIH relied on purported evidence about the safety of rDNA research to shift the burden from the proponents of the technology to those who sought to regulate it. Further, most people (including most of those in Congress) were unaware that the "evidence" was nothing more than unsubstantiated conjectures that were floated at the Falmouth and Ascot conferences.

However, within less than a decade, that NIH policy required broadening to suit the needs of the GE venture. When it was implemented, the main safety issue involved microorganisms employed in biomedical research. But as the range of recombinant technology expanded, biotech proponents wanted the range of the shifted burden to expand along with it. As we saw in Chapter 2, the Coordinated Framework established by the Reagan Administration in 1986 effected this expansion by instructing the administrative agencies to apply the shift to the new varieties of plants being created through genetic engineering. But this policy change lacked scientific legitimacy. As with the preceding shift in 1978, although there was great pretense of scientific backing, none existed.

Further, this chapter has revealed that when the shift was extended beyond the environmental safety of GMOs to impact the issue of food safety, not only did it lack scientific authority, it had no legal authority either. In fact, it violated explicit mandates of the Food, Drug and Cosmetic Act. Accordingly, the FDA grappled with the challenge of how to institute it without revealing that the law had been transgressed in the process. In crafting a conducive policy, administrators not only had to sacrifice the law to political demands, they also had to sacrifice science, subordinating the opinions of their own experts to the directives from the Bush White House. And to pull it all off, they had to cover up the facts and issue a string of lies – deceiving Congress and the public into once more believing that a government policy on GMOs was based on overwhelming scientific consensus and solid evidence when it actually opposed the judgments of many scientists and had no evidentiary backing at all.

Moreover, in this case people were also led to believe that responsible oversight was being exercised and that strict testing was being conducted. The FDA engendered these illusions, and they have been progressively strengthened by a stream of misrepresentations from government officials acting as "cheerleaders for biotechnology" (in the words of Bill Clinton's Secretary of Agriculture, Dan Glickman).[82] Glickman himself had been one of the biggest cheerleaders – and uttered some of the boldest falsehoods. He declared that "test after rigorous scientific test" had proven that GE foods were safe;[83] and he proclaimed that every one on the market had been so certified: "Without exception, the biotech products on our shelves have proven safe."[84] Other officials have asserted that the tests were actually performed by the FDA. For instance, an undersecretary at the department of agriculture announced: "The Food and Drug Administration tests all genetically modified foods before they go on the market. . . . We're doing everything to protect our food."[85] And some high-ranking officials have boasted that government-run tests have not merely proven that GE foods are safe, but proven it *absolutely*. Thus, Tommy Thompson, Secretary of Health and Human Services in George W. Bush's administration, proclaimed: "GM (genetically modified) food is absolutely safe, our experts have done tests and found it completely safe."[86]

Further, although these officials fed the confusion, they were also its victims; and it seems that none realized the falsity of his statements. As Glickman later admitted: ". . . I pretty much spouted the rhetoric that everybody else around here spouted; it was written into my speeches."[87] The falsehoods that he and the others were expressing had become conventional wisdom throughout the federal government, and even presidents were taken in. For instance, Bill Clinton assured the nation: "We have confidence in the findings of our Food and Drug Administration that these [biotech] foods are safe. And if we didn't believe that, we wouldn't be selling them and we certainly wouldn't be eating them. . . . I would never permit an American child to eat anything I thought was unsafe."[88] Had he learned that the FDA's "findings" were not based on scientific evidence or even on sound scientific reasoning but were merely unfounded presumptions that countered the judgment of its scientific staff, his confidence would have collapsed.

Over the years, the FDA not only cultivated the confusion, it intensified the misrepresentation. Although the presumption that GE foods are GRAS technically implied that their safety had been scientifically demonstrated, the 1992 policy statement refrained from explicitly propounding such a false assertion and instead relied on theoretical arguments.[89] But the agency eventually grew bolder. Instead of merely proffering such arguments in support of safety, it proclaimed that safety had been positively demonstrated. For example, on May 3, 2000, Commissioner Jane Henney declared:

"FDA's scientific review continues to show that all bioengineered foods sold here in the United States today are as safe as their non-bioengineered counterparts."[90] But the previous year, the agency had acknowledged that it was *not* performing scientific reviews, stating: "FDA has not found it necessary to conduct comprehensive scientific reviews of foods derived from bioengineered plants . . . consistent with its 1992 policy."[91] Further, as we've seen, the information that manufacturers have chosen to submit to the agency is incompetent to establish that even one GE food is as safe as its conventional counterpart, let alone that they all are.

Nonetheless, the FDA has persisted in its bogus claim about the demonstration of safety. One notable instance occurred in October 2002, when the agency made what *USA Today* called "an unusual move" and sent a well publicized letter to the governor of Oregon that helped defeat a ballot initiative for the mandatory labeling of GE foods in that state.[92] Among the letter's falsehoods was the renewed assertion that "FDA's scientific evaluation of bioengineered foods continues to show that these foods, as currently marketed in the United States, are as safe as their conventional counterparts."[93]

The agency similarly deceived Congress. On June 14, 2005, an FDA official once again delivered misleading testimony about genetic engineering to a legislative committee by declaring: "Over the last ten years, FDA has reviewed the data on more than 60 bioengineered food products. . . . To date, the evidence shows that these foods are as safe as their conventional counterparts."[94]

Through chronic exposure to such disinformation, most Americans have had no clue that the industry was essentially granted self-regulation. A poll conducted in 2004 by the non-profit Pew Initiative for Food and Biotechnology revealed how wide-spread the delusion had by then become. As reported by the IPS News Agency, "According to an expert familiar with the poll, Americans have tremendous faith in their regulators, but wrongly believe GE foods have been approved and tested by the FDA. 'They're under the false impression there is thorough testing like . . . for drugs,' said the expert, asking to be unnamed. When people learned that GE foods are not tested, they were very uncomfortable and indicated they want mandatory, uniform testing and evaluation of GE foods, noted the expert."[95]

The IPS report also observed: ". . . the U.S. public does not want to take risks with its food. . . . Indeed, 81 percent of those surveyed by Pew believed the FDA should approve the safety of GE foods before they come to market, even if that would mean 'substantial delays.'"[96]

The Centrality of the FDA's Fraud to the Survival of GE Food

As previous pages have shown, although there is *not* a seamless continuum between genetic engineering and conventional practices (as its proponents claim), for more than three decades there *has been* an essentially seamless continuum between the preferences of the biotech industry and the agenda of the US government – which has caused a drastic *discontinuity* between FDA policy and the law. So close is the connection between industry and government that numerous individuals have smoothly transitioned between the two sectors, sometimes repeatedly. The most striking example is Michael Taylor, who after serving for five years as an FDA attorney, became a private-practice lawyer representing Monsanto, then returned to the FDA as Deputy Commissioner for Policy to oversee the policy on GE foods, and, after giving Monsanto what it wanted, joined the company as Vice President for Public Policy. Then, in 2009, he again returned to the FDA, this time as senior advisor on food safety – a position that's been referred to as "food czar."[97] And in January 2010, he was again elevated to the rank of deputy commissioner: this time as Deputy Commissioner for Foods, a new position that he was the first individual to hold.[98]

Due to industry influence, GE foods have entered the US market, not through a transparent, science-based process, but through sleight of hand. And the consequences have been anything but slight. For more than a decade, the majority of processed foods in the US have contained ingredients derived from engineered organisms (with the current percentage close to 90%); and the number of crops that have been genetically restructured keeps growing. Moreover, the deceptions not only allowed GE foods to pervade the United States but to permeate much of the world.

Without the FDA's fraud, the GE food venture would not have expanded but imploded. The effect of the facts on consumers, legislators, and investors would have quickly doomed the entire enterprise. This is evident to anyone who understands the socio-economic realities. For instance, when I met in 2001 with the chief scientist of the food safety authority for Australia and New Zealand (who had a favorable attitude toward GE foods), I presented the information about the FDA's cover up and then asked what she thought would have happened if, in 1992, the agency had announced that its scientists had concluded that these novel products entail unusual risks and that each should undergo extensive toxicological testing. She promptly replied that it would have "killed" the whole industry.[99]

In taking stock of the first five chapters, it's clear that the venture to genetically engineer our foods has been chronically dependent on systematic suppression of facts conjoined with the persistent spread of misinformation – and could not have survived without either. It's equally obvious that, besides the necessity to exceed the bounds of truth, there's been a need to breach important regulatory boundaries on both the natural and societal levels. To create organisms with functional foreign genes, biotechnicians had to surmount a series of regulatory mechanisms that maintain the structural and operational integrity of genomes. Then, to market such products without proper testing, they had to induce public officials to let them evade the regulations that preserve the integrity of the food supply.

And, just as nature's regulatory safeguards were sundered throughout the various biological kingdoms, so the breach of society's regulatory safeguards in the United States set the stage for similar infractions by officials in many other nations – accompanied by a similar stream of double talk to convey the impression that nothing irregular had happened.

CHAPTER SIX

GLOBALIZATION OF REGULATORY IRREGULARITY

How Food Safety Officials in Canada, the EU, and Other Regions also Sidestepped Science and Sound Policy

People discouraged by how poorly GE foods are regulated in the US might have hoped that the regulatory systems in Europe and other regions would provide a back-up, forcing products that receive an essentially free pass through the American system to undergo adequate safety testing before they're marketed elsewhere. However, any such hopes have been misplaced. Although the EU and most other industrial nations have imposed some testing requirements on these novel products, they've been too feeble to furnish even a modest assurance of safety.

The Insubstantiality of 'Substantial Equivalence'

The central concept that has underlain the international regulatory system is referred to as *substantial equivalence* – and it's been accompanied by substantial confusion. For one thing, most people think that it is also a basic feature of the US system when, in reality, it's quite foreign to the official American approach. As we saw in the last chapter, US food safety law requires that all new food additives be regarded as unsafe until proven safe; and it imposes the burden of proof on the manufacturers. Further, in order for an additive to qualify for the Generally Recognized as Safe (GRAS) exemption, and be excused from undergoing testing, its safety must already have been demonstrated via rigorous testing rather than through theoretical reasoning. Moreover, an overwhelming consensus must exist within the scientific community that such proof has in fact occurred. However, as we also saw, in order to put GE foods on the market, the FDA illegally (and fraudulently) presumed that they are GRAS, even though none of them fulfilled either of the GRAS requirements.

In contrast to the US system (as it exists on the books), the approach based on the concept of substantial equivalence does not demand solid proof of safety and significantly relies on theoretical assumptions and reasoning. In this approach, if a bioengineered food organism can be ascertained to be "substantially equivalent" to its conventional counterpart, it will be considered as safe as that non-engineered organism, even without the kinds of tests that are necessary to establish that it actually is.

But there's great uncertainty as to what the concept actually means. In further contrast to US food safety law, where the terms and concepts are strictly defined, "substantial equivalence" as a regulatory principle has remained quite vague. It was introduced in 1993 by the Organization for Economic Cooperation and Development (OECD), and in 1996, the United Nations endorsed it through the Food and Agriculture Organization and the World Health Organization.[1] However, numerous experts have criticized the degree to which it has stayed loosely delineated. For instance, three scientists writing in the journal *Nature* noted: "Given the weight the concept has been required to carry, it is remarkable how ill-defined it remains. . . ." They pointed out that at the time they were writing (1999), the following statement from the OECD was the closest thing to an official definition: "The concept of substantial equivalence embodies the idea that existing organisms used as foods, or as a source of food, can be used as the basis for comparison when assessing the safety of human consumption of a food or food component that has been modified or is new."[2]

Such vagueness has worked in favor of the biotech industry. As the authors of the letter in *Nature* observed: "The adoption of the concept of substantial equivalence by the governments of the industrialized countries signalled to the GM food industry that as long as companies did not try to market GM foods that had a grossly different chemical composition from those of foods already on the market, their new GM products would be permitted without any safety or toxicology tests."[3] In practice, this has allowed the assessments to focus solely on the intended or expected effects of the gene insertion, based on the assumption that there will be no unintended effects that need to be monitored. So a crude (and incomplete) chemical profile of the engineered organism is compared to that of the non-engineered parental line, and if no major discrepancies are noted, the former is deemed substantially equivalent to the latter. Of course, even though it's obvious that the two organisms do differ in regard to the expression product of the inserted gene, if analysis supports the idea that this substance is not dangerous, then the entire engineered organism is regarded as having been shown to be safe.

Thus, when *substantial equivalence* has been the operative principle, it has relieved manufacturers of any obligation to perform thorough testing to discover whether unexpected changes have occurred that can't be detected through simple chemical analysis. In such circumstances, the kinds of tests that the FDA experts said are necessary (toxicological tests employing the whole food) have not been done – and the testing has been uniformly lax. For instance, in 2000, Professor E. Ann Clark of Guelph University, Ontario reviewed the GE crops that had been approved in Canada and found that 70% (28 of 40) had not been subjected to any actual lab or animal toxicity testing. In the remaining 30%, although some types of animal studies were performed, the animals were *not* fed the whole GE organism. Instead, they were fed the isolated protein that was expressed by the foreign gene. Further, this protein was not even produced within the engineered plant. It was derived from laboratory bacteria into which the gene had been inserted.[4] Accordingly, such testing was incapable of detecting any unintended, deleterious effects within the plant caused by disruptions associated with the genetic manipulation – the kinds of unexpected hazards about which the FDA experts repeatedly warned.

And this deficient approach was universally utilized. The year after Dr. Clark's review, a scientist at Iowa State University published a paper verifying that to the extent toxicological studies had been performed on GE crops, they did not involve the whole plant but were limited to the known foreign proteins. She stated that other forms of safety testing were not considered necessary.[5]

Also in the year following Dr. Clark's review, an expert panel of the Royal Society of Canada published a more extensive examination of the Canadian regulatory regime – and leveled some harsh criticism. The Society's report had especially disparaging words for the way the concept of substantial equivalence was being employed. It branded the concept as "scientifically unjustifiable and inconsistent with precautionary regulation of the technology"– and noted that it was being used to excuse manufacturers from performing full risk assessments.[6] Although the regulators had contended that their reviews were rigorous, the Society's experts rejected their claims. As reported in *The Toronto Star*, they declared that the review system was "fatally flawed . . . and exposes Canadians to several potential health risks, including toxicity and allergic reactions."[7]

In stark contrast to both the FDA's official (but fraudulent) policy and the suppositions underlying the concept of substantial equivalence, the Canadian experts stated that the "default presumption" for every GE food should be that the genetic alteration has induced unintended and potentially hazardous side effects, encompassing "a range of collateral changes in expression of other genes, changes in the pattern of proteins produced and/or changes in metabolic activities." [8] They declared that approvals of GE foods should no longer be based on the loose approach associated with the concept of substantial equivalence and instead "should be based on rigorous scientific assessment." [9]

Moreover, it's obvious that the actions of the Canadian government cannot be excused as innocent oversights – and that its attitude has been not only irresponsible but reprehensible. For instance, although the Royal Society undertook its investigation of GE foods at the request of the government, the government eventually became uncooperative. As reported by the *Toronto Star*, it even "barred" the Society's expert panel "from seeing evidence that safety tests had actually been done on genetically modified foods." [10] Further, although the experts criticized the government for excessive secrecy, to this day, Canadian regulators continue to conduct safety assessments on GE foods in a clandestine manner that precludes external scrutiny.[11]

Continuation of Shoddy Safety Assessments

Despite the public scolding it received, the Canadian government held to its course; and it failed to implement the essential reforms the Royal Society had called for. Further, the concept of substantial equivalence continued to reign in most other nations as well; and the flimsy research on which the Canadian approvals were based was similarly accepted by regulators throughout the world. Consequently, for many years GE foods entered the market on the basis of safety assessments that were sorely defective.

A striking example of the extent of the deficiencies is provided by the experience of the eminent food safety researchers Arpad Pusztai and Susan Bardocz (his wife) when they were at the UK's Rowett Institute of Nutrition and Health. As reported by investigative journalist Jeffrey Smith, in April 1998 the Institute's director asked the two scientists to evaluate a large stack of documents (totaling around 700 pages) that comprised six or seven requests for approvals of specific GE foods (including varieties of tomatoes, soy, and corn).[12] He explained that an important EU meeting on bioengineered foods would soon convene in Brussels and that the head of the British Ministry of Agriculture, Forestry and Fisheries (MAFF) was going to attend and wanted a scientific basis for recommending these submissions. He then added a jolting piece of news. The minister needed the evaluations within two and a half hours.

In order to meet this demand, the team honed in on the critical parts of each submission: the research design and the data. What they discovered was deeply disturbing. "As a scientist, I was really shocked," said Pusztai, in relating his experience to Smith. "This was the first time I realized what flimsy evidence was being presented. . . . There was missing data, poor research design, and very superficial tests indeed. . . . And some of the work was really very poorly done. I want to impress on you, it was a real shock." [13]

Although he and his wife had initially assumed that two and a half hours would be insufficient, well before the time was up, they were prepared to definitively state that the research fell far short of demonstrating that any of the foods was safe for human or animal consumption. But when Pusztai phoned the minister of the MAFF to inform him, he received another shock. The minister told him that all of the submissions had already been approved in the UK. The review that he and Susan performed had not been intended to assist the UK government in deciding whether to approve those particular GE products but was merely supposed to provide the minister with scientific assurances about them that he could employ in the EU meeting. Within the UK, the populace had already been consuming those foods for close to two years. Moreover, they had been eating them unknowingly, because the approvals had all been made in secret.

Unfortunately, the studies that Pusztai and Bardocz discredited were not rarities. Time and again, investigation of the regulatory process (in whatever the nation) has revealed research that was deficiently designed, poorly conducted, and irresponsibly reviewed. One of the more unsettling episodes involved three of Monsanto's applications to the Australia New Zealand Food Authority (ANZFA) for approval of GE plants (varieties of herbicide resistant corn (maize) and canola, and a pesticide-producing corn).[14] Prior to ANZFA's final action, the Public Health Association of Australia (PHAA) reviewed the applications and discovered some troubling data. As explained in the comments the Association sent to ANZFA in October 2000, in *all* three cases, the GE plant differed significantly from its parental line in amino acid composition, and there were significant differences in other areas as well.[15] The PHAA experts pointed out that the variations in amino acid profiles alone warranted rigorous toxicological testing of the whole GE foods. They noted that these alterations were of such magnitude that they could not be attributed solely to the known products of the inserted foreign genes. They further explained that since amino acids are the building blocks of proteins, either the concentrations of some of the proteins naturally present in the plants had been altered or else one or more new proteins had been produced that do not naturally occur in the plant. They cautioned that in either case, harmful effects to consumers could result and that additional testing was required to demonstrate the plants were in fact safe.

Monsanto had attempted to minimize the importance of these alterations by arguing that the levels of amino acids in the GE plants still fell within the range of previously reported values for conventional varieties of the particular plants involved. But the PHAA countered that it was illegitimate to compare the GE plant to an average range compiled from non-GE plants cultivated in widely differing growing conditions. It stated that the proper comparison was between the GE plant and the plants that had been used as controls in Monsanto's field trials. Those plants belonged to the line from which the GE plant was derived (the parental line) and were grown at the same time as the GE plants under the same conditions. Therefore, the statistically significant differences between the GE plants and the controls could be assumed to have resulted from the genetic engineering process itself, not from naturally occurring factors. Accordingly, a shift in amino acid concentration could be indicative of a unique protein composition that does not ordinarily occur in conventional plants – and that might be hazardous to human health. The PHAA noted that using data gathered from plants in widely varying conditions would undermine the very purpose of having used controls, since it would allow entry of all the variables that are excluded by the controlled experiment – such as differences in climate, soil, and cultivation techniques, which are unrelated to the engineering process and can only serve to obfuscate collateral changes it may have caused.[16]

Amazingly, ANZFA sided with Monsanto instead of the PHAA. Instead of upholding the integrity of the controlled experiment, the agency circumvented the controls and compared the GE plants to plants grown under widely varying conditions. Only in that way was it able to rule that the three plants were substantially equivalent to their conventional counterparts – and therefore suitable for human consumption.

Besides disregarding the warning signs generated by controlled experiments, ANZFA also overlooked the absence of key elements of standard scientific investigation. For example, the PHAA noted that Monsanto's statistical analyses reported only a few values while omitting several pieces of information that are necessary to enable other scientists to assess the data – and that are required by peer-reviewed journals. And when these experts then examined four other GE food applications submitted to ANZFA, they discovered similar omissions.[17] For this reason alone, none of the seven submissions was suitable for publication in a standard scientific journal. Moreover, the sample sizes used in comparisons were surprisingly small. As the PHAA noted, "With such low numbers it is almost a foregone conclusion that a statistically significant difference will NOT be found between the GE food and the non-GE food for most analyses, even if one exists in nature."[18] Further, it pointed out that because in several cases potentially important statistical differences had still been detected, even with such small sample sizes, those differences might well be "substantial indeed."

Nonetheless, ANZFA approved the other four submissions as well. Moreover, it was not the only regulatory body willing to accept such shoddy research. Six of those seven submissions had already been approved in the EU.

Thus, manufacturers have regularly declared substantial equivalence in the face of substantial differences; and the regulators have gone along with it. One of the more striking examples of such inaptly proclaimed equivalence involves the world's most widely planted GE crop: Monsanto's Roundup Ready® soybean, engineered to tolerate the potent herbicide *glyphosate*. As described in the letter in *Nature* cited previously, regulators began "by assuming that the known genetic and biochemical differences" between the engineered beans and their counterparts "are toxicologically insignificant." Then they focused on a "a restricted set of compositional variables, such as the amounts of protein, carbohydrate, vitamins and minerals, amino acids, fatty acids, fibre, ash, isoflavones and lecithins."[19] However, the comparison was deeply flawed. For one thing, conventional soybeans cannot be sprayed with glyphosate because it would kill them. In contrast, the GE beans would be subjected to substantial doses of glyphosate in the process of destroying surrounding weeds. Further, as the *Nature* letter pointed out, it had been known for many years that applying glyphosate to soy beans "significantly changes their chemical composition." Yet, instead of comparing sprayed GE beans to the conventional, unsprayed beans, some key compositional tests employed engineered beans that had *not* been subjected to the herbicide, even though people and livestock would be consuming the sprayed beans.[20]

But the defects with the assessments of Monsanto's Roundup Ready® soybean do not stop there. Investigation by scientists at Japan's Nagoya University of the data submitted to regulators in that nation revealed that the GE beans employed in the animal feeding tests had not been sprayed with glyphosate either.[21] Further, these independent investigators reported that although a difference in body weight between rats fed the GE soy and those fed the conventional type was described as "statistically significant" in the data sheet, the company's conclusion declared that "no statistical significance is observed." At least as troubling, these scientists discovered striking discrepancies in chemical composition after the beans were toasted in the standard manner for turning them into animal feed (108 degrees centigrade for 30 minutes). Not only were major components like water content, protein, fat, fiber and ash different in the GE beans compared to the non-GE ones, the GE beans contained significantly higher concentrations of three specific proteins that are known to be harmful (trypsin-inhibitor, lectin and urease). These proteins remained active in the GE beans while in the non-GE beans they were denatured and inactivated. Moreover, their levels were above accepted standards for animal feed.

According to the Nagoya team's report, rather than acknowledging that a problem existed, Monsanto claimed that the beans had been "insufficiently" toasted and instructed the lab that had performed the test to toast them at 220 degrees centigrade for 25 minutes. Although this is considerably higher than normal processing temperature, it actually widened the difference in the activity between the two strains, with GE beans showing a high level of heat resistance. Yet, instead of admitting that the beans were substantially different, Monsanto claimed that this second toasting was still insufficient. So it had the beans subjected to two more rounds of toasting at increasing levels of temperature until all the proteins were denatured and inactivated. As the investigative report points out, only by putting the GE beans through such an extraordinary series of heat treatments did Monsanto render the harmful proteins as inactive as those in the non-GE beans. But because such extreme measures are not employed when the GE beans are processed for animal feed, it raises doubt about the product's safety for that use.

In all, the Japanese investigators found so many irregularities in the safety assessment of the GE soy they concluded it was "inadequate and incomplete." Their report concludes: "The safety assessment of the Monsanto Roundup Ready soybean needs to be reassessed." But regulators have not done so, even though their initial reviews were shown to have been exceedingly sloppy. Accordingly, the product continues to be ingested by people and livestock the world over.

Regulators have rarely distinguished themselves in other respects either. For instance, besides allowing many of the feeding tests on GE foods to employ only the expression product of the foreign gene rather than the whole GE food, they've approved tests on the whole foods that were wholly deficient. For instance, during tests on Aventis's bioengineered T25 corn, twice as many chickens died in the group that ate it than in the control group fed the non-engineered parental stock.[22] Despite this result (and several defects in the way the study was conducted), EU authorities approved the product. The chair of a UK government advisory committee later admitted that the chicken study should have been reanalyzed. He also admitted that at the time approval was given, his committee had only seen a summary of the study.[23] Other experts who subsequently reviewed the study emphasized its flaws – with one university scientist stating it was "not really good enough to base a student project on, let alone a marketing consent for a GM [genetically modified] product."[24]

Substantial irregularities also occurred in connection with Syngenta's application for EU approval of Bt11 sweet corn (maize). This high profile case was intensely debated during 2004 because at that point, there had been a six-year moratorium on approvals of GE foods in the European Union, and approving Bt11 would end it.

But even knowing there would be a spotlight on the application, Syngenta cut corners. It undertook no long-term toxicological tests using the whole plant, and the more superficial nutritional feeding studies it performed with cattle and hens did not employ the sweet corn under consideration (that was intended for human consumption) but a variety of engineered field corn intended for livestock – which the French Food Safety Agency (AFFSA) noted had "significant genetic differences" from the sweet corn.[25] The AFFSA warned that "unforeseen effects" from the sweet corn "cannot be discounted," and it called for new tests. The report of the Austrian agency was also critical. It noted that not only were the allergy tests insufficient, but that Syngenta's claim of safety was primarily based on hypotheses rather than direct evidence – and that several of its presumptions were false.

Nonetheless, the European Commission, the EU's main executive body, approved the product. Moreover, it justified its action through misrepresentative statements, with the Health and Consumer Protection Commissioner declaring that Bt11 "has been scientifically assessed as being as safe as any conventional maize [corn]."[26] And although there was no evidence to support such a bold pronouncement, he implied that there was, proclaiming that Bt11 "has been subject to the most rigorous pre-marketing assessment in the world."[27] But this is false. It implies that Bt11 successfully passed every type of safety test that's been applied to a food or food additive, and that each was administered to the highest standards, which clearly never happened.[28]

Further Inadequacies of the Testing

The GMO-Generated Proteins Are Not Directly Tested

Besides dispensing with essential feeding tests and tolerating slack performance of the remainder, the current regulatory system relies on analytic tests that don't provide adequate data. For instance, the tests it accepts for assessing the safety of the foreign protein cannot fully do so. As previously noted, the main reason is that the tested proteins are normally not the ones that are synthesized within the engineered plant. Rather, they're produced by inserting the related gene into bacteria, because it's much easier to garner a sufficient amount of the protein in this manner. But even if these bacterially-derived surrogates possess exactly the same sequence of amino acids as their plant-produced counterparts, they can yet be significantly different – and more dangerous.

That's because a protein's effects are not solely determined by the arrangement of its constituent amino acids. They're also a product of other factors, and these factors can be unexpectedly altered as genes are transferred between species that cannot breed through natural means.

The New Proteins Made Within GMOs Can Have Deleterious Additions

One factor is whether the protein gains add-ons – and the specifics of what gets added. The adjoined substances can be sugars, fats, or other types of molecules; and whether and to what extent they're attached to a protein depends on the conditions in the cells where the protein is formed. According to David Schubert, a molecular biologist and protein chemist with the Salk Institute, although we know that such modifications can render an otherwise harmless protein toxic or allergenic, we don't know enough to predict how and when such malefic modifications will occur.[29]

However, we do know that plant cells can induce such modifications in a way that bacterial cells cannot. The particular process is called *glycosylation*, and it involves the addition of sugar chains to a protein. While this process does not occur in bacteria, it does in plants and animals. Consequently, when a bacterial gene is transferred to a plant via genetic engineering, the resultant protein could become glycosylated; and although sugar-coated, it could cause effects that are not at all sweet.

Moreover, if such a harmful alteration did take place, it would never be detected merely by testing the effects of the bacterially-produced, unsugared protein. That protein could pass every test, while the plant-built version could make people pass away. In light of the fact that most of the

GE plants on the market possess genes transferred from bacteria, this regulatory deficiency is serious.

Further, when the inter-species transfer instead occurs between plants, the glycosylation pattern of the associated protein could still be adversely altered within the foreign environment – even if the plants are closely related. This was discovered when a protein normally produced in a kidney bean was synthesized within a pea. Although the protein as made by the bean is safe for humans when fully cooked, something changed for the worse when the source gene was inserted in peas – despite the fact the two species belong to the same biological family (and are members of the sub-group referred to as *pulses*). Tests on mice brought this change to light.[30] All the animals were fed a standard diet for four weeks; and twice per week, one group was also fed beans, another was fed non-GE peas, and the third was given the GE peas. The mice then underwent immune response tests that are supposed to indicate whether a substance will be allergenic for humans. Surprisingly, although the protein produced in the altered peas and the protein produced in the beans had identical amino acid sequences, only the former provoked an immune response. When injected into the footpads of the mice that had consumed it, significant swelling occurred. Their lymph nodes also reacted against it. And when their tracheas were exposed to it, tissue inflammation and mild lung damage resulted.

Thus, after the entry of this pea-produced protein into the animals' diet, their immune systems became primed to repel it. In trying to account for the dramatic difference between the effects of this aberrant protein and its bean-built counterpart, the researchers employed an advanced test that probes the patterns of sugar chains that have been added to proteins. This revealed small changes had occurred when the protein was made in peas instead of beans. The scientists concluded this tiny shift in sugar pattern was the most probable factor underlying the shift in allergenicity – and that it explained why one pulse-produced protein did not rouse the rodents' defense mechanisms while the same protein made by another pulse was repulsed.

What's more, the effects of this shift were broad. The altered protein not only induced the animals' immune systems to react against it, it predisposed them to mount a response against other concurrently consumed proteins as well. In contrast, the mice fed the normal version of the protein showed no such inappropriate sensitivity. Only the mice that ingested the modified version were induced to react against other ordinarily inoffensive proteins as if they too were allergens.

In a further surprise, the allergenic properties of the modified protein persisted even after the GE peas had been boiled for 20 minutes. Although this denatured the protein, and curtailed its usual effects, it did not deactivate the extraordinary effect on the mouse immune system – which refutes the widely held assumption that when GE plants are cooked, any allergenic attributes will disappear through protein denaturing.

This underscores the importance not only of testing the protein that's actually produced by the engineered plant, but of testing it thoroughly. However, as noted, the plant-produced version of the protein is rarely tested; and even in the exceptional cases when it is, the measures are not sufficiently sensitive. Although the test that revealed the allergenic nature of the pea-produced protein is commonly used in assessing medicines, it had not been previously employed in the screening of commercialized GE plants – and it's still largely ignored. Nor is the special test that detected the subtle change in glycosylation regularly used. In those cases when a protein's structure is examined, a less sensitive test is employed that does not provide detailed information about glycosylation patterns – and could not have detected that those of the pea-made and bean-made proteins were different.[31] Moreover, because that inferior type of test *had* been employed when the GE pea was initially developed (several years prior to the more extensive testing), scientists were led to believe that the two versions of the protein were identical.

So if that genetically altered pea had been subjected only to the usual modes of testing, it would have been cleared for human consumption – and might have caused substantial suffering.

The Shapes of the GMO-Generated Proteins Could Be Dangerously Altered

But even if nothing gets added to the foreign protein after insertion in a new organism, it could still cause unexpected harm. One way is through change of shape. A protein's function is primarily determined by its unique three-dimensional structure; and that structure, although very complex, is tightly organized. Further, this organization is not achieved when the protein is synthesized; and a newly synthesized protein has a largely two-dimensional, ribbon-like form. This two-dimensional form must then be folded into the correct three-dimensional configuration.

The cellular biologist Barry Commoner has noted that scientists used to think the protein "always folded itself up in the right way once its amino acid sequence had been determined." But, as he pointed out, this notion changed in the 1980s when scientists discovered that some proteins "are, on their own, likely to become misfolded – and therefore remain biochemically inactive – unless they come in contact with a specific type of 'chaperone' protein that properly folds them."[32] Accordingly, Commoner cautioned that when a protein that has co-evolved with a particular chaperone is transferred to a foreign environment, it might not fold in the proper manner. Without the assistance of its own chaperone, misfolding might occur; while if it's instead influenced by an alien chaperone, it could also be misshaped. Further, although in some cases misfolding would deactivate the protein, in others it could cause the protein to act in a dangerous way. For instance, Mad Cow Disease is caused by an errantly folded protein.[33]

Moreover, like proteins in GE foods that are harmfully glycosylated, those that are malignantly misfolded would likewise slip through the current regulatory system undetected. And there's at least one case in which the evidence suggests that the foreign protein in a GE plant may indeed have become misshapen.[34]

The Amino Acid Sequences of the Proteins Can Be Changed

There's yet another way in which a protein made from a gene inserted in a plant can significantly differ from the protein produced when that gene is instead inserted in a bacteria. Not only can it become misshapen or altered by harmful add-ons, even the sequence of its amino acids can be unintentionally changed. This can happen because, in contrast to the mode in which biotechnicians add new genes to bacteria, the insertion process in plants is routinely unruly – and often induces the deletion of DNA that's supposed to be within the added gene or the addition of some that's not supposed to be there.

Further, besides being common, some of these alterations are difficult to detect. According to Doug Gurian-Sherman, a plant biologist who had performed risk assessments on GE crops for the US Environmental Protection Agency, "Rearrangement of the nucleotide sequence of a gene often occurs during the insertion of that gene into the genome of the recipient plant . . . . Most of those random changes impair or even eliminate the function of the protein coded by the gene and may be easily detected by bioassays. Some changes, however, may be more subtle and less easily determined. Even single nucleotide changes can alter a protein's amino acid sequence and affect the protein's properties."[35]

However, despite the known risk of such subtle yet potentially harmful changes, regulators were lax in responding. As Gurian-Sherman pointed out: "FDA provides little guidance for assuring that potentially deleterious changes have not occurred in the transgene, and consequently to the GE protein, due to transformation of the plant." And he noted that these changes can "alter the properties of the GE protein" and that "detection of many plant characteristics of health concern require specific testing."[36]

Yet, as he also noted, there was still a need for determining how to assess such changes in the proteins added to plants via the engineering process. And that was in 2003. So although GE crops had been entering the food supply for more than eight years, regulators had not yet devised a protocol for detecting whether such potentially deleterious alterations had occurred within them.[37]

Additional Proteins Can Be Accidentally Added

Moreover, besides the risk that the intended proteins will be altered, there's a risk that an unintended one can be introduced. As Chapter 11 explains in more detail, the same section of DNA can be involved in the production of more than one protein; and this entails that pieces of unwanted protein could be generated by an inserted segment of DNA. Further, even regions of DNA that appear to be non-protein-coding can contain protein-coding sequences; and these potentially problematic sequences can go unnoticed by regulators for many years. Indeed, this has actually happened – not merely in one GE crop, but in the majority of those that have entered the market.

This unsettling situation arose due to the widespread reliance on the powerful 35S promoter from the cauliflower mosaic virus to drive the expression of the inserted genes. Because in its viral home this promoter comes in front of the gene whose expression it influences, the regulatory agencies assumed it didn't contain any protein-coding sequences itself. But they were wrong. In fact, there's another gene on the *other* side of the promoter ("upstream" of it) known as Gene VI; and its coding sequence extends *into* the 35S promoter. As a result, many of that promoter's nucleotides also encode a segment of the Gene VI protein.

But this unpleasant reality wasn't discovered by any regulators until 2012; and the discovery was not made by the FDA (because its reviews of GE foods are so superficial), but by another regulatory agency: the European Food Safety Authority (EFSA).[38] And because it was already known that similar fragments of Gene VI can express active proteins,[39] the EFSA scientists had to acknowledge that the presence of this gene fragment in a GE plant has the potential to induce changes in the plant's traits. Nonetheless, they tried to downplay the risks – despite the fact that the protein produced by Gene VI is not only toxic to plants[40] but interferes with a basic mechanism of protein synthesis relied on by humans as well as plants[41] – while also interfering with another important biological mechanism that's common to both (RNA silencing).[42]

Accordingly, other experts have not been as comfortable with these circumstances as have the regulators. And two who've lucidly expressed their concerns are Jonathan Latham, a molecular biologist and plant virologist, and Allison Wilson, a molecular geneticist. In January 2013 they published an article cautioning that the presence of this gene fragment in GE plants poses human health concerns, especially because many viral proteins work to disable their host. And they asserted, "The data clearly indicate a potential for significant harm."[43] Therefore, they advocated a recall of all GE crops that harbor a piece of Gene VI – which besides pulling all those with the 35S promoter from the cauliflower mosaic virus off the market, would have removed another set that contained a promoter derived from the figwort mosaic virus.

Unsurprisingly, the EFSA was not enamored with this call for a recall; nor was the Australia/New Zealand food safety agency. So each countered with arguments purporting to show that the partial presence of Gene VI doesn't entail significant risk.[44] But in rebuttal, Latham and Wilson presented an incisive analysis that exposed several flaws in these defensive arguments.[45] And, based on this analysis, they asserted that the regulators' arguments are not only "inadequate to meet a potentially major food crisis" but "scientifically misleading." They further alleged that the arguments "do not address the key agronomic and human safety concerns raised by Gene VI," are "irrelevant and illogical," and "rest on scientifically unverified or unsupported assertions." Moreover, they issued a strong admonition: "It is potentially acceptable for regulators to condense or simplify complex scientific information to educate or inform a lay public. What is not acceptable, however, is to 'inform' the public with misinformation."

Their rebuttal is well-reasoned and well-worth reading and can be freely accessed at the link provided in endnote 45.

Regulators Refuse to Recognize These Risks – and Have Made No Effort to Control Them

Yet, despite the known potential for the foreign proteins to be altered and the inserted sequences to mutate, those responsible for the regulatory system have remained unresponsive to these additional risks – and caustic toward calls for greater caution.

Throwing Caution to the Wind

Throughout the ages, societies have recognized the wisdom of exercising caution before rushing into new activities that could entail substantial risk. This understanding has been conveyed in several adages, such as: "look before you leap," "better safe than sorry," and "an ounce of prevention is worth a pound of cure." In the late 20th century, as policy makers and experts attempted to intelligently deal with several technologies that, despite their apparent promise, also appeared to have significant potential to damage human health or the health of the environment, they endeavored to express this folk wisdom in a more formal manner. The result became known as the *precautionary principle.* Although this principle has been articulated in several ways, one group of experts points out that all of them contain the following concepts:

1) When we have a reasonable suspicion of harm, and
2) When there is scientific uncertainty about cause and effect, then
3) We have a duty to prevent harm.[46]

One way of stating the principle in the context of food safety is that when there is significant disagreement among experts about whether an additive might pose an unacceptable level of risk, the proponents of the additive have the burden of proving that it's safe.

As we saw in Chapter 5, long before anyone started talking about a precautionary principle by name, the US Congress established a strict mandate for precaution as the central feature of the nation's policy on food additives. Moreover, that 1958 law is *more* stringent than any version of the precautionary principle articulated thereafter. Whereas such formulations specify that there must be reasonable suspicion of harm (along with significant expert disagreement about risk) before the principle is triggered, the US law lays down no such conditions. It categorically presumes that *all* additives introduced after 1958 are dangerous until proven safe, and it places the burden of proof squarely on the manufacturers.[47] However, as Chapter 5 revealed (and as was noted at the beginning of this chapter), in order to promote GE foods, the FDA illegally exempted them from the precautionary requirements; and it continues to pretend that they're all Generally Recognized as Safe, despite overwhelming evidence to the contrary. (The Royal Society of Canada's report alone clearly establishes that they are not GRAS).

Sadly, the situation has not been much better in other nations. Although many purport to apply the precautionary principle to food safety, in the case of GE foods, they have tended to turn their backs on it. The example of the EU is instructive. On April 30, 1997, the European Commission issued a paper on consumer health and food safety stating that it "will be guided in its risk analysis by the precautionary principle, in cases where the scientific basis is insufficient or some uncertainty exists."[48] On March 10, 1998, the European Parliament issued a resolution affirming this precautionary approach. It stated that EU food law is "based on the preventive protection of consumer health . . . founded on a scientifically-based risk analysis supplemented, where necessary, by appropriate risk management based on the precautionary principle . . ."[49] And the next year, The Joint Parliamentary Committee of the European Economic Area (which comprises the EU and three neighboring nations), adopted a resolution reaffirming "the over-riding need for a precautionary approach" in regard to the approval of GE foods.[50] Then, in 2000, the European Commission issued a major document on the precautionary principle stating that its implementation should start with a scientific evaluation that is "as complete as possible."[51]

However, it is not possible to sustain this precautionary approach while permitting the concept of substantial equivalence to hold sway. The two approaches are so incompatible that applying the latter undermines the former. Yet, EU officials have continued to assess GE foods according to the relaxed standards of substantial equivalence – even while professing that they're imposing stricter criteria.

This pretension eventually entailed a change in terminology. Because the concept of substantial equivalence had received so much criticism, the chair of the GMO Panel of the EU's European Food Safety Authority co-authored an article in 2003 advising that the term be replaced by the phrase "comparative safety assessment" in order to evade the "controversy" associated with the older term. However, he and his co-author admitted that although the words would be new, the underlying regulatory approach would remain the same.[52]

Further, not only have the EU regulators continued to give undue weight to the superficial comparisons of the substantial equivalence paradigm, they still sanction comparisons that are scientifically illegitimate, permitting producers to compare a GE crop to varieties that not only differ genetically from the non-GE parental line but that were grown many years before in widely varying locales. Some of the data allowed to skew the studies even pre-date World War II.[53]

However, despite the infirmities in the EU's approach, officials claim that it's more precautionary than any alternative. For instance, on March 17, 2011, the Commissioner for Health and Consumer Policy (John Dalli) delivered a presentation in which he deplored the "misunderstandings" that have led people "to wrongly believe that the potential risks of GMOs . . . are not adequately assessed. . . ." But his subsequent remarks revealed that it was he who had fallen into misunderstanding. He asserted that the "thorough comparison between a GMO and a conventional safe counterpart" which occurs via the EU's approach "allows the identification of all the differences created by the genetic modification." And he alleged: "All these differences are then investigated in detail with respect to possible toxicological, environmental, allergenic or nutritional aspects." He then declared: "There is no stricter alternative available to this comparative approach."[54]

Somehow, Mr. Dalli failed to realize that the comparisons being made do not by a long shot identify "all the differences created by the genetic modification." Even if they were restricted to controlled studies employing the non-GE parental strain, they still could not detect the full range of unintended effects. For one thing, although many GE crops have now undergone 90-day rat feeding studies employing the whole food, a 2011 article in a peer-reviewed journal by six scientists emphasized that such medium-term tests are "insufficient to evaluate chronic toxicity."[55] The authors pointed out, based on solid scientific principles, why only long-term studies that include reproductive and multigenerational analyses can adequately screen for the potential deleterious effects of the genetic manipulations. But neither the EU nor any other regulators require such tests. For this reason alone (not to mention several previously discussed), the assessments conducted by EU regulators cannot be deemed the strictest alternative.

Moreover, the EU did not even require the 90-day feeding tests until 2013, and when such tests were voluntarily performed prior to that, it was usually not until at least 10 years *after* the GE crops they assessed had been approved by various regulators and had entered the food supply in numerous nations – which, according to the authors of the above article is "a matter of grave concern."[56] These experts also noted that the sample sizes in the tests have in several instances been too small to ensure reliability, a fact that didn't deter the EU regulators from accepting those tests.

Even more troubling, when these scientists reviewed the data from 19 of the feeding studies (on soy and corn varieties comprising 83% of the GE foods that people have been regularly eating), they found that 9% of the measured parameters, including blood and urine biochemistry and organ weights, were significantly disrupted in the animals that ate the GE feed. Moreover, the greatest disturbances were to the kidneys of the males and the livers of the females; and the scientists emphasized that because livers and kidneys "are the major reactive organs" in cases of chronic food toxicity, these results should be viewed as danger signs – something the regulators have not seen fit to do.[57]

In light of these facts, it's clear that Commissioner Dalli has been seriously misinformed about the risk of unintended effects. And there are pro-GE scientists well-positioned to keep him and his co-commissioners confused. One is Anne Glover, who (at the time of this writing) is the EU Commission's Chief Scientific Advisor. This influence-wielding scientist recently declared: "There is no substantiated case of any adverse impact on human health, animal health or environmental health, so that's pretty robust evidence, and I would be confident in saying that there is no more risk in eating GMO food than eating conventionally farmed food."[58] She went on to announce that, as a consequence, the precautionary principle no longer applies. But, as this chapter has documented (and as Chapter 10 will demonstrate more thoroughly), there *is* substantial evidence of adverse impacts on the health of animals that have consumed GE foods. Moreover, even if such evidence had never been generated, Dr. Glover would not be justified in treating its absence as "robust evidence" for the safety of GE foods – and as grounds for abandoning the precautionary principle. Unfortunately, the commissioners she advises are unaware of her disregard for facts and her looseness with logic.

Yet, despite the significant influence of misinformation, the regulatory irregularities have not always stemmed from it. There are regulators who *have* been properly informed and yet have brushed the information off. For instance, by February 2001, the Australia New Zealand Food Authority (ANZFA) had been informed about several problems of GE foods by the Public Health Association of Australia, and they were aware of the recently released report by the Royal Society of Canada. And when I met with the agency's chief scientist, biotechnology manager, and general manager of standards on February 15 of that year, I emphasized the key points conveyed by both of those sources, especially the conclusion of the Canadian experts that the "default presumption" for every GE food should be that the genetic alteration has induced unintended and potentially hazardous side effects. I also informed them about the memos written by the FDA's scientists and emphasized how they had repeatedly cautioned about genetic engineering's unusual potential to generate unintended side effects. But the chief scientist summarily dismissed such thinking as "mere speculation."[59] This attitude is reflected in her agency's published acknowledgement that it regards GE foods to be safe until proven dangerous.[60]

Through such behavior, this agency that's duty-bound to protect the citizens of Australia and New Zealand from unsafe foods has provided yet another case study in regulatory irresponsibility.[61]

Thus, by inadequately monitoring for unintended harmful effects, and by ignoring significant signs of problems, regulators in the EU and several other regions have retreated from the precautionary path that they were supposed to have been following. As the reviewers of those 19 disquieting feeding studies have asserted, it's "unacceptable" that billions of consumers worldwide are being subjected to GE foods on the basis of the substandard testing that's performed under the current regulatory system.[62] It's also unacceptable that in order to perform adequate independent reviews of this shoddy testing (on behalf of the public interest), these scientists had to get court orders to obtain all the necessary data for three of the GE foods – despite the fact it had been submitted to government regulators and should have been readily accessible.[63]

Conflicted Missions: Regulation is Incompatible with Promotion

The irresponsible behavior of the various regulatory agencies is largely due to the fact that (like the US FDA) although they are supposed to regulate GE foods, they also have a mission to promote them. This has engendered serious conflicts of interest.

Commenting on the pervasiveness of these conflicts, Suzanne Wuerthele, a toxicologist with the US Environmental Protection Agency, stated: "This technology is being promoted, in the face of concerns by respectable scientists and in the face of data to the contrary, by the very agencies which are supposed to be protecting human health and the environment. The bottom line in my view is that we are confronted with the most powerful technology the world has ever known, and it is being rapidly deployed with almost no thought whatsoever to its consequences."[64]

The expert panel of the Royal Society of Canada also emphasized the corrosive effects of the government/industry collusion. Its report observed: "In meetings with senior managers from the various Canadian regulatory departments . . . their responses uniformly stressed the importance of maintaining a favorable climate for the biotechnology industry to develop new products. . . . The conflict of interest involved in both promoting and regulating an industry or technology . . . is also a factor in the issue of maintaining the transparency, and therefore the scientific integrity, of the regulatory process. In effect, the public interest in a regulatory system that is 'science based' – that meets scientific standards of objectivity, a major aspect of which is full openness to scientific peer review – is significantly compromised when that openness is negotiated away by regulators in exchange for cordial and supportive relationships with the industries being regulated."[65]

In several cases, the connections between regulators and industry have not merely been close but unsavory. According to a report by Friends of the Earth, as of 2004, one of the scientists at the European Food Safety Authority (EFSA) had direct financial links to the biotech industry, several had indirect links, and two had appeared in promotional videos sponsored by the industry.[66]

It's therefore not surprising that this agency has issued several highly questionable opinions supporting the safety of GE foods. One eyebrow-raiser involved Monsanto's YieldGard® Rootworm Corn (MON 863 maize), a variety designed to express the Bt pesticide. As reported by Friends of the Earth, the reviewers from EU member states raised "a large number of concerns about the quality of the assessment," and some of the strongest related to a feeding study on rats.[67] They were especially concerned about the significant difference in white blood cell counts for rats fed the GE corn compared to the animals that ate the non-GE variety. But the EFSA dismissed the differences as not "biologically meaningful." The agency also dismissed every other concern, including those about differences in additional blood cell parameters, kidney weights, and kidney structure.[68] But the other experts were not so dismissive. For instance, the French Commission for Genetic Engineering determined that the data fell short of demonstrating an absence of harm; and the Director of a French national research body declared: ". . . what struck me in this file is the number of anomalies. There are too many elements here where significant variations are observed. I never saw that in another file."[69]

Many other reviewers were similarly struck by the variations, and at a regulatory committee meeting in September 2004, only a few countries supported the YieldGard® corn. (Friends of the Earth was informed it was four out of 25.) Accordingly, the European Commission asked the EFSA to consider a new evaluation submitted by the German national authorities that advised an additional feeding test better designed to detect whether any of the unintended effects were linked to unintended alterations in the corn. But the EFSA countered that this would only be "worthwhile" in a case with "indications of the occurrence of unintended effects" and that there were no such indications in regard to this corn. So it reaffirmed its original conclusion that the product is safe – which confirms that when the agency is assessing a GE crop, even the presence of an unusual number of significant variations between the experimental and control groups of rats is still not considered indicative of unintended effects.[70]

The concerns that the EFSA dismissed were subsequently proven to have been well-justified when a team of independent experts obtained the raw data from the 90-day feeding study Monsanto had conducted and subjected it to the kind of rigorous analysis that this regulatory body should have performed itself. The three members of that team were also part of the five-member group that later performed the telling review of the 19 feeding studies; and this particular Monsanto study was also included in that set. Further, it was one of the studies for which the researchers could not obtain sufficient data without first obtaining a court order to compel its release from the regulators. And this closely-guarded data was illuminating. When the scientists published the results of their re-analysis in the *Archives of Environmental Contamination and Toxicology*, they reported several differences between the rats that ate the GE corn and the conventionally fed controls, including chemical changes that "reveal signs of hepatorenal [liver and kidney] toxicity."[71] If the EFSA had been more intent on regulation rather than promotion, such differences would have initially been detected and fully acknowledged – and prevented the product from being approved.

Through its prolonged promotional efforts (and concomitant dereliction), the European Food Safety Authority's actual authority has been so degraded that even the European Commission lost trust in it – while concurrently gaining awareness that there are good grounds for greater caution. This was disclosed when Friends of the Earth obtained documents the EC had submitted to the World Trade Organization (WTO) in response to charges by the United States that its restrictions on GE food constitute an illegitimate restraint on trade. In defending the need for its modest (and inadequate) food safety regulations against a nation that argued there was no scientific justification for them, the EC asserted that there *are* reasonable scientific concerns about the safety of GE foods, that the research conducted by the biotech industry tends to be of low quality, and that industry's applications for product approvals are frequently flawed. Moreover, besides criticizing the submissions of industry, the commission criticized the safety assessments of the EFSA.[72]

However, because the commission is also motivated to promote these products (though perhaps less strongly than is the EFSA), its public statements have imparted an opposite impression. According to the Friends of the Earth, the EC has consistently assured the public that the GE foods it has approved are "completely safe;" and it has "continually used EFSA opinions to justify its decisions" – one of which granted approval for Monsanto's hotly debated (and highly dubious) YieldGard® corn to enter the EU food supply.[73]

Through such duplicity, the commission has duped a lot of people – and helped sustain a regulatory regime that's overdue for an overhaul.

Thus, in addressing the risks posed by genetic engineering, there has been regulatory irresponsibility on a global scale. Further, as we shall see, the dereliction has not been confined to agencies entrusted with the preservation of food safety. It has also infected those that are supposed to protect the environment. And, once again, many of the most egregious offenses have occurred in the United States.

CHAPTER SEVEN

Erosion of Environmental Protection

Multiple Risks, Minimal Caution

The two US federal agencies responsible for protecting the environment from hazards of bioengineered organisms, the Environmental Protection Agency (EPA) and the Department of Agriculture (USDA), have not lagged far behind the FDA in subordinating the standards of science and the demands of the law to the interests of the biotech industry. Although Phil Regal's efforts during the 1980's had halted the release of several high-risk organisms, the overall behavior of the EPA and USDA has not been encouraging. While the agencies did implement some guidelines in response to environmental concerns that had been raised by Regal and other scientists, they tended to be minimal. They demanded little in the way of genuine biosafety analysis, and they provided scant protection.

The US government did not even undertake a full environmental impact study of a GMO (under the guidelines of the National Environmental Protection Act) until 2005. The level of regulation is so lax that permits for performing field tests on GE crops are usually issued without asking the manufacturer to do anything more than providing notice of intent to do such a test. And oversight rarely approaches rigor thereafter.

Mishandling a Hazard: The Alarming Case of K. Planticola

The depth of the system's deficiency is revealed by its obtuse handling of a genetically engineered soil bacterium that turned out to be far less innocuous than expected. The altered bacterium was developed by a German corporation; and in 1994, preparations were underway for its release. Hopes were high, because it had been designed to increase ethanol production within fermentors that process agricultural waste. This system would produce a two-fold benefit. Besides converting field waste into an automobile fuel that's cleaner burning than gasoline, the process would yield a nutrient-rich sludge that could be spread on the soil as a fertilizer.

However, the engineered organisms also had significant downside potential. They were derived from a particular strain of a soil-dwelling bacterium named *Klebsiella planticola*. And *K. planticola* do not only congregate on plant matter that's dead and decaying, they also inhabit plants that are alive. Further, their residence is the root system, to which they adhere by exuding a sticky layer of slime. Moreover, although their presence on the plant is strictly localized, their geographical dispersion is global. They've attached themselves to plant roots the world over and there is not a single type of plant tested for them in which they have not been found.[1]

Consequently, because viable *K. planticola* would be present in the sludge spread on the fields, it would have been prudent to carefully assess the effects of these engineered organisms on plants that are living as well as those that are already dead. This is especially so in light of research that indicated bioengineered microorganisms can disturb microbial populations in natural environments, in some cases inducing loss of a fungus that's a normal component of the soil.[2] Further, scientists had observed that even a low level of ethanol can negatively affect some biological systems.[3] But the tests being conducted on *K. planticola* were not rigorous enough to provide an adequate assessment, even though most were occurring at the Oregon laboratories of the EPA.

Fortunately, a few soil scientists at Oregon State University recognized that a bioengineered bacterial strain with the potential to continually effuse ethyl alcohol amid the roots of plants should be more thoroughly studied before it was allowed to proliferate in nature. So they designed an experiment to do so. They put wheat seedlings into jars filled with identical samples of soil and then added the engineered variety of *K. planticola* to some of the jars and the strain from which they had been derived to an equal number of others.

All the plants grew well for the first week. But then, every plant in the soil containing the engineered bacteria wilted and died. In contrast, all the plants growing amid the normal version of the bacterium remained healthy. The research design was healthy too, and the ensuing report was eventually published in a professional scientific journal.[4]

The data indicated that the engineered bacteria altered the population of soil microorganisms in a way that could have affected the nutrient cycling processes wheat plants rely on. Further, because the evidence also indicated that the engineered bacteria persisted in the soil as well as did the unaltered strain, it raised the possibility that if released, they could have become established in the fields.[5] Moreover, if they were able to survive in a variety of environments, given the rapid rates at which bacteria multiply and migrate, they could have been virtually impossible to control.

Accordingly, several experts believe that these engineered microbes posed a major risk. Elaine Ingham, who, as an Oregon State professor, participated in the research that discovered their lethal effects, points out that because *K. planticola* are in the root system of all terrestrial plants, it's justified to think that the commercial release of the engineered strain would have endangered plants on a broad scale – and, in the most extreme outcome, could have destroyed all plant life on an entire continent, or even on the entire earth. This in turn would have wiped out animal life, including humankind. Whereas bacterial life and some higher species could have continued, the biosphere as we know it would have been drastically depleted.[6] Another scientist who thinks that a colossal threat was created is the renowned Canadian geneticist and ecologist David Suzuki. As he puts it: "The genetically engineered Klebsiella could have ended all plant life on this continent. The implications of this single case are nothing short of terrifying."[7]

However, other scientists think there's scant likelihood that such a devastating outcome could have occurred. They note that the experiment only employed one particular type of sandy soil and that there's no evidence the engineered bacteria could have induced similar affects in other soil types. They further argue that it's unlikely the bacteria could have become broadly established throughout a wide range of varying eco-systems.

Unfortunately, there's not enough data to definitively resolve these issues. It's not even clear whether the ethanol was a factor in harming the plants or whether they were felled by an unintended effect of the bioengineering process – induced by an unplanned perturbation of the bacterium's normal mode of function.

But it *is* clear that the engineered *Klebsiella* caused the plants in that particular experimental set-up to die – and that more research should have been conducted to ascertain exactly how it had happened, and whether it would happen again in different soil types. Phil Regal emphasizes that the study "raised a red flag" and that it in effect "demanded further detailed research." He says that without more evidence, "there was no way of knowing how the engineered bacteria might behave in the field."

He continues:

> The EPA had partly funded the research conducted by the Oregon State scientists, and I thought the agency should sponsor more studies

to find out exactly what was happening so that we would build our knowledge base and provide a firmer foundation for assessing subsequent engineered organisms. That would have been the proper thing to do in science. But instead, someone up the EPA ladder, or possibly in the White House, cut the funding altogether.

Apparently, those empowered to set the agenda did not want more light shed on this issue and were hoping it would quickly subside. It seems that they even instructed their subordinates to stop talking about it, because the middle-level agency people with whom I had been communicating abruptly informed me, with shaky and apologetic voices, that they could not discuss the matter any further – even though I had known them for years and had been on the EPA's Science Advisory Board.

The genetic engineers had enormous political clout, and those at middle levels were afraid to stick their necks out.

Moreover, not only did the EPA shut down further research involving the engineered *Klebsiella*, it initially resisted the critical evidence and took an adversarial attitude toward the scientists who brought it out – despite the fact it had funded their research. Elaine Ingham reports: "When the data first started coming in, the EPA charged that we couldn't have performed the research correctly. They went through everything with a fine tooth comb and they couldn't find anything wrong with the experimental design – but they tried as hard as they could. At that time, some EPA researchers did not understand 'ecologically-based' testing systems designed to look at the microbial interactions and nutrient cycling processes that would occur if *Klebsiella planticola* were released into the environment. The fact is that the regulatory system as it stands today is totally inadequate to catch these kinds of unexpected effects. If we hadn't done this research, the *Klebsiella* would have passed the approval process for commercial release."[8] And it seems that the revelation came just in the nick of time. In Ingham's understanding, the impending release was "mere weeks away" when the data was presented to the EPA.[9]

Besides the *Klebsiella* incident, Ingham had additional first-hand experience on which to base her judgment about regulatory inadequacy. As she explains: "I've worked with folks in the Environmental Protection Agency and I know the tests the EPA performs on organisms. They often begin their tests with 'sterile soil.' But if it's sterile, then it's not really soil. Soil implies living organisms present. If you use 'sterile soil' and add a genetically engineered organism to that sterile material, are you likely to see the effects of that organism on the way nutrients are cycled, or on the other organisms in that system? No, you're not likely to. So it's probably no surprise that no ecological effects are found when they test genetically engineered organisms in sterile soil. They really need to put together testing systems which assess the effects of the test organism on all of the organisms present in soil."[10] She emphasizes the importance of such an approach because real soil is densely packed with diverse forms of microscopic life, most of which play an important role in fostering favorable conditions for plants to flourish.

However, despite the deficiencies of their experimental system, the regulators seemed set on sticking with it as tenaciously as *Klebsiella* cling to plant roots.[11] They not only appeared unconcerned by its inability to detect the full effects of GMOs, but undeterred by the problems that had come to light. For them, promotion rather than precaution would remain the priority. Further, though unperturbed by the prospect that some GMOs could be highly harmful, they *were* upset by investigators who revealed actual dangers – not (it seems) from concern about the damage that might befall the biosphere but from irritation over the injured image of GMOs. And this provocation colored their behavior toward those researchers.

Thus, despite the fact that Ingham and her colleagues had performed a highly valuable service – and may have averted a massive catastrophe – rather than honoring her accomplishment, the EPA apparently tried to punish her for it. Although she had regularly received significant funding from the EPA prior to her work on *Klebsiella*, the situation suddenly changed. After submitting several research proposals that were consistently rejected, she finally gave up and decided to stop wasting her time.[12]

Before moving on, it's important to address the bottom line question: *Even though we lack detailed data about many aspects of the engineered Klebsiella, can we justly conclude that they posed a meaningful risk?* To answer it, we need to be clear about what risk is. As we'll examine more thoroughly in Chapter 9, risk is technically defined as the product of the multiplication of two quantities: (1) the probability that a potential problem could actually result in harm and (2) the amount of the harm that would result. Thus, if the probability of harm is high but the harm that would be caused is miniscule, then the risk is low. Conversely, even when the likelihood of harm is low, if the resultant damage would be colossal, then the risk is significant. Therefore, even if the worst-case scenarios envisioned by Drs. Ingham and Suzuki were extremely unlikely, because they're catastrophes of incalculable proportions, the risk should be deemed significant.

An Imbalanced System: Promotion Tops Prevention

Although the risk posed by the engineered *Klebsiella* was far from trivial, the regulatory response was all too typical; and the incident is symptomatic of the ills of the system, ills stemming from a rigid political agenda to promote bioengineering. One of the more dramatic displays of the ardor with which this agenda is advanced was prompted by a speech Philip Regal delivered at a conference on GMOs held in Washington, D.C. on January 31, 1989. The event was co-sponsored by the National Wildlife Federation and the Corporate Conservation Council and was well-attended by representatives from both industry and government. By that date, Regal had seen more than enough evidence to convince him that the government's promotional bias was hindering the implementation of rational regulation. So he decided to shake things up by boldly calling attention to the problem. At the end of his speech, he forthrightly described the realities as he saw them and emphasized how the EPA was "bending over backward" to accommodate the wishes of the biotech industry.

After he had finished, and as the ensuing break began, one of the top EPA officials, accompanied by his assistant, advanced toward him in what appeared to be a state of excitation. Bracing himself for a scorching glare and a harsh harangue, he was astonished to instead encounter eyes beaming with pleasure and a voice erupting in praise. The man told Regal that he and his colleagues had been laboring to convince the industry that the EPA was on their side and would accede to their plans, and he commended him for so clearly making this case. As Regal's surprise subsided, he realized that by receiving his critique as a compliment, the official not only confirmed its validity but revealed that the agency's integrity had degraded even farther than he had thought.

The promotional prejudice of the government agencies, and the deficiencies it breeds, have remained constant features of the US regulatory regime from the advent of genetic engineering to the present day. Defects were apparent to Regal in 1983, and serious defects were still readily discernible by the government's General Accounting Office in 1988 when it reviewed how the USDA, the EPA, the FDA, and the NIH were dealing with GMOs – and faulted all of them for making safety determinations without an adequate scientific basis.[13] Moreover, despite continual pronouncements from government officials that genetic engineering is being carefully regulated, the situation has not improved.

By January 2000, the system had become so biased toward promoting this technology, and so averse to establishing sensible safeguards, the

Public Employees for Environmental Responsibility (PEER) felt the need to re-release a report that it described as "a devastating critique of the risk assessments employed by the United States prior to the widespread, commercial release of genetically modified organisms (GMOs) into the environment." [14] The report, titled "Genetic Genie," had first been issued in September 1995 in an attempt to deter the EPA from approving the release of an engineered bacterium (*Rhizobium meliloti* RMBPC-2) that numerous agency scientists regarded as posing unreasonable risks to both human and environmental health. It was written by several of these EPA experts and reviewed by faculty members of major universities. Because the EPA had so often been found in violation of whistleblower protection statutes, the authors had to remain anonymous to avoid retaliation by their supervisors. Their report charged that besides the fact the EPA's risk assessment was seriously deficient, the assessment of benefit was also fatally flawed. Although the organism had been designed to increase nitrogen fixation in alfalfa and thereby boost yields, the data indicated that yields might not significantly improve. However, the EPA ignored the evidence that the farmers buying the bacteria might receive no benefit and instead only took into account the projected economic benefit the corporation would gain from sales to the farmers. And it decided that this benefit was sufficient to outweigh the risks – which, according to the PEER report, were drastically downplayed.

Like the internal experts who wrote the critique, the EPA's outside advisory panel readily recognized the weaknesses of the agency's work; and five of the six members refused to endorse its proposed approval of the bacterium.[15] When the agency indicated that it planned to proceed with the approval anyway, one member resigned in protest.[16] The EPA approved the bacterium for commercial release in 1997 and, according to one of the agency's own risk-assessment experts, as of July 2001 it had not performed any follow up studies to detect adverse impacts – or even to determine if crop yields had actually increased.[17]

PEER re-released its report in 2000 because it decided the situation had not improved. Its accompanying statement declares: "The passage of time has not diminished the accuracy or the urgency of the scientific critique contained in Genetic Genie. The same commerce-driven dynamics described in Genetic Genie are occurring today as American regulatory agencies struggle with overwhelming uncertainties while assuring an increasingly skeptical world that everything is safe." [18]

Unfortunately, the behavior of the Department of Agriculture has been just as dismal as the EPA's. Its handling of biotechnology has stayed so shoddy that in December 2005, twenty-one years after Regal organized the first major scientific workshop on the ecological risks of GMOs at the renowned Cold Spring Harbor Laboratories, nineteen years after the first permit was issued for releasing one into the environment, and after more than 10,000 field trials had been authorized, the agency's inspector general issued what the *New York Times* termed a "stinging" [19] report rebuking it for failure to properly oversee such trials. The document declared that the USDA "lacks basic information" on where the tests are or what's done with the crops after they are harvested, and it pointed out that agency regulators have too frequently either failed to notice violations of the rules or committed violations themselves. It charged: "Current (USDA) regulations, policies and procedures do not go far enough to ensure the safe introduction of agricultural biotechnology." [20]

This chronic regulatory infirmity, coupled with recurrent corporate carelessness, has resulted in a string of serious accidents. For instance, although a variety of GE long grain rice developed by Bayer never gained approval for commercial planting, it was widely found in the US rice crop. This large scale contamination occurred despite the fact the engineered rice was only field tested in Louisiana – and only between 1999 and 2001. Moreover, the contamination went on for five years before it was discovered.[21]

Further, the errant rice was not the only GE crop mismanaged during 2001. In that year, problems arose at an Iowa test field when the manufacturer of an experimental variety of corn engineered to produce pharmaceutical medicines failed to take adequate measures to prevent the gene-altered pollen from riding the wind and transforming the corn on surrounding farms – necessitating the destruction of 155 acres of produce.[22] And across the Missouri River in Nebraska, another test plot of that pharmaceutical crop also caused problems when some corn accidentally commingled with soybeans, ultimately adulterating 500,000 bushels.[23]

Moreover, among numerous other incidents of contamination, pollen from an unapproved variety of GE herbicide-resistant bentgrass escaped an Oregon test plot and altered other varieties of grass. This escape went unreported for several years, and some experts suspect that by now there's been substantial contamination of the commercial grass seed supply, 70 percent of which is cultivated in Oregon.[24] The EPA has discovered several other instances in which pollen from herbicide resistant GE bentgrass escaped field trials and transferred its reconfigured genes to native grasses several miles away – including a case where a national grassland was contaminated.[25]

Keeping Score: How the Main GE Crops are Not Helping the Environment but Harming It

Proponents of GE crops have routinely claimed that the two types most prevalently grown (those that produce their own pesticide and those that are resistant to one or more commercial herbicides) are, on the whole, beneficial for the environment. But extensive evidence refutes this claim – and reveals that the net effect of these crops on the environment is harmful.

The Hazards of Pesticide-Producing Plants

Biotech advocates have boasted that GE crops would reduce pesticide use because a large percentage would be designed to produce their own insect-repelling pesticides. To date, such insect-killing crops have all been engineered to express a pesticide that's made by a common soil bacterium. The pesticide is called "Bt;" and in its natural form, it has long been used as a spray to control insects. Because it is generally safe for mammals and non-target organisms, including beneficial insects, it has even been approved for use in organic agriculture.

However, when a copy of a gene that synthesizes Bt pesticidal protein is engineered into corn or cotton (the two basic commercialized Bt crops), the results don't replicate the natural situation. For one thing, the plant-generated proteins do not always have the same structure as the natural ones, and their effects are different.[26] Natural Bt protein breaks down rapidly in daylight and does not accumulate in soil or waterways. In contrast, because every cell of a Bt crop expresses the Bt toxin in active form, the roots continually exude the toxic proteins into the soil. Further, when residues of Bt corn are ploughed under at harvest, the Bt in their tissues doesn't readily break down and persists in the soil for months.[27] This has negative effects. Researchers have found that mycorrhizal fungi (which colonize the roots of plants and help them to absorb nutrients, resist disease, and tolerate drought) are less abundant in the roots of corn engineered to express Bt than on corn that is not.[28] Scientists have also discovered reduced concentrations of another beneficial fungus (*Arbuscular mycorrhizal* fungi) in the roots of Bt corn.[29]

Further, unlike natural Bt proteins, those produced in engineered plants have harmful impacts on waterways and aquatic life. One study (conducted in Indiana) found that corn-generated Bt proteins were polluting 25% of the streams that were tested.[30] Another study determined that the biomass of Bt corn is toxic to aquatic organisms.[31] And when Bt corn was fed to water fleas (an organism often used as an indicator of environmental toxicity), toxic effects were observed, including reduced fitness, higher mortality, and impaired reproduction.[32]

Moreover, researchers have found that Bt crops exert toxic effects on non-target insect populations such as butterflies and that they may even impair the ability of bees to find nectar sources.[33] Additionally, the evidence indicates that they harm beneficial insects that prey on plant pests.[34]

Bt crops pose other threats as well. One of the biggest is that they might ultimately undermine the effectiveness of Bt itself. This could happen through the development of widespread resistance to Bt within the target pest populations.

Just as overuse of antibiotics has given rise to drug-resistant supergerms, so excessive use of Bt toxin could lead to Bt-resistant insect pests. However, until the advent of genetic engineering, there was never a threat of such an overload. In its natural form, Bt is not extensively employed in large-scale agriculture because its rapid breakdown necessitates intensive management (including precise timing in application) that is ill-suited to most industrialized farms.[35] And although Bt has been utilized on many organic farms (and on farms employing integrated pest management, a system that uses less synthetic pesticides than the typical industrial operation), it has been applied sparingly, providing no pressure for resistant insects to develop. But the widespread planting of Bt-endowed crops has caused insects to be continually exposed to the toxin. Accordingly, those that possess a mutation conferring resistance to it gain a big competitive advantage over their vulnerable species mates; and, over time, they become the dominant population.

The development of Bt-resistant pests is not just a hypothetical possibility. For instance, it took only six years for the western corn rootworm to evolve resistance to the Bt corn that was specifically designed to curtail it.[36] And populations of Bt-resistant rootworms have been observed in Iowa and Illinois.[37] As such resistant strains proliferate, farmers must resort to pesticides to control them.

Further, even when the Bt crops initially knock out most of the pests against which they were intended, other pests often occupy the resulting gap in the ecosystem.[38] Because these troublesome species are not inherently vulnerable to Bt, they also induce the application of pesticides.

Consequently, although Bt crops have so far accounted for a reduction in pesticide use, the drop has been far less substantial than anticipated. Moreover, the reduction can be expected to steadily shrink as Bt-resistant pests continue to propagate – and as pests that succumb to Bt provide opportunity for impervious pests to take their place. Further, if Bt resistance becomes so prevalent that natural Bt sprays are rendered ineffective, it would be a major blow to organic agriculture and to farms that employ integrated pest management. As several experts have observed, it's likely that no other naturally produced pesticide combines the efficacy of Bt with the same degree of safety.

Herbicide Overdose

Two major kinds of pesticides are those that kill insects (*insecticides*) and those that poison weeds (*herbicides*). Even though Bt crops have to date enabled a decrease in use of insecticides, overall, GE crops have caused an increase in pesticide use. That's because the bulk of commercialized GE crops don't produce insecticides but have been engineered to withstand herbicides; and 84% of the land planted in GE crops contains these herbicide-resistant varieties.[39] Consequently, herbicide use has sharply escalated. Employing official US Department of Agriculture data, agronomist Charles Benbrook found that during the first thirteen years of their cultivation in the United States (1996 to 2008), GE crops induced an increase of 383 million pounds of weed-killing chemicals.[40] According to Dr. Benbrook, "This dramatic increase . . . swamps the decrease in insecticide use attributable to GE corn and cotton, making the overall chemical footprint of today's GE crops decidedly negative. . . ."[41]

Further, most of this additional usage was due to a phenomenon about which ecologists had repeatedly warned – and regulators and biotech proponents had shown scant concern: the emergence of herbicide-resistant superweeds. Such weeds develop in two ways. The first is through herbicide overdose, with superweeds arising as do Bt-resistant insects. Although the steady onslaught of poison kills most weeds, mutations eventually develop in a few of them that confer resistance to it. Those resistant weeds then rapidly multiply, since there are virtually no weeds left within the sprayed areas to compete with them for nutriment.

In contrast, the second route does not involve mutation but cross-pollination. Because canola can fertilize some of its wild and weedy relatives, canola pollen containing the genes for herbicide resistance can endow those weeds with the engineered trait.

Superweeds don't emerge immediately, but once they do, their spread accelerates. It took around four years after the initial plantings of Monsanto's Roundup Ready® (RR) soy (which resists the herbicide glyphosate) before the first species of glyphosate-resistant weeds appeared; and their presence was originally confined to the state of Delaware. But within another 10 years, there were 10 resistant species in 22 states infesting up to 10 million acres in the US alone.[42]

And the impact has been enormous. According to the president of the Arkansas Association of Conservation Districts, "It is the single largest threat to production agriculture that we have ever seen."[43] For one thing, the data demonstrates that superweeds not only emerged due to excessive use of herbicide, but that their subsequent proliferation has led to abundant additional spraying. Dr. Benbrook has determined that this superweed-induced spraying is "the primary cause" of the increased volume of herbicide attributed to GE crops.[44] In other words, more herbicide has been deployed in trying to control these novel weeds than was dispensed during the prolonged process through which they were given birth and caused to widely spread.

Moreover, because such weeds have become resistant to glyphosate, farmers have been forced to use an older generation of herbicides to deal with them – herbicides that are generally considered to be more harmful than glyphosate. But one of the big selling points of Roundup Ready crops was that they would shift use from these older herbicides to the more environmentally friendly glyphosate. Thus, not only have GE crops caused a major increase in herbicide use, they have stimulated a significant return to the harsher herbicides whose use they were supposed to have minimized.

One of these herbicides is 2,4-D. It was an active ingredient in Agent Orange, a defoliant used extensively by the US military during the Vietnam War. 2,4-D is very potent, and its vapor causes damage to most broadleaf plants (plants that are not grasses) at extremely low levels.[45] Two surveys of state pesticide regulators indicate that its drift has caused more instances of crop injury than has any other pesticide.[46] Some species that are especially sensitive to it include grapes, soybeans, sunflower, beans, tomatoes, and cotton.[47] Further, besides being more toxic than glyphosate in general, 2,4-D is especially brutal on such sensitive species: at least 300 times more toxic than glyphosate to the emerging seedlings and nine times more toxic to the plants.[48]

Another herbicide that farmers have been turning to in the battle against superweeds is *dicamba*. Like 2,4-D, it packs a more harmful environmental punch than glyphosate. In fact, chemically speaking, dicamba is a close cousin of 2,4-D.

Moreover, despite the fact that 2,4-D and dicamba are harsher on the environment than glyphosate, the latter is hardly benign. For instance, researchers have demonstrated that glyphosate caused birth defects in frog and chicken embryos at doses far lower than those used in agricultural spraying.[49] The malformations in the embryos were similar to human birth defects observed in areas of South America with heavy planting of RR soy. Monsanto's herbicide "Roundup®" (which has glyphosate as its active ingredient) has also been found to be a potent endocrine disruptor at levels up to 800 times lower than residue levels allowed in food and feed. And it was observed to be toxic to human cells and also to damage DNA at doses far below those used in agriculture.[50]

Among the many other downsides of glyphosate is its tendency to stimulate the growth of harmful fungi in the soil. One of the worst is *Fusarium*, a fungus that causes wilt disease and sudden death syndrome in soy plants.[51] Further, *Fusarium* not only affects plants, it generates toxins that can enter the food chain and harm humans and livestock.[52] In pigs, it has been found to impair reproduction and increase stillbirths.[53]

How Industry Compounds the Problems – While Amply Profiting from Them

Faced with the emergence of glyphosate-resistant superweeds, and farmers' increasing need to employ herbicides that are much more potent than glyphosate to deal with them, the biotech industry has responded in a predictable, but regrettable, fashion. Although the problems were caused by genetic engineering, Monsanto and the other giant corporations have decided to try to solve them through additional genetic engineering. Thus, to enable farmers to apply stronger herbicides more profusely in a way that won't kill their crops, the industry has developed a new line of GE plants that are resistant to those toxic chemicals. For instance, the Dow corporation has created a soybean that's resistant to 2,4-D; and, for good measure, it also engineered the bean to tolerate glyphosate and glufosinate (an herbicide that can be used against plants that are glyphosate-resistant). It has also produced corn that can be doused with 2,4-D and several other herbicides as well.[54] Not to be outdone, Monsanto has developed a soybean that's resistant to both dicamba and glyphosate – and designed corn and cotton endowed with triple herbicide resistance.

While this makes sense from a purely commercial standpoint, since the corporations profit handsomely from selling farmers premium-priced seeds along with the herbicides to which they're resistant, from a scientific perspective, it's unwise and irresponsible. As Bill Freese, a science policy analyst with the Center for Food Safety has noted, "Increasingly toxic herbicide cocktails will be used on multiple herbicide-resistant (HR) crops, spawning weeds with multiple resistances. The chemical arms race with weeds triggered by these HR crops entails an ever-escalating spiral of pesticide use and pollution, and attendant adverse impacts on public health and the environment."[55]

Besides stacking more herbicide resistant traits onto their GE seeds, the manufacturers are steadily expanding the number of crop species endowed with such resistance. One of the recent additions is Roundup-resistant alfalfa. This was a major event because alfalfa is a major crop. It's the fourth-largest one grown in the US, and it ranks number three in value. Further, it is primarily used as feed for dairy cows and beef cattle; and it's heavily relied on by organic dairy operations.

But only organically grown alfalfa can be used in organic dairy production; and because alfalfa is pollinated via the action of far-ranging bees, organic alfalfa fields will be increasingly encroached by pollen that's been genetically engineered. And the contaminated crops that emerge will be unsuitable for use by organic dairy farmers.

When a lawsuit was brought in US federal court in 2006 to halt the use of the GE alfalfa, the judge determined that significant contamination had already occurred, stating: "Such contamination is irreparable environmental harm. The contamination cannot be undone."[56] He also ruled that in approving the crop, the US Department of Agriculture (USDA) had violated environmental laws by failing to analyze such risks. Accordingly, he banned further use of the crop until the agency performed a thorough environmental impact statement (EIS) that would include analysis of the effect on farmers trying to grow GE-free alfalfa.

In its EIS, the USDA admitted that contamination of conventional and organic alfalfa could occur and that the economic interests of the growers could be harmed. Nonetheless, it yielded to intense pressure from the industry and *deregulated* GE alfalfa, an action that superseded the court's ruling and permitted the crop to be planted without any restrictions.[57]

Unfortunately, the short-sightedness displayed by the industry in addressing the predicaments caused by herbicide resistant GE crops also characterizes the way it has dealt with the difficulties of Bt crops. The response to the problems of Bt corn has been especially unenlightened. Because these gene-altered varieties have failed to control a large range of pests, manufacturers have altered their seeds in an additional way. They have coated them with powerful insecticides – and they have chosen ones that are highly controversial.[58]

These potent chemicals are called neonicotinoids, and they are aggressive. As the plant grows, they spread throughout all its tissues, even entering the pollen and nectar. Neonicotinoids differ from sprayed insecticides (as do the Bt toxins expressed within GE crops) because they endure within the growing plant and are ever active. Accordingly, there's a much greater probability that pests will become resistant. There's also greater likelihood that beneficial insects will be harmed, especially since neonicotinoids are toxic to a wide variety of them. Further, the toxic effects are substantial even at very low doses because the chemicals persist over long periods in soil and water.[59] Moreover, evidence suggests that neonicotinoids may play a role in bee die-off and colony collapse.[60]

A Crucial Choice: To Keep on Hitting the Wall, or to Hit Upon a Better Way Forward

In surveying the evidence, it's clear that GE crops have not delivered environmental benefits as was promised but have instead created serious problems. Rather than reducing pesticide use, they've substantially elevated it. And by inducing the growth of superweeds, they're also inducing continual increase in the volume of pesticide needed – as well as reliance on chemicals more damaging than the glyphosate with which they were designed to be used. Further, the failure of Bt crops to control pests without application of synthetic pesticides has caused the manufacturers to intensify the toxicity of the entire plant. In their initial phase, every cell expressed the Bt toxin in active form, which entailed a substantial range of hazards. But in their current phase, the plants are even more hazardous, because besides containing the Bt toxin, every cell now also contains a toxic neonicotinoid.

Due to these unintended outcomes, other anticipated benefits have not materialized either. For instance, herbicide resistant crops were supposed to have greatly expanded the employment of "no-till" practices that had been on the rise for decades, because the ability to kill surrounding weeds even while the soy or other crop was growing would presumably eliminate the need to inhibit the weeds by plowing the land prior to sowing. In turn, reduction in plowing was supposed to reduce soil erosion and the harmful runoff of fertilizer and pesticide.

However, one research study indicates that even when no-till practices have been adopted by growers of GE soy, the net environmental impact is still negative, largely due to the high use of herbicide.[61] Further, even assuming that some farmers have produced net positive impacts through combining GE crops with the no-till system, the rise of superweeds is impeding their ability to continue the practice. In May 2010 the *New York Times* reported that a retreat from the no-till approach was occurring throughout America's East, Midwest, and South; and it cited the case of a Tennessee farmer who had adhered to the no-till system for 15 years but who was starting to plow again in an effort to combat superweeds. Reflecting on the situation, the farmer remarked: "We're back to where we were 20 years ago." [62]

But this statement is only partly true. While many farmers have been forced to re-adopt practices they employed two decades previously, that doesn't mean that the problems of that time were as great as those of today. Due to the effects of GE crops, today's dilemmas are much worse. Twenty years ago, farmers did not face superweed infestations, and they were using a lot less herbicide. Further, no plants had pesticidal chemicals genetically infused within every cell. Accordingly, the various risks currently posed by such plants were non-existent; and there was no threat that Bt spray would cease to be an effective tool in organic farming and integrated pest management. Nor was there a threat that organic dairy farming might no longer be viable because truly organic alfalfa might no longer be available. In even less than 20 years, agricultural bioengineering has produced all these problems – and several more.

Thus, through the widespread use of GE crops, agriculture has essentially hit a wall. In response, the biotech industry has made the wall much thicker – while encouraging farmers to keep driving into it. However, another approach is available, one that offers an open and much more promising road. In such an approach, genetic engineering and its attendant problems would no longer be features of farming. And agriculture would be safe, sustainable, and productive. We will examine this attractive (and realistic) alternative in the book's final chapter.

As we've seen, although some genetically engineered organisms have significantly imperiled the environment, and although the GE crops currently in use impose environmental impacts that are predominantly negative, the US Department of Agriculture and the Environmental Protection Agency have largely ignored the risks and continued to promote (and approve) these crops as if they entailed no unusual problems at all. Along with the Food and Drug Administration's fraudulent claims that GE foods have been proven safe to eat, the pronouncements and practices of these agencies have left the American public deeply confused about the facts – and more complacent about GE foods than citizens in most other industrialized nations.

However, not all the confusion and complacency can be attributed to the actions of government agencies, even in combination with the extensive deceptions of the scientific establishment that have been documented in previous chapters. A significant role has also been played by the mainstream American media.

CHAPTER EIGHT

MALFUNCTION OF THE AMERICAN MEDIA

Pliant Accomplices in Cover-up and Deception

As I walked to lunch on May 27, 1998, I was elated. The Alliance for Bio-Integrity and the International Center for Technology Assessment had just held a press conference at the National Press Club in Washington, D.C. announcing the filing of their lawsuit against the FDA; and it seemed there was good reason to be buoyant. The conference had been well-attended, with numerous print reporters and camera crews from the major national TV networks. I and other speakers had described the many flaws of the FDA's policy on GE foods and emphasized that, despite its pretensions the agency was not regulating these products in the slightest degree. We had also driven home the fact that among the plaintiffs were nine well-credentialed scientists, whose participation refuted the FDA's claim that GE foods are "Generally Recognized as Safe."

Consequently, I expressed great optimism about the kind of media coverage I was expecting to the friends who had attended the conference and were accompanying me to a restaurant. But one of them didn't share my optimism. She had extensive experience with the press and during the conference she heard a sobering statement from a member of the media that she felt I needed to hear as well. She had been sitting next to a correspondent for one of the national TV networks.[1] He regularly provided reports during the national news about important stories originating from Washington, and his camera crew was taping the conference. Toward the end of the session, he turned to her and remarked: "This is an important story. It should be widely told. But it won't be. I'll file my report this afternoon, but it's not going to go any further. It won't make it onto the evening news, and it won't be on the morning news either."

When I heard this, I found it hard to believe. Why would such an important story not be broadcast? After all, it was vitally relevant to all

Americans because they were regularly consuming GE foods without their knowledge. Didn't they have a right to know that, contrary to the assertions of their proponents, these products have not been carefully tested and that the claims about their safety were based exclusively on dubious assumptions? Moreover, shouldn't the sham about general recognition of safety be exposed? Shouldn't citizens be informed that, in reality, there was not a consensus among experts that GE foods are safe – and that nine were so concerned about the risks that they were suing the FDA?

So, while my optimism was somewhat tempered, I maintained a belief that although forces at that particular news network might obstruct the reporting of our story, conditions would be different at the others – and that they and the rest of the media would dutifully convey the key facts to the public.

But I was wrong. Despite the presence of their crews at the press conference, none of the national television networks reported on our lawsuit. Nor was it mentioned in the *New York Times,* the *Washington Post*, or the *Wall Street Journal* – the nation's three most influential newspapers. National Public Radio didn't even refer to it. Further, although reports on the suit did circulate through some news services and appear in several newspapers, they furnished no grounds for celebration. While they noted that scientists were included among the various plaintiffs, they didn't reveal that there were nine of them, and they failed to point out that the involvement of so many experts undermined the FDA's claim about general recognition of safety.

In fact, the articles did not even report the basic message that our scientists were communicating, even though it was amply conveyed by speakers at the press conference and the supplementary documents we provided. Consequently, readers had no idea that these experts had branded the FDA's policy as scientifically unsound, warned about the unusual potential of GE foods to cause harmful unintended effects, and called for rigorous safety testing. Moreover, in blacking out our scientists' assertions, some dramatic ones had to be disregarded. For instance, during the question and answer session, the molecular biologist Liebe Cavalieri was asked to comment on the fact that many eminent scientists declare genetic engineering to be substantially the same as traditional breeding. As noted in Chapter 4, his answer was not timid. He denounced their behavior as "disgraceful" – and their claim as a "sham."[2] He then added, "And you can quote me on that."

But none of the articles did. Instead, they quoted several spurious assertions from proponents of GE foods issued in response to our suit. One of the most outrageous was from Stephen Ziller, vice president of the

Grocery Manufacturers of America, whose members produce most of the name brand foods and beverages sold in the US In extolling the safety of GE foods and the soundness of FDA policy, Ziller painted the plaintiffs as "opponents of progress and science-based research."[3] In light of the fact that the plaintiffs were actually suing the FDA for ignoring science-based research, and were demanding that more research be performed, this accusation was absurd. But due to the deficient reporting, readers could not discern its absurdity – and many were probably being taken in by it. For the same reason, many may have also been deluded by another absurd assertion, made by an FDA official to trivialize the differences between GE foods and conventional ones that our call for labeling was like demanding that labels be placed on grapes picked by non-union workers.[4]

A Long-Standing, Pro-Bioengineering Bias

As I eventually learned, the media's bias toward bioengineering has been long-standing. For instance, when researchers analyzed media coverage of a major GMO-related controversy in the early1980's, they found significant imbalance in how risks were reported. As they noted, "with very few exceptions, the primary bearers of risk information . . . formed dichotomous groups."[5] Even though many scientists had issued cautions, when warnings were cited by the media, they came from the mouths of non-experts, while scientists were almost always presented as assurors of safety.[6]

This portrayal of disputes about the safety of bioengineering as a conflict between scientific experts and non-expert "critics" has remained a general feature of US media coverage of genetic engineering. Extensive data indicates that, on the whole, the American media have been highly selective in the scientists they cite – and so reflective of the life science establishment that they've only been willing to associate concerns about GMOs with scientists so long as such concerns were voiced by prominent members of the pro-biotech mainstream. Accordingly, when the establishment stopped expressing concerns and began to consistently assert safety, the media started portraying such concerns as phenomena found only among non-experts, despite the fact that numerous well-credentialed biologists continued to warn about risks.[7]

The study cited above is not an anomaly. Research has consistently detected a bias toward genetic engineering in the reports of the US media. For instance, three researchers at Cornell University conducted a survey of media coverage of biotechnology from 1970 to 1996 and determined that it has been "overwhelmingly positive."[8] And a study by researchers at Texas A&M

University indicated that newspaper coverage of biotechnology is heavily dominated by the individuals in universities and industries who promote it.[9] Eight years after that study, one of the researchers noted the continuing imbalance in coverage, stating: "The bottom line is that the voices promoting (biotechnology) are more prominent than those that object."[10]

Further, it would be a mistake to blame the bulk of the biased reporting on the reporters themselves. To large extent, it's attributable to their editors – and the executives who set editorial policy. Phil Regal recounts that when he engaged reporters in extensive conversations, they often told him that they had "to be very careful" about what they submitted because their editors "were very pro-biotech." He surmises that this in part reflects the fact that several media companies have been acquired by massive corporations with substantial interests in sectors that would be adversely impacted by negative news about bioengineering. Further, as ingredients from GE crops became increasingly integrated into the processed foods on American supermarket shelves, all media corporations developed a financial stake in the public image of biotechnology, even if they did not own biotech-related businesses. That's because a substantial portion of their advertising revenue comes from food producers; and the US food industry not only promotes GE products, it aims to squelch reports that could mar their reputation. Accordingly, it's not surprising that the assertions by the Grocery Manufacturers of America, whose members account for so much of the media's advertising income, have been routinely quoted, regardless of their accuracy.

But media efforts to favor biotechnology have not been limited to linking safety concerns with ignorance and to providing a venue for the unchallenged presentation of propaganda. There has also been systematic suppression of contrary views. This practice was already well-established, and well-documented, by 2002. In that year, Food First, an institute that focuses on food policy released the results of a review it conducted of 11 major newspapers (including the *New York Times*, the *Washington Post*, and *USA Today*) and three weekly news magazines (*Time*, *Newsweek*, and the *Economist*) between September 1999 and August 2001. According to the researchers, these publications "have all but shut out criticism of genetically modified (GM) food and crops from their opinion pages."[11] They pointed out that "an overwhelming bias in favor of GM foods" was revealed not only within the editorial pages, but also on the op-ed pages, a forum usually reserved for a variety of opinions.

Commenting on these findings, the institute's co-director stated: "It is a great disservice to the American public when the media filters out critical viewpoints on issues that are central to our times. This is an issue where there is significant difference of opinion among both scientists and the general public, and those differences must be represented in the media if the public is to be able to exercise its democratic right to make informed decisions about new technologies." [12]

Moreover, media magnates have not merely promoted the GE agenda by slanting the content that's conveyed through the various outlets they control. They have sometimes wielded their considerable influence in person. A striking instance occurred during the Spring of 1998, when (as was later reported by *The Guardian*) there was a series of meetings at which the Clinton Administration put "intensive pressure" on the UK government to open up British and EU markets to America's GE foods. At two of the meetings, the American team was not limited to government officials but included top executives from the private sector, including the CEO of the US international media company, Warner-Lambert (which is now Time-Warner).[13]

Capitulation to Industry Threats

Even when the media have endeavored to report unflattering facts about bioengineering, industry threats have often deterred them. In some instances, they not only cowered, but were downright cowardly. Moreover, to achieve full appeasement, their cowardice coupled with deviousness.

A glaring example occurred in 1997, when a Fox-owned TV station in Tampa, Florida (WTVT) became so intimidated by Monsanto that it tried to fraudulently alter a report on one of that company's genetically engineered products: recombinant Bovine Growth Hormone. This product, commonly referred to as rBGH, is a drug designed to boost milk production in cows. Although the FDA allowed rBGH to enter the US market, most industrialized nations (including Canada and those in the EU) have banned it due to concerns that it could cause harmful changes to the composition of milk. One of the biggest concerns, expressed by several scientists, is that milk from rBGH-treated cows contains significantly elevated levels of a hormone called "insulin-like growth factor 1" (IGF-1). Research suggests that pre-menopausal women under age 50 with high levels of IGF-1 are seven times more likely to develop breast cancer – and that men with high levels are four times more likely to develop prostate cancer.[14]

During 1996, a husband and wife team at WTVT began to investigate the rBGH story. They both had distinguished backgrounds. Jane Akre was a former CNN anchorwoman and reporter. Steve Wilson had won three Emmy Awards for exposing hazardous defects in Chrysler and Ford vehicles. WTVT had hired them to add punch to its reporting. For three months, they dug into the facts surrounding rBGH, compiling sufficient information for a powerful four-part series. One of its startling features was an excerpt from a Canadian television program in which a government official recounted how a Monsanto representative bribed her committee by offering a big sum of money if they recommended that rBGH be approved with no further studies.[15] (In presenting Monsanto's version of the story, a spokesman stated that the officials had misunderstood the situation and that the offer was merely to furnish "research" funds.)

WTVT was enthusiastic about the series, which was due to air on February 24, 1997; and it paid for extensive radio broadcasting to promote it. But on the 21st, a Monsanto attorney faxed a letter to Roger Ailes, the head of Fox News, at his New York office. The letter was designed to dampen enthusiasm about the series, and it succeeded. It alleged that the series was biased and unscientific; and it warned, "There is a lot at stake in what is going on in Florida, not only for Monsanto, but also for Fox News and its owner."[16] In Akre and Wilson's opinion, this was the part of the letter that Ailes found most troubling. Not only was Monsanto a major national advertiser on Fox, it used the services of the Actmedia advertising agency – which was owned by Rupert Murdoch, who also owned Fox.[17] Thus, if Monsanto withdrew its advertising from Fox and also switched ad agencies, Murdoch would take a double financial hit.

WTVT was abruptly notified that its heralded series was postponed pending "further review," despite the fact it had already passed a thorough review by attorneys. But the station's general manager, a former investigative reporter, was in no mood to retreat. He and the station's lawyers scrutinized the episodes and determined that they were accurate and fair. So he re-scheduled the series for the following week.

This prompted an even stronger letter to Ailes from the Monsanto attorney. It threatened that airing the report "could lead to serious damages to Monsanto and dire consequences for Fox News."[18] This was too much for Fox, and the series was once again put on hold. Further, within a short time, WTVT's general manager and news manager were fired.

Not surprisingly, the new general manager was far less sympathetic to what Akre and Wilson were trying to accomplish; and, according to Wilson, he was deeply concerned about the advertising revenues that would be sacrificed if the series were broadcast. Besides the big losses that Fox would incur on a national scale if Monsanto pulled its advertising from the corporation, he calculated that the dairy industry and many supermarkets would reduce their advertising on WTVT.[19]

Akre and Wilson report that the manager tried to find out what they would do if the story was killed. "Would you tell anyone?" he inquired. "Only if they ask," replied Wilson.[20] The reporters relate that he subsequently offered them a substantial sum of money if they would leave the station. But he imposed an onerous condition. They would have to agree not to publish any details they'd learned about rBGH – or to disclose how Fox had dealt with the story.[21] Shocked and repelled by this offer, the reporters refused to gag themselves and deny the public the important information they had accumulated. Instead, they offered to re-write their documentary to make it more acceptable.

Although they were permitted to try, their scripts were scrutinized by attorneys at Fox's regional office in Atlanta, who routinely required that the content be rendered more favorable to Monsanto. Among other demands, the reporters were told to cut the information that the FDA had approved rBGH based only on short-term testing, to delete all mention of IGF-1, and to refrain from using the word "cancer."[22] But after six months and 73 rewrites, the attorneys still wanted to modify the language so as to mollify Monsanto. Moreover, they did not merely instruct Akre and Wilson to revise or remove statements; they ordered them to add some false ones by Monsanto's dairy research director. They even commanded them to include an assertion that milk from rBGH-treated cows was the same as milk from untreated ones. According to a London newspaper, "Monsanto insisted that this statement be aired."[23] And according to the reporters, they were threatened that they'd be terminated if it wasn't.

But they would not comply because they knew that the statement was not true. Further, they pointed out to management that Monsanto's own data demonstrated there were differences, including elevated levels of IGF-1. Yet Fox would not risk running the series without falsifying the facts, so it suspended the reporters for "insubordination," and ultimately fired them. Then, in an extraordinary step, it hired another reporter to prepare a broadcast with Monsanto's statement intact. Wilson emphasizes the aberrant nature of Fox's action: "It's no secret in journalism that stories are sometimes killed. What is so unusual and egregious about our case is that this is the first time I know of that a newspaper or broadcaster has opted not to kill a story but to mold the story into a shape that the potential litigant and advertiser would like."[24]

Akre then sued the station alleging that her dismissal was an infraction of Florida's whistle-blower statute. The jury agreed, and it found that Fox had fired her for threatening to inform the Federal Communications

Commission (FCC) about the attempt to falsify a news report. It also awarded her $425,000.[25]

Fox appealed the verdict, claiming that no infraction of the Florida statute had occurred. It noted that the whistle-blower protection only covers employees who try to report violation of a law, rule, or regulation; and it argued that there was no rule or regulation forbidding a broadcaster from falsifying the news. Akre countered that the FCC's policy against intentional distortion of the news constitutes a rule within the statute's purview.

The Florida appellate court sided with Fox, and it overturned the award of damages. It declared that the FCC policy does not qualify as a rule or regulation under the state statute and that Akre was therefore not entitled to the whistle-blower protections.[26] Thus, in effect, the judges held that while it's illegal for Florida broadcasters to retaliate against employees for reporting a minor infraction of the building code, they have an unbridled right to fire anyone who blows the whistle on deliberate falsification of the news.

Self-Censorship and the Chronic Suppression of Facts

Although it may be especially egregious in regard to its particulars, Fox's stifling of reporters to protect GE foods is *not* exceptional in terms of general editorial policy. The magnates who steer the mainstream American media have routinely sided with the promoters of agricultural biotechnology and have consistently imposed restraints that were not merely selective, but suppressive. As a result, any efforts of their subordinates to fairly inform the public have been systematically stymied. Further, while it's hard to gauge how much suppression of negative news has stemmed from overt industry threats, it *is* clear that every act of restraint was ultimately a case of media self-censorship. Each ensued from either commercial calculation or ideological commitment, not from government compulsion – the main driver of censorship in most other parts of the world.

Such voluntary muzzling on behalf of GE foods has been a constant phenomenon in the United States and has squelched stories that would ordinarily have been jumped on as journalistic gold mines. It even blocked transmission of the most momentous biotech news ever handed to the American media: the revelation that the FDA fraudulently concealed the warnings of its scientific staff about the abnormal risks of GE foods.

I'm intimately familiar with this case of suppression, because I was the one who gave the media the incriminating facts. In June 1999, at a well-attended press conference in Washington, D.C., at which several experts spoke about the problems of GE foods, I exposed the FDA's cover-up and made copies of key memos from its files available to the media. Despite the disappointing coverage of the 1998 conference that had announced the lawsuit, I was confident that things would be different this time. After all, these clear-talking, hard-hitting memos were written by FDA experts, and they refuted the basic claims the agency had been making about GE foods for the last seven years. They also proved that it had been lying. Therefore, I assumed that within a few weeks, the public would be awakened to the risks, their trust in the FDA would dissolve, and their voices would be raised in an overwhelming demand for the reform of its policy and the removal of GE foods from the market.

So did many other experienced observers. For instance, because word of my revelations reached other countries, I was quickly contacted by a reporter in Brussels who had been covering biotech issues for years. After asking me to fax him copies of some of the memos, he remarked that if they were indeed genuine, within a month the entire GE food venture would be well on its way to extinction.[27]

But he and I were mistaken. Our expectations were based on the assumption that the mainstream American media would communicate the important information it had been given. However, this assumption was too optimistic. Even though the conference was attended by reporters from the nation's major newspapers and news magazines, none of these publications informed their readers about what the FDA scientists had said – and how agency administrators had covered up the warnings and lied about the facts. And this blackout was not due to lack of interest on the part of the reporters in attendance.

For instance, Rick Weiss, a science reporter with *The Washington Post*, called me shortly after the conference and interviewed me extensively. He requested that I fax him copies of the key memos and that I also tell him how to contact several of our scientist-plaintiffs. As he prepared his story, we spoke several more times; and I had high hopes that his report would initiate a major breakthrough. But when the article finally ran, I was shocked – and deeply disappointed. There was no mention of the FDA memos, no quotes from our scientist-plaintiffs, and no indication that many experts had serious concerns about the potential toxicity of GE foods.[28] Equally egregious, although the article noted that a lawsuit had been initiated against the FDA to compel safety testing, it termed the plaintiffs "activists," with no hint that the group included nine knowledgeable life scientists.

I was just about to phone Weiss and demand to know why he had failed to include the critical information he had gathered when he phoned me. He said that he knew I was very disappointed, and he wanted me to understand that he was disappointed too. As he explained what had happened, I began to feel sorry for him. The article he wrote had exposed the FDA fraud, quoted from the memos of the scientific staff, and also quoted scientists who were plaintiffs in our lawsuit. But his editor refused to let it stand – and demanded deletions and revisions. Weiss objected, but the editor was adamant. So, with the editor's active participation, substantial excisions and revisions were made; and the article that the public read was far from the one Weiss had intended to produce.

Further, even when the first ranks of editors and executives have been eager to report the crucial facts, individuals with greater authority have quashed their attempts. A striking example occurred a few months after the press conference when a producer at CBS national news phoned me requesting an interview that would be run on an upcoming evening broadcast. He was so interested that he was not deterred when he recognized there would be significant problems in setting things up. Because I was to be interviewed by a newscaster at the national studio, the exchange needed to take place at a CBS affiliate station so that a video link could be established enabling him and me to see and hear one another. But the nearest station to where I lived was in Cedar Rapids, Iowa, almost 95 miles away. Further, a crew would have to be flown from Chicago to Cedar Rapids to coordinate everything and to tape me.

When the producer asked if I'd be willing to drive to Cedar Rapids to do the interview, I was frank. I explained that I had grown tired of spending time and energy to do interviews about the FDA fraud that were never communicated to the public. So I told him that I would only drive the 185 total miles and consume the better part of a day on the condition that quotes from the FDA scientists would be aired, along with a demonstration that the FDA had lied. He agreed to this condition. However, he pointed out that although the interview would go for about 15 minutes, it would be substantially edited down and that because the entire segment about the GE food controversy would only be on air for around five minutes, with other interviewees as well as me, I could not expect to be on screen for even that amount of time. I told him that I understood; but I made it clear that if I did the interview, I planned to spend almost all of the time reading from the memos of the FDA experts, and I wanted an assurance that some of that footage would be broadcast. He guaranteed that it would be. So I agreed to do the interview.

While I was in the Cedar Rapids studio, the newscaster who conducted the interview allowed me to quote extensively from the FDA memos and to make the other points I intended. Only his last question shifted the focus. He asked me to name some of the religious denominations that were represented in our lawsuit. So I gave a brief one-sentence answer specifying several of them.

I drove home in an upbeat mood. Because CBS had invested so much in the interview, and because of the producer's guarantee, I was confident that there would finally be national exposure of how the FDA's policy on GE foods did not conform to the findings of its scientists but instead countered them. However, my confidence was once again misplaced. When the story aired, the only statements about food safety came from the proponents of genetic engineering, and they contained the standard assurances that GE foods are as wholesome as natural ones and that the FDA had certified their safety. None of the statements from the FDA scientists made it into the broadcast; and the only time I appeared was in answering that last question about the religious denominations. Even then, my full one-sentence reply was not used. Rather, it was cut to about three seconds in which only a few of the denominations were mentioned.

It's obvious that the decision to delete all the significant information I provided did not come from the producer who set the interview up. He knew what I planned to say, he guaranteed that a significant part of it would be included, and he expended substantial resources in order to make the interview possible. Yet, despite his desire to include statements from the FDA scientists, one or more individuals in CBS's upper echelons apparently could not abide the broadcast of such potent information and forbid him from doing so, even though it entailed a substantial waste of resources – and the breaking of a promise.

Further, although a few excerpts from memos by FDA scientists were finally disseminated by a major US media outlet (17 months after the media had first been informed of them), the impact was minimal because their significance was muted. The excerpts appeared within a *New York Times* article chronicling the history of GE foods, and their appearance was remarkable because the reporter who started the article, Melody Petersen, had encountered sustained resistance from her editors.

She first interviewed me in the summer of 1999, while I was working on the lawsuit in the offices of the ICTA in Washington, D.C. She also interviewed ICTA attorneys. She was keenly interested in the FDA memos, and as she learned the facts, she recognized that the agency had effected a major cover up and that its policy on GE foods was based on dubious claims. Further, it was apparent that she wanted her article to clearly convey this information. She also wanted to contact some of our scientist-plaintiffs and present their concerns.

But her editors did not share her inclinations. In fact, they seemed intent on hindering her from producing the exposé that she envisioned. This became obvious over the following months as she and I communicated by phone. She informed me that she was repeatedly instructed to do additional research so the piece would be more balanced. And in response to her inclusion of statements from our scientist-plaintiffs, she was told to instead find scientists who were "objective," as if by expressing scientific reasons for questioning the safety of GE foods, one's objectivity was forfeited. Because, when I had last heard from her, she was still frustrated, I assumed that the article would never make it into print – or that if it did, the critical information would be absent.

Phil Regal relates that Petersen also had some long phone conversations with him, and that she contacted several others "in the know." He says that she struck him and the others as somewhat naïve because she really seemed to believe that the *Times* was going to publish the information she was garnering. By then, he had learned to expect that little of what he and other scientists said about the risks of GE food would make it past the layers of editorial review. As he recounts: "One just got used to reporters gathering material that never got published. If you didn't take it in stride it would have been very discouraging. So you just sloshed along. I came to regard it like the sand paintings of the Hopi and the Tibetans. It was a ritual that just had to be performed, even if you knew the wind would blow your efforts away. The tradition of truth-telling had to be preserved – even if the words that one spoke were not."

So both Regal and I were surprised when, on Thursday, January 25, 2001, the article was actually published.[29] And I was amazed to see that a few excerpts from FDA memos had survived the editing process. But I was not surprised by the article's basic thrust, which downplayed the safety issue and instead focused on how Monsanto's marketing and public relations mistakes had provoked so much consumer opposition abroad, and enough resistance from activists at home, to prevent the GE food venture from realizing its full potential. There were no statements from Regal or any of our other scientist-plaintiffs, and the quotes from the scientists whom the editors presumably regarded as "objective" were generally supportive of bioengineering.

Even more telling, the FDA memos were not given prominence. The article was exceptionally long, and they weren't mentioned until about three-quarters of the way through it. As is well known, newspaper articles are ordinarily structured to provide the most important information near the beginning. So if the editors had wanted to emphasize the import of the memos, they would have featured them much earlier, and quoted them more extensively. But that was not the aim of the article. It was titled: "Biotechnology Food: From the Lab to a Debacle." It was not titled: "FDA Fraud on Bioengineered Foods Exposed." There was no mention of the fact that the FDA had covered up its scientists' warnings, or the fact that it had issued lies, or the fact that the agency's files (in conjunction with our plaintiff group) demonstrated that GE foods are not *Generally Recognized as Safe*.

Although Melody Petersen had seemed to want to highlight those facts, she was eventually replaced by a journalist who was prepared to take a different tack. When the article ran, she was not listed as the author but was merely credited as having contributed reporting. If the editors had desired to shake things up by making full use of the information our lawsuit had generated, the article would have been the one that she had envisioned; and it would have had a profound effect. But as written, its impact was insubstantial; and the *Times* did nothing more to enhance it. The paper never followed up on the revelations in the FDA files and (to the best of my knowledge) printed no further reference to what agency scientists had written.

This behavior substantially differs from how the *Times* has handled other instances of FDA malfeasance. For instance, the front page of the Sunday edition for July 15, 2012 featured an article headlined: "Vast F.D.A. Effort Tracked E-Mails Of Its Scientists."[30] And it was the top story in the emailed newsfeed for that day.[31] In this incident, several agency scientists claimed that faulty review procedures had resulted in the approval of particular medical imaging devices for mammograms and colonoscopies that exposed patients to dangerous levels of radiation. And they communicated their concerns to journalists and members of Congress. The *Times'* article revealed that the FDA had used "so-called spy software" to monitor emails sent by five of the scientists; and it pointed out that while government agencies have substantial discretion to monitor employee computer use, the agency may have gone too far (and broken the law) "by grabbing and analyzing" legally protected confidential information such as attorney-client communications and whistle-blower complaints to Congress.

Moreover, the *Times* did not let the story drop. It published three follow up articles within the next two weeks. Due to such prominent and persistent coverage, the issue grabbed the attention of the public and those on Capitol Hill. Thus, the paper appears to have no penchant for protecting the image of medical imaging – and no hesitation about fully disclosing disputes between FDA administrators and scientists over the dangers of some of the devices.

But in the case of GE foods, not only was the *Times'* coverage of the warnings by FDA experts subdued and devoid of follow up, its subsequent reporting on the products has largely projected a positive picture, often presenting the industry's promotional claims as established facts. A typical piece, written by the paper's personal health columnist, was titled: "Facing Biotech Foods Without the Fear Factor." In the section headed "Ignorance vs. Progress," she dismissed food safety concerns as due to the publics' lack of knowledge. And in a following section she portrayed genetic engineering as preferable to traditional breeding. She indicated that when one plant pollinates another, that's a form of "genetic modification;" and she emphasized that it entails the transfer of "dozens, hundreds, even thousands of genes of unknown function." In contrast, she described recombinant DNA technology as "the most refined, precise and predictable method of genetic modification because the function of the transferred gene or genes is known." [32]

Since 1897, the motto of the *New York Times* has been: "All the News That's Fit to Print." But for the last few decades, this slogan has become inaccurate in at least one key respect. It would better reflect reality if it were revised to state: "All the News That's Fit to Print – Unless It Casts Doubt on the Safety of GE Foods."

Of course, to compensate for the commercial media's cowardice, National Public Radio could have dutifully stepped forward. But it did not; and, despite being given ample opportunity, it displayed at least as much reluctance to inform the public about the FDA fraud as its commercial counterparts. Perhaps such timidity stems from the fact that this nonprofit organization must seek substantial funding from private corporations – and that for many years, Monsanto, General Mills, and other businesses with a big stake in GE foods have been sponsors.

Watergate and the Pentagon Papers vs. GE Foods: A Striking Contrast in Media Attitude

The American media's willingness to suppress key facts about GE foods is even more astounding when compared to the stance it took when addressing two other major instances of government deception: (1) the Watergate cover-up and (2) the misdeeds revealed by a set of official documents about the Vietnam War.

Watergate

In the Watergate affair, which began in June 1972, operatives in league with key advisors of President Richard Nixon were caught in the act of burglarizing the National Democratic Committee headquarters; and the White House undertook a sustained effort to conceal its incriminating connections to them. But through relentless investigation, the *Washington Post* systematically dug out the facts that were being suppressed and exposed them in a series of hard-hitting articles. Further, the paper persisted in the face of repeated denunciations by the Nixon administration – and also despite its threats and harassment.[33] The *Post* also pushed on even though the rest of the media, as well as most of the country, doubted the soundness of its reports.[34] And through such courage and perseverance, it ultimately brought down several top White House officials – and even the president himself.

The contrast between the newspaper's determination to expose government deception in the case of Watergate and its distinct disinclination to do so in regard to GE foods is dramatic. In 1972, it was so strongly committed to uncover government wrongdoing, even to the point of bringing down a US president that it pressed ahead despite the danger to its reputation and its very survival.[35] But in 1999, it chose to be complicit in the continuation of a cover-up rather than expose government delinquency that would endanger the image of the biotech industry and the novel products it had placed on the nation's supermarket shelves. In order to disclose the truth about Watergate, it had exerted extensive effort and taken major risks; yet it refused to publish key truths about GE foods, even though that would not have entailed investigative effort, or been open to doubt. Despite the fact a reporter had been handed rock-solid evidence from the FDA's own files establishing that GE foods were not "generally recognized as safe" as defined by law – and that the FDA had been concealing this through fraudulent misrepresentations, the *Post* prevented him from bringing these facts to light – and excised all reference to them from the article he'd written.

The Pentagon Papers

Although the case of GE foods may mark the only time the *Washington Post* has deliberately abetted the cover-up of government wrongdoing, Watergate was *not* the only instance in which the paper incurred significant risk in order to expose it.[36] In the preceding year, it had also acted courageously in that cause; and on that occasion, it was not alone. The *New York Times* and several other newspapers were part of the same endeavor.

These news organizations were determined to publicize important information contained in a 7,000 page study of US involvement in Vietnam between 1945 and 1967. The documents comprising this study came to be known as "the Pentagon Papers" because they were prepared by a task force within the Defense Department. They were produced during the Vietnam War, and they were classified as "Top Secret – Sensitive." Fifteen copies of the study were produced, and two were given to the RAND Corporation, a think-tank that had often provided research and analysis to the Defense Department.

The study contained many unsettling facts. Not only did it reveal the government had realized at an early stage that there was little likelihood the war could be won, it also exposed extensive government deception. As a *New York Times* editor later observed, the documents "demonstrated . . . that the Johnson Administration had systematically lied, not only to the public but also to Congress, about a subject of transcendent national interest and significance."[37]

An analyst at RAND who had contributed to the study, Daniel Ellsberg, decided to make several photocopies and release them in order to hasten an end to what he considered an immoral war. In March 1971, he gave out the first copy of the study – to a *New York Times* reporter, who brought them to the attention of his superiors. Although these executives recognized the study's importance and wanted to publish some of those that were most revealing, they had doubts about the legality of doing so. Not only were the documents top-secret, they had been stolen. Further, they contained sensitive information that, if publicized, might weaken the government's capacity to conduct a major war.

So the executives sought the advice of the law firm that served as their outside counsel. What they wanted to know was: *Even though the transmission of the documents to the Times had been illegal, could publishing them nonetheless be lawful?* And what they hoped to receive was a go-ahead. But the answer they got was not encouraging. The lawyers advised them to hold off because they might be prosecuted for violating the Espionage Act of 1917, which made it a crime to transmit classified information to unauthorized persons or to publish such information "in any manner prejudicial to the safety or interest of the United States."[38]

However, their in-house lawyer saw things differently. He argued that through the First Amendment of the US Constitution, the press had the right to publish information providing citizens crucial insight into government policy. And his words won out. Instead of opting for the cautious (and comfortable) stance counseled by the outside experts, the paper's executives went with the bolder thinking of their staff attorney; and on June 13, despite the serious risks, the *Times* began publishing key excerpts from the stolen study.

An infuriated Nixon administration promptly demanded that it stop; but it refused. So the Justice Department sought a temporary restraining order from a federal court, asserting that further publication of the classified documents would cause "immediate and irreparable harm" to national defense interests. The court issued the order, and the *Times* was forced to cease after only three articles had been published.

However, that did not prevent the sensitive documents from appearing in a major newspaper. Ellsberg had also given some of them to the *Washington Post*, and it began to publish excerpts on June 18.

The Justice Department again obtained a restraining order. Meanwhile, the *New York Times* sought to remove the one that had been issued against it by filing an appeal. On June 26, the US Supreme Court agreed to hear its case – in conjunction with the case involving the *Washington Post*. Four days later, the Court ruled that the government could not enjoin publication because it had failed to meet the "heavy burden of showing justification for the enforcement of such a restraint."[39] However, five of the nine justices stated that even though the government could not restrain the newspapers from publishing the secret documents, it could nevertheless criminally prosecute them for having done so.[40]

Moreover, during the period the *Times* and *Post* had to desist because of restraining orders, other newspapers showed no inclination for restraint. Ellsberg had passed copies of the study to several, and they too began to publish. The *Boston Globe*, the *Chicago Sun-Times*, and the *St. Louis Post-Dispatch* soon got into the act; and the beleaguered Justice Department could not stay on top of things. According to a report in *Air Force Magazine*, "As soon as one newspaper was enjoined, the next one picked up publication."[41] In all, fifteen papers in addition to the *Times* and *Post* joined the effort.[42]

As with Watergate, there are glaring contrasts between the case of the Pentagon Papers and the case of the FDA's GE food files. Although both sets of documents revealed that the government had lied about matters of great importance, the similarity stops there. The former were top-secret classified documents that were stolen and then passed to newspapers in violation of federal statute. Not only was there reason to think that publishing them could hinder the nation's war effort, it seemed that any newspaper executives who did would be technically guilty of treason. Nonetheless, despite the risk of criminal prosecution, two of the nation's most influential papers started publishing the secret report; and when the government restrained them through the courts, fifteen other newspapers leaped in to keep the documents rolling off the presses – even though the legality of their actions was still in doubt and the threat of prosecution loomed lively.[43]

In contrast, the FDA files were unclassified and irrelevant to national security. And instead of having their dissemination blocked by a court order, they were *obtained* through a court order and presented to the media in a legal manner; and there was never any doubt they could be lawfully published. Moreover, although the Pentagon Papers had no direct bearing on the health or well-being of any citizens outside the military, the FDA papers did, since they revealed that foods being regularly consumed by most Americans pose unusual risks.

Yet, the managers of the media consistently prevented the content of these documents from being conveyed to the public. Even though their actions would have been perfectly lawful and highly laudable, they refused to inform the citizenry that the government had been defrauding them – not when the fraud had been essential in getting GE foods on the market and was still essential for keeping them there. In 1971, the nation's newspapers had courageously published the Pentagon Papers despite the risk of criminal prosecution; but from 1999 onward, they (along with the other organs of the mainstream media) have balked at disseminating the documents that would expose the government's GE food frauds – for motives presumably far less noble than those operative during that earlier episode.

It's also noteworthy that, like the deceptions revealed in the Pentagon Papers, the Watergate cover-up did not pose a threat to public health either. Yet the *Washington Post* was committed to uncover the deceptions – while it would later remain complacent in the face of a fraud that *did* put the health of the nation's citizens at risk. It's also noteworthy that, in prelude to the 40th anniversary of Watergate, Leonard Downie, Jr., the *Post's* executive editor during the period when its apparently protective (and restrictive) stance on GE foods was adopted, wrote an article warning that due to the pressures of the digital age, the future of investigative reporting is "at risk." Unaware of the inherent irony in his effort, the man who directed editorial policy when Rick Weiss's attempted exposé was quashed sought to enhance appreciation for the very kind of responsible reporting that the *Post* had thwarted in that instance – asserting that America would be "best served . . . [through] widespread recognition of the importance of accountability journalism in our democracy – and the need to ensure that it survives and flourishes."[44]

Although one might question the consistency of Downie's practical position on the matter, his declaration about the importance of "accountability journalism" in a democracy is beyond dispute – and the centrality of its role has been widely and regularly recognized. One of the most forceful acknowledgements was issued as part of the Supreme Court's ruling that the government could not place prior restraint on the publication of the Pentagon Papers. In his oft-cited opinion, Justice Hugo Black declared: "Only a free and unrestrained press can effectively expose deception in government."[45] In that case, the deception was exposed because the media heroically fought off the imposition of government restraint. But in the case of genetically engineered foods, the media have voluntarily (and spinelessly) imposed restraints upon themselves. While such self-censorship would have been at home in Stalin's Soviet Union, it's incompatible with the needs of an open and democratic society. And because of its persistence, one of the biggest and most dangerous deceptions ever perpetrated by the United States government has been allowed to maintain its force more than fifteen years past the point it should have been fully brought to light.

Another Major Contrast: America's Media vs. Europe's

Besides the dramatic contrast between how the American media covered Watergate and the Pentagon Papers and how they've dealt with critical information about GE foods, there's an enormous difference between the latter and the corresponding performance of the media in Europe. In Chapter Two, we saw that the European media were much less prone to print unsubstantiated promotional claims as established truths, and more willing to scrutinize and challenge them. And when I started giving press conferences on that continent, I quickly learned that the media there were also more willing to unstintingly report the facts, regardless of the effect on the image of GE foods. I had become so accustomed to the suppression prevalent within the US that it was initially somewhat overwhelming to see stories about the FDA's misbehavior featured on the front pages of major newspapers and in the prime-time broadcasts of mainstream television. As such coverage continued it became clear that the free functioning of the European media on the GE food issue had significantly facilitated the informed consumer resistance that had kept most of these foods out of their supermarkets. And I was increasingly struck by the irony that although revelations about this major fraud were being openly disseminated within countries minimally affected by it, the citizens of the nation that was most directly affected – and adversely impacted – were being kept in the dark by the irresponsible policies of the press and the broadcasters.

Thus, due to the malfunction of the American media, false notions about GE foods continue to prevail. Even though the managers of mass communication have been given incontestable evidence of a government fraud that's a reckless gamble with public health, they've consistently failed to report it. While they have accommodated the biotech proponents and uncritically communicated their misrepresentations as fact, they have recoiled from conveying what the FDA scientists said. And they've given scant coverage to other scientists who have likewise expressed doubts about GE foods.

Accordingly, most Americans have remained under the illusion that GE foods are substantially the same as conventional ones – and that they are generally recognized as safe. And they're still not aware that the FDA's experts determined that these products pose unusual risks.

Moreover, the publics' confusion about risks runs far deeper than this; and it is not merely the result of deficient media reporting. Over the years, the proponents of GE foods in industry and academia have persistently misrepresented the risks that are associated with them – and done it so methodically that they've even distorted the very concept of what risk is.

CHAPTER NINE

Methodical Misrepresentation of Risk

Oversights, Anomalies, and Delinquencies

As the previous chapters have revealed, the risk that has most troubled the proponents of GE foods is the risk these products will be perceived as abnormally risky. Consequently, this is the risk they have primarily sought to manage. And in the process, they've consistently contorted both logic and the basic facts.

This tendency toward distortion has persisted from the birth of bioengineering to the present. As we saw in Chapter 1, during the early years of the technology, when the range of GMOs was limited to microorganisms, the molecular biology establishment calmed public concerns and fended off regulation by imparting false impressions about risks. Not only did they feign there was an overwhelming expert consensus that research employing engineered microbes is safe, they claimed that this alleged consensus was based on new evidence, when in fact no such evidence had been generated and the claims of safety were essentially based on conjectures. Further, one of the founders of rDNA technology declared he had demonstrated that it merely mimics what natural processes are already doing, despite the fact the purportedly natural process in his experiment was highly unnatural – and even then failed to replicate the radical results of genetic engineering. Moreover, when research with actual bearing on risks was eventually conducted, the biotechnicians suppressed the outcomes that were inconvenient – or significantly misrepresented them.

Then, as we saw in Chapter 2, when genetic engineering expanded to the plant kingdom, the range of misrepresentation also expanded. As the technology's advocates strove for unregulated release of engineered organisms, they argued that virtually all such entities would be environmentally safe. But although these arguments were advanced in the name of science, they relied on assumptions that were scientifically unsound. Further, most advocates continued to assert these arguments even after scientists who did have the relevant expertise had solidly refuted them.

Adding to the corrosion of science, a 1987 report by the National Academy of Sciences declared that GMOs pose no unusual risks, despite the fact this declaration was far more political than scientific. Unabashed, the Academy issued a follow-up report in 1989 that again made broad pronouncements about the safety of GMOs that were scientifically unwarranted. Exacerbating these excesses, biotech proponents presented the unjustified generalities as authoritative conclusions about all aspects of GMOs, even though the scope of the reports was restricted to environmental risks of field trials conducted within the continental US.

Chapter 3 revealed how misinformation became even more crucial to the advance of bioengineering after a food supplement produced through the technology was linked to a deadly epidemic that swept the US in 1989. Although the evidence implicates the engineering process as the most likely cause of the contamination that turned the supplement toxic, the proponents of the technology have spun the impression that it's been fully absolved. Moreover, they've even been able to becloud the fact that it played a role in the lethal supplement's production.

Yet, although obfuscating the facts about the disaster a GE product caused was essential in paving the way for their commercialization, it was not sufficient; and Chapter 4 showed how biotech proponents have needed to distort the story of how engineered foods are created. Through such misrepresentations, and through misstating basic facts of biology, they've imparted the illusion that these novel products are essentially as natural as – and no riskier than – their conventional counterparts.

We next saw, in Chapter 5, how in order for GE foods to be allowed onto the market, the United States Food and Drug Administration had to re-assert the claims, first issued by biotech proponents during the 1970's, that there's an overwhelming scientific consensus about the safety of genetic engineering – and that this consensus rests on solid evidence. Further, although the earlier claims related to research with microbial GMOs while those in 1992 involved the wholesomeness of engineered plants as food, the latter were just as bold – and bogus – as were their predecessors. In fact, they were more brazen, since even the FDA's own files attested their falsehood.

Chapter 6 went on to demonstrate that regulatory irregularity in regard to GE food was not confined to the Unites States. We saw how officials in Canada, Europe, Australia, and Asia employed superficial reasoning to justify superficial testing – turning their backs on sound scientific knowledge and standard scientific protocols in order to grant approvals to GE foods and project the impression that they're no more dangerous than conventional ones.

Chapter 7 then showed how the environmental risks of GE crops have been routinely underestimated and mismanaged. And Chapter 8 revealed how the American media have, through suppression of key information and slanted reporting of the remainder, presented a significantly imbalanced picture of both the health and environmental risks posed by engineered organisms.

Now it's time to take our examination deeper. It's time to more clearly understand why there has been, and continues to be, major disagreement about the risks of GE foods within the scientific community. We need to discover how numerous esteemed scientists and scientific institutions can declare that the risks of GE foods are similar to those of conventionally produced foods while many equally qualified experts maintain that the risks are greater. Have both groups been examining the same set of evidence according to the same scientific protocols and with an equal degree of logical rigor? If so, it will indicate that the available evidence does not yield clear-cut answers and can be interpreted in different ways by equally earnest experts. Or has one group overlooked significant facts, ignored important standards, and applied loose logic? If this is the case, it would show that the disagreement stems less from a lack of adequate data than from a lack of scientific integrity.

But before proceeding, it's important to recognize that there are many more experts counseling caution – and far fewer reports declaring safety – than the public has been led to believe.

Misrepresenting the Degree of Consensus that GE Foods Are Safe

GE proponents present themselves as champions of science, and they promote the impression that scientists and scientific institutions have uniformly concluded that GE foods are safe. But to do so, they've needed to misrepresent the record – to the extent that reports from scientific conferences and esteemed institutions have frequently been twisted to make it seem they contain definitive conclusions about the safety of GE foods even if they do not. Further, the mischaracterizations are frequently propagated by organizations wielding significant authority.

For example, in October 2012, the board of directors of the American Association for the Advancement of Science (AAAS) issued a statement in opposition to the labeling of GM foods that, as part of its argument for their safety, asserted that the World Health Organization has determined GE foods to be "no riskier" than conventionally produced ones.[1] But in reality, the WHO has stated that "it is not possible to make general statements on the safety of all GM foods" and that their safety should therefore be assessed on a case-by-case basis.[2] Moreover, the WHO noted that while safety assessments are not required for "traditional" foods, most national authorities require them for bioengineered products – and that one of its objectives is to assist in this process.

Truth notwithstanding, the AAAS statement went on to proclaim that "every respected organization" that has examined the evidence has likewise determined GE foods to be "no riskier" than conventional ones. But in fact, several respected scientific organizations *have* concluded otherwise.

For instance, in 2001 the expert panel of the Royal Society of Canada (that nation's academy of science) issued an extensive report declaring that (a) it is "scientifically unjustifiable" to presume that GE foods are safe and (b) the "default presumption" for every GE food should be that the genetic alteration has induced unintended and potentially harmful side effects.[3] In describing the report's criticism of the current approach to regulating GE foods, the *Toronto Star* stated: "The experts say this approach is fatally flawed . . . and exposes Canadians to several potential health risks, including toxicity and allergic reactions."[4]

The British Medical Association is another respected organization that has expressed reservations about the safety of these novel products.[5] As described in the *British Medical Journal*, the Association released a 2004 report stating that "more research is needed to show that genetically modified (GM) food crops and ingredients are safe for people and the environment and that they offer real benefits over traditionally grown foods."[6] And, as Chapter 6 noted, the Public Health Association of Australia (PHAA) has expressed concerns about the risks posed by the bioengineering process and has strongly criticized what it regards to be the "flawed" regulatory process through which GE foods have gained approval.[7]

The eminent medical journal *The Lancet* has also criticized the presumption that GE foods are no riskier than conventional ones. In an editorial titled "Health risks of genetically modified foods," it stated that there are "good reasons to believe that specific risks may exist;" and it declared that "governments should never have allowed these products into the food chain without insisting on rigorous testing for effects on health."[8]

Moreover, as Chapter 5 demonstrated, it's appropriate to include the US Food and Drug Administration among the organizations whose scientific reviews have determined that the risks of GE foods should not be equated with those of conventional ones. On the basis of a thorough analysis, the scientists on the agency's biotechnology task force concluded that these new products pose an unusual degree of risk. However, they were not empowered to express their conclusions in the agency's policies on, and statements about, GE foods. Instead, as we saw, those policies and statements were issued by administrators and were skewed by political and economic factors divorced from science. Consequently, the FDA's official statements about GE foods have not only disregarded the conclusions of its own scientists, but contradicted them. Therefore, although it's common practice for biotech proponents and media personnel to assert that the FDA has determined GE foods to be safe, because such assertions imply that the determination was a scientific one, they're inaccurate. In fact, they're doubly mistaken, since the agency has not even made a formal determination about GE foods. It has merely made a "rebuttable" presumption that they're generally recognized as safe – a presumption that has always clashed with reality, and has grated ever more glaringly as cautionary reports like those just cited have been released.

Of course, the reports by the FDA scientists were suppressed, so the AAAS board can be excused for not knowing about them. But those of the WHO, the Royal Society of Canada, the British Medical Association, the PHAA, and *The Lancet* were public, and quite noteworthy. Yet, they were carelessly overlooked – or callously misrepresented. Further, as we shall see, the AAAS statement is peppered with other falsehoods.

Although such fact-fracturing is regrettable, especially when committed by the directors of one of America's premier scientific organizations, by now it should not be surprising. As we've repeatedly seen, pro-GE scientists tend to be blinkered by their biases – and to boldly express them via assertions that are overly broad or utterly bogus. Moreover, the AAAS board was chaired by one of the most biased of biotech boosters: Nina Fedoroff. As Chapter 4 demonstrated, when it comes to GE foods, Dr. Fedoroff apparently harbors several illusions, and has promulgated several delusions, one of which is that only a "few" scientists have serious doubts about their safety – and that "they are rarely those who know this new science well."[9] Evidently, this erroneous notion engendered the board's invalid claim that "every respected organization" has sided with the position it espouses.

The Canadian Experts vs. the NAS: Assessing the Asymmetries, Disclosing the Anomalies

So, despite what the proponents of GE foods would have us believe, it's clear that there is not now, nor has there ever been, a scientific consensus about the safety of these novel products – which brings us to our investigation into how the substantial split in expert opinion has come about.

Two prominent reports embody that split: the 2001 report by the Royal Society of Canada and one issued in 2004 by the US National Academy of Sciences.[10] Although each sets forth an extensive case, the former maintains that GE foods should be treated as riskier than their conventional counterparts, while the second asserts that the risks are essentially the same.

Comparing the procedures through which these contrasting reports were generated should be illuminating. It can reveal whether the processes of analysis have been substantially similar and the only differences lie in how the final judgment calls were made (as when a judicial panel examines a particular case and issues a divided opinion), or whether instead the main differences reside in the process of analysis itself (as if two separate panels of judges had tried to determine an issue by considering related, yet distinct, sets of evidence according to dissimilar judicial standards). Through this inquiry, we can clarify how strongly each of the clashing conclusions is backed by sound scientific thinking – and to what extent (if any) the scientific basis of one or the other is illusory.

The Canadian Report

The report issued by the Canadian expert panel took issue with the common characterization of genetic engineering as more "precise" than traditional breeding; and it pointed out that the implication of safety imparted by the term is unwarranted. In the process, it critiqued the way regulators in Canada and other nations have been assessing and approving GE foods.

The panel's reasoning can be summarized as follows:

- Genetic engineering can only be deemed precise in cases where its effects are limited to those that are intended (and predicted) – and do not perturb the activities of the organism in unexpected ways.
- However, while the technology's advocates advance a "simple linear model [in which] the action of one gene and its products will have no significant effects on other genes, gene products, or metabolic functions . . . , empirical evidence suggests that linear models are not good predictors of complex biological systems, which involve extensive interactions between cellular components at all levels."[11]
- "It is clear that living cells are exquisitely tuned to both their internal and external environments. Perturbations in either will typically induce a spectrum of changes in gene expression, protein synthesis and metabolic patterns. . . ."[12]
- Moreover, "[m]utations in single genes have long been known usually to produce multiple effects . . . within the mutated organism."[13]

- There are several ways in which "collateral impacts" could arise from bioengineering. One is via the "strong" viral promoters that are affixed to the inserted genes in order to induce a high level of expression. Among their drawbacks, these promoters impose a "metabolic cost to the plant of having to accumulate products unnecessarily." [14]
- In light of all the facts, "[t]he default prediction for the impacts of expression of a new gene (and its products) within a transgenic organism would therefore . . . be that this expression will be accompanied by a range of collateral changes in expression of other genes, changes in the pattern of proteins produced and/or changes in metabolic activities." [15]
- Consequently, genetic engineering cannot be considered more precise than conventional breeding, because it's more likely to cause a greater extent of unintended alterations in the organism's cellular processes – some of which could be harmful.
- Extensive experience indicates that the products of conventional breeding have almost always been safe to eat. In contrast, not only do we lack such an experiential basis with GE foods, we also lack a sound theoretical basis for presuming that they'll be safe – but do have good reason to view them with caution.
- Accordingly, we can only deem a particular GE food safe if comprehensive testing has demonstrated it to be. And such tests must be able to detect unintended results that might not be evident through either visual observation or the kinds of chemical analysis that are currently relied on. There must be "direct testing for harmful outcomes" employing the whole food, not merely an extract of the expected new substance(s). Tests should include those for "short and long-term human toxicity, allergenicity and other health effects." [16]
- Only when a GE food has successfully completed this comprehensive testing should it be declared "substantially equivalent" to its conventional counterpart. Absent this testing, such a declaration is "scientifically unjustifiable." [17]
- But such testing is not currently required. Further, regulators are not even prepared to implement it. There are no "validated study protocols currently available to assess the safety of GM foods in their entirety . . . in a biologically and statistically meaningful manner." Thus, it's imperative to develop "practical and scientifically robust approaches for the safety assessment of such foods." [18]

The NAS Report

The report released by the National Academy of Sciences in 2004 stood in stark contrast to its Canadian predecessor. Whereas the Canadian experts had concluded that the "default prediction" for every engineered food is that "collateral changes" have occurred that could make it riskier than its conventional counterpart, the NAS panel asserted that the engineering process is not inherently riskier than conventional breeding. Therefore, although the Canadian scientists stated it's "scientifically unjustifiable" to presume that a GE food is substantially equivalent to its counterpart without establishing equivalence through rigorous testing, those of the NAS declared that it's "scientifically unjustified" to require such testing merely because the food was produced through genetic engineering.

Discerning How the Differences Developed

How did these opposite outcomes arise? Both groups purport to have reached their conclusions based on careful examination of the best available evidence. Accordingly, if each group had actually done so, it would imply that one had taken account of evidence not considered by the other. And because the NAS report was published three and a half years after the Canadian one, it would suggest that during that period, important data had been generated indicating that GE foods are not in fact more likely to cause harmful side effects than are conventional ones.

Is this the case? As it turns out, one report *does* discuss substantially more evidence with direct bearing on the risks of GE foods than does the other. However, it's not the one that came latest. Contrary to reasonable expectations (and as we shall see), the Canadian report is bolstered by more such evidence, even though it was produced much earlier than the other.

Moreover, not only does the NAS report disregard a considerable store of data that counters its main contentions, it proffers scant evidence to prop them. Equally odd, besides its evidentiary deficiencies, it lacks sound logic. As we'll see, its main arguments are in key respects incoherent and even inconsistent – sometimes to the extent of severe self-contradiction. Moreover, in trying to establish that GE foods present minimal risk, they contort the very meaning of the term.

Because this semantic shift is pivotal, and because it occurs covertly, it's important to examine it carefully. And to do that, we first need to get a firm understanding of what risk is.

Covertly Contorting the Concept of Risk

As we discussed in Chapter 7, calculating risk is a process of identifying potential problems, assessing the probability that they will occur, and gauging the degree of harm that would result. In performing such analyses, *hazards* are distinguished from *risks*. Although in common parlance the two terms are used interchangeably, in technical contexts, they have distinct denotations.[19] A *hazard* is a specific aspect of an object or situation that has potential for harm, while the *risk* posed by the hazard is a function of (1) the likelihood that the potential will be actualized, multiplied by (2) the severity of the consequent harm. For instance, walking in the open spaces of Arizona entails the hazard of receiving a lethal snake bite. So does strolling the fields of Ohio. But despite the presence of the same hazard, the risks are hardly the same, because a far greater number of venomous snakes are slithering about in Arizona. Therefore, an Arizona amble entails a higher risk of fang-induced death.[20]

Therefore, in assessing risk, it's necessary to consider not only the hazards, but the probability they will actualize. This is underscored by the fact that one methodology can entail every hazard posed by another, and even some additional ones, and yet be much safer. A comparison of flying with driving provides a striking example.

On first impression, air travel appears more perilous. Not only does it entail virtually every hazard inherent in automobile travel, it includes *additional* dangers. As do cars, airplanes can collide with land-traveling vehicles (the vans and trucks that traverse the runways, and other planes taxiing them), and such on-the-ground collisions have become the biggest problem.[21] Also like cars, they can suffer the consequences of brake failure, blow a tire and skid out of control, and so on. But, unlike cars, they can plummet to earth from great altitudes due to a variety of mishaps that are uniquely aeronautic (such as hitting a flock of birds). So, if we focus only on the mere presence of hazards, because air travel entails more of them than does traveling by car, we'd be compelled to conclude that it's more dangerous. (Or else we'd have to believe that, on average, the hazards of aviation cause less harm when they manifest than do those of driving – which is obviously false.)

However, we know from extensive data that air travel is much safer – to an astounding degree.[22] And that's because, per vehicle mile, there's a much lower probability that any of its hazards will actualize. In contrast, although the hazards posed by automobile travel are fewer in number, the likelihood that one or another will result in concrete catastrophe is significantly greater. Therefore, although the hazards of flying are more numerous (and in some cases much scarier) than those of driving, flying entails far less risk.

Thus, merely identifying hazards (potential problems) does not in itself reveal what risk is posed. For that, we need to know the probability that the potential problems will actually do damage. And, as earlier noted, we also need to gauge how extensive that damage is likely to be.

However, although the concept of risk is easy to grasp, the NAS has been chronically incapable of correctly applying it in the case of genetic engineering. This deficiency dates from the academy's first formal statement about the technology: its position paper issued in 1987. As presented in the document's overview, there were two "key findings."

1. "There is no evidence that unique hazards exist either in the use of R-DNA techniques or in the movement of genes between unrelated organisms."
2. "The risks associated with the introduction of R-DNA engineered organisms are the same in kind as those associated with the introduction of unmodified organisms and organisms modified by other methods."[23]

Then, to ensure these findings were implanted in public awareness, they were restated at the end of the document, topping the list of conclusions.[24]

Yet, despite being issued by an eminent scientific body, the combined claim of these conclusions is illogical. Even if we grant that the hazards of genetic engineering are the same as the hazards of conventional breeding (which, as will be shown, is not true) it provides no basis for declaring that the risks are the same. As we've seen, merely establishing that two situations entail identical hazards does not entail that the risks are also identical. To gauge the relative risks, deeper analysis is required.

Of course, it's possible that the NAS committee might have engaged in this type of analysis. But the facts weigh against it. First, examination of the document provides no indication that a careful scientific analysis was made. Second, as we saw in Chapter 2, when Phil Regal confronted one of the authors about the way scientific issues had been mishandled, the man refused to engage in a discussion on that level because, as he explained, the paper was not primarily scientific but "political." Third, the very language of the conclusion about risks reveals that the committee had not seriously considered them.

That conclusion declared the risks of GE organisms to be "the same in kind" as those of other organisms. But such a declaration is not scientific, because it speaks of risks as if they're hazards, despite the difference between the two concepts. As we've seen, while hazard identification involves recognition of various kinds of dangers, risk assessment does not primarily deal with *kind* but with *degree*. It seeks to measure the probability that one or another kind of potential danger will actually manifest. It doesn't focus on quality but on quantity. It does not merely ask "What kind?" but "How much?"

So, by proclaiming that the risks are the same "in kind," the committee demonstrated that it hadn't ventured to move beyond hazard identification and engage in genuine risk assessment. It also disclosed that either it was seriously confused about the concept of risk – or was intent on confusing others.

The Academy increased the confusion with its next report on GMOs in 1989. Whereas the first report had conflated hazards and risks, the second blurred the boundaries between genetic engineering and conventional breeding. In fact, it attempted to equate the processes, asserting that there's "no conceptual distinction" between them.[25] As we saw in Chapter 4, although this statement is logically absurd, it has been extensively employed by biotech advocates – and passed off as an authoritative scientific conclusion.

The Academy could have set things straight in the report it released in 2000. But instead, that report endorsed the loose thinking of the initial one – and even expanded its scope. Because that 1987 report (as well as the one issued in 1989) was limited to the environmental risks of field trials within the continental US, its statement that the risks of engineered organisms "are the same in kind" as those of conventional ones was, as a technical matter, similarly restricted. But the authors of the 2000 report not only repeatedly (and approvingly) cited that flawed statement, they extended it to the issue of food safety.[26]

However, although the scope of the 2000 report was broader than either of its predecessors, it was nonetheless limited. It only dealt with plants engineered to resist pests, such as those endowed with the Bt gene (which were discussed in Chapter 7). So there was still a need for an NAS report that applied to GE crops in general.

The one issued in 2004 addressed that need; and it was requested and funded by three federal agencies: the FDA, USDA, and EPA.[27] Yet, although its range was wider than the 2000 report, its thinking was just as narrow.[28] And its reasoning about risks was even more muddled.

Straining to Uphold the 'Product Not Process' Doctrine

Such muddling was the result of a sustained (and strained) attempt to establish that the products of genetic engineering should not be regulated any differently than those of conventional breeding. This notion was the report's central message, and it was unequivocally expressed in the first sentence of the press release heralding the document's publication.[29] That sentence announced, as the key conclusion of the authoring committee, that all new varieties of food, whether derived through conventional breeding or genetic engineering, should be assessed for safety "on a case-by-case basis," not on the basis of what technique produced them. The next sentence provided the rationale for that conclusion: ". . . because even traditional methods such as cross-breeding can cause unexpected changes."

To drive this message home, the second paragraph of the release elaborated it in a statement from the scientist who chaired the committee:

> "All evidence to date indicates that any breeding technique that alters a plant or animal – whether by genetic engineering or other methods – has the potential to create unintended changes in the quality or amounts of food components that could harm health. The possible impact of such compositional changes should be examined on a case-by-case basis to determine whether and how much further evaluation is needed."

A few paragraphs later, the theme was again sounded in highlighting the report's lead finding: that subjecting GE foods (as a class) to a higher level of safety testing than conventionally produced ones is "scientifically unjustified."[30]

And, to assure that the basic message was clearly disseminated, it was also repeated when committee members spoke with the media. For instance, in a telephone news conference, one member declared: "The most important message from this report is that it's the product that matters, not the system you are using to produce it."[31] Major news outlets such as the *New York Times* gave this statement prominent placement in their stories.

Thus, the 2004 report vigorously upholds the maxim, first issued as a directive to federal agencies by the Reagan Administration in 1986 (and routinely voiced by biotech proponents thereafter), that regulation should focus on the product, not the process. However, as we saw in Chapter 2, while this phrase is well-worn, it is not well-founded; and although it was presented as a scientifically backed principle, it did not derive from sound scientific analysis but from a purely political and economic agenda.

Nonetheless, many influential documents have subsequently espoused it, to the detriment of their purportedly scientific nature. One of the most important was the FDA's 1992 policy statement, which presumed that all

GE foods are so safe they need no testing. As we saw in Chapter 5, an FDA compliance officer protested the agency was "... trying to fit a square peg into a round hole ... [by] trying to force an ultimate conclusion that there is no difference between foods modified by genetic engineering and foods modified by traditional breeding practices." And she blamed this forcing on "the mandate to regulate the product, not the process."[32]

Although the committees that wrote the various NAS reports on genetic engineering were not officially mandated to embrace this '*product not process*' mantra (as was the FDA), they have nevertheless chosen to do so – while endeavoring to enhance its aura of scientific authority. The maxim was propounded as the "fundamental principle" of the Academy's first report on GE crops in 1987, and the authors of the one that followed in 1989 indicated their intent to "reemphasize" it – which they did unstintingly.[33] It was emphasized again in the report of 2000; and the one released in 2004 firmly upheld the tradition. Yet, as we shall see, that report could not uphold the maxim while also upholding science – or logic; and the committee that produced it floundered in the face of inconvenient facts.

For instance, although the authors of the 2004 report stressed that every method of breeding can produce unintended effects, they also had to acknowledge that some are more likely to do so than others – and that genetic engineering has one of the greatest tendencies. Nonetheless, in order to support the notion that this erratic technology should not be subject to stricter standards, they maintained that it does *not* pose greater risks – even compared to the simplest, most natural process employed since ancient times.

But in striving to sustain this contention, their arguments became circuitous, and sometimes ludicrous.

Loose Language and Lame Logic

In trying to make their case, they produced a chart presenting their conclusions about how the various modes of plant breeding differ in propensity for unintended effects.

See Figure 9.1, which is a reproduction of the chart. It can also be viewed at: http://www.nap.edu/openbook.php?record_id=10977&page=64

Figure 9.1 COMPARATIVE CHART FROM THE NAS REPORT OF 2004

Relative likelihood of unintended genetic effects associated with various methods of plant genetic modification.

1. Selection from a homogenous population
2. Selection from a heterogeneous population
3. Crossing of existing approved plant varieties* (includes all methods of breeding)
4. *Agrobacterium* transfer of rDNA from closely related species
5. Conventional pollen-based crossing of closely related species
6. Conventional pollen-based crossing of distantly related species and/or embryo rescue
7. Somatic hybridization
8. Tissue culture (somaclonal variation)*
9. Biolistic transfer of rDNA from closely related species
10. *Agrobacterium* transfer of rDNA from distantly related species
11. Biolistic transfer of rDNA from distantly related species
12. Mutation breeding via chemicals or ionizing radiation*

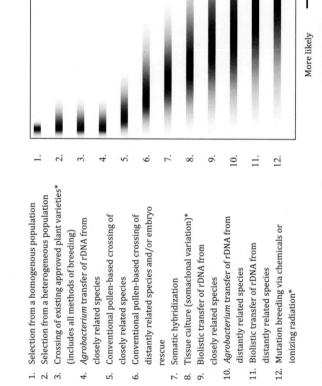

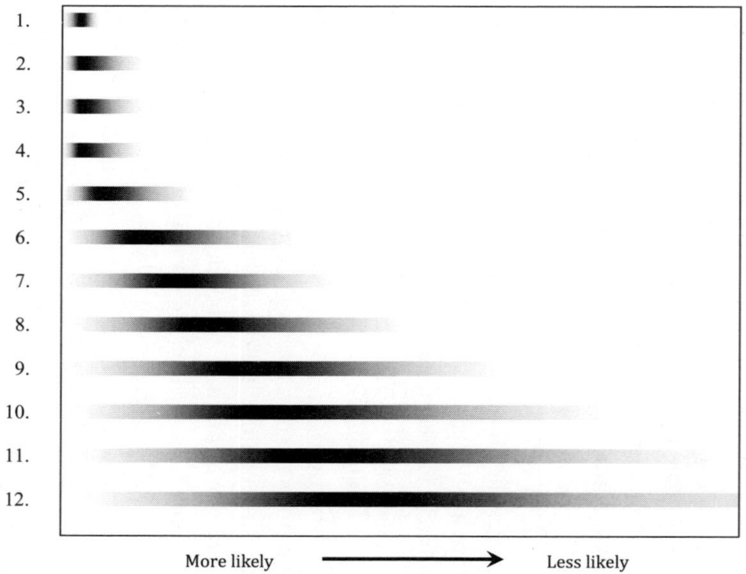

* These three entries have been slightly modified for purposes of clarification. In #3, the original wording has been maintained, and the only change involved repositioning the text in parentheses so it's easier to see.

Numbers have been added along the vertical axis for the sake of clarity.

Reproduced with permission from "Safety of Genetically Engineered Foods: Approaches to Assessing Unintended Health Effects," National Academy of Sciences Courtesy of the National Academies Press, Washington, D.C. (2004), p. 64.

© Copyright 2004 National Academy of Sciences

The modes are listed along the vertical axis, descending from lowest to highest propensity, with a shaded bar alongside each to indicate the magnitude of its likelihood for unintended effects. Consequently, each successive bar is longer than the one above it. According to the report, the "gray tails" indicate "the range of potential unintended changes" while the dark sections indicate "the relative degree of genetic disruption."[34] Although the meaning of these statements is not crystal clear, it seems the committee is saying that the entire length of the line represents the number (and kinds) of effects that the method is likely to induce while the dark areas represent the net severity of the various disruptions. But there's no need to belabor this issue, because the longer the composite bar, the greater the dark area within it; so one can focus on overall length as the key distinguishing feature.

The mode at the top of the list has the shortest bar – and the longest history. It's simple selection: the time-honored practice of letting plants propagate naturally and then choosing those that are most desirable to provide seed for the next generation.[35] According to the report, this method "is least likely to express unintended effects, and the range of those that do appear is quite limited." At the list's bottom is a category that includes two techniques of inducing mutations: one via radiation and the other via chemicals. These techniques were assigned the longest (and darkest) bar because, in the committees' words, they are "the most genetically disruptive and, consequently, most likely to display unintended effects from the widest potential range. . . ."[36]

Genetic engineering appears at more than one position because it can be performed in more than one manner; and, in the committees' opinion, each has a distinct disruptive potential. As Chapter 4 explained, there are two main methods for transferring recombinant DNA into plants. One employs pathogenic bacteria to insert the rDNA by infecting an isolated array of plant cells; the other blasts the rDNA into the cells with a gene gun.[37] Because research has shown that the latter tends to cause more extensive side effects, the committee ranked it as more disruptive than the former. Moreover, they also deemed that the degree of biological disparity between the organisms involved in the genetic transfer makes a difference. Therefore, they classified bacterial transfer between closely related species as less disruptive than bacterial transfers between distantly related ones, while recognizing a similar distinction for gene-gun transfers. Thus, four modes of genetic engineering appear along the vertical axis, at positions 4, 9, 10, and 11 (specified by method of insertion as well as distance between the participant species), with gene-gun transfer between remotely related organisms ranked as the most disruptive of the four – and second most disruptive overall, just short of radiation and chemical-based breeding. Bacterial transfer between distant species comes in a close third, with an adjoining bar that's almost as long as that of its ballistic-based counterpart.

However (as we shall see), despite the committees' pretension of precision, the rankings for genetic engineering are deeply flawed, and all of its modes are depicted as substantially less disruptive than is actually the case. But even if these rankings *are* accepted as accurate, so long as the standards of reason are adhered to, the accompanying claims cannot be. In one of the most astounding, the committee tried to allay concerns their chart may have raised by assuring the reader that even if a technique increases the likelihood of unintended effects, it does *not* thereby increase risk. As they put it: "Placement along this continuum has no bearing on risk of adverse outcomes, but only on the probability of unintended changes, which need not be hazardous." [38]

However, while it is true that not all unintended changes in plant breeding will be harmful, it's absurd to assert there's no connection between a process's propensity to induce them and the degree of risk that's posed by its products. A little analysis makes this obvious.

There are various kinds of unintended alterations that can adversely affect consumer health. For instance, some can reduce the quantity of one or more nutrients. Or nutrient intake could be impeded by inadvertently increasing the concentration of substances that impair the digestion or assimilation of particular nutrients – or by creating anti-nutritive substances that were not already there. Further, other changes could make the food directly dangerous by introducing new allergens or by perilously elevating the level of toxins that normally exist at low levels. Or new toxins could even be created.

Further, these types of changes involve different degrees of risk. For instance, a slight reduction in a particular nutrient would ordinarily pose less danger than the creation of an allergen that could severely sicken tens of thousands of people. And making a food toxic could be far more dangerous than that, because toxins tend to affect everyone, while an allergen is only problematic for a small portion of the population. Moreover, elevating a toxin that's already in the plant would generally induce less overall harm than creating one that had never been there before, because most native toxins are known, enabling breeders to detect the problem and prevent commercialization. In contrast, unknown toxins could easily go unnoticed. The potential for total harm also depends on how quickly a toxin acts.

Those that cause an immediate and noticeable reaction can subsequently be avoided, while those that induce incremental harm over many years are difficult to discover, even if their cumulative effect is huge. That's why tobacco has killed millions more people than have toxic toadstools. Although the latter are far more lethal per unit consumed, because they're dramatically deadly, humans have long known to shun them. Conversely, because no one drops dead after his first (or even thousandth) cigarette, it took centuries before their aggregate effects were finally recognized.

In this regard, it's important to understand that, like tobacco-induced disease, most food-borne illnesses are not acute, but chronic. They develop slowly through repeated exposure, and they're more difficult to detect than those inducing a rapid and noticeable reaction. Therefore, they entail greater total risk.

Thus, we again see that the risk entailed by an unintended change is a function of both the likelihood it will cause harm and how great the harm is likely to be. But the committee lost sight of this fact – and also lost touch with common sense. Just consider the implications of their claims.

On the one hand, they acknowledged that the main forms of genetic engineering (as well as radiation and chemical breeding) entail a far greater likelihood of unintended effects than does pollen-based breeding, while on the other, they alleged that these more disruptive techniques do *not* necessarily pose higher risk. Therefore, if these high-tech methods induce unintended effects more frequently than pollen-based breeding and yet do not engender higher risk, then the effects they do generate must, on average, be *less* dangerous than the less frequently induced side effects of that natural method. Or, to look at it the other way around, if the potential side effects of both the invasive and the natural methods pose the same degree of risk, but the latter are ten times *less* likely to manifest, they must be ten times *more* dangerous.[39]

This result is clearly ridiculous; and it attests the Academy's ongoing willingness to mangle the meaning of risk on behalf of genetic engineering.

Yet, the snubbing of reason did not stop there, and as the committee pressed on, they descended more deeply into logical dysfunction. For example, to buttress their basic claim, they asserted that *none* of the various methods of breeding, whether based on simple selection, radiation, or genetic engineering, is "inherently hazardous."[40] Not only is this pronouncement momentous, it's curious; and one may well wonder on what grounds it rests. Here's their explanation: "If a particular method were inherently hazardous, all products resulting from its use would be potentially harmful. However, it is known that each method can provide safe products. . . ."[41]

This rationale is dubious in multiple respects. For one thing, just because a method can *sometimes* yield safe products does not imply it's inherently nonhazardous. For instance, if a technology cranked out 100 harmful products for every safe one it delivered, we would not regard it as innately innocuous, despite the fact it can occasionally yield a non-injurious result.

Further, it's confusing (and misleading) to assert that none of the processes is inherently hazardous. After all, the report itself repeatedly emphasizes that each method can induce unintended effects. Accordingly, it would be more accurate (and honest) to acknowledge that *all* of them entail inherent hazard – and to then engage in an earnest assessment of comparative risk by examining the probabilities that the respective hazards will manifest.

But it appears that the members of the committee were less concerned with accuracy and honesty than with portraying rDNA technology as no riskier than conventional processes. Otherwise, they wouldn't have appended their curious assertion with the clause: "so the key for breeders and regulatory agencies . . . is to identify the relatively rare, potentially hazardous products resulting from any method." [42]

This statement could win a medal for length of logical leap. It's a prodigious feat to start with the proposition that every method can produce *some* safe products and jump to the conclusion that each produces them so routinely that its potentially hazardous products are "relatively rare."

Thus, in the space of two sentences, the committee ostensibly demonstrated that genetic engineering (as well as radiation and chemical breeding) entail negligible risk, because if it's indeed the case that these methods rarely produce a food that's even "potentially" hazardous, they can hardly pose any danger. But this purported demonstration was carried off without citing any evidence – and by wrenching the rules of logic.

So, just as the Reagan Administration established the '*product not process*' maxim by edict rather than evidence (or earnest analysis), the committee tried to establish the safety of genetic engineering in the same manner. Accordingly, their assertion is more imperial than empirical – and more laughable than logical. Although their convoluted verbiage and pretensions to scientific process tend to obscure it, in essence they declared, "Genetic engineering is safe because we say it is."

The assertion's dubiousness has yet another dimension: it is *not* "known" that bioengineering (or radiation breeding and chemical breeding) can provide safe products.[43] That's because *knowing* entails a high degree of certitude. For instance, we know that 2 plus 2 equals 4 and that the area of a rectangle equals its height times its length. Further, although philosophers debate the extent to which we can actually know things outside the realm of truths that can be proven through pure deduction, in our common understanding, when a group of scientists state that something is *known*, they're implying it's backed by a lot of solid evidence – to a degree that leaves little room for doubt. They are thus indicating that their alleged knowledge is based on far more than a reasonable presumption, an informed opinion, or even an earnest assessment of facts that, while suggesting a particular conclusion, are not fully conclusive.

But the committee's claims about radiation breeding and genetic engineering have no hard evidence behind them. As previous chapters have discussed, and as the Canadian report confirms (and the next chapter will more thoroughly demonstrate), none of the products of bioengineering has been proven safe via adequate testing.[44] And in the case of foods created via radiation, no safety tests have even been conducted.[45] Furthermore, the committee acknowledged "it is almost certain" that plants bred via radiation contain mutations in addition to those that are selected for – and that these changes could remain undetected and induce "unknown effects."[46]

The best the committee can say in regard to radiation is that its products are "widely used and accepted" and that no harm has been linked to any.[47] But the same could have been said about tobacco prior to 1960. Further, tobacco was finally linked to disease only through long-term epidemiological studies, while no such studies have even been started for the products of radiation or bioengineering – and no records kept that could enable them.[48] So if any of the foods produced via radiation or bioengineering have been causing common ailments such as cancer or colitis, experts could not have discerned it. Further, as David Schubert (a professor at the Salk Institute) has noted, even if adequate monitoring *were* in place, any increase in a common disease induced by a novel food could still not be detected unless it at least *doubled* the ordinary frequency of new cases.[49]

Moreover, at least one of the products of genetic engineering (a tryptophan supplement) *has* been clearly linked to harm. And the harm was not a minor annoyance but a deadly epidemic that harmed thousands of people. However, as we saw in Chapter 3, GE advocates tend to obscure this fact – and try to justify their refusal to report it by arguing that because the engineering process itself was not shown to have caused the disease, they're excused from acknowledging that the disaster arose through one of its products. Yet, as that chapter demonstrates, this argument is not only logically invalid it's empirically off-base, because the evidence indicates that the process *was* the key cause.[50]

The tryptophan disaster underscores the fact, pointed out by the FDA experts (and many others), that in assessing whether GE foods cause harmful effects, absence of evidence is *not* evidence of absence – even if no problems are observed over many years. After all, if (back in 1989) the engineered tryptophan supplement had harmed all those people via a commonplace malady instead of a highly unusual one, scientists would still not be aware it was toxic – and it would probably still be on the market, continuing to deal death and disability.

Yet, as have so many others who wish to protect the image of genetic engineering, the committee neglected to mention the epidemic, while also neglecting to face another unpleasant reality. They expressed confidence in the current regulatory system and contended that it has adequately screened the GE foods on the market for harmful changes – despite the fact the Canadian experts had charged that the system is seriously ill-suited to do so.[51]

Thus, it's clear that the committee overstated the state of our knowledge. Contrary to their unequivocal claim, we do *not* know that genetic engineering, or radiation breeding, can provide safe products, because there's no solid evidence to confirm that any they've created is actually safe – and, as we've seen (and will soon more fully appreciate), there's good reason to think that they could be harmful.

Nonetheless, biotech advocates have been emboldened by the committees' pronouncement, and have even unwarrantedly amplified it. For instance, although the committee said we know that genetic engineering is capable of producing *some* safe foods, Pamela Ronald (the University of California scientist whose distortions were discussed in Chapter 4) relies on their report to declare she knows that *all* the GE foods on the market are just as safe as the rest of the foods in her refrigerator.[52] And because this unconditional assertion appears in an ostensibly authoritative (and highly influential) book, it has colored the thinking of many intelligent people, despite the fact it's false. As the foregoing discussion has demonstrated, although Ronald can *think* that these foods are safe, and may even fervently *believe* it, short of divine revelation, she cannot actually *know* it.[53]

Due to the committees' devotion to the *'product not process'* doctrine, their arguments continued to be clumsy. For instance, in attempting to establish that the process of production is not a reliable predictor of problems, they classed all such processes (including simple pollination) as "methods for genetic modification" and then declared, "The potential for hazard resides in specific products of the modification. . . ."[54] This implies that the production process is disconnected from hazard – a notion that's clearly mistaken. Although consumers are not directly injured by the production process (as laboratory or factory workers can be) and are instead harmed by whatever deleterious foods it may produce, if a particular process has a high likelihood of churning out poisonous products, it's fair to say that it is hazardous. After all, a hazard is a condition with the potential to cause harm. But according to the committee's strange logic, such a process cannot even be said to entail the *potential* for hazard.

However it's expressed (and the committee expresses it in several ways), the notion that the process of production is essentially neutral in regard to risk, and that regulation can be primarily based on the specific products themselves, is fallacious. As the Canadian report explained, we can only determine what initial level of regulation is appropriate by assessing the risks posed by each process. We cannot do so by assuming the risks are essentially equal simply because each process has the potential to induce injurious outcomes – while ignoring the fact that some will do so far less frequently than others. If the products of genetic engineering are much more likely to contain harmful ingredients than are those of traditional methods, then it's reasonable (and necessary) to require that each be tested more rigorously. This is especially so because a number of harmful effects would be difficult to detect without extensive testing. So if superficial analyses are relied upon instead, and deeper testing is not performed *unless* such analyses happen to discover signs of problems, many actual problems could escape notice.

Thus, to focus on the product without considering the risks of the process through which it emerged is to run the significant risk of ignoring noxious substances it may contain. And because the NAS committees' own chart acknowledges that the modes of genetic engineering behind most of the crops currently consumed are much more likely to produce unintended effects than almost every other form of breeding, it's clear that the regulatory path they endorse is not only unsound, but irresponsible.

What's more, it's *illegal* – at least in the United States, whose government requested the report. As Chapter 5 demonstrated, according to US law, every GE food must be presumed unsafe until proven safe by standard scientific procedures – even if it is claimed to be "generally recognized as safe." However, in common with most Americans, the committee members were confused about how the law relates to GE foods – and were apparently under the illusion (fostered by the FDA) that the manufacturers have no burden to demonstrate their safety.[55]

Understating Hazards by Overlooking Facts

The committee's commitment to the '*product not process*' maxim skewed another important facet of their report as well. Not only did it sully discussion of how the comparative chart relates to risk, it caused every bar associated with bioengineering to be drawn too short. And the inaccuracy of the lengths is attested by the chart itself. A bit of analysis makes this clear.

When the report describes the genetic engineering process, it gives the impression that after an isolated cell has incorporated the cassette of transferred rDNA, it's a simple matter to regenerate it into a mature plant.[56] However, as Chapter 4 noted, such regeneration could not occur without substantial human intervention, usually in the form of the artificial process called tissue culture – which imparts what experts have referred to as a "genomic shock" that triggers substantial mutations.[57]

Accordingly, plant breeders have sometimes employed tissue culture in the hopes of creating useful change, just as they've done with radiation. Yet, like radiation, tissue culture generates a broad range of unintended disruptions – some of which could be harmful to the people and animals that consume the altered product.

In describing the various modes of plant production, the NAS report devotes two paragraphs to tissue culture.[58] However, not only do these paragraphs fail to mention its essential role in the genetic engineering process, they imply that the latter has no need of it. The reader is told that although some breeders still use tissue culture to generate mutations, the practice is now largely confined to developing countries because it "has largely been supplanted by more predictable genetic engineering technologies."[59] And when, on the following two pages, the committee described how GE plants are produced, there was no mention of tissue culture.[60] But these omissions belie the fact that although most breeders no longer employ tissue culture in the quest for favorable mutations, they still must use it if they want to transform genetically engineered cells into full plants – and that in sticking with it for that purpose, they're also stuck with undesirable disruptions it creates.[61]

However, they're not stuck with all of them. That's because after a regenerated plant has been produced with tissue culture, it's usually put through repeated rounds of sexual crossing through which mutations that result in easily observed abnormalities can be gradually removed. And this is done by bioengineers as well as by breeders who employ tissue culture to produce useful mutations in non-engineered plants.

Yet, while this process reduces the number of collateral changes, it would usually fail to eliminate them. According to the authors of an extensive review of the changes that tissue culture induces, "there is a high probability" that regenerated plants (even when market-ready) will bear alterations to their genomes as well as to areas that are trans-genomic. Consequently, they asserted that this reality "should be carefully considered" in *all* of the technique's practical applications; and they noted that one of these is the production of GE crops.[62]

Nonetheless, although tissue culture is widely employed in creating GE plants, and although it's highly likely to induce unwanted alterations that will remain in the final product, the committee failed to account for these effects when, in devising their comparative chart, they decided how long the bars affixed to the modes of bioengineering should be.

And that was not because they failed to recognize that the final products of tissue culture will almost surely harbor such effects. They recognized this reality; and this is reflected in the bar they assigned to the process, which marks it as one of the more disruptive techniques. In fact, it's portrayed as far more disruptive than one of the modes of genetic engineering (bacterial rDNA transfer between closely related species), with a bar more than four times longer than the one accompanying that GE-based process. (Compare bars 4 and 8 in Figure 9.1)[63]

But that's bizarre, considering that the latter process *relies* on tissue culture to turn the engineered cells into adult plants. This anomaly implies the committee took no account of culture-induced effects when estimating the length of that method's bar. Accordingly, one might quickly conclude that the bar should be increased by a length equal to the *entire* bar for tissue culture – and that the bars associated with the three other modes of bioengineering should be likewise elongated.

If this were done, important changes would occur in the rankings. Although the bars depicting the disruptive tendencies of the modes of GE-based methods would have lengthened (and darkened), the one belonging to radiation would remain unchanged. That's because the radiation is directed at seeds rather than isolated cells, and the seeds grow into plants without assistance from tissue culture. Consequently, three of the applications of genetic engineering would become the *most* disruptive methods in the chart – significantly surpassing radiation.

Yet, it's probably unwarranted to readjust the rankings to such a degree. That's because, although both bioengineers and the breeders who actually aim to cause mutations via tissue culture try to reduce unwanted changes in the same way *after* the culturing process has resulted in a regenerated plant, they tend to perform the process that gives rise to the plant in different ways. And the differences can reduce the initial number of mutations.

There's flexibility because the number and severity of the mutations are influenced by factors that can vary from case to case, and several of these can to some extent be adjusted by the breeders. For instance, if cells are subjected to the culturing process for shorter times, mutations tend to lessen. Lowering the level of applied hormones can have the same effect.[64] Accordingly, genetic engineers try to manage conditions so as to avoid mutations, whereas the breeders who want to induce them require conditions that favor their formation. Therefore, the plants these breeders regenerate would be expected to contain, on average, more unwanted culture-induced mutations than do engineered plants, which increases the likelihood that more of them would remain after the subsequent breeding cycle has ended.[65]

In light of this difference, is it plausible that the committee accounted for the role of tissue culture in genetic engineering after all? In their assessments of the latter's mutational propensity, did they assume that the efforts of the biotechnicians have significantly softened the perturbational punch of tissue culture – and then adjust their calculations accordingly? *It's clear they did not.* Although they provided scant explanation of the reasoning behind their rankings, it's almost certain that they didn't take the effects of tissue culture into account at all. For instance, the bar associated with bacterial-based transfer of rDNA between different, but closely related, species (bar 4) is quite small, and almost identical to the one affixed to selecting from a group of non-identical plants within the *same* species – plants that were propagated via the natural process of pollination (bar 2). So if merely 5% of that GE-related bar represented the effects of tissue culture, it would entail that randomly wedging a chunk of DNA from one species into the DNA of a different (though closely related) species via bacterial infection is, in itself, *less* likely to induce unintended changes than is the flow of pollen between two zucchinis.[66]

It's extremely unlikely that such a differential exists; and it's also unlikely the committee intended to imply that it does, because that would have contradicted a premise they consistently applied in the chart: the premise that the greater the biological distance between the organisms involved in gene transfer, the greater the tendency for unexpected changes. So it's evident they disregarded the effects of tissue culture.

Accordingly, their chart should be adjusted to properly account for these effects. However, it's difficult to determine how to do so, because several variables are involved. Not only do different biotech companies probably have their own methods of doing the cultures, it's more challenging to mitigate mutations in some species than in others; and additional factors can also come into play. So, while it would likely be mistaken to amend the chart by increasing the bars associated with the modes of bioengineering by the full length of the bar assigned to tissue culture (when employed as a purposely mutational process), failing to enlarge them by at least one-third that length would probably also be errant.

So, if we stay conservative and stretch the bars associated with GE by adding just one-third the length of the bar the committee assigned to tissue culture (bar 8), the two modes that have produced most of the engineered crops on the market no longer lag behind radiation breeding in propensity for unintended effects. One becomes equal (if not slightly greater), and the other, the one responsible for the bulk of the GE foods sitting on grocery shelves, exceeds radiation by a noticeable margin.[67]

Therefore, according to the committee's own calculations (when conservatively adjusted to correct for a crucial oversight), genetic engineering stands as the most disruptive technique on the chart: with one of its predominant commercial modes surpassing the perturbational power of blasting seeds with DNA-damaging radiation, and the other equaling it. Moreover, if the rankings are instead adjusted by increasing the bars depicting the GE modes by half the length of the bar for tissue culture (which may be more appropriate), radiation breeding decisively drops to third place, well behind each of those two GE modes in disruptive potential.[68]

Consequently, setting the rankings straight fully undermines one of the main arguments the report promoted: the contention that because radiation-based breeding is even more disruptive than bioengineering, and because it's safe, the latter must be safe as well. We've already seen that the second premise of this argument (that radiation breeding is known to be safe) is false; and now we know that the first premise is also wrong – even within the confines of the committees' own thinking, because when their chart is adjusted to accommodate commonly recognized facts (which they overlooked), the premise is contradicted.

Further, even if one accepts the assertion about the safety of radiation, the larger argument still fails – which is highly significant, considering how heavily biotech advocates have relied on it. For instance, in *Tomorrow's Table*, Pamela Ronald cites the 2004 report as evidence that mutation breeding via radiation "is considered very safe" despite the fact it poses "even greater risks" than genetic engineering.[69] Had she realized that the report actually shows the mode of bioengineering behind most GE foods on the market to be substantially *more* disruptive than radiation breeding, she would have also realized that even if the latter *had* been proven safe, it would not confirm the safety of the former.

Moreover, in addition to disregarding the effects of tissue culture, the committee overlooked several other ways in which genetic engineering causes unwanted changes. And, as we shall see, when all relevant facts are properly assessed, it's obvious that the aggregate food safety risk posed by this newest mode of breeding exceeds the risk posed by every other mode by a far greater margin than would be indicated by merely adjusting the chart to account for tissue culture – and that, if the chart were re-drawn to fairly reflect reality, the bar associated with *every* mode of bioengineering would be significantly *longer* (and *darker*) than the bar of *any* other form of breeding, including radiation.

So let's discover why those bars should be re-drawn.

How the Risks of Genetic Engineering Were Consistently Misrepresented

The NAS committees' aversion to a forthright confrontation with the unintended effects of bioengineering was so strong that they ignored several key facts – while misrepresenting others that they did mention. These faults will be obvious as we examine how they dealt with some important issues.

a. The Use of Viral Promoters

The committee was especially cavalier regarding the promoter from the cauliflower mosaic virus (the 35S promoter) that's present in most GE foods. As discussed in previous chapters, this powerful promoter is attached to the inserted genes and impels them to constantly express the proteins they encode outside the controls of the plant's intricate regulatory system. This abnormal activity can disrupt complex biochemical feedback loops and induce adverse outcomes. Such disruptions can also result from the continuous drain of energy to power functions that the plant doesn't need. Moreover, the promoters can induce erratic expression of nearby (and even distant) native genes or can activate biochemical pathways that are ordinarily inactive – either of which could lead to the production of harmful substances.

Yet, although at the time the report was written such hyper-active promoters were operating within the DNA of virtually every GE food on the market, and although many experts (including the Canadian panel) had raised concerns about the above-described risks, the committee did not deign to so much as mention even one of them. In fact, only a single sentence in their report refers to viral promoters within GE plants, and besides being unrelated to the above concerns, it provides no inkling about what the promoters are, how they function, and why they're needed.[70] So a person who did not already have such knowledge would still be uninformed after reading the full report.

Further, although some GE advocates do at least acknowledge the various promoter-related concerns, they dismiss them by arguing that many conventionally produced and widely consumed vegetables have been infected by the cauliflower mosaic virus – and that the activity of the 35S promoter in those cases is no riskier than in the case of GE products. But this argument disregards an important fact: the 35S promoters in naturally infected plants are in a different location than are those that enter the plants via genetic engineering. And, as in real estate, location is of prime importance in this case too – and it must be taken into account.

In genetic engineering, the recombinant cassettes and the 35S promoters within them wedge directly into the plants' DNA, while the promoters within an invading cauliflower mosaic virus do not. Instead, they remain outside the plants' DNA. Consequently, the promoters in naturally infected plants don't exert the same disruptive influence on the native DNA as do those that that are artificially inserted; and, according to the plant virologist Jonathan Latham, the latter are "almost certain" to cause different effects – and can alter expression of the plants' genes in ways that the naturally introduced promoters won't.[71]

So the promoters that drive expression of the inserted genes pose different food safety risks than those tucked within the viruses, and it was a major oversight for the committee to ignore this reality.

b. Disruptions Caused by the Insertion Process

In the same section of the report in which it dodged a discussion of viral promoters, the committee also evaded an earnest examination of the disruptions caused by the insertion of recombinant DNA. While they did discuss alterations that can occur when one section of DNA inserts into another, their discussion was unduly focused on the disruptive potential of a natural phenomenon: the ability of some DNA segments to become mobile and re-insert into a new region of the genome.

These mobile segments are technically termed *transposable elements* or *transposons* – and they're commonly called "jumping genes." And, although the committee didn't refer to them by the latter term, they apparently *did* want to make people feel jumpy about them. Thus, the section on insertional disruptions begins, not with a description of those that are known to result from genetic engineering, but with a discussion of disturbances that transposons can cause. It notes that when transposons move, they sometimes re-insert into genes and disrupt their function. It also points out that even if a gene is not directly disrupted, the insertion disrupts the

DNA sequence that's involved – and can also induce rearrangements of surrounding DNA.[72]

The committee then tried to show that the insertion of DNA cassettes via bioengineering poses less risk. They stated that it's "rare" for transferred DNA to insert into important genes – and that even if it does, the resultant undesirable change will be noticed during subsequent screening and that the plant line containing it will be discarded.[73] But they failed to note that transposons don't ordinarily pose any greater insertion-related risks. In the absence of extraordinary stresses, these potentially mobile elements rarely mobilize; so most of their current positionings occurred in the ancient past.[74] This means that over the vast stretch of biological time, any insertions that impaired the health of the plant would have been eliminated. And those that impaired the health of the plant's consumers would have also been discarded. Further, to the extent that any transposon-related problems remain, they're just as salient in a GE plant as in the non-engineered variety from which it's developed.

In fact, a bioengineered plant carries greater transposon-related risk than its parent because the engineering process tends to activate transposons and get them jumping. But not only did the committee ignore this fact, they implied the opposite. Similar to their portrayal of tissue culture, their discussion of transposon activation imparts the false impression that it's a phenomenon unconnected to (and more worrisome than) the bioengineering process.

Yet, in reality, that process can induce transposon movement in three ways. First, the insertion of the cassette can itself cause transposons to shift locations.[75] Further, as previously noted, the engineered cells are then subjected to tissue culture, which imparts a "genomic shock." Due to this jolt, transposons often jump.[76] Additionally, the 35S promoter can stir nearby transposons into a mobile mode.[77]

Conversely, the process of pollination rarely causes transposons to mobilize.[78] Therefore, contrary to the impression the committee created, whatever risks are entailed by transposon activation are inherent in bioengineering but largely absent from most pollen-based modes of breeding.[79]

Moreover, in trying to depict the relocation of a transposon as a riskier event than the insertion of an rDNA cassette, the committee obscured an important fact. It failed to acknowledge that, like a re-entering transposon, the process of implanting a cassette can likewise alter the surrounding DNA. Although this is a well-documented effect of bioengineering, from the committees' wording, one would never know it. While they linked such effects to transposon shifts, they ignored them in regard to cassette insertions – implying that the latter don't induce any.

But induce them they do – and in profusion. In one study, researchers examined 112 lines of Thale cress into which rDNA cassettes had been inserted via *Agrobacterium* infection and discovered insertion-related aberrations in almost every plant they examined. In most cases, there were small deletions of the plant's DNA at the site the cassette had entered. And in 21% of the plants, the alterations were large-scale, with sizable sections of the cress DNA either excised or rearranged. In two of these cases, a section from one chromosome had re-located to another chromosome – a major change.[80]

Further, not only was DNA that should have been in the plants deleted, unintended sequences were inadvertently added. Eight of the plants had large insertions of extraneous DNA that came from either the plasmid that had conveyed the cassette into the *Agrobacterium* or from the cassette itself. And most of the rest contained smaller insertions of DNA of undetermined origin. In none of these cases were the superfluous sequences supposed to be there.

Such messy insertions are not the exception but the rule. Three biologists who conducted an extensive review of the scientific literature observed: "It is apparent that small and large-scale deletions, rearrangements of plant DNA, and insertion of superfluous DNA are each common occurrences at *Agrobacterium*-mediated transgene insertion sites."[81] And they noted that particle bombardment seems to make an even bigger mess. As stated in their report, inserting genes through that explosive method "is usually or always accompanied by *substantial* disruption of plant DNA and insertion of superfluous DNA."[82]

As we've seen, the NAS committee also recognized the greater perturbational power of the ballistic method. Yet, they failed to acknowledge its full disruptive range. They only noted that it ups the odds the insertion will disrupt crucial sequences of DNA, but they said nothing about its propensity to disrupt the regions surrounding the insertion site – or to introduce unintended sequences within the site.[83]

Each type of alteration, whether caused by bacterial-mediated insertions or those resulting from the gene gun, can adversely impact the plant's function. As the review article points out, when native DNA is deleted, important sequences can be lost or impaired; and unwanted effects can also ensue from alteration of the surrounding DNA or the addition of superfluous sequences. Further, even if there were no deletions or other unintended structural changes, the coding sequences and promoter regions that are *intentionally* added could exert a disruptive influence on the operation of native genes.[84] This influence can be extensive – and can alter the expression of genes that are thousands of base pairs away from the insertion site.[85]

Moreover, bioengineering's potential for such alterations entails a *unique* hazard. As the molecular biologist David Schubert has stated, the insertion process "generates unpredictable changes in gene expression that are going to be different in *kind* from those produced by traditional breeding."[86] Thus, besides the committees' multiple misstatements about risks, even their basic contention about hazards is false, because, contrary to their claim, bioengineering *has* in this case introduced hazards that are not posed by other modes of breeding. And, as we shall see, it has wrought other new ones as well.

So great is the disruptive power of the plant bioengineering process that it can cause disruptions throughout the genome. The majority of these are apparently caused by tissue culture, which is usually employed in a "particularly mutagenic form" when regenerating engineered plants.[87] But it's probable that some are caused by the process of *Agrobacterium* infection – and possible that particle bombardment can also act as a cause.[88]

These mutations are multitudinous. Several studies indicate that GE plants typically contain hundreds or even thousands of them.[89] Moreover, their magnitude is probably much greater than was measured. Three scientists who reviewed the studies observed that because the researchers employed analytical techniques that cannot reliably detect point mutations and small deletions, it's "likely" they missed most of them.[90]

Although a GE plant is commonly out-crossed or back-crossed with other non-GE lines several times, which can eliminate many mutations that are distant from the insertion site, it's difficult to remove all of them (as was previously noted when discussing tissue culture).[91] Many experts think that any of these remaining mutations could potentially render a plant unsafe to eat – and that this risk has not been adequately reduced by the regulatory system.[92] For one thing, the biotechnicians don't even perform genomic analyses that could detect whether (and how many) wide-spread mutations remain.[93]

Further, multiple rounds of crossing can't remove the mutations in the areas adjacent to the insertion site; and the assessments routinely fail to examine these regions properly. Moreover, they usually don't even examine the insertion sites with adequate care.[94] The system has been so lax that significant disruptions have gone undetected in several GE crops that have been cleared for sale and extensively marketed.[95]

Thus, although the disruptive effects of inserting rDNA are well-documented, and so substantial that some scientists have referred to them as "genome scrambling," the committee disregarded most of them and instead projected the impression that they don't exist.[96] It's difficult to view this as an inadvertent oversight – especially since the committee freely described the insertional disruptions that transposons cause. Further, they didn't refer to any of the relevant publications cited by the research review mentioned above, despite the fact many were readily available to them.[97]

Aggravating the imbalance of their account, not only did the committee ignore some well-known hazards of cassette insertion, they minimized the risks of those that they did mention, stating that breeders would be able to notice plants with deleterious disruptions of important genes and screen them out. However, this ignores the fact, emphasized by the Canadian experts, that many alterations might not cause noticeable effects but could still induce unexpected toxins that the regulatory system is not designed to detect. Many other experts have also discussed the system's inability to ensure the safety of the engineered products passing through it.[98] But, as the foregoing analysis reveals, the committee was apparently more intent on maintaining the pace of that passage than on acknowledging evidence that could imply there's a need to control the flow with greater care. The existence of such an intention is more firmly implicated by the fact that, in portraying radiation-bred plants as the type most prone to bear unintended effects, they were quite willing to point out that these products can harbor unwanted mutations that have eluded the screening process.[99]

c. Production of Unintended Harmful Substances

Besides arising via insertional disruptions, unsuspected (and even novel) toxins could be formed within bioengineered plants through several other avenues. And this hazard was a central concern of the Canadian expert panel. Although the NAS committee never specifically acknowledged any of the concerns raised by that panel, they did acknowledge that GE plants can pose such a problem. Yet, they argued that conventional crops can do the same – in an apparent attempt to dilute this GE-related concern by stretching it to encompass traditional breeding as well.

But their attempt fell flat. For one thing, they could only point to a few instances (involving some varieties of potatoes and celery) in which harmful levels of a toxin resulted from conventional breeding.[100] Further, none of the substances was novel or unexpected.[101] Instead, toxins that are ordinarily present in low concentrations had been dangerously elevated.

Moreover, breeders have been long aware of these toxins and usually monitor the plants to ensure their levels stay within reasonable limits.

However, the committee did not keep their argument within such limits. For instance, they claimed that conventional breeding is not only liable to elevate existing toxins but to generate some that are totally novel. Yet, they could cite no more than one example – and besides being solitary, it was spurious. In their account, crossing a domesticated potato with a particular wild variety of the species ". . . produced not only the usual glycoalkaloids, but also the toxin demissidine, which is not produced in either parent."[102] And to prop the importance of this point, they repeated it several pages later, asserting the toxin was "novel to both parents."[103]

But what was novel was the claim itself. That's because potatoes produce significant quantities of demissidine as they develop. Yet the process normally stops before they ripen.[104] So the extraordinary feature of the case the committee cited was not the production of the substance, but its continuation in plants that were ripe. Therefore, the assertion that the toxin had not previously appeared in the parents was blatantly false – and highly misleading.

And it was still more misleading for the committee to impute outsize significance to this singular incident – and would have been even were it accurate. Besides claiming the incident "shows that non-genetic engineering breeding methods can have unintended effects and generate potentially hazardous new products," the committee went on to warn that such dangerous new chemicals can emerge "any time" these methods are employed.[105] But in reality, there's essentially no risk of this happening through conventional methods. This fact is underscored by eight experts who co-authored an article published in 2013 in the journal *Plant Physiology*. In their words: "Although breeders recombine tens of thousands of genes with virtually infinite potential interactions, to our knowledge, there has never been a report of a completely novel toxin or allergen appearing in a genus as a result of conventional breeding." They therefore concluded that because hundreds of thousands of varieties have been generated without the emergence of such novel substances, the likelihood that any will occur through conventional methods "is virtually zero."[106]

Thus, the threat of novel toxins is itself novel – and *unique* to bioengineering. So it's yet another hazard that's absent from other methods.

The illegitimacy of the committees' contention becomes more glaring in light of the fact that, as Chapter 4 described, there are numerous cases in which genetic engineering has caused weird and totally unexpected results, including the production of a novel toxin in a reconfigured plant. As we saw in Chapter 5, FDA experts were well aware that GE plants have a unique potential for inducing such oddities and warned their superiors about it. Further, Chapter 3 demonstrated that a toxic food supplement produced through genetic engineering contained a strange, never-before-seen substance which was significantly linked to a deadly epidemic – and that this contaminant most likely emerged through the engineering process. Moreover, the altered organisms that spawned this queer contaminant had not been endowed with any foreign genes but merely with extra copies of some of their own, one of which was fused to a viral promoter that forced it to hyper-express. As the Canadian panel, the FDA experts, and many others have pointed out, the genetic engineering process *itself* can produce such unexpected and extraordinary dangers, regardless of the specific gene that's inserted; and they cautioned that this likelihood should not be taken lightly.

Additionally, as Chapter 6 described, bioengineering can even turn an ordinarily harmless protein produced by the inserted gene into an injurious agent. As we saw, even when the sequence of amino acids that comprise a protein remains the same, other factors that affect its structure can be unexpectedly (and dangerously) altered as genes are transferred between species that cannot breed through conventional means. And we examined published research demonstrating that when a bean protein was produced within a pea, it not only became abnormally allergenic itself, it even induced the laboratory mice to develop allergic reactions toward unaltered proteins that were consumed along with it.

Therefore, the potential for protein reconfiguration represents another hazard unique to genetic engineering; and it poses a serious risk, because the harmful changes that could occur would most likely escape detection by the current regulatory system – in Europe as well as America. David Schubert has emphasized the magnitude of the risk by asserting that the research on the GE pea "is probably the single most important study published to date showing the potential dangers of GE crops because allergens can be dangerous at incredibly low levels, and most if not all GE proteins have antigenic properties that are different from their normal counterpart."[107]

Chapter 6 also discussed how the shape of a protein could be deleteriously changed when it's produced within an alien environment. This too is a hazard of bioengineering that's absent from conventional plant breeding. Moreover, there's indication that such misfolding has in at least one instance occurred.[108]

Yet, the committee remained unmindful of the critical facts that bear on the threat of unintended substances.[109] Although *every* instance of genetic engineering entails some risk of inducing novel toxins or allergens, and although the probability that any will arise through conventional means is essentially nil, they pushed the notion that the two approaches are equally prone to produce them.

In light of the foregoing discussion, it's clear that the bars associated with genetic engineering in the report's comparative chart must be adjusted to reflect far more than the unintended effects of tissue culture, because those effects are not the only ones the committee ignored. As we've seen, they not only overlooked the range of disruptions that could occur through both the viral promoters and the insertional process itself, they failed to comprehend bioengineering's *unprecedented* capacity for forming harmful substances – and due to such oversights, they would surely have underestimated its full perturbational power.

This has important implications. As we've already seen, conservatively revising the chart to reflect a measure of influence from tissue culture boosts the rank of one major mode of bioengineering beyond radiation breeding (and well off the chart) – rendering it the most disruptive method of all. And the technology's other main commercial mode gets lifted into a virtual tie with radiation breeding. So if the committees' conclusions are further recalibrated to register an additional (and sizable) group of neglected effects, the bars associated with both these forms of bioengineering will have to be stretched a lot farther. And, although it's not clear exactly how far the stretching should go, it *is* obvious that the bars for both of these modes should surpass the one for radiation by so wide a margin that each extends way beyond the border of the chart, decisively branding them the most perturbational and unpredictable techniques. And because (as was previously demonstrated) the length of the bars ultimately depicts the degree of danger to human health, these techniques also stand forth as the riskiest.

Further, when a careful comparison is made between genetic engineering and the mutative methods based on radiation and chemicals, it's evident that the revised chart proposed above fairly reflects the risk differential. Such a comparison appears in Appendix D, which supplements Chapter 11 and should be read in conjunction with it.

However, even after making the adjustments prescribed above, the chart would still need further revision – not to the length of the GE-associated bars, but to their number. That's because there should only be two of them, not four; and they ought only to register the mode of gene transfer, not the biological distance between the species involved.

I realize that this assertion may seem questionable, considering that biological distance *is* highly relevant in conventional contexts, where crossing plants that are closely related is safer than crossing those that are distant. But genetic engineering is different. Although the products of genes from distinctly foreign species could cause problems due to their foreignness, equal (or greater) risks could also result when the species are similar.[110] After all, the uncontrolled hyper-expression of a protein that ordinarily interacts with a plant's native chemicals in a highly regulated and well-coordinated manner could cause riskier side reactions than the overexpression of one that's completely foreign to the cellular system – and may have scant impact upon it. As we saw in Chapter 3, when bacteria were engineered to over-produce L-tryptophan, a substance they routinely make, metabolic disruptions occurred that induced formation of at least one novel substance – which may well have been the source of the toxicity that caused an epidemic. Further, the over-expression of protein is *in itself* problematic; and impelling a plant to produce *any* protein (from *any* source) in abnormal abundance could cause imbalances that lead to undesirable, and potentially harmful, results.[111] This is yet another hazard that's essentially unique to genetic engineering, and the risk it entails will be more thoroughly discussed in Chapter 11.

Thus, there's good reason to think that dangerous unintended effects could accompany *all* instances of agricultural bioengineering; and given the current state of our knowledge, we cannot accurately determine if there's a differential in risk between the average case involving distantly related species and the average instance in which they are close. Accordingly, it's reasonable to conclude that there should be a single bar for bacterially mediated rDNA transfer and a single one for bioballistic transfer as well – and that, while the latter will be longer than the former, *both* will be significantly longer than the bar for radiation breeding.

Final Conclusions: How the Disagreement about Risks Has Arisen

Early in this chapter, an important question was raised about how the disagreement within the scientific community regarding the risks of GE foods has arisen. In particular, it asked: *Does the disparity primarily stem from a lack of adequate evidence or from a lack of scientific integrity on the part of one faction?* To answer this query, we wanted to discover whether both groups have been examining the same set of facts according to the same scientific standards and with an equal degree of logical rigor. And the reports issued by the Royal Society of Canada in 2001 and the National Academy of Sciences in 2004 were selected to serve as examples.

Now we have a sound basis for providing a definitive answer. Our investigation has revealed that the NAS report reached an opposite conclusion from that of its Canadian predecessor due to recurrent dereliction and a dearth of scientific integrity. We've observed how the committee that produced that report systematically disregarded significant facts, misrepresented several of those they did mention, and repeatedly breached the laws of logic. We've also seen how they even distorted the very concept of risk – and how these various delinquencies were apparently driven by a determination to uphold both the image of bioengineering and the presumptions on which the lax regulatory policy of the US government is based. Further, we've seen that when the committees' own calculations are cured of inconsistency and adjusted to accommodate neglected facts, they designate genetic engineering as the riskiest form of breeding, with *every* manner in which it's applied posing a much larger degree of danger than does any other method – confirming the Canadian report's conclusion that it is "scientifically unjustifiable" to declare any GE food safe unless its safety has been confirmed through rigorous testing.

Thus, it's clear that the purported scientific basis of the NAS report is largely illusory. And a revealing measure of the degree to which it's estranged from science, and divorced from reality, is how it treats its Canadian precursor. Although that report had been issued a mere three and a half years earlier, and although it was produced by an eminent group of scientists, nowhere does the NAS report so much as mention it – thereby dodging the difficulty of confronting its contrary arguments, while projecting the impression that no scientific body has reached a conclusion at odds with its own.

Yet, despite its deep defects, this report has been extolled and extensively relied on. For instance, in their influential book, *Tomorrow's Table*, Pamela Ronald and Raoul Adamchak repeatedly cite it as the primary support for their claims about the safety of GE foods, and many of the products' other proponents have done the same. Further, due to the Academy's status, the media and the public have accepted the report's conclusions, assuming they derived from an assessment that was aligned with science when the process had in fact dishonored it.

Other Esteemed Organizations Have Likewise Released Defective Reports

Sadly, the 2004 NAS report is far from unique. Many respected organizations have also released favorable evaluations of GE foods that are remarkably shoddy – like the 2012 report of the American Medical Association (AMA) that's so strangely reasoned it prompted a professor of public health at New York University to declare that it "doesn't make sense."[112] However, although the statement may not make sense from a scientific standpoint, it does from a political one; and it's in perfect accord with the AMA's avowed intention, first announced in a 1990 policy statement on agricultural biotechnology, "to endorse or implement programs that will convince the public and government officials that genetic manipulation is not inherently hazardous. . . ."[113] Further, besides announcing a commitment to belittle the risks, the organization pledged "to actively participate in the development of national programs to educate the public about the benefits of agricultural biotechnology." In furtherance of this educational endeavor, it declared an intent "to encourage physicians, through their state medical societies, to be public spokespersons" for the technology. Therefore, it's not surprising that the Association has remained more dedicated to promoting GE foods than to sustaining balance and accuracy in reporting on them.

In a prime instance of such errancy, the 2012 report begins the section that discusses (and minimizes) the potential human health effects with the assertion that during the nearly 20 years GE foods have been marketed, "no overt consequences on human health have been reported and/or substantiated in the peer-reviewed literature."[114] But this is significantly misleading because the authors neglect to note that a few years prior to their selected time frame a bioengineered food supplement caused a well-documented epidemic. Further, besides failing to acknowledge this inconvenient fact, they avoided acknowledging another one: the fact that even a complete lack of evidence of harm cannot count as evidence for its absence because (1) no clinical toxicological studies with human subjects have been performed and (2) no epidemiological studies to detect adverse chronic effects have been conducted either – nor could they have, given the lack of labeling and of other means for monitoring.

Even worse, not only has the AMA refused to admit that the lack of labels has impeded proper assessment, and not only has it refused to insist on the mandatory labeling that's been instituted in most industrialized nations, it has actively *opposed* such a step, despite the fact that labeling is widely demanded by consumers and necessitated for the responsible monitoring of public health.[115] So, the Association has on the one hand worked to thwart the gathering of data that could help assess whether GE foods are inducing adverse chronic effects, while on the other, it points to the lack of data as an indication that such effects don't exist.

It's disquieting that America's premier medical organization would promote a group of distinctly new and unproven foods in such an unseemly manner, especially since this blatant promotional policy stands in stark contrast to the earlier stance of the medical community in regard to tobacco. During the 1950's, although a substantial percentage of American doctors smoked, they were barred by their ethical code from appearing in cigarette advertisements. So tobacco companies had to pay actors to pose as doctors in their ads in order to project the impression their cigarettes were endorsed by this esteemed class of health professionals. But when GE foods were developed a few decades later, the AMA *itself* unreservedly championed them – and even encouraged individual members to endorse them as agents of its official policy.

Equally remarkable, the intensity of its intent drove it to initiate its effort when the GE food venture was still in an early stage. The AMA first proclaimed the inherent safety of agricultural biotechnology, and declared that the benefits "greatly exceed any risk posed to society," two years before the FDA issued its policy statement on GE foods, four years before one was brought to market, and long before safety tests had even been conducted, let alone published.[116] The evidential deficit is underscored by the fact that, two years *after* the Association's bold pronouncements, an FDA official acknowledged there was no scientific data that could support a determination that GE foods are safe.[117] But this evidentiary void did not deter the AMA from issuing assertions that required an evidentiary base – and thus projecting the false impression that one actually existed.

Like the NAS and the AMA, several other eminent institutions have behaved more as promoters of agricultural bioengineering than as objective evaluators of its risks; and their reports have likewise been slack. Appendix B discusses two of them: one from the UK's Royal Society and one from the Institute of Food Technologists. And these are only a representative sample. Had the appendix examined every report that propounds flawed arguments for the safety of GE foods, it would have run for many more pages.

Thus, although the NAS and other prestigious organizations present their assertions about the safety of bioengineered crops as science-based, analysis reveals that they're *not* based on sound scientific assessment but on its circumvention. And they all befog the fact that, compared to every form of conventional breeding, genetic engineering entails unique hazards and imposes higher risks.

What's more, besides misstating the risks, they misrepresent the benefits.

The Benefits: How They're Inflated – and Why They're Irrelevant

Proponents of GE foods routinely argue that the risks not only must be considered in context of benefits, but be offset by them. And just as they illegitimately minimize the former, they liberally exaggerate the latter. Moreover, in addition to overstating the benefits that will directly derive from whatever GE crop is being considered, they further inflate them by linking that product with the aggregate boons that are expected to accrue from the technology as a whole. Thus, although the first generation of GE crops were widely recognized as generating scant benefit for the public, they were touted as integral parts of a monumental process whereby yields would be bountifully increased, the environment would be better protected, and food would become more nutritious.[118] For example, despite the fact that the most extensively planted GE crop (Monsanto's Roundup Ready® soybean) was actually reducing yields, expanding herbicide use, and increasing herbicide residues in food, it was allowed to ride the wave of these expectations – and be adjoined with attractive outcomes it was actually obstructing.[119]

So, although biotech advocates are devoted to the doctrine that the process of genetic engineering should not itself be associated with risks – and that any risks are separate attributes of one or another of its individual products, they reverse this approach when it comes to the technology's benefits. In that case, every benefit that *any* GE product might eventually bestow is deemed an aspect of the engineering process *per se*, and every food it produces is somehow connected with a plethora of marvels that are supposedly soon to manifest.

But even if the evaluation of benefits *were* performed in a legitimate fashion, it would be *irrelevant* to the question of food safety. And if it impacted the approval process, it would also be *illegal* – at least in the United States, the nation in which GE foods are most widely consumed. That's because US law requires that new additives to food must be demonstrated safe; and in this context, "safety" is defined as "a reasonable certainty . . . that the substance is not harmful under the intended conditions of use."[120] Consequently, potential benefits are irrelevant, and it's unlawful to consider them.

Many people may find this surprising (and confusing), including those in the American Medical Association, which (as we've seen) promotes GE foods and has a stated goal to "convince the public and government officials . . . that the health and economic benefits of recombinant DNA technology greatly exceed any risk posed to society."[121] The confusion stems from the fact that benefits *are* weighed against risks in determining the safety of prescription drugs. But it's appropriate to do so in that context, because the drug is intended for situations in which the individual is already at risk and a decision must be made as to whether the benefits of the drug will outweigh the risk posed by the disease in combination with the risk posed by the drug's side effects.

However, although the regulatory system accepts the side effects of drugs as necessary risks in the curing of disease, food is a different case. It's supposed to be safe and free of harmful side effects. Therefore, it is reasonable to require that a food additive must not impose a significant risk of harm, regardless of any benefits it might also confer. This is especially important in light of the fact that, in contrast to prescription drugs, foods are sold "over-the-counter" and are consumed by everyone, not just by individuals in need of special physician-monitored interventions.

Thus, the notion that in order to assess the safety of GE foods, it's necessary to weigh their benefits is *way* out of line with both reason and US law.

When the facts are thoroughly examined, it's clear that the title of this chapter is not an exaggeration – and that the risks of GE foods have indeed been methodically misrepresented. And the complicit parties include many eminent scientists and scientific institutions. Of these, the US National Academy of Sciences has played the most prominent role; and although its influential report of 2004 has been hailed as a paradigm of scientific risk assessment, analysis reveals it to be more parody than paradigm. Despite its aura of authority, it is empirically deficient and logically dysfunctional; and its case for the safety of genetic engineering is overly reliant on the overlooking of data and the wrenching of reason. Further, it misrepresents a lot of the information that it does discuss, and the weight of the evidence runs counter to its main contentions. Moreover, when its own calculations are aligned with reality, they attest the high-risk nature of genetic engineering, distinctly designating it the most dangerous form of breeding.

On the other hand, the report issued by the Royal Society of Canada in 2001, which presents a case for regarding GE foods as abnormally risky, displays none of the defects of its NAS counterpart – or of the other reports with a similar slant. Furthermore, because it counters the illusion of consensus that GE promoters project, most of those reports have ignored it; and none has refuted its reasoning. Nor has any evidence appeared that demonstrates its precautionary stance is no longer appropriate.

However, as we shall next see, extensive evidence *has* accumulated confirming the need for precaution and the necessity of strengthening rather than loosening the regulatory requirements. In light of the risks described in this chapter, the emergence of such evidence is not surprising. And, considering how biotech proponents have consistently misrepresented these risks, it's also not surprising that they've just as vigorously distorted this portentous evidence – and even vilified the scientists who published it.

CHAPTER TEN

A Crop of Disturbing Data

How the Research on GE Foods Has Failed to Show They're Safe – and Instead Confirmed They Should Be Off the Market

As the preceding chapter demonstrated, when risks are assessed in a manner that upholds logic and registers the relevant facts, genetic engineering looms forth as the most dangerous method for producing new varieties of food. Moreover, as we've seen in other chapters, and will more fully comprehend in this one, the risks have not remained solely theoretical; and when one looks beyond promotional claims to focus on concrete evidence, it's clear that the edible output of recombinant DNA technology has often been linked with adverse effects.

These unsettling outcomes date from the earliest phase of food-centered bioengineering. As Chapter 3 documented, long before the technology had been able to develop any market-ready grains, fruits, or vegetables, it had produced a food supplement of the essential amino acid L-tryptophan; and that seemingly innocuous supplement, the GE venture's first ingestible offspring, spawned a major epidemic. Further, the epidemic was caused by a toxic contaminant that most likely arose from the engineering process itself. Moreover, as the chapter also documented, the catastrophe was of such magnitude that biotech proponents were driven to thoroughly obfuscate what had happened in order to prevent the GE food enterprise from being halted. Absent such obfuscation, it's doubtful there would have been a need to write this book because, even if the venture had been allowed to proceed, public pressure would almost certainly have compelled it to be conducted far more responsibly than has been the case.

However, although skirting the facts of this initial disaster enabled the GE food venture to continue, it has *not* been possible to skirt the technology's continuing ill effects. And years after the toxic tryptophan first hit the market, when bioengineers had finally developed a plant that was destined for dinner tables, it too entailed problems.

The Flavr Savr™ tomato: A Shaky Start for the Produce of Agricultural Bioengineering

This first bioengineered whole food was a tomato. And, like the injurious tryptophan supplement that had gone before, the primary gene inserted into its DNA was not a foreign one. In fact, this tomato had arguably been altered to a *lesser* degree than had the tryptophan-producing bacteria because, in contrast to them, its cells weren't even engineered to produce a higher volume of a native protein. Instead, they were modified so as to *suppress* the production of an innate protein.

This targeted protein was an enzyme (called the *PG enzyme*) that causes the fruit to grow increasingly soft. And by diminishing its output, the developer (Calgene, Inc.) aimed to enable the fruit to be picked *after* it had ripened and yet remain firm throughout the shipping process – setting it apart from most commercial tomatoes, which are picked prematurely and reach final ripening away from the vine. In this way, the new tomato was supposed to be more flavorful when harvested but not too squishy by the time it reached supermarket shelves. Accordingly, it was dubbed the Flavr Savr™ tomato.

In order to achieve their goal, Calgene's biotechnicians made a copy of the PG gene (the one responsible for the fruit-softening enzyme) and reversed the order of its sequence. In this way, they created a new gene that generates an RNA transcript that they hypothesized would bind to the messenger RNA of the native PG gene and prevent it from being translated into an enzyme.[1] But in order to get the new gene inserted, they had to employ *Agrobacteria* to wedge the gene into the tomato's DNA; and in order to get that new gene to express RNA at a high enough level, they didn't rely on any of the plant's native promoters but instead boosted the output through the insertion of the powerful 35S promoter from the cauliflower mosaic virus. And to get the reconfigured cells to grow into whole tomato plants, they had to put them through tissue culture.[2]

So there were ample avenues through which adverse alterations could have occurred. And it appears that some did.

But this only became apparent because Calgene did more than the FDA required it to do. Well before the agency had issued its 1992 policy statement that illegally presumed GE foods are *Generally Recognized as Safe* (GRAS) – and falsely informed manufacturers they weren't required to test them – the corporation had requested the FDA to issue an advisory opinion regarding the tomato's GRAS status. And, at that time, the FDA apparently thought it would aid the image of GE foods if there were tests to support the safety of the first bioengineered plant that was headed for market. So the agency encouraged Calgene to perform 28-day feeding studies with rats.[3] And, although this encouragement was not a command, the corporation decided to have an outside laboratory put three of the most promising tomato lines it had developed through such short-term testing.[4]

However, although this voluntary decision was, from the public's perspective, commendable, from the side of Calgene, it was probably regrettable.

Initially, two studies were conducted. In the first, only one of the three Flavr Savr™ lines was employed; and in the second, the other two were used. To the corporation's dismay, some of the rats consuming one of the engineered lines in the second study displayed disturbing abnormalities. The lab detected that 10 percent of them (4 out of 40) developed gross lesions in their stomachs (which meant their stomachs were bleeding).[5] In contrast, no lesions were observed in any of the rats in the control group that ate natural tomatoes.

These lesions presented reasonable cause for concern; and the food safety expert Arpad Pusztai has pointed out that comparable erosions in humans could result in "life-threatening" hemorrhage – particularly among the elderly and individuals regularly taking aspirin. Accordingly, he noted it's "not legitimate" to call such lesions "mild" in the case of human pathology – and that because the rats were being used as models for humans, the term could not be legitimately used in the context of these studies.[6]

Faced with the emergence of such problems, Calgene decided to run another test on the line that was linked with them. It then hired a team of pathologists to assess the results and to reevaluate those of the first two studies as well. However, instead of finding that the initial observations of lesions in the second study were flawed (as Calgene presumably hoped would happen), this team discovered lesions in four *additional* rats that were fed the problematic variety of engineered tomato. This boosted the percentage of lesion-afflicted rats consuming that product to 20 percent, a highly unwelcome outcome.[7]

But the results for the third study (in which lesions occurred in some rats not fed the Flavr Savr) were more encouraging – to the extent that Calgene, along with other outside evaluators it employed, believed that when the data from all three studies were assessed, a reasonable case could be made that the lesions were "incidental" and not induced by any change related to the engineering process. Accordingly, when they submitted the data and their arguments to the FDA, they were optimistic the agency would agree that the Flavr Savr was safe.

The FDA Experts Are Not Convinced

However, the FDA's scientists were not won over – as is clearly revealed by documents pried from the agency files by the Alliance for Bio-Integrity lawsuit. For instance, a memo written on June 16, 1993 by a toxicological pathologist that was also signed by both the Chief of the Pathology Branch and the Leader of the Diagnostic Pathology Section noted that "the criteria for qualifying a lesion as incidental were not provided;" and it also observed there was "considerable disparity" between the various studies that "has not been adequately addressed or explained."[8] The memo consequently concluded there was insufficient evidence to rule out the role of the engineered tomato in causing the lesions.

The corporation responded and tried to allay these concerns, but the FDA experts deemed the response unsatisfactory. Another memo signed by the same three pathologists stated that "Calgene failed to adequately address . . . some of the major issues raised by the Pathology Branch. . . ." It further asserted that the unresolved issues "leave doubts as to the validity of any scientific conclusion(s) which may be drawn from the studies' findings."[9] Moreover, the pathologists did not stand alone. A scientist with the Additives Evaluation Branch concurred that Calgene's responses "were insufficient" and noted that ". . . unresolved questions still remain." He also stated: "Until these questions are answered, Pathology will be unable to conclude whether or not there is a treatment related effect in rats consuming Flavr Savr tomatoes."[10] The Director of the Office of Special Research Skills was not persuaded either. He wrote that the data "raise a question of safety"– and that they "fall short" of satisfactorily resolving it.[11]

Accordingly, it was clear that these scientists would not be convinced of the tomato's safety without additional data; but by that time, it was also clear that Calgene would not have to provide any. Despite the fact the FDA experts had concluded that, as a matter of science, the data were inadequate, the FDA administrators had by then decided that, as a matter of policy, it was adequate to have inadequate data – or even no data at all. So they didn't ask Calgene to do more testing, and Calgene was content to stand pat.

Déjà vu: The FDA Once Again Misrepresents the Conclusions of Its Scientific Staff

However, although the FDA decided to proceed without legitimate evidence of safety, it needed to project the impression that such evidence was in hand. It had previously publicized the fact that the tomato was undergoing testing; and if it now conceded that the resultant data raised an unresolved safety issue, the GRAS presumption it had recently issued on behalf of GE foods would have been called into question. Consequently, as it had done in 1992 when issuing a policy statement that clashed with the opinions of its own experts, the FDA once again suppressed those opinions. But whereas the previous cover-up had enabled it to declare that GE foods can be *presumed* safe, this one enabled it to proclaim that the first bioengineered whole food had actually been *proven* safe.

To provide backup for this action, Commissioner David Kessler decided to hold a meeting of the agency's Food Advisory Committee, a standing group of scientists occasionally consulted on important matters.[12] Kessler had probably appointed many (if not most) of the members; and for this particular review, he added some temporary members and consultants.[13]

Further, it appears that the committee was constituted in a manner conducive to the agency's desired outcome – and that Kessler and his colleagues seemed highly confident of obtaining it. This is evidenced by the fact that a week *before* the meeting commenced, the agency released a document asserting it had determined that "Flavr Savr tomatoes are as safe as other tomatoes."[14] This bold move broke with the agency's policy of not commenting on products that were under review,[15] and it signaled the administrators' certainty that the committee would uphold their position. After all, if they had believed there was even a modest chance the committee would oppose them, they would almost certainly have refrained from issuing that statement, since being rebuffed by their advisors at a public meeting would have created an extremely awkward predicament. That's why this pre-meeting pronouncement prompted one Calgene executive to regard the upcoming event "primarily as a PR exercise for the FDA."[16]

But the document's timing was not its only curious feature. It also purported that the determination of safety had been made by the scientific staff, even though several agency experts who examined the feeding studies had asserted that they did *not* support such a determination – and that within this group (which apparently comprised the majority of those who had scrutinized the studies) were the Director of the Office of Special Research Skills, the Leader of the Diagnostic Pathology Branch, and even the Chief of the Pathology Branch, the branch that specializes in the assessment of adverse changes. Consequently, the media and the public were seriously deceived, as attested by a report in the *Sacramento Bee* headlined: "FDA scientists find Flavr Savr safe."[17]

The advisory committee was deceived as well. During the meeting (held from April 6th through 8th, 1994) none of the agency scientists who called for more testing was selected to speak, and no mention was made of their concerns. Instead, the FDA picked two scientists who sided with the administrators to represent the position of agency experts on the food safety issue. Further, not only did their views differ from those of the scientists who expressed doubts about the Flavr Savr, they clashed with the opinion of the experts on the agency's biotechnology task force about the need for rigorous testing of GE foods in general. Thus, the one who led off said the FDA had made the "correct" decision in leaving animal studies optional rather than mandatory.[18] Moreover, when he discussed the Flavr Savr, he downplayed the significance of the stomach lesions; and although (as we've seen) it was illegitimate even to term them "mild," he declared they were "very mild."[19] In all, he conveyed the impression that, in the view of the scientific staff, there was no unresolved issue and no need for further studies; and the second scientist did the same.[20]

Any committee member who heard these presentations would have been surprised to learn that a substantial number of agency scientists (including the Chief of the Pathology Branch) were not so sanguine about the lesions and had concluded that additional testing must be done. Only by covering up this fact and preventing the concerns of those experts from being presented could the agency have been so confident its desired outcome would be achieved.

And the subterfuge succeeded. By the end of the meeting's third day, most of the members of the advisory committee were in such accord with the FDA's position that one attendee described the wrap-up as a "love fest."[21] If the facts had been fairly reported, and the committee had learned that the agency's pre-meeting proclamation was deceitful, the fest would almost surely have involved some festering.

Regrettably, the FDA was not prepared to amend its ways, and after the meeting ended, it continued (and even deepened) its deceptions regarding the Flavr Savr. In a subsequent press release, Commissioner Kessler declared: "We have approached our review of this product with scientific rigor and a commitment to full, public disclosure of that science."[22] But the agency's style of disclosure fell so short of "full" it was fraudulent. Thus, the document went on to assert that, according to the agency's assessment, "all relevant safety questions about the new tomato had been resolved," when in reality, key members of the scientific staff had insisted that "major issues" were not adequately addressed and that "unresolved" safety questions still remained.

And to further give the lie to the assurance of full disclosure, the FDA not only solidified the illusion that its scientists had determined the tomato was safe, it obfuscated the embarrassing feeding studies. For instance, in a release issued the same day as the one boasting a commitment to full disclosure, the agency provided a purportedly comprehensive summary of "all the data" submitted by Calgene that had been reviewed by its experts. But while it noted there were analyses of the tomato's composition, the identity and stability of the inserted genetic material, and issues regarding the antibiotic resistant marker, it said nothing about the testing in which rats suffered stomach lesions.[23] Another official account published the next year was similarly mute in this regard.[24] And this sanitized story of what evidence was examined, which omits mention of the most crucial (and troubling) tests, has become the one most widely circulated.

So, although when the FDA initially encouraged Calgene to conduct the feeding tests there was a mutual feeling that "everyone would breathe a little easier" if they were carried out,[25] after the results were in, the agency apparently felt that easy breathing would instead be fostered by pretending that the tests had never happened.

Déjà vu II: The FDA Breaks the Law Again Too

What's more, besides being deceitful, the FDA's behavior was illegal. Just as its general presumption that GE foods are GRAS was contrary to law (as we saw in Chapter 5), so were its actions regarding the Flavr Savr. In fact, they were illegal in more than one respect.

The FDA was supposed to be evaluating the tomato according to the GRAS standard, which (as Chapter 5 discussed) requires an overwhelming consensus about safety among experts – and also requires that the consensus be based on adequate technical evidence. But neither requirement was met. It's clear there was not a consensus in favor of safety within the FDA's expert staff, and it's clear that the requisite technical evidence was likewise lacking. For instance, the previously cited memo of the Director of the Office of Special Research Skills (Dr. Robert Scheuplein) stated that ". . . the data fall short of 'a demonstration of safety' or of a 'demonstration of reasonable certainty of no harm' which is the standard we typically apply to food additives."[26]

However, in an attempt to ease the tomato's passage through the review process, the FDA had instructed its scientists to disregard this standard and apply a less rigorous one. As Dr. Scheuplein wrote in regard to Calgene's submission: ". . . it has been made clear to us that . . . the safety standard is not the food additive standard. It is less than that but I am not sure exactly how much less."[27]

Of course, the reason he was unsure about how much looser the standard was supposed to be is that no lower standard exists. According to US law, all substances added to food fall into one of two categories in regard to testing. In the first are those that don't have to be backed by any testing because they have a history of safe use prior to 1958.[28] All others fall into the second category; and their safety must be demonstrated through scientific procedures, whether they seek entry to the market via the formal food additive process or by virtue of being GRAS. And, as Chapter 5 has made clear, the standard for testing is the same in each instance.[29] There's no special category for which a lower standard applies, and the FDA's ad hoc effort to create one was illegal.

Yet, the FDA (through its administrators) was not prepared to let the requirements of the law impede the introduction of the first GE whole food, and in its endeavor to circumvent them, it eventually decided to deny any legal significance to the problematic feeding studies. So, although the agency had originally instigated the testing, it ultimately argued that testing hadn't actually been needed anyway, that the results could be ignored, and that the status of the tomatoes could be adequately assessed through analytical means alone. And, in a final flourish, it asserted that, according to its own analysis, these tomatoes "have not been significantly altered" in a legally meaningful way.[30]

But it was the argument itself that was not legally meaningful; and its nonsensical nature is exposed in Appendix C, which elucidates just how arbitrary and capricious the FDA's actions in regard to the Flavr Savr, and to GE foods in general, have been. Further, even if we humor the agency by assuming that there had not been an initial requirement for testing, the fact is that feeding tests *had* been done (at the FDA's encouragement) and that, according to its own scientists, they raised a safety issue that was not satisfactorily resolved. Thus, it was clear that a significant number of qualified experts did *not* regard the tomato as safe according to standard criteria – and that it therefore could not be GRAS.

Another Wrinkle: There Were Ill Effects Worse than Lesions

Moreover, besides being linked with the lesions, the Flavr Savr was associated with outcomes that were far more severe. Seven of the original 20 rats that ate one of the GE lines died within two weeks. But in each of the other groups (in which the rats were fed the other GE line, a natural tomato, or a water control) only one death occurred.[31] Further, the death-linked line was not the one linked with lesions;[32] so each of the GE lines in that particular study was connected with adverse effects.

The data on deaths was contained in an endnote in one of the research reports; and it was only brought to my attention because in 2000, Arpad Pusztai contacted me and requested copies of Calgene's submissions.[33] It was through his diligence that the portentous endnote was discovered and eventually reported more widely.[34]

This discovery shocked him, not merely because of the number of rats that died, but because the response of the researchers had been so abysmal. In their crucial endnote, they cursorily attributed the cause of death to a husbandry error while furnishing no evidentiary back-up for this determination – and no further explanation. Moreover, they replaced the dead animals with new ones and kept the study going as if nothing had happened. As Dr. Pusztai explained to the investigative journalist Jeffrey Smith, these actions violated standard procedures in several respects. Smith reports: "He told me emphatically that in proper studies, you *never* just dismiss the cause of death with an unsupported footnote. He said that the details of the post mortem analysis *must* be included in order to rule out possible causes or to raise questions for additional research. Furthermore, you simply *never* replace test animals once the research begins."[35]

Dr. Pusztai and his two colleagues found several other flaws in the research as well. For one thing, the range of the rats' starting weights was "unacceptably wide" – much wider than is permitted by "high-quality" nutritional journals.[36] Moreover, they faulted the researchers for failing to probe more deeply, especially after having discovered the disruptions to the rats' stomachs. Such results, they said, "clearly . . . should have prompted more experimentation," not only to investigate the effect of the GE tomatoes on stomach histology, but "even more important," to look for effects on the small and large intestines.[37]

Overall, these experts found it "regrettable" that the feeding studies were "poorly designed" – especially considering that "so much rested on the outcome." And they asserted that due to the various defects, the FDA's conclusion that the Flavr Savr was as safe as ordinary tomatoes "does not therefore appear to rest on good science and evidence which could stand up to critical examination."[38]

Of course, that flimsily founded conclusion was not made by the specialists who had evaluated the data according to the legally mandated standards. Like Dr. Pusztai and his colleagues, those scientists determined that the data was insufficient to serve as a demonstration of safety. The specious conclusion was instead made by administrators, who were not acting as agents of science and the law but as servants of a directive to foster biotechnology – and were therefore more intent on promoting its products than on protecting the public.[39]

Déjà vu III: The Administrators Spurn Yet Another Set of Expert Warnings

Moreover, not only did agency administrators snub science and infringe the law by dismissing the conclusions of their experts about the feeding studies, they also did so in regard to another important feature of the Flavr Savr: its possession of an antibiotic-resistance marker gene.

As Chapter 4 explained, because most cells subjected to gene implantation techniques fail to incorporate the cassettes of recombinant DNA, a large number must be targeted; and there must be some way of recognizing the tiny fraction that have taken it up. Consequently, the cassettes are almost always endowed with a marker gene that expresses a protein which renders the cell resistant to a particular antibiotic. This enables biotechnicians to select the cells that have acquired the new cassette by exposing all of them to that antibiotic and eliminating those that were not transformed by killing them off.[40]

Calgene had chosen a marker gene conferring resistance to kanamycin, a broad-spectrum antibiotic with a significant medical use. And the wide deployment of this gene raised the risk of widely spreading such resistance – which, in the view of many experts, was a risk not to be taken lightly.[41] So it was important for the prospects of the GE food venture that this marker be approved, not only because Calgene planned to use it in additional GE crops, but because other manufacturers intended to employ it in producing theirs.

Accordingly, the FDA believed it was imperative not only for the kanamycin-resistance gene to gain agency approval, but public confidence – and that this would best be achieved if it was certified as safe via the standard food additive petition process. Therefore, the agency asked Calgene to submit a separate food additive petition for the enzyme the gene produces.[42] And although the petition ultimately gained approval, the story of how it happened is a sordid one – as is graphically exposed by memos obtained through the lawsuit.

Months before filing the petition, Calgene had submitted several documents in support of its case for the safety of the marker gene system; and the FDA had requested its Division of Anti-Infective Drug Products to evaluate them.[43] It was the division with greatest expertise in assessing the specific risks posed by the presence of this gene within every cell of a widely consumed food. And this prospect roused its concerns. Accordingly, agency experts once again did not deliver the response the administrators wanted; and the Division's scientists instead expressed strong reservations about the marker gene's use. Further, perhaps because they knew how the administrators had disregarded previous memos from expert staff that described risks of bioengineering, they emphasized the depth of their concerns by capitalizing all the letters in the key sentence of their conclusion: "IT WOULD BE A SERIOUS HEALTH HAZARD TO INTRODUCE A GENE THAT CODES FOR ANTIBIOTIC RESISTANCE INTO THE NORMAL FLORA OF THE GENERAL POPULATION."[44]

Because this document could have severely undercut the agency's agenda to promote GE foods, the Division director sent it to another FDA official with a cover letter titled, "The tomatoes that will eat Akron." In it he stated: "You really need to read this consult. The Division comes down fairly squarely against the kan gene marker in the genetically engineered tomatoes. I know this could have serious ramifications."[45]

But such ramifications could only ensue if the warnings were heeded by the administrators – or else became known by the public. And neither of these conditions occurred. In what was becoming a standard procedure, the administrators again spurned the warnings of their experts, even though the main admonition had been conveyed in an emphatic, and dramatic, manner. Further, also in line with their previous practice, they made sure the concerns were covered up.

However, in pursuing this callous course, they not only had to brush off the initial report from the Division of Anti-Infective Drug Products, they had to ignore a strongly stated follow-up from its Supervisory Microbiologist. In an effort to drive home the reality of the risks, he delineated the various weaknesses in Calgene's arguments and explained why its submission failed to demonstrate that use of the kanamycin-resistance gene would be safe. He next pointed out that although other markers were available that did not pose appreciable risk, the industry preferred the antibiotic resistant ones because they're easier to use. Then, in forthright words that he hoped would be given the attention they deserved, he declared it was wrong to let industry convenience trump public safety. "In my opinion," he wrote, "the benefit to be gained by the use of the kanamycin resistance marker in transgenic plants is out weight [sic] by the risk imposed in using this marker and aiding its dissemination nation wide. If we allow this proposal, we will be adding a tremendous quantitative load of genetic material to the environment which will probably assure dissemination of kanamycin resistance."[46]

But his efforts were in vain. The administrators ultimately shunned all the cautionary input that he and his colleagues had provided and instead gave Calgene formal approval to use the antibiotic-resistance marker gene not only in its Flavr Savr tomatoes, but in the GE canola and cotton it planned to produce in the future. And to justify their action, they projected the impression that the agency's scientists had no qualms about the marker gene and were uniformly convinced of its safety.

Moreover, as in the case of the feeding studies, effecting this fraud entailed hoodwinking the Food Advisory Committee – and falsely purporting there was consensus about safety among agency experts. Thus, as had happened during that earlier deception, when the marker gene was discussed, none of the experts with reservations was selected to appear; and the discussion was instead conducted by a scientist who was well-attuned with the agency's agenda. Indeed, she belonged to the Biotechnology Policy Branch, which was keen to promote GE foods and had helped craft the agency's lax and illicit policy statement.[47] Accordingly, in asserting that the agency deemed the gene safe,[48] she skirted the fact that the entire Division of Anti-Infective Drug Products (comprising the scientists with the greatest expertise in evaluating the relevant risks) had determined that its safety had *not* been demonstrated – and that it posed an unacceptable threat.

Within a few weeks of the meeting's end, the FDA formalized this deception by approving Calgene's food additive petition – and spuriously certifying that using the kan marker gene in the production of the corporation's bioengineered tomatoes, canola, and cotton was, from the standpoints of both science and the law, safe.

Final Take on How the FDA Put the Tomato on the Market: Multiple Frauds and a Four-fold Breach of the Law

The FDA announced the approval of the marker gene on May 18, 1994, the same day it notified Calgene that it had no objection to the marketing of the tomato itself.[49] Further, the agency expressed no reservations about either of the lines that had been linked with harmful effects; and it appears that the one Calgene commercialized was the line associated with the curious deaths.[50]

But probably no one outside the FDA knew that this first bioengineered whole food was reaching the nation's dinner tables through the organization's systematic misrepresentations and repeated acts of law-breaking – and that without such malfeasance, this novel fruit would have essentially died on the vine.

The frauds were crucial to the law-breaking, and the law-breaking was crucial to commercialization. There were four distinct transgressions; and the first was the worst, since it laid the groundwork for the marketing of *all* GE foods. It occurred in 1992, when the FDA brazenly breached the law by pretending that GE foods were GRAS when it fully knew they weren't. The next year, the agency broke it again by telling its experts to assess the Flavr Savr according to a looser standard than the law prescribes. And when the agency subsequently declared the tomato problem-free, that was another breach, because it did so in the face of agency experts who maintained that safety had *not* been demonstrated – and that an important issue was unresolved. But that transgression was not in itself sufficient to enable commercialization. In order for the tomato to appear on supermarket shelves, the food additive petition for the kanamycin resistance gene had to be approved; and the FDA could only do so by breaking the law one more time – and certifying the additive safe even though a key division of agency experts had concluded it entailed unreasonable risk.

Further, because so much confusion has been created about what the law requires, it's important to point out not only that the tomato's entry to the market was illegal, but that it would have remained illegal even if the advisory committee *had* been fully informed of the suppressed staff memos but supported the administrators' position anyway. That's because general recognition of safety cannot be achieved by getting one group of experts to disagree with another group – and even two experts have sufficed to defeat a claim of GRAS in federal court.[51] Therefore, the fact that a significant number of agency experts concluded that the safety of the tomato had not been demonstrated nullifies the notion that it was GRAS. Although the administrators could befog that fact, they could not expunge it – and due to that obstinate reality, the product reached the market through a breach of the law.

Additionally, the approval of the marker gene's food additive petition was not (and could not have been) legitimized by the opinion of the committee either. The experts in the Division of Anti-Infective Products were especially concerned about the risk that kanamycin resistance would be spread to human gut bacteria; and they noted that Calgene had not presented a concrete demonstration to rule out this problem but only theoretical arguments.[52] They further stated that even though some of them were "plausible," Calgene needed to "demonstrate these arguments as fact."[53] Accordingly, they advised that a controlled animal study should be conducted to test them.

In his follow-up memo, the Division's Supervisory Microbiologist emphasized the unreliability of Calgene's assumptions. He pointed out that several were based on the belief that scientists have "adequately" understood important mechanisms at play in the bacterial world, and he then declared: "In my opinion, nothing could be further from the truth."[54] He additionally emphasized that there was no solid basis to support Calgene's calculations about the effects of deploying the marker gene; and he asserted: "We can not predict what the consequences of this action will be."[55]

However, although these experts concluded that safety could not be established on the basis of Calgene's assumptions and arguments, and that concrete experimental data was required, Calgene did not submit any.[56] Therefore, it's obvious that the marker gene failed to meet the legal standard of safety, which requires "a reasonable certainty in the minds of competent scientists that the substance is not harmful under the intended conditions of use."[57] And even though the FDA succeeded in getting some other scientists on its staff to say they saw no problems with using the gene, and even though it eventually was able to convince most of the scientists on its advisory committee to concur with this viewpoint, that could not overcome the fact that the group of scientists most skilled at assessing the gene's safety remained dubious.[58]

The Failure of the Tomato's Basic Hypothesis – and Its Failure on the Market

Not only were the potentially complex consequences of employing the antibiotic-resistance marker gene unpredictable, even the seemingly straightforward effect of the enzyme-inhibiting gene was not accurately anticipated. The central hypothesis on which the tomato's success had been staked was the assumption that decreasing the crucial softening enzyme would enable the fruit to ripen on the vine and yet stay firm enough to resist damage during shipping. But this assumption did not hold up – a failure which, in combination with poor business planning, induced the product's demise only a few years after its debut.[59]

Moreover, this collapse was in no way connected with public resistance to genetic engineering.[60] So deft had been the FDA's deceptions that even prominent opponents of the technology were convinced the tomato's safety had been established.[61] Thus, the fruit failed, not because of opposition to bioengineering, but due to the failure of bioengineering to yield the desired results.[62] Of course, had the facts been fully aired, and the public informed of what the FDA experts had actually said, they would almost surely have reacted differently – and not only rejected this first fruit of bioengineering, but served notice that no others would be accepted either.

Thus, the commercialized phase of the GE food venture had a remarkably marred beginning. And the facts connected with its first food supplement and first whole food were so unsettling that, if they'd been honestly reported instead of stubbornly obfuscated or systematically misrepresented, the entire enterprise would have been halted – and you would not be reading this book because there would have been no need to write it.

But you *are* reading it; and as you continue, you will learn how, as the venture continued, it continued to yield troubling results – and how its proponents were continually driven to keep on obfuscating or distorting them.

The Caliber of Testing Slackens

The safety-related studies on the Flavr Savr that Calgene submitted to the FDA were never published as research papers in peer-reviewed journals, and, as we've seen, they did not meet the minimal standards for mainstream scientific journals that publish such studies. Yet, despite their flaws, they were of a significantly higher caliber than most of the testing of GE foods for many years thereafter. In the United States, the FDA professed that the tests on the tomato had so successfully demonstrated safety that tests on other GE foods would not be necessary – and that the soundness of its policy to not require any had been confirmed.[63] But even if the tests *had* actually shown the tomato to be safe, it would still have been illegitimate to so boldly generalize, because (as will later be explained in more detail) being able to demonstrate that one GE food is safe would *not* establish that any others are. And although most other industrialized nations did not adopt the FDA's hands-off policy, none required the kind of testing to which the Flavr Savr had been subjected; and they all settled for a standard that was much lower.

As Chapter 6 explained, this standard, which is termed "substantial equivalence," is far *less* substantial than the GRAS standard, which is the one by which the Flavr Savr was supposed to have been assessed. In fact, it's so lax that for many years, only a few studies on GE foods were published in peer-reviewed journals because (as Chapter 6 also revealed) the regulatory requirements were far below the standards the journals upheld.

Accordingly, most GE foods that entered the market had scant hard evidence to attest their safety.

So deficient was the system of assessment that in 2001, five years after the wave of post-Flavr Savr GE foods had begun spreading throughout North American markets, a report by the Royal Society of Canada (discussed in the previous chapter) not only lamented the lack of adequate testing of these foods, but noted that there weren't even any validated test protocols to assess their safety "in a biologically and statistically meaningful manner."[64] Thus, all of the many claims up to then alleging that safety had been proven were, from a genuinely scientific standpoint, baseless.

How an Effort to Improve the Means for Detecting Ill Effects Was Viciously Attacked for Actually Detecting Some: The Astonishing Story of Arpad Pusztai

An Award-Winning Research Design

But the Society's lament was not the first time the dearth of meaningful test protocols was prominently acknowledged. Six years earlier, before any bioengineered soy, corn and canola appeared on the market, the Scottish Agriculture, Environment and Fisheries Department (SOAEFD) recognized the lack and endeavored to remedy it by stimulating the development of a model procedure that could reliably verify a GE food was safe to eat. So it called for research proposals, and out of 28 submitted from throughout Europe, it selected the one from a group headed by Arpad Pusztai, whom, as noted earlier, was a renowned food safety expert.[65] This proposal, which had won out over so many competitors, was awarded a 1.6 million-pound grant, with the expectation it would establish better methods for reliable risk assessment of GE crops and lead to sound published studies, none of which were then in existence.[66]

This promising project was coordinated through the Rowett Research Institute in Aberdeen, Scotland, one of the world's most prestigious nutritional research centers, and the place where Pusztai had been employed for 32 years. Other participants were the Scottish Crop Research Institute and the University of Durham School of Biology. The GE crop the project aimed to test was a potato engineered to fend off predators by producing a pesticide. This would be achieved by endowing the spuds with a gene derived from the snowdrop plant that expresses a type of protein (called a *lectin*) toxic to aphids and several other insects. The researchers hoped that, when inserted in potatoes, it would provide the same protection it affords the snowdrop.

Another variety of pesticide-producing potato (produced by Monsanto) was soon to be marketed in America, containing a different kind of insecticidal protein that was derived from a soil bacterium instead of a plant. But this was not the most important difference between the two types of altered potatoes. Although Monsanto's would be consumed by a large number of people, they would not be meaningfully safety-tested, whereas those developed by Pusztai's group would undergo the most stringent testing yet applied to a GE crop.

This testing was designed not only to detect ill effects, but to discern if any were attributable solely to some aspect of the bioengineering. To that end, there were two groups of control rats against which the rats fed the engineered potatoes were compared. One was fed potatoes from the natural-state parental line from which the GE lines had been derived, and the other consumed potatoes from the parental line that had been spiked with lectin at the same level produced within the GE lines. And to ensure that no extraneous differences existed between the GE and parent-line potatoes, they were grown alongside each other in a tunnel that was isolated from the environment and were otherwise exposed to identical conditions as well.[67]

Further, to provide fuller insight, there were two distinct GE lines, each created from the parent line through a separate insertional event. In all, four separate studies were conducted, with two employing one of the GE lines and two employing the other. In each case, the details differed. For instance, one study tested raw potatoes for 10 days while another tested cooked ones for 30.

However, despite the fact the set-up was rigorous, Pusztai was confident it would not detect any adverse effects from the lectin. He was the leading authority on lectins, and his prior research had demonstrated that, although the snowdrop's version wreaked havoc on the innards of insects, it was safe for mammals – even at levels hundreds of times higher than the GE potatoes would contain. So he assumed that if any problems were discovered, they would be side effects of the genetic engineering process. Moreover, because like so many scientists at that time, he had been led to believe that this process was essentially safe, he also assumed that it wouldn't be linked to problems either – and that these GE potatoes would become a beneficial addition to agriculture. This latter assumption was also held by the Rowett Institute, which planned to commercialize the product.[68] But the assumption was wrong.

Unexpected, and Troubling, Results

When the tests were finally completed, and the results registered, they were disquieting in several respects. Chemical analysis revealed that neither line of GE potato was substantially equivalent to the parent line and that there were statistically significant differences in several constituents that are of major nutritional importance. It also revealed that the two GE lines were not even substantially equivalent to one another – which was probably due to the fact that the bacterial-delivered cassettes had been wedged into different regions of the genomes and had therefore induced disparate effects.[69]

What's more, the feeding studies revealed that the chemical changes correlated with substantial physiological changes. The rats eating the GE potatoes differed from those fed the parental line in many measures of general metabolism and organ development, and their immune systems were weakened.[70] In all, 39 statistically significant differences were found (by independent multivariate statistical analysis), of which no more than five could have been the result of random error.[71] Most of these differences were observable after a mere 10 days.[72] Further, they were found even when the GE-fed rats were compared to those that ate the parental potatoes spiked with the lectin.[73]

One of the more worrisome outcomes was that the GE potatoes induced abnormal proliferative cell growth in the middle section of the small intestines (the *jejunum*).[74] Such growth can be a precursor of cancer.[75] Further, *both* lines of GE potatoes were associated with intestinal abnormalities, and both were also linked to altered organ growth and diminished immune response.[76]

Responding to a Sensed Duty and Providing the Public the Facts

In Pusztai's assessment, the data clearly indicated that the troubling changes had not been caused by the mere presence of the lectin but instead had been induced by some aspect or aspects of the genetic engineering. And this concerned him, because he knew that these perturbing potatoes not only could have effortlessly entered the US market but would have passed through the superficial regulatory regimes in rest of the world without a hitch. And he further realized that the various GE foods already being consumed by multitudes of people might be harming them in a similar way that his potatoes had harmed the rats – and that even if adverse effects were progressively cumulating, there was little chance they'd be linked to the bioengineered products that were causing them.[77]

Moreover, he was disheartened by the fact it would take a long time to bring out the facts – *if* he stuck to standard procedures. That's because, according to convention, a scientist is not supposed to communicate research findings prior to presenting them at a conference or via a published article. And observing this convention would create a substantial delay. As he later explained, "I had facts that indicated to me there were serious problems with transgenic food. . . . It can take 2 to 3 years to get science papers published and these foods were already on the shelves without rigorous biological testing. . . ."[78]

So when the British TV show, "World in Action," requested to interview him about his research and his views on GE foods, it offered an opportunity to provide the public with important information he believed they had a right to know – especially since the research that produced it had been funded by their taxes. However, he was reluctant to take such a bold step without gaining the permission of the Rowett Institute's director, Professor Phillip James – permission that was readily granted.

Pusztai's was only one of several stories in the program, and his interview was edited down to 150 seconds for broadcast. But those 150 seconds were potent. In them, Pusztai noted some of the adverse effects that had been induced by the GE potatoes and emphasized the importance of tightening up the testing standards. Moreover, he expressed his concerns about GE foods and strongly criticized the fact they were on the market even though their safety had not been adequately demonstrated. As he put it, "I find that it is very, very unfair to use our fellow citizens as guinea pigs. We have to find the guinea pigs in the laboratory."[79]

Colossal Controversy and Cruel Reprisals

Not surprisingly, when the interview aired on August 10, 1998, it created a colossal stir; and immense attention was directed onto Pusztai and the Rowett institute. Professor James was initially delighted with this attention, and he spoke of Pusztai and his research in glowing terms.[80] When Pusztai met with James before leaving for home the following afternoon, the latter's attitude toward him had not noticeably changed.[81] But things took a drastic turn on August 12th. Pusztai was abruptly fired, the research was terminated, and all the data was confiscated.[82] And to aggravate the insult, Pusztai was put under a gag order that forbid him from speaking about the research – and threatened legal action if he did.

Such a dramatic reversal cries out for an explanation. How did Pusztai morph from hero to outcast in less than 24 hours, and why did his research so suddenly turn repugnant? It appears that this abrupt attitudinal shift did not originate within the Rowett Institute – and had no basis in science. Instead, evidence points to the intervention of an outside influence that was not only powerful, but political. According to a 2003 article in *The Daily Mail,* Pusztai alleged he was separately informed by two employees at Rowett that the day after the broadcast (which was the day before he got sacked) Professor James received two calls from the office of the Prime Minister, Tony Blair. He also alleged that a senior manager at the institute told him and his wife that Blair interceded after being called by then US President, Bill Clinton, whose administration was heavily promoting GE foods – and pressuring other nations to accept them.[83]

The *Daily Mail* article further reported that the story of Clinton's role was corroborated by two other eminent researchers at Rowett, one of whom stated he was informed of it by a senior official at the institute. The article also pointed out that although Professor James strongly denied he was contacted by Blair, it was evident that the Blair government was allied with pro-GE forces and even participated with them in "a coordinated counter-attack" against Pusztai.

Intense, Unjust, and Hypocritical Attacks

This attack was vehement, and often vicious, and it berated his behavior as well as his research. Further, besides being unfair, it was hypocritical. After all, consider that Pusztai was scolded for violating protocol by speaking prematurely despite the fact no protests were raised when (as we saw in Chapter 1) Stanley Cohen announced his research results well ahead of publication in order to derail legislative attempts to regulate bioengineering. And, in contrast to Pusztai's case, his main claim wasn't even true and was at odds with his actual study. Nor were there outcries when the FDA broke with a long-standing policy by prematurely (and fraudulently) declaring the Flavr Savr's safety. So it's clear that biotech proponents approve of premature pronouncements that serve to protect the GE venture and only object when the aim is to instead protect the public from its routinely-denied risks.

And the hypocrisy didn't stop there. Although the scientists who praised GE foods had previously ignored the protocols of science so they could proclaim their safety on the basis of testing that was glaringly deficient, now that one had been prominently linked with problems, they suddenly became ardently devoted to the principles they had formerly been willing to scrap. Moreover, although they demanded that this unwelcome research be subjected to the highest standards, their rediscovered scruples were not strong enough to deter them from decrying defects that didn't even exist.

Consequently, they derided its design for several alleged flaws, despite the fact it actually *did* conform to the rigorous standards they now professed to champion. And it's difficult to see how they could have honestly done so. The study's essential design had already been employed in dozens of published studies conducted at Rowett, and it had won out over 27 other proposals in a competition held by a department of the Scottish government. Moreover, according to Pusztai, it had also been approved by the UK's Biotechnology and Biological Sciences Research Council.[84]

One of the most deplorable attacks was mounted by the institution that had for centuries been regarded as a paragon of scientific rectitude: the Royal Society. This august organization (which is the UK's national academy of science) was founded in 1660 and is the oldest scientific academy in continuous existence.[85] And, although for most of its history the Society had refrained from taking sides on issues or from even expressing an official opinion on a topic,[86] by the mid-1990's, it had become a partisan defender of GE foods and embraced a proactive policy on their behalf.[87] Pursuant to this policy, it endeavored to quell the unnerving influence of Pusztai's interview by impugning him and his research. As part of this effort, 19 of the Society's fellows wrote an open letter attacking his work.[88] And a month later, in March 1999, the Society itself initiated a major, and unconventional, phase in the campaign.

Up until then, it had never operated as a peer-reviewing body, leaving that function to journals and other institutions. Yet, as Chapter 9 and Appendix B have shown, the Society was prepared to stray from its standard practices in order to protect GE foods, so it broke with its long tradition and undertook the first peer review in its history.

And this peer review was truly peerless. Not only was it unprecedented in regard to the Society's own past, it was without peer in regard to other peer reviews. However, its uniqueness was not due to exceptional merit, but distinctive dereliction.

Ordinarily, a peer review examines a complete data package, because that's the only way research can be properly assessed. But the Royal Society did not review all of Pusztai's data, because the Rowett Institute only sent an incomplete report that had been prepared for use by scientists on the research team who were familiar with the basic details. Accordingly, that abbreviated report didn't describe several key facets of the design – and was not what Pusztai and his colleagues would have submitted to a refereed journal. However, although a standard journal would never have accepted such a submission, the Royal Society did. And it even refused Pusztai's offer to provide adequate data, which he had finally obtained in late 1998 when Rowett was obliged to release it to him so he could testify before a Parliamentary committee.[89] Thus, instead of obtaining a proper package, this premier scientific institution saw fit to review one that was clearly unfit – and to then roundly critique the lack of fitness for which it itself was largely responsible.[90]

Moreover, although the research design was *not* deficient, the composition of the review panel *was*. According to Pusztai, none of the members had expertise in nutritional studies, and therefore none was properly qualified to assess some important aspects of the research.[91] This led to error. For instance, one of the reviewers claimed that there were too few rats in each group to obtain reliable results, unaware that the number was indeed sufficient, that Pusztai had previously conducted more than 40 nutritional studies in which this number was employed, and that all those studies were published in respected journals because they had satisfied reviewers who *were* properly qualified.[92] Further, the same referee made several other erroneous statements, one of which, according to Pusztai, was "not only outrageous," but revealed that individual "had no idea how such an experiment is conducted."[93] He also noted that the "poor" quality of another reviewer's comments implied that he or she was "out of touch with present day nutritional science."[94] And not only did yet another reviewer display ignorance of an important fact about nutritional research that should have been known by scientists reviewing this nutritional study, he or she apparently read the documents carelessly (if at all), because every fact about the study that was recited was wrong.[95]

Therefore, all the reviewers' criticisms were baseless. Most pertained only to the substandard submission, not to the research as actually designed and executed; and the rest were due to ignorance about important facts that should have been known by a competent review team.

However, although this review more closely resembled the proceedings of a kangaroo court than those of an objective scientific panel, the critical verdict that was issued in June 1999 nonetheless served to strongly discredit Pusztai and his research. But those who knew the ugly details realized that the Society had actually discredited itself. As the editor of the prestigious journal *The Lancet* declared, the Society's action was "a gesture of breathtaking impertinence to the Rowett Institute scientists who should be judged only on the full and final publication of their work."[96] And he subsequently branded it a "reckless decision" that abandoned "the principle of due process."[97]

But the Royal Society remained undaunted; and having unjustly damaged Pusztai's reputation, was determined to squelch any endeavor that might rehabilitate it. Consequently, it tried to prevent his research from being published, and its actions were once again unsavory.[98] What's more, they "intensified" when the Society learned that *The Lancet* was planning to print some of it.[99]

How Publication in a Premier Journal Was Denied Its Proper Effect

That journal had received a paper co-authored by Pusztai and Dr. Stanley Ewen, a pathologist at the University of Aberdeen, describing one of the studies that detected abnormal cell growth in the intestines. Given the volatility of the topic, the editor selected a team of six reviewers to scrutinize the paper, twice the usual number. And it survived this increased degree of scrutiny, with only one reviewer (who worked at a government-funded institute) siding against publication.[100] Accordingly, it was slated for inclusion in the issue to be published on October15, 1999.

This was bad news for the boosters of biotechnology. If the quality of Pusztai's research, and the validity of his findings about the intestinal abnormalities, were vindicated by such an eminent journal, it would cast reasonable doubt on the safety of bioengineered food in general. It would also cast doubt on the reliability of the scientists and scientific institutions that had so savagely disparaged the research. Accordingly, several GE proponents urged the journal to abandon its plans. And the Royal Society was among them. The journal's editor, Richard Horton, told the *Guardian*, "[T]here was intense pressure on *The Lancet* from all quarters, including the Royal Society, to suppress publication." [101]

A lot of the pressure was exerted in a phone call Horton received on October 13 from a senior member of the Society that he said began in a "very aggressive manner." [102] The caller told him he was "immoral" for planning to publish a paper which he "knew to be untrue" – and subsequently told him that if he proceeded, it would "have implications for his personal position" as editor.

However, unlike the scientists who were pillorying Pusztai, Horton maintained his integrity, and he published the paper as planned. Yet, this important occurrence did not induce the significant effects it should have. And that's because the pro-GE brigade intensified their attack, relentlessly distorting the facts about the research and the circumstances of its publication.

The unrepentant Royal Society remained at the forefront, adding mightily to the barrage of disinformation. In 2002, its Biological Secretary falsely asserted in the Society's journal, *Science and Public Affairs*, that the

Lancet published Pusztai's research "in the face of objections by its statistically-competent referees." [103] And in the same year, it released a report that more thoroughly misrepresented the research.

The Deplorable Power of Disinformation

That's the report on GE foods critiqued in Appendix B; and its foul treatment of Pusztai, which is not described in that appendix, was worse than any of the offenses that are. Not only did the report relegate his research to one short paragraph, that paragraph packed a big deception. It purported that the Society had published a review in 1999 that scrutinized the research as published in the *Lancet* and found it flawed, with no convincing adverse effects demonstrated.[104] But that publication was of the shoddy peer review previously discussed, and it appeared more than three months *before* Pusztai's research was published.[105] Accordingly, the members of that review panel had not even seen the complete data package submitted to the journal and had instead passed judgment on the incomplete one the Rowett Institute had provided. Thus, the paragraph deceitfully implied that the Society had analyzed the published study and determined it was defective when, in reality, its assessment had not even considered that study and had instead focused on an abbreviated summary it knew was deficient even before the assessment began.[106]

Further, the 1999 review was at least as devious as the 2002 report that fraudulently misrepresented it. For instance, after declaring that ". . . we have reviewed all available data related to work at the Rowett Research Institute," the document then claimed that "Dr. Pusztai indicated to us that further information existed, but did not provide it." [107] Yet, as previously revealed, Pusztai *did* offer to provide it and was rebuffed – apparently so the Society would have a package with ample fodder for critique.

By persistently spreading such disinformation, the Society and its allies in the scientific establishment misled and emboldened GE proponents who were not scientists. For instance, Lord Dick Taverne, an influential member of Parliament and founder of the organization Sense About Science, displayed a peculiar scientific sensibility in regard to Pusztai's research by declaring that the rats had been fed "harmful lectins inserted in potatoes" – thereby implying that the problems were not caused by basic features of bioengineering.[108] Ironically, Taverne's grossly inaccurate comment came in context of a complaint about the "irresponsible and reckless disregard for fact and evidence which has characterised the reporting of many scientific issues. . . ." [109]

Of course, the irresponsible and reckless disregard for fact and evidence had actually been displayed by the Royal Society and other influential voices of science, and it was through the confusion they intentionally fostered that the media fell into erroneous reporting – with the errors prejudicial not to biotechnology, but to Pusztai. In one of the most telling examples, the technology editor of *The Independent* demonstrated his failure to stay independent of such deceptive influence by stating that when *The Lancet* evaluated Pusztai's work, "the reviewers refused it for publication, citing numerous flaws in its methods – notably that the rats in the experiment had not been fed GM potatoes, but normal ones spiked with a toxin that GM potatoes might have made."[110] The fact that the technology editor of a major British newspaper was under the impression the research hadn't been published and that the rats had not even been fed engineered potatoes shows how stunningly successful the disinformation campaign had been.

Further, the notion that the research never got published in a peer-reviewed journal had gained considerable traction well before being bolstered by *The Independent.* For instance, it was prominently promoted by the report of New Zealand's Royal Commission on Genetic Modification issued in 2001. In commenting on defects that it purported to have recognized in Pusztai's research, the report states: "It was unfortunate that the process of peer review was pre-empted by premature media release, thus preventing further scientific assessment."[111] But the research *was* adequately assessed in *The Lancet's* peer review, and it was the Commission's assessment that got obstructed – not by any flaws in Pusztai's work, but by its own preference for false claims from GE proponents over the genuine facts obtained directly from him. Thus, although the Commission received extensive written briefs from Drs. Pusztai and Ewen and also received oral testimony from both, its report on their research nonetheless misrepresents it in several additional (and significant) respects.[112] Moreover, it omits several important facts of which the Commission had been apprised; and consonant with its claim that peer review was prevented, one of the omissions is the fact that the research was published in *The Lancet.* Whereas many references are listed for the report's discussion of Pusztai's research, nowhere in the main text or the reference section is there any mention of *The Lancet,* which is strange indeed. That a government-appointed, blue-ribbon commission could have become so seriously befuddled about the basics of the Pusztai research is in itself compelling proof that, even though its soundness is solidly supported by the facts, the forces that desire to discredit it have substantially triumphed.

So powerful has been the influence of disinformation that the main omission in some reports has been, not the failure to mention that the research was published, but the failure to mention the research at all. For instance, the 2004 report by the National Academy of Sciences that was analyzed (and rebutted) in the preceding chapter says nary a word about it. Presumably, the committee that wrote the report thought the research had been so thoroughly discredited that there was no need to note it. More disturbing, even the 2001 report by the Royal Society of Canada (also discussed in Chapter 9) says nothing about it either. Although that report examines a substantially larger amount of risk-relevant information than does its NAS counterpart, and although it cautions about the risks instead of trying to minimize them as the other one does, the authors nonetheless appear to have fallen under the illusion that Pusztai's research was so devoid of merit it could be justly ignored. Accordingly, it's not surprising that Pamela Ronald's book, *Tomorrow's Table*, is also mute regarding Pusztai.

And while Nina Fedoroff did discuss his research in *Mendel in the Kitchen*, her discussion *augments* the inaccuracies. Although she initially provided a fuller and more accurate account than have most other pro-GE commentators, she ultimately could not steer clear of error, and she went on to make a misstatement that's not only major, but novel – and stands as her unique contribution to the ongoing confusion. Her false report came in the form of a critique. Despite the fact Pusztai and his group had taken several steps to ensure that the control and GE potatoes had been grown under the same conditions, Fedoroff claimed there was a big fault in their set up; and she faulted Pusztai for allowing it, and for failing to recognize that it could have caused the differences between the two sets of spuds. In particular, she alleged (a) that he had "jumped to the conclusion" the differences were due to some aspect of the bioengineering, (b) that this conclusion was "unwarranted," and (c) that it was likely due to ignorance of a critical fact.[113] As she explained, "What he probably didn't know – because he was neither a plant breeder nor a plant biologist – was that the very process through which the plants are put during the introduction of the transgene – culturing through a callus stage and then regeneration of the plant – can cause marked changes in both the structure and expression of genes."[114] She therefore asserted that this comparison of GE potatoes with inappropriate controls was the "central flaw" in the experiment; and she marveled that, despite all the attention given to the research, no one before her seemed to have appreciated this crucial point.[115]

But what *she* failed to appreciate is that no one had seen this "central flaw" because it didn't exist – and that she'd 'seen' it only because she failed to see all the facts. In reality, Pusztai was well aware that generating plants via formation of an amorphous mass of cells known as a *callus* tends to induce substantial alterations, and he designed the experiment to avoid it. That's one reason he chose potatoes, because they can naturally propagate in an asexual manner and do not require the extreme form of callus-inducing tissue culture that ordinarily must be employed to regenerate a whole plant from a single cell. Therefore, unlike most GE plants, the engineered potatoes he developed had not gone through this disruptive process. Further, he had taken additional steps to minimize any differences that might have arisen during the gentler culturing they did endure. Thus, the troubling disparities between his GE potatoes and the controls were most likely due to some aspect of the bioengineering other than tissue culture. Further, even if they *were* attributable to the tissue culture, that wouldn't vindicate the bioengineering process, because tissue culture is one of its essential components.

However, as had the NAS committee in 2004, Fedoroff pretended the tissue culture process is so separate from bioengineering that its risks are irrelevant when assessing the latter's safety. But, as Chapter 9 explained, the risks posed by tissue culture are *inherent* to bioengineering; and they *must* be considered when gauging its overall risks. Moreover, Pusztai's research underscores just how substantial those aggregate risks are. That's because, by minimizing the disruptive influence of tissue culture, and reducing the probability it would be the source of adverse effects, the study strongly enhanced the odds that those observed were caused by basic features of the GE process distinct from the culturing – especially since it had shown that the lectin produced by the foreign gene had not itself caused harm. Accordingly, because most GE crops pose a higher set of culture-related risks (due to the more disruptive form of tissue culture employed in their development), it's reasonable to regard them as, in principle, of even higher risk than were Pusztai's potatoes.

But the proponents of GE foods were not about to acknowledge the logical implications of the research; and they instead tried to strip it of its rightful relevance. The Royal Society was in the thick of this endeavor. Its 1999 statement took pains to point out that: "The work concerned one particular species of animal, when fed with one particular product modified by the insertion of one particular gene by one particular method." It then grandly declared: "However skillfully the experiments were done, it would be unjustifiable to draw from them general conclusions about whether genetically modified foods are harmful to human beings or not. Each GM food must be assessed individually."[116]

Characteristically, not only was this assertion unsound, it was duplicitous. Although the Society was, on the one hand, anxious to *restrict* the significance of the problematic findings to that one unique set of circumstances, on the other, it was eager to *expand* the applicability of any study that could be used to refute those findings – even if its particulars were substantially dissimilar to those of Pusztai's. Thus, the Society's 2002 report tried to discredit Pusztai's research by alleging that subsequent feeding studies employing GE sweet peppers, tomatoes, and soya had tested "clearly defined hypotheses focused on the specific effects reported by him" and had not found any adverse effects.[117]

But this claim had no credible backing – and was a carefully crafted deceit. Although the Society cited a published document in a manner that implied it was a report of the primary research, it was not. It was merely an opinion piece that briefly alluded to two research studies. And neither of them came close to refuting Pusztai's findings. For one thing, they weren't reliable. One, performed by Chinese scientists on both GE sweet peppers and tomatoes, was not even published and had not yet undergone peer review.[118] And although the other, which tested GE soybeans, had been published in a Japanese journal, its methodology was deeply flawed. For instance, the soybeans used as the conventional comparator were not from the parental line that had given rise to the GE beans, so there were too many genetic differences between them. Further, it appears they were grown under different environmental conditions than had the GE beans, which additionally diminished their fitness to serve as controls. Moreover, all the beans were heated to such a degree that any unintended toxic proteins in the engineered ones would have been denatured.[119] Perhaps even worse, the study so drastically departed from standard procedures that, according to Pusztai, had it been conducted in the UK, "the researchers would have lost their animal licence and the research would have been forcefully terminated."[120] That's because the animals were essentially starved – which in itself destroyed the study's reliability.[121] In all, the shoddiness of this study contrasts so sharply with the thoroughness of Pusztai's that the Salk Institute biologist David Schubert called it "a joke."[122]

However, there's nothing funny about the outsize and utterly unwarranted influence these substandard studies have wielded. Their impact was established in 2001 by an opinion piece in the prestigious journal *Nature Reviews* – the one the Royal Society's 2002 report artfully cited as if it were original research. That document declared that the studies on engineered peppers, tomatoes, and soybeans had "tested" Pusztai's claim but hadn't detected adverse effects – implying that his claim had thereby been refuted.[123] Later that year, the report by New Zealand's Royal Commission on Genetic Modification also disparaged Pusztai's study by stating: "Extensive testing carried out by Chinese researchers, similar to that described by Drs. Pusztai and Ewen, has not replicated their results."[124] Thus, even before the Royal Society weighed in, the impression that Pusztai's research had been solidly discredited had been instilled within the scientific community and the public mind.

Yet, what's most unsettling – and amazing – is not that the Royal Society and other ostensibly authoritative commentators have tried to discredit Pusztai's research with studies on peppers, tomatoes, and soybeans that were substandard, but that they cited studies on those species in the first place. After all, Pusztai's research involved GE potatoes, so it can't be refuted by tests on different species of plants. And this holds even more strongly for the studies that have been cited for that purpose, since even the transferred foreign genes in those cases were different. While a lectin-producing gene from a plant had been inserted in the potatoes, the soybeans were endowed with an herbicide resistance gene from a bacterium, and the peppers and tomatoes were transformed with a gene from a virus that expresses its coat protein.

The Irrefutability of Pusztai's Research

Moreover, even if the other studies had employed potatoes from the same species as did Pusztai and had transformed them with the same cassette that he'd inserted in his, and even if they had precisely replicated his research design, they could not have refuted his results – no matter how many times they may have failed to reproduce them. That's because, as the Royal Society had emphasized when trying to restrict the relevance of Pusztai's study, it involved "one particular product," and that product contained a unique set of alterations induced by a unique insertional event (and to some degree by a unique transit through tissue culture). In fact, because the insertion of an rDNA cassette is such a singular event, with a singular set of effects, the two GE potato lines Pusztai created substantially differed from one another, despite the fact they derived from the same parental stock, had been transformed with identical cassettes, and were grown under identical conditions. Accordingly, any deleterious attributes of the GE varieties could be the results of where the cassette had lodged within the DNA, or what kind of disruptions it had caused, or how the viral promoter was affecting surrounding genes – or many other factors that were distinctly associated with the specific insertional events.

Thus, the reliability of Pusztai's findings could only be tested by employing the *same* lines of GE potatoes that were used in his experiments; and that was rendered impossible when the government and the Rowett Institute shut down his research and destroyed all the potatoes. Consequently, his findings about the adverse effects of those GE potatoes can *never* be refuted by further testing, no matter how well designed, and the fact that the Royal Society and prominent scientists appear oblivious to this reality is further proof that they do not adequately comprehend the basic workings of rDNA technology – or else are willing to misrepresent them in order to protect its image and promote its products.

Given the intensity of the controversy and the degree of disinformation that's been generated, it's important to be clear about the imperviousness of Pusztai's research. The paper Ewen and Pusztai published in *The Lancet* only discussed abnormalities in the rats in that study and made no attempt to generalize the results to other GE food crops. Consequently, their findings in that experiment, and the conclusions they drew, cannot be impugned by research on *any* other GE plants (including potatoes) – and they are thus, in regard to further testing, irrefutable.[125] They can only be challenged by discerning flaws in the way the study was designed or executed; and, as we've seen, there are no solid grounds on which to do so. The situation is similar regarding the other three studies within the research project, some of which have been published as book chapters. Their findings cannot be refuted by further research either.

And although in other contexts Pusztai opined that some of the effects he observed might be attributable to broader features of the engineering process – and that similar problems might therefore be induced by transforming other plants with the technology – he did not assert that this would invariably happen. Therefore, even his broader statements cannot be refuted by two or three (or even ten) tests on other GMOs, no matter how rigorously conducted. That's because demonstrating that one or another particular insertion event is not linked with adverse results would not demonstrate that every other insertion will likewise be problem-free. Of course, if the vast majority of well-conducted tests on GE crops were to find no ill effects, that would weigh against the idea that a wider range of GE foods might also be harmful. However, as subsequent sections demonstrate, no such evidence has accumulated, and the bulk of the rigorous testing has resulted in adverse outcomes.

Nonetheless, despite its strengths, and although its basic facts are beyond reasonable dispute, not only is Pusztai's research still shrouded by confusion, the impression prevails that it has been discredited. Moreover, the proponents of bioengineering have not been content merely to repeat their falsehoolds but have augmented their stock of them. For instance, Derek Burke, a former President of the Society of General Microbiology and a co-author of the 2001 opinion piece that unjustly impugned the research, jacked the distortion to a new level in 2014. He asserted that Pusztai had claimed in press releases and on a TV program "that GM potatoes caused cancer when fed to rats." He then embellished that bogus assertion by alleging: "A claim that this was true of all GM foods was then made, but never sustained." And in his final flaying of fact, he declared that "the claims were disputed and could not be reproduced."[126] But in reality, Pusztai had said nothing about cancer; nor did he extend the claims that he actually did make to all engineered crops.[127] Further, because the potatoes he used had been destroyed, there was never a legitimate attempt to reproduce his results – and it was deceptive to imply that studies on different species employing different recombinant cassettes were valid endeavors to do so.

Even worse, besides inventing new falsehoods, GE proponents have maintained them even after being alerted to their inaccuracy. An especially egregious example involves Nina Fedoroff. As we've seen, in 2004 she claimed to have discovered a "central flaw" that no one else had detected, although in reality the flaw was in her own misunderstanding of the facts. That claim appeared in her influential book; and in February 2006, she repeated it when she posted an expanded version of the book's discussion of Pusztai on an influential website. Shortly after her piece was posted, Pusztai sent her comments that pointed out her error. But he didn't receive a reply, and seven years later, her erroneous accusation was still standing on that website – and still spreading the false notion that the research was fatally flawed.[128]

A Perturbing Trend: As Adverse Outcomes Mount, Cover-ups Continue and Precautions Diminish

Perhaps you've noticed the growth of a persistent, and perturbing, trend. Bioengineering's first edible product caused a major epidemic. Its first whole food was linked with harm to lab animals that (in the eyes of the FDA's pathologists) cast its safety in reasonable doubt. And the first GE food to undergo thorough, industry-independent testing was found to cause significant adverse effects in the rats consigned to dine on it. Further, in each instance, governmental entities covered up key evidence and misrepresented key facts; and in the two most volatile cases, numerous eminent scientists and scientific institutions aided and substantially expanded the government's effort, turning it into a slick and systematic disinformation campaign.

Moreover, if the facts about either of the first two incidents had been fairly reported to the public, the GE food venture would have been brought to a stop – and probably couldn't have continued. And if it had restarted, full dissemination of the facts about the third incident, in combination with revelations regarding the earlier two, would almost surely have ended it. However, such dissemination was prevented. Although, thanks to Pusztai's bravery, the essential information was transmitted throughout the UK, and eventually spread through Europe, the American media kept the US citizenry in the dark. And this crucial black-out, conjoined with the adroit disinformation campaign mounted by biotech advocates, has robbed the research of its rightful influence. While the revelations in Europe significantly contributed to the growth of widespread resistance to GE foods that has largely kept them out of that continent's supermarkets, the ag-biotech venture has continued full force in the US and Canada and is still being ardently pushed by several governments and the mainstream scientific establishment.

Equally appalling, despite their pretensions to the contrary, the proponents of genetic engineering and the governments that abet them have no wish to follow up on research that raises doubts about its safety – and have consistently thwarted efforts to do so. In fact, instead of spurring increased scrutiny and tightened protocols, the adverse effects that were detected during the 1990's actually induced a reduction in oversight and a loosening of standards.[129]

Thus, as we saw in Chapter 3, although the Showa Denko Corporation endeavored to give the FDA the bacteria used in producing the toxic tryptophan supplements so studies could be done to determine the specific cause of the contamination, the agency would not cooperate. So the corporation finally destroyed the microbes, curtailing any chance of reaching a definitive conclusion. And a few years after the tryptophan-induced epidemic, when faced with problematic results of tests on the Flavr Savr tomato, the FDA again chose to curtail rather than encourage additional research. Although its experts had called for further testing to clarify the extent of risk, the agency not only refused to demand any, it declared there wasn't even a need for the testing that already had been done, that the product could be marketed without reliance on it, and that future GE crops did not have to undergo any testing at all. Around the same time, (as Chapter 7 revealed) the US Environmental Protection Agency was stifling follow-up studies on research that had shown a gene-altered soil bacterium was lethal to vegetation. Even though the associated risk was enormous (prompting Phil Regal to assert that further detailed research was "demanded"), and even though the EPA had funded the initial study, the agency not only refused to fund any follow-up research, it treated the university professor who participated in that revelatory study as a pariah.

Four years later, in its handling of the Pusztai incident, the UK government demonstrated it also desired the disabling of research that could damage the image of biotech – and was just as willing as the US government to effect this aim in a ruthless manner. Accordingly, although Pusztai's research had been funded by the Scottish government, and was supposed to establish a sounder protocol for future studies on GE food, when it produced embarrassing results, the central government swiftly shut it down and made sure those engineered potatoes would yield no further distasteful discoveries. Further, not only did the government foil any follow-up studies, it refused to implement the research protocol (into which so much public money had been invested) as a new standard – and instead retreated to the comfortable confines of the *substantial equivalence* doctrine, enabling GE foods to remain free from the scrutiny that could detect their unintended side effects.

Another Continuing Trend: Research that Produces Disturbing Results Provokes Nasty Attacks

Hence, at the dawn of the 21st Century, more than a decade after the first edible offering of bioengineering entered the market, the venture still rested on a precariously feeble footing. Not only had some of its products caused discernible problems, the Royal Society of Canada reported that testing was inadequate and that validated protocols were lacking. And the footing is even feebler today. There are still no requirements for adequate testing, and several of the tests that *have* been conducted augment the store of data that raise reasonable doubts about safety.

Among the unsettling results are the following:

- Male rats fed a variety of Bt maize developed by Monsanto for the Egyptian market differed from those fed the non-GE control maize in organ and body weights and in blood chemistry, despite the fact the control plants were the parental variety and were grown next to their engineered relatives.[130] The differences were detected after 45 days; and after 91 days, several toxic effects were measured, including abnormalities in liver cells, excessive growth of intestinal membranes, congested blood vessels in the kidneys, and damage to cells that are essential for sperm production.[131]

- Feeding another type of Bt maize to both young and old mice was associated with a marked disturbance of the immune system and of biochemical activity.[132]
- When mice were fed for five consecutive generations on GE triticale (a hybrid of wheat and rye) their lymph nodes enlarged and the number of some important immune system cells significantly decreased.[133]
- Rabbits that consumed GE soybeans had adverse changes in enzyme function in their hearts and kidneys.[134]
- Mice that ate GE soybeans for two years displayed indications of acute liver aging in comparison to those fed on non-GE soy.[135]

A more detailed summary of these and several other troubling studies is available at: http://earthopensource.org/gmomythsandtruths/sample-page/3-health-hazards-gm-foods/3-1-myth-gm-foods-safe-eat/

The Maltreatment of Malatesta

Further, not only have the results of testing often been unsettling, so is the way researchers who've generated them have been mistreated. And in some cases, the inflicted indignations have been all too similar to those heaped upon Pusztai. For instance, Manuela Malatesta, a professor at the University of Urbino in Italy, led a team that conducted long-term research on Monsanto's glyphosate-resistant soy and found that the mice that ate this GE food had disturbed functioning of their livers and pancreases – and that the males also had altered function of their testes.[136] Although colleagues advised her to refrain from publishing her results, she decided to do it anyway – and reaped the unfortunate consequences her friends had anticipated. Besides being forced from her post at the university, she was unable to get funding to do follow-up research. As she related: "I lost everything: my laboratory, my research team. I had to begin again from scratch at another university."[137]

The Searing of Séralini

Moreover, as in the case of Pusztai, besides punishing the researchers, biotech proponents have sometimes savagely assaulted the research. And the more important the findings, the more intense have been the attacks. Thus, one of the most important, and alarming, studies was subjected to special abuse. Like Malatesta's research, it was a long-term study on a glyphosate-tolerant GE crop – but in this case, the product was Monsanto's NK603 maize, which, like the soybean she tested, was designed to survive application of the company's Roundup® herbicide. It was conducted by a research team led by Gilles-Eric Séralini, a Professor at the University of Caen, France, and it was published in a peer-reviewed journal in 2012.[138]

The research grew out of a prior study the team had conducted on the same GE maize. It re-analyzed the raw data Monsanto had generated during a short 90-day feeding trial to convince regulators that the product was safe. But the European Food Safety Authority (EFSA) didn't need much convincing. Although differences were detected between the rats eating the GE and non-GE maize, Monsanto's researchers discounted their importance, stating they weren't "biologically meaningful."[139] And the EFSA readily accepted this assessment.[140]

However, when Séralini and his colleagues got their hands on the data (which required a legal action and a court order), they discovered symptoms of liver and kidney toxicity in the rats on the GE diet; and they published their findings in a standard journal in 2009.[141]

They then conducted their new study, feeding rats the engineered maize for two years instead of three months, to determine if those effects were truly insignificant in the long run. Further, not only was their study longer than Monsanto's had been, it was more comprehensive. It was also better designed, because it could distinguish between effects of the Roundup herbicide that would be applied to the crop and the effects of the maize itself – the first study to achieve such discrimination.

The results were highly damaging to the image of Roundup-resistant crops because they demonstrated a high degree of damage to the rats that ate them. They revealed that Roundup and the GE maize each independently caused serious injury to the livers and kidneys, abnormal onset of large tumors, and increased mortality. Such problems had not been seen in Monsanto's study because they take longer than 90 days to develop. For instance, the first tumors didn't form in male rats until the study had gone a month longer than Monsanto's, the first tumors in females didn't appear until the 7th month, and most of the tumors weren't apparent until 18 months had elapsed.

Further, the kidneys and livers weren't the only organs adversely impacted. Statistically significant damage was also detected to pituitary glands and mammary tissues; and all the negative effects were observed for the three basic categories of experimental rats: those that ate GE maize sprayed with

Roundup, those that ate unsprayed GE maize, and those that ate no GE maize and instead were given a small amount of Roundup in their drinking water (an amount similar to the amount that would have been ingested in a dose of sprayed maize).

Because regulators have often not required any toxicological feeding studies, and have never required any that are longer than 90 days, and because Séralini's study showed that a bioengineered product that had gained world-wide approval based on a 90-day trial could nonetheless induce severe and comprehensive harm when consumed for a longer term, it threatened to discredit the entire regulatory system – and to undermine the entire GE food enterprise. So the proponents of that enterprise quickly set out to discredit it instead.

As usual, their attack was vigorous, venomous, deceptive – and effective.

Their key contention was that the study had been defectively designed; but this complaint rested on the notion that its object was to detect cancer – despite the fact that clearly had not been its aim. Consequently, their critique was way off-base. In reality, the study was not intended to monitor signs of cancer but to detect long-term toxicity; and the two types of trials have different design protocols. Further, the study Séralini conducted not only satisfied all the criteria for such a toxicity study, it *exceeded* them in some respects. So all its measures of toxic effects were reliably obtained – and, even without taking the tumors into account, they were more than sufficient to cast reasonable doubt on the product's safety.

Further, it *is* legitimate to also take the tumors into account. Although the critics complained that cancer studies are supposed to employ more rats per group than did Séralini, that guideline is meant to make the studies more sensitive to the abnormal incidence of tumors. In technical language, it's a precautionary measure to avoid false negatives, not to prevent false positives. In other words, the purpose of using more rats is to decrease the likelihood that an unusual rate of tumor incidence will go undetected, not to guard against the wrongful imputing of significance to differences in tumor rates between groups of rats that aren't actually meaningful.[142] Therefore, as several experts have emphasized, because Séralini's study was *less* sensitive than the standard tumor-detecting trial but nonetheless detected numerous tumors, its results are even more portentous than if a higher number of rats had been used.[143]

Thus, the critics are in effect claiming that because the study employed fewer rats than are ordinarily needed to detect tumors, the tumors it detected don't really count – and that this somehow also nullifies the multiple findings of toxicity that were obtained via the standard procedures of toxicity tests. Obviously, this argument is not only false but ridiculous. And so are the others they've mustered.

For instance, the study has been attacked for using a strain of rats especially prone to tumors – which, it's alleged, would lead to such growths even in the absence of the GE maize and the Roundup®. But the researchers used the same strain that Monsanto had used in its 90-day study on the maize and in its rat studies with glyphosate. And this strain is a standard one used in long-term toxicity studies and in cancer studies as well. So if the use of that strain invalidates Séralini's study, it also invalidates those Monsanto studies and the all the other studies in which it's been employed – studies that include many other GE foods.

Moreover, the absurdity of the argument looms larger in light of the fact that the rats consuming the GE maize (or the Roundup alone) displayed a quicker onset of tumors, and a higher incidence of them, than did the control rats – which demonstrated that something besides their pedigree exerted a tumor-inducing effect.[144]

Yet, although their criticisms were utterly unwarranted, a host of GE advocates doggedly maintained the effort to discredit the study. Especially galling to them was the fact it had been published in a respected peer-reviewed journal, *Food and Chemical Toxicology* (FCT), which endowed it with a credibility that they could not abide. So they put prodigious pressure on the journal; and, more than a year after the study had been published, the editors finally succumbed and took the extraordinary step of retracting it – a step that may well have been facilitated by the appointment of a former Monsanto scientist to the journal's editorial board.

And, just as the Royal Society's peer review of Pusztai was without peer, so was the retraction of the Séralini study from a peer-reviewed journal. According to the Committee on Publication Ethics (COPE), the only valid reasons for retracting an article are unreliable findings (due either to misconduct or honest error), plagiarism, redundant publication, or research that's unethical.[145] Yet, in his initial statement about the retraction, FCT's editor-in-chief, A. Wallace Hayes, didn't cite any of these reasons and instead lay the sole blame on "inconclusive" outcomes regarding the rates of tumor incidence and mortality. And he alleged that the inconclusiveness was due to the use of too few rats and also to the particular strain of rat that was employed – allegations which, as we've seen, were without merit.

However, after receiving several letters criticizing his failure to follow the COPE guidelines, and belatedly realizing that "inconclusive" findings are improper grounds for retraction, he abruptly augmented his argument. Because the guidelines permit retraction if there's "clear evidence that the findings are unreliable" due to misconduct or "honest error," he made an audacious, and awkward, attempt to turn the alleged inconclusiveness into a case of error-based unreliability. He gamely asserted that because the data are "inconclusive . . . the claim (i.e., conclusion) that Roundup Ready maize NK603 and/or the Roundup herbicide have a link to cancer is unreliable." And he then attributed this "unreliable" claim to "honest error."[146]

But this attempt to rehabilitate the retraction, although game, was lame – because it relied on a gross misrepresentation. And, although Dr. Hayes attributed Séralini's alleged blunder to "honest error," it's difficult to do the same regarding his. That's because nowhere in Séralini's paper is there a claim that either the GE maize or the Roundup is linked to cancer. In fact, the word "cancer" does not even appear.[147] Yet, despite this reality, Dr. Hayes was not content to rest with the false accusation that Séralini and his co-authors had claimed a link between NK603 and cancer. He went on to exacerbate the falsehood by declaring that their paper contained "the claim that there is a definitive link between GMO and cancer" – thereby painting them as having irresponsibly extended their claim to all engineered organisms.

But the only irresponsibility on display was his own. Not only had Séralini's team behaved responsibly, they would have been irresponsible if they hadn't mentioned the tumors. That's because, according to standard protocols, researchers performing chronic toxicity tests must report the presence of tumors, even if their studies aren't designed to detect them.[148] And that is all that Séralini's team did. They diligently reported the data on the tumors without making any claims about links to cancer. And in trying to cast this conscientious behavior as a delinquency, Dr. Hayes had to strenuously twist the truth.

Moreover, even if there *had* been a legitimate basis for rejecting their discussion of the tumors, it would not have provided valid grounds for retracting the entire article. The COPE guidelines state that "if only a small part of an article reports flawed data," the "best" course is to rectify it via a correction. And they emphasize that: "Retraction should usually be reserved for publications that are so seriously flawed . . . that their findings or conclusions should not be relied upon." But Séralini's findings regarding the multiple toxic effects that were linked to both the GE maize and the Roundup were not only solid, they were the central focus of his study; and his discussion of the tumors was unconnected with them. So even if that discussion had been inappropriate or unreliable, it still would not have weakened those findings in any way. In fact, Hayes acknowledged that the number of rats was adequate to support those findings,[149] and he also acknowledged that the raw data were accurate.[150] So the retraction violates basic standards and offends logic.

It's also starkly at odds with the course of science. Numerous scientists have protested the retraction, emphasizing that inconclusive research can nonetheless be important – and cannot be dismissed solely on that basis. Reflecting this view, David Schubert wrote: "The editors claim the reason [for retraction] was that 'no definitive conclusions can be reached.' As a scientist, I can assure you that if this were a valid reason for retracting a publication, a large fraction of the scientific literature would not exist."[151]

Nonetheless, although they had neither fact, logic, nor science on their side, the forces that promote the GE food venture were once again able to discredit a well-designed study that solidly linked a bioengineered crop with adverse affects on health – despite the fact it had been peer-reviewed and published in a respected journal. Further, in some respects, their attack on Séralini's work was even more successful than the one they'd mounted against Pusztai's. In the latter case, they failed in their attempt to prevent the research from being published, and although the disinformation they spread deluded many people into believing that it was published contrary to the decision of the experts who reviewed it – and many others into believing it had not been published at all – it nonetheless does stand as a peer-reviewed study in a premier journal. But in Séralini's case, they pressured the journal into retracting the study, which stripped it of the distinction of publication and formally branded it as unreliable.

Moreover, although both studies had broad implications, due to the disinformation, those implications have been essentially ignored. Pusztai's research indicated that the harms it detected could have been caused by one or more general features of the bioengineering process,[152] while Séralini's indicated that the Roundup® herbicide heavily sprayed on many varieties of GE crops poses a disturbing degree of risk – and that a variety of maize engineered to tolerate the Roundup does as well (even when unsprayed). Thus, the former casts reasonable doubt on most of the GE crops on the market, while the latter shows that Roundup itself (which is sprayed on the majority of GE plants currently being consumed) and at least one variety of Roundup-resistant maize in itself *do* cause harm when fed to rats – while casting doubt on the inherent safety of all other Roundup-resistant crops as well.[153] However, not only has there been scant recognition of these realities, because the impression gained hold that Séralini's study focused solely on cancer, few people are even aware that it found severe toxic effects in the kidneys, livers and pituitary glands – and that the journal's chief editor had not contested these findings.

This dearth of awareness is due to the critics' consistent complaints about the reporting of tumors – and utter disregard of the other findings. Like a magician who misdirects peoples' attention so they won't see something that would ordinarily be obvious, the study's enemies concentrated their critiques on the tumor-related findings and created the illusion there weren't any others. And the illusion was so strong even seasoned journalists were taken in. For instance, in his report on the retraction, a *New York Times* reporter who had regularly covered biotech issues discussed *only* the contested findings regarding tumors and made no mention of the others, even though they were solidly established and beyond reasonable dispute.[154]

A Heartening Outcome: The Study's Restoration to Publication

Fortunately, the Séralini story now has a happier ending. On June 24, 2014, his unjustly maligned study was republished in yet another peer-reviewed journal: *Environmental Sciences Europe* (ESEU).[155] Because it had already passed the peer review process twice (once to gain publication in FCT and a second time when that journal performed a special review that confirmed there was nothing "incorrect" in its reported results), ESEU concluded that it deserved a place in the published literature.

But it remains to be seen whether the proponents of genetic engineering will finally afford it the respect it deserves – and openly acknowledge its serious implications. In light of their past behavior, such a response, though long overdue, and highly beneficial, would be highly surprising.

Entrenchment of Hypocrisy, Duplicity, and Audacity

Regardless of how the pro-GE forces treat the Séralini study in the future, their prior treatment has been shameful – like most of their previous practices. And it's clear that although that study and the one conducted by Pusztai are of far higher quality than any on which claims of safety have been based (two geneticists have called Séralini's "the most detailed and thorough study ever carried out on a GM food crop."[156]), the proponents of the bioengineered food venture have unfairly attacked them for non-existent faults while overlooking the serious deficiencies in the studies on which the venture rests.

Unfortunately, such hypocrisy has become the norm – and a double standard has become standard.[157] Even when the proponents have not fabricated flaws in the studies that reveal risks of GE foods, they've demanded that such studies conform to criteria far higher than any employed when approving these foods for sale. Consequently, when independent researchers have been able to reassess the data on which approvals have been based, they've routinely discovered adverse effects that the regulators either missed or misinterpreted. This chapter has already discussed one important example (the reassessment of Monsanto's NK603 maize) and Chapter 6 has discussed several others. A more comprehensive discussion is available at: http://earthopensource.org/gmomythsandtruths/sample-page/3-health-hazards-gm-foods/3-1-myth-gm-foods-safe-eat/

In light of all the preceding facts, one may marvel at the certainty that's constantly expressed about the safety of GE foods by those who promote them, because the evidence clearly puts their safety in reasonable doubt – and, at minimum, raises a presumption that every claim asserting that it's been proven is false. Further, when even the most ostensibly authoritative of these assertions are subjected to careful analysis, their falseness is conclusively confirmed.

For instance, in October 2012 the Board of Directors of the American Association for the Advancement of Science (the AAAS), issued a statement in support of bioengineered foods which declared that extensive testing has shown they're safe to eat.[158] And to support this assertion, they relied on an apparently impeccable source: a report issued in 2010 by the European Commission (EC) reviewing 131 research projects the EU has funded.[159] But they failed to note that only twenty-two of them related to food safety. Moreover, when a team of independent investigators (which included two molecular biologists) analyzed the ten most recent of those twenty-two, they concluded: "Within those ten projects, there is astonishingly little data of the type that could be used as credible evidence regarding the safety or harmfulness of GM foods."[160] They found that only one of the projects resulted in published studies on food safety – and that those three studies "do not show the safety of GM food but rather give cause for concern." What's more, because no bioengineered crops that are on the market were involved, even if the tests *had* demonstrated safety, the results could not be extended to any of the products that people are actually consuming.

Further, none of the other twelve projects demonstrated food safety either. In fact, their titles indicate that they were primarily focused on evaluating test methodologies or gauging consumer attitudes.[161] Thus, to the extent that the EU report the AAAS directors cited actually reflects on food safety, it goes against their claim.

Moreover, the other supposedly solid source they cited (a review of 24 animal feeding studies) doesn't support their claim either;[162] and the same investigative team that exposed the defects in the EC report demonstrated that this review is deeply flawed as well. According to their analysis, the authors dismissed statistically significant differences between GE and non-GE crops for "scientifically unjustifiable" reasons; they applied a double standard, rejecting the reliability of studies that detected harm while accepting the soundness of those that didn't, despite the fact the latter displayed the same design features that triggered their dismissal of the former; and several of the purportedly favorable studies were on animals (such as cows and fish) with digestive systems so different from ours that they're not deemed suitable for assessing human health effects.[163] Further, the authors even accepted studies as support for safety in which too few animals were employed (in some cases only six per group) to enable reliable conclusions – apparently oblivious to the fact that GE proponents had lambasted Séralini's study for using 10 per group, even though that number *did* comply with accepted standards. Accordingly, the investigators determined this review to be so "fatally biased . . . [that] no valid conclusions can be drawn from it."[164]

An additional review that the AAAS directors would surely have noted if it had been published prior to their statement, and that GE proponents have profusely cited since it appeared in 2013, likewise fails to establish that GE foods are safe. It was conducted by Alessandro Nicolia and three colleagues and purportedly reviewed 1700 studies (over 600 of which related to food safety and the rest to environmental safety). But despite its extolled comprehensiveness, it left out many important studies with inconvenient results. For one thing, it only reviewed those published between 2002 and October 2012, so it automatically excluded Pusztai's and several others with unsettling findings.[165] For another, even within their chosen time frame, the authors admittedly "selected" the studies for their review – but inexcusably neglected to describe the criteria by which they did so.[166] Moreover, according to the team that discredited the two reviews mentioned above, many relevant studies "are simply omitted" from the list while others are mentioned without any discussion of their findings, despite the fact they're "seminal to any discussion of GMO safety."[167]

Among those mentioned but then ignored are the studies conducted by Malatesta and her colleagues, even though (unlike any of the studies allegedly demonstrating safety) they performed long-term monitoring of a glyphosate-tolerant GE crop.

Even the Séralini study was omitted, despite the fact it had not been retracted when the review was published. In trying to excuse this crucial omission, the authors alleged that the study is "of no significance." However, not only did they fail to provide an explicit definition of "significance," they failed to explain why the study's statistically significant findings of organ damage and hormonal disruption were somehow insignificant – a most significant dereliction.[168]

Thus, by unjustifiably excluding the Malatesta and Séralini papers, the authors of this ostensibly thorough review disregarded two seminal long-term toxicological studies on the type of plants that comprise more than 80% of the edible GMOs on the market (the plants resistant to glyphosate) – essentially thumbing their nose at the notions of thoroughness and fairness.[169]

Accordingly, due to the above-noted defects alone, the document deserves no deference. Furthermore, the previously-cited investigative team has detailed several others that more decisively discredit it.[170] In all, these defects reveal that the paper is not the balanced scientific assessment it purports to be but is instead a partisan, unfair, and unconvincing effort to prop the image of GE foods.

A Crucial Transgression: The Illicit Reversal of the Burden of Proof

The previous chapter established that there has never been a consensus within the scientific community that GE foods are safe to eat; and this one shows that the experts who support the course of caution, and who assert that the safety of these foods has not been demonstrated, are in the right. It further shows that those within this camp who take a stronger position, and assert that the evidence does not merely fail to establish safety but casts considerable doubt upon it, are still standing on solid empirical ground.

It additionally demonstrates that the continued claims of safety made by the scientists who support GE foods significantly stem from ignoring, misinterpreting, or misrepresenting key facts. But even the claims that don't derive from such cognitive lapses or intentional deceptions are based on a major misconception – one on which the other claims are to a substantial part founded as well. It's the mistaken belief that those who question the safety of these products have the burden of proving that they're dangerous – and that unless they do, the products can be deemed safe.

As prior chapters have documented, according to US food safety law, the opposite is true, and those who introduce a novel additive have the burden of demonstrating that it won't be harmful. They've further shown that every GE food is subject to this requirement. Moreover, it's a standard legal principle that the party who bears a burden of proof must establish its case through a preponderance of the evidence – and that unless this happens, the other party wins. So the party without the burden need not produce any evidence at all in order to prevail, while the other side must not only produce it, but produce it to a dominant degree.

However, as earlier chapters have also shown, in the case of bioengineering this burden has been steadily and illegally shifted. As we saw in Chapter 1, the first illicit shift occurred in 1978, when deceptions generated by successive conferences of pro-GE scientists persuaded the US National Institutes of Health to declare that those who advocated the regulation of research with genetically engineered organisms would thereafter bear the burden of demonstrating danger. Chapter 2 then described how the US Department of Agriculture extended this shift to the environmental effects of GMOs, and Chapter 5 revealed how the nation's Food and Drug Administration made it applicable to food safety, even though transferring the burden contravened explicit provisions of the relevant statutes and regulations. Despite the FDA's pretensions of propriety, this shift was so clearly illegal, and so utterly unjustifiable, that when FDA officials have been publicly challenged about it, they've had to lamely deny that it's happened.[171]

Nonetheless, the developers and proponents of GE foods are either oblivious of the legally-imposed burden they bear, or else, like the FDA, in denial about it. Consequently, their arguments routinely rely on the notion that those who question the safety of GE foods must prove that they're harmful – and that they themselves can carry the day by persistently carping at threatening research. Thus, they deem it sufficient merely to cast doubt on any doubt-raising data, while failing to recognize that they're obliged to provide a separate and solid demonstration of safety for every GE food that's headed for market.

Moreover, not only do the manufacturers bear the burden of proving that each GE food is safe, their burden is extraordinarily heavy. In most non-criminal trials, it's sufficient for the party with the burden to establish that the weight of the evidence is on its side – and that its argument is more likely than not true. But US law imposes a much higher burden in the case of new additives to our food. A manufacturer cannot prevail merely by showing that the additive is more likely to be safe than harmful. It must instead demonstrate that there's "a reasonable certainty" it won't be harmful. Although this burden is probably not as heavy as the one borne by the prosecution in a criminal trial (where the defendant's guilt must be proven beyond a reasonable doubt), it is clearly supposed to be more onerous than the one laid on plaintiffs in non-criminal litigation.

This is obvious from how the FDA experts, who routinely work with this rigorous standard, applied it when evaluating the Flavr Savr. For instance, the Director of the Office of Special Research Skills acknowledged that he thought Calgene had made a "strong" case that the feeding studies "do not show harm." [172] Therefore, if he'd employed the standard of proof operative in an ordinary trial, he would have decided that because the evidence indicated the tomato was more likely than not to be harmless, its safety had been adequately established. But he applied a stricter standard. He emphasized that Calgene was required to provide "a positive demonstration of safety;" and he then explained why he thought it had failed to do so. He asserted that because the data "raise a question of safety," they "fall short of 'a demonstration of safety' or of 'a demonstration of reasonable certainty of no harm' which is the standard we typically apply to new food additives." [173] He then stated that in order to provide such a demonstration, it would be necessary for Calgene to conduct "a stronger study that resolves the safety question raised by the current data." [174]

So when the agency administrators subsequently permitted the product's commercialization without requiring any additional testing, they were not following the law but flouting it. And the same exacting standard against which the agency's experts assessed the Flavr Savr is legally mandated in the case of all other GE foods as well.

Yet, despite this stark reality, the products' proponents not only insist that the burden of proof must be borne by the other side, many demand that the standard of proof be extraordinarily strict – essentially arguing that a GE food can be deemed safe as long as no one has demonstrated there's a reasonable certainty it's harmful (which turns the legally mandated standard on its head). For instance, the pro-biotech scientists Bruce Chassy and Wayne Parrott have asserted that even if a study on a GE food that detects an adverse health effect is published in a peer-reviewed journal, it still must be "accepted by a consensus of the scientific community," [175] a condition that would legitimate the capacity of baseless attacks to delegitimize it. Moreover, they state it is also "necessary" that the results be verified by follow-up studies.[176] Of course, as is all too typical, their scruples are inconsistently applied, and they seem quite willing to accept studies in support of safety that don't come close to meeting such stringent criteria.

Therefore, these scientists are, in essence, advocating that challenges to the safety of bioengineered foods be subjected to the strictest evidentiary standards – and demanding that each be granted the special evidentiary protection extended to defendants in a criminal trial, with its safety presumed and accepted unless its harmfulness has been proven beyond a reasonable doubt. And not only do other pro-biotech scientists propound such preposterous measures, some take a position that's even more extreme. For example, according to Kevin Folta, a plant scientist at the University of Florida, "Those that support the hypothesis that GM crops are dangerous need to have the cleanest experiments, perfect controls, [and] massive numbers." Why? Because, as he puts it, "They are trying to overturn a paradigm, a scientific consensus." In his version of the facts, this "paradigm" regarding the safety of GMOs is "as intuitive as gravity for most plant scientists;" and he draws a parallel between challenging this paradigm and trying to disprove that gravity exists. He accordingly asserts that those who mount such a challenge face a "very different evidence threshold" than do those conducting research that supports the paradigm.[177] Thus, this prominent biotech proponent would have us believe that, from a scientific standpoint, the safety of bioengineered foods is so well-established that most experts accept it in the same way they accept the existence of gravity – an assertion so absurd it's difficult to believe it came from a credentialed scientist. After all, while no rational person would presume gravity doesn't exist, the law presumes that GE foods are not safe unless proven to be; and hundreds of scientists regard them as risky.

The Developers of GE Foods Have Fallen Far Short of Meeting Their Burden of Proof

So shifting the burden of proof, with its concomitant focus on whether the evidence has proven GE foods dangerous, has shifted attention from where it should be placed: the question of whether any GE food has been proven safe. And to properly answer this question, we need to understand what kinds of tests are needed to truly satisfy the requirements of US law – tests that could collectively demonstrate there's "a reasonable certainty" that a particular GE food won't be harmful.[178]

In 2014 the molecular biologists John Fagan and Michael Antoniou described a set of procedures that they consider minimally necessary to support such a demonstration.[179] In their prescription, the initial phase of safety assessment would still comprise comparative analysis, but in contrast to the present system, it would not be superficial. Instead, it would include the full range of available molecular profiling techniques (genomics, transcriptomics, proteomics, and metabolomics) that provide a far more sensitive assessment of whether potentially problematic changes have occurred in the bioengineered organism. The next phase would aim to determine whether that product causes troubling changes in laboratory animals that eat it. And, in further contrast to the current system, the testing would entail long-term feeding studies that not only study the animals throughout their lifetime, but monitor their offspring – and the next generation as well. In this way, researchers could gauge not merely whether the food is harming them, but is adversely affecting their capacity to reproduce – or adversely affecting the health of their progeny. Moreover, unlike so much currently permitted testing, all the studies would uphold proper scientific standards, with appropriate controls and rigorous adherence to all other basic protocols.

Additionally, the tests would include "comprehensive anatomical, histological (microscopic examination of body tissues), physiological, and biochemical analysis of organs, blood, and urine."[180] There would also be "molecular profiling of selected organs from test animals to evaluate effects on gene expression, proteins, metabolites, and RNA interference, which could underlie any negative health effects observed."[181] And, after such thorough feeding studies with lab animals, there would be similar testing of farm animals followed by long-term dose escalation trials with human volunteers.

Obviously, when even the most rigorous tests that have been employed in the testing of a GE food are compared to the set of tests prescribed above, they appear dismally deficient; and even if a less stringent set of requirements were adopted as a standard, as long as they were capable of reliably monitoring for long-term, multi-generational effects, they would tower far above the type of testing that's been used up to now.

Moreover, not only have the tests been deficient and the manufacturers wrongly relieved of their burden to demonstrate that GE foods are safe, they've been allowed to ignore the troubling results that even this inadequate testing has so often generated. And the extent of this dereliction is striking. As Michael Antoniou has pointed out:

> If the kind of detrimental effects seen in animals fed GE food were observed in a clinical setting, the product's use would be halted and further research instigated to determine the cause and to find solutions. However, what repeatedly happens in the case of GE food is that despite increasing evidence of serious adverse health test results, government and industry continue unabated with the development, endorsement, and marketing of these foods as if nothing has happened – to the point they even seem to ignore the results of their own research! There is clearly a pressing need for independent research

into the potential ill effects of GE food – and this research *must* include extensive animal and human feeding trials.[182]

Therefore, in light of all the above considerations, it's eminently reasonable to conclude that no GE food has successfully borne its evidentiary burden – and that none has been proven safe.

Facing Up to Reality: GE Foods Are Illegally on the Market

The preceding sections have amply demonstrated that if the facts about the earliest GE foods had been fairly reported, none others would have come to market – and that, according to the dictates of the law, none *should* have. They've further shown that if those requirements *had* been followed, not even the Flavr Savr would have been commercialized.

Moreover, it's by now evident that even if the relevant US laws were as lax as the FDA has pretended them to be, adherence to the accepted standards of science would in itself have largely kept the products of the GE food venture out of commerce – thus fatally deflating its prospects.

Additionally, a substantial body of research has raised serious doubts about safety; and the studies that have detected adverse effects are, on the whole, significantly more solid than those purporting to have found no problems. The strength of this disconcerting research is attested by the rank unfairness of the attacks to which it's routinely subjected, since if it were truly as flawed as its detractors allege, there would be no need to distort it. Consequently, those who've attempted the unjust discreditation have ultimately discredited themselves.

Although many will surely contest the above assertions, the irrefutable fact remains that there is extensive and intensive controversy about the status of the research within the scientific community. Therefore, it's patently clear that there is not an expert consensus about the safety of GE foods; and due to this resolute reality, it's also clear that all those on the US market have not only entered illegally, but illegally persist, because none of them is generally recognized as safe (GRAS), and none has been approved via a food additive petition.

And this conclusion, no matter how unpalatable, is inescapable. The law requires not only that there be solid evidence of safety, but that this evidence be widely accepted by experts. But neither of these conditions has been satisfied. There's too little sound evidence of safety, too much evidence raising doubt, and too much scientific controversy about what the evidence indicates. Although biotech proponents like Chassy and Parrott assert that studies reporting adverse effects of GE foods are, even after publication in peer-reviewed journals, still of no account unless they and their cohorts are willing to accept them, the law actually prescribes the opposite. In truth, the studies that purport *not* to find problems are the ones bound by special strictures, and it is they that lack legal effect unless they're generally recognized and accepted within the scientific community. Moreover, it doesn't take many doubters to delegitimize them. A federal court has ruled that the testimony of five experts is sufficient to defeat a claim that such recognition and acceptance exist – and in one case, even two sufficed.[183]

On the other hand, because the law is precautionary, and demands that the safety of novel additives be so well established there's a reasonable certainty they won't be harmful, any published study that raises doubts must be given legal weight; and its significance cannot be snuffed by the attacks of scientists who dislike its findings – especially if the attacks appear unjustified in the eyes of other equally competent experts.

Consequently, it's obvious that GE foods are not legally GRAS – and that the FDA's rebuttable presumption that they are has been solidly and repeatedly rebutted ever since it was announced in 1992. It's also obvious that the degree of refutation that has by now occurred is colossal. Previous chapters have documented numerous instances of unequivocal rebuttal, and the evidence that eviscerates the claim of consensus continues to mount. Thus, as of January 2014, almost three hundred scientists had signed a statement asserting that there is not a consensus about the safety of GE foods, that their safety has not been adequately demonstrated, and that some studies "give serious cause for concern."[184] And one of them, the Salk Institute biologist David Schubert, had, in the year prior to signing, clearly shown how well-founded are the concerns about the evidentiary deficiencies. In a letter published by the *Los Angeles Times*, he asserted: "As a medical research scientist who published a comprehensive, peer-reviewed critique of genetically modified food safety testing, I can state confidently that it is false to say such foods and the toxic chemicals they require are extensively tested and proved safe."[185]

Accordingly, those who persist in their claims of consensus demonstrate either how thorough is their insulation from reality, or how staunch is their resolve to misrepresent it. And, when we remain in reality, it's indisputable that dispute exists – and certain that the 'reasonable certainty' standard has not been satisfied. After all, a substantial number of experts regard GE foods as abnormally risky, do not think their safety has been demonstrated, and doubt that it will be within the foreseeable future. Therefore, the legally mandated burden of proof has not been met and almost surely won't be within any acceptable time frame – which means that as long as GE foods are on the US market, their presence will be illegal.[186]

Thus, when we take a hard look at the hard evidence, it's hard to accept the assurances that the safety of GE foods has been scientifically demonstrated. In fact, it's well-nigh impossible. And our investigation up to this stage has revealed that on the empirical as well as the theoretical plane, there's good reason to regard genetic engineering as the riskiest form of food production.

Moreover, this evaluation has stemmed solely from the standpoint of biological science. As we'll see, from the perspective of computer science, there's an expanded basis for viewing these foods as dangerous – and enhanced reason to regard the enterprise that's producing them as one of the most reckless in history.

CHAPTER ELEVEN

Overlooked Lessons from Computer Science

The Inescapable Risks of Altering Complex Information Systems

DNA is commonly compared to computer software, and it's routinely described as the program of instructions that drives the processes of life.[1] Further, the alterations that bioengineers effect in DNA are often likened to the programming refinements implemented by software engineers. For instance, Pamela Ronald, one of the most prominent biotech boosters, has written: "Over the last 20 years, plant breeding has entered 'the digital age of biology.' Just as software engineers tinker with computer codes to improve machine performance, scientists and breeders are altering the 'DNA software system' of plants to create new genetically engineered crop varieties . . . ."[2]

However, such analogies are inapt – and far too simplistic. Not only do they misrepresent how exquisitely intricate the information systems within living organisms actually are, they misrepresent the role of DNA within them.[3] Moreover, reputable software engineers would never revise computer programs in the way biotechnicians alter genomes; and, besides acting with far greater insight, they exercise much greater caution. As we'll ultimately see, when the fundamental facts are examined, and the lessons of software engineering are carefully considered, it's clear that reprogramming life's information systems through recombinant DNA technology is an inherently high-risk process that cannot be practiced in conformity with the essential safety standards of software engineering – or any other branch of engineering.

The Basic Features of Human-Derived Software

Today, computer use is widespread and most people are familiar with several of the basic terms and concepts. They know that hardware comprises the parts of the system that are concrete and physical, while the software is a component that's abstract and essentially immaterial. They know that the former is formed from matter and that the latter is a pattern of information – that the one can be grasped by the hand and the other primarily by the mind. But just as the majority of computer owners don't know the intricacies of the gadgets and circuits within their unit's motherboard, they're only superficially aware of what software is and how it gets developed. And without this knowledge, it's difficult to fully appreciate the dramatic differences between genetic engineering and software engineering.

Instructions vs. Data

There are two basic categories of software: instructions and data. The first prescribes actions, the second is acted upon. For instance, a set of encoded rules for making numerical computations are instructions; the information representing the numbers that get computed is the data. A word processing program is also a system of instructions, while the document files it processes are data.

Generally, the term *program* refers to a set of instructions; and such programs are executable. But that doesn't mean programs never contain data. For instance, the spell checker in a word processor is a sub-set of instructions that includes a dictionary of words against which the processed data can be compared. So that dictionary serves as data.[4]

In most contemporary computers, instructions and data are both stored in the computer's memory – but usually in separate regions. However, although they tend to be stored separately, they're both stored in the same encoded form. The information in each is represented via merely two digits: zeroes and ones (0's and 1's). Thus, the code is *binary*, because it utilizes only two symbolic units.

Distinct Levels of Code

Yet, although a set of programmed instructions is fed into the computer as a pattern of 0's and 1's, the program never starts out that way. The binary pattern is the final phase of the programming process, and the information it conveys was manifested in different modes at earlier stages. In the earliest, the program exists as ideas within the mind of the programmer. This is the most abstract phase of the program – and also the most important, since it's at this level that the system gets created and coordinated. The programmer then expresses the program in a specialized language, called a *programming language*. Through such a language, the specific instructions can be set forth and their interconnections worked out. Many different programming languages have been developed; and the trend is toward increasing their abstraction – which entails diminishing the concrete correspondence between the language and the electrical operations through which it will be applied. Such abstraction is valued because the greater it is, the more powerfully and efficiently the language can be employed.[5] For instance, as the language becomes more abstract, each of its statements can represent a larger number of individual instructions.

Moreover, although programming languages are not composed with 0's and 1's, they consist of codes. They generate an intricate array of symbolic statements laying out the logical structures of the various instructions and establishing their web of interrelations. Accordingly, such an array is called the program's *source code*. However, while this code can be read by humans who are familiar with the particular language employed, it cannot be read by computers. So it must be translated into a format suitable for these machines.

The translation is carried out by another program called a *compiler*. It compiles the encoded instructions of the higher level into a code that the machine can read and execute. This is the *machine code*, and it's the level at which the instructions are expressed in 0's and 1's. Each of these instructions is directly executed by the computer to perform a discrete operation. So at this level there's a close correspondence between the elements of the software and the actions of the hardware.

It's important to note that while most programs are not initially expressed in machine code, it is possible to do so. And programmers have sometimes taken this direct route because it can maximize the efficient use of computer resources. However, because working at this level is a difficult and error-prone process, it's only feasible when the program is very small.

Enhanced Reliability Through Reduced Interactivity

One of the biggest problems for software engineers has been the propensity of their programs to generate unintended effects, a propensity that increases with the program's size and complexity. Most of these effects arise when different parts of the program interact with one another in unanticipated ways.

So programmers have endeavored to isolate modules of coded instructions that are designed to function in coordination and insulate them from the instructions with which they are not supposed to interact. They've increasingly aimed to develop segmented programs in which the units self-interact but minimally affect one another – and then only in a tightly controlled manner.[6] The overall goal is to develop a system that's as linear as possible: where each command or discrete series of commands will yield a specific, predictable outcome without also inducing results that are unpredicted.[7]

A common way of depicting the types of programs that are desired, as well as those that are not, is through analogies to pasta. Software engineers generally want to avoid programs that are structured like a mound of spaghetti, in which the various strands are so entangled that not only is it difficult to follow the course of any one of them, but pulling on one affects several others. Thus, they disdainfully refer to programs in which the logical flow is complicated and the elements are extensively interconnected as *spaghetti code*. What many instead aim to create are programs patterned like a serving of ravioli – programs in which the constituents of the various modules are essentially as independent from one another as the cheese and vegetables enclosed within separate packets of pasta.[8] However, they still have to keep the modules sufficiently interconnected to function as an integrated whole – a capacity with which no plate of ravioli has ever been endowed, no matter how sticky the sauce.

How the Software of Life Dramatically Differs from Human-Derived Software

High Complexity, Low Comprehensibility

As we've seen, human-fashioned software can grow very complex, and as its complexity grows, so does the tendency for unintended interactions between its components. However, despite these unpredictable events, such highly complex systems are still highly comprehensible. Programmers can clearly discern what the components are and can comprehend almost all the rules through which they operate. There's a written record detailing all the elements and how they're intended to function together; and it's precisely known where in the computer the program is stored and how its commands are transferred to the hardware.

But even the most complex human-made system seems simple compared to nature's software. And the complexity of this naturally formed system is so great that even some of its basic contours can't be clearly determined. We do not know what all the components are, nor do we fully understand how they're arrayed. Although DNA is commonly regarded as the locus of a cell's information system, the evidence indicates that the system is not fully localized within that molecule – and that some of its most important parts reside elsewhere. Further, not only are many of the components outside the genome, the rules through which all the components interact are to a large degree outside our understanding.

Evidence of a dispersed program was already strong by the last decade of the 20th century, and its significance was examined in an influential article published in *Nature Biotechnology* in 1997 by Richard Strohman, a professor of molecular and cell biology at the University of California, Berkeley. In it, Strohman explained that although mainstream molecular biology had for more than forty years portrayed genes as the "ultimate" agents controlling life's processes, exerting their control by issuing the key commands within the cellular program, this portrayal was highly inaccurate.[9] He pointed out that "the real secrets of life" cannot be found at the level of the genetic agents but instead at the level of "the rules and constraints that organize genetic agents into functional arrays."[10] And he noted that not only is this level of "gene management" a higher level than the one at which gene expression occurs, it is not confined within the DNA but is "coextensive with the cell itself."[11] Moreover, he emphasized that the dynamics operating at this level are different than those at the lower one and that the interactions are far more complex – to the extent of being ultimately "transcalculational," which, as he noted, is "a mathematical term for mind boggling."[12]

In a subsequent paper, he elaborated on these themes, emphasizing the importance of the factors that are *not* determined by DNA sequences (termed *epigenetic* factors) – and our deficient understanding of how they operate. He noted that "many biologists, world wide, have known for decades that genetics alone is not sufficient to explain life's complex outcomes, and that another kind of information management system must be present. . . ." He continued: "This second informational system is coextensive with the cell itself [and] consists of many interconnected signaling pathways. . . ." And he emphasized: "The key concept here is that dynamic/epigenetic networks have a life of their own: they have network rules not specified by DNA, and we do not understand these rules."[13]

Evelyn Fox Keller, a professor of the history and philosophy of science at the Massachusetts Institute of Technology, is another expert who has repeatedly pointed out that the cell's informational program extends well beyond the gene – and far beyond our comprehension. In her book, *The Century of the Gene*, she quotes a statement made by the president of the National Academy of Sciences in 1998 that "[w]e always underestimate the complexity of life, even of the simplest processes;" and she shows how a substantial amount of such underestimation has occurred by *overestimating* the role of genes. Like Strohman, she emphasizes that an organism's development and coherent functioning are primarily coordinated not by the genes themselves but by "the complex regulatory mechanisms that, in their interactions, determine when and where a particular gene will be expressed."[14]

However, the pattern of these interactions is difficult to apprehend, especially since the system is not fixed, but fluid. As Keller points out, in contrast to the sequence of the genome, which is largely "static," this regulatory system is "dynamic."[15] Moreover, she equates this dynamic system with "the developmental program;" and she asserts that ". . . an understanding of its dynamics needs to be sought at least as much in the interactions of its many components as in the structure or behavior of the components themselves." She continues: ". . . the program consists of, and lives in, the interactive complex made up of genomic structures and the vast network of cellular machinery in which those structures are embedded."[16] She then remarks: "It may even be that this program is irreducible – in the sense, that is, that nothing less complex than the organism itself is able to do the job."[17] Accordingly, she notes that, instead of being bound by the genome, the program is essentially "everywhere."[18]

Nonlinearity and Ambiguous Agency

Thus, in the analyses of both Strohman and Keller, a biological information system is spread throughout the organism, and this substantially obstructs its comprehensibility. Such dispersion sharply contrasts with the well-defined contours of human-shaped programs, the structures of which are far better understood. Further, our capacity to comprehend these natural systems is further constrained by the fact that, besides being nonlocalized, they're nonlinear.[19] In such complexes, distinct actions induce system-wide effects in a significantly unpredictable fashion. This also contrasts with the human-made systems, which are substantially linear because they're designed so that discrete operations yield discrete, predictable results.

Further, as Strohman pointed out, one of the big cognitive complications posed by life's software is the elusiveness of its rules. Not only are we unable to fathom how they interact as a network, we have little understanding of what the various rules actually are; and it's difficult to discern how any of them is embodied within (and distributed among) the cell's multifarious components.

Compounding the conundrums, the distinction between instructions and data is significantly blurred. In human-made systems, there are usually clear boundaries between the two, and it's easy to differentiate the parts of the program that are active agents from the passive parcels they manipulate. We can distinguish between that which commands and that which is commanded.[20] In contrast, nature's software is not merely ambiguous, but enigmatic. As Strohman and Keller have noted, not only has it been wrong to view genes as the *ultimate* agents, whatever agency they do exert is limited, and they're often the elements that are acted upon.

Moreover, some experts say they don't exercise agency at all. The theoretical biologist Michael Conrad points out that DNA does not prescribe behavior as do the instructions in a computer program, noting that while in the latter programs, "each unit . . . reacts to defined outputs of other units" in a sequential manner, the components of a bio-based information system interact in such a holistic and non-sequential fashion that each is essentially responding to "global properties" of the entire system.[21] Accordingly, he says that such a system "cannot be programmed like an ordinary computer."[22] And he emphasizes that, contrary to common opinion, discrete command functions are not localized within DNA. As he explains, DNA does not provide prescriptions for specific cellular behaviors but rather contains symbolic descriptions of the "primary structure" of many of the cell's important molecules.[23]

From this perspective, besides their incapacity to prescribe distinct behaviors, the sections of DNA referred to as genes do not even dictate the production of particular proteins. Rather, they serve as repositories of information that are used by the cell in producing proteins according to needs that are registered, expressed, and responded to through the operations of the system as a whole. And these operations are so complexly coordinated that discrete outcomes cannot be reliably predicted.

Conrad's conclusions were published in 1972, and over the following four decades, the grounds for them have grown increasingly compelling. An abundance of evidence has amassed not only confirming the soundness of his analysis, but revealing that the complexity of bioinformation systems is far greater than was then known – and that their dynamics are even more global. For instance, in 2003 BBC news reported that a team of scientists who tracked more than 20,000 interactions between 7,000 of the genes in the fruit fly discovered "a hidden level" of organization between "apparently disconnected proteins," and hence also between the genes that express them.[24]

The profound extent of the complexity and holistic coordination that characterizes cellular life can be glimpsed by considering the mechanics of how genes are expressed, which also demonstrates the passivity of the role that they play.

As discussed in Chapter 4, the sequential information within a gene becomes expressed as a particular protein through a multi-stage process.

In the first phase, a specialized enzyme transcribes the information into a strand of another (but related) type of nucleic acid called ribonucleic acid, or RNA. This RNA serves as a messenger, and it conveys the information to an intricate structure called a ribosome that translates it into a chain of amino acids that subsequently folds into a protein. But in plants and animals, before the RNA travels to the ribosome, it must be edited. And the editing is done by a set of enzymes that remove the sections of DNA that do not code for amino acids (the introns). However, that's often not the end of the editorial process. A group of other enzymes frequently intervenes to rearrange the information so as to code for a different protein than would have derived from the initial sequence. And in many instances, the range of alternative proteins is substantial. Some genes can give rise to dozens of different types; and the mechanics by which the enzymes determine which protein is going to be produced is neither prescribed by the gene whose associated RNA they reconfigure nor even influenced by it. Further, although these enzymes are proteins that are coded by other genes, those genes do not direct the details through which the enzymes operate either. Many additional factors come into play; and although some of them derive from yet other genes, those genes also lack the capacity to prescribe the complex ways in which their products interact with other genes and with the substances they produce.

Further, not only is the full process of gene expression dependent on a number of factors outside the gene, most genes can't even get the process started in an autonomous manner. A transcribing enzyme will only attach to a gene's promoter region when that promoter is in a receptive mode, whereas the default condition for most promoters is to be unreceptive.[25] And a promoter in its closed-down default state only becomes receptive through the agency of specific molecules that are usually independent of its associated gene. Most of these molecules are regulatory proteins produced by other (often distantly located) genes that are not influenced by the genes their expressed proteins regulate.

The interplay between such regulatory proteins and a promoter's activation sites can be astounding. Consider the case of the promoter attached to a gene (dubbed *Endo 16*) that encodes a multi-function protein critical for the development of the sea urchin embryo. The expression of this gene is controlled by 14 proteins that bind to its promoter in a manner that enables transcribing enzymes to convert its information into RNA at the appropriate times and rates. Further, not only are the binding sites for these transcription-enabling proteins highly specific, there are far more of them than proteins: at least 50. Moreover, 20 of them are tailored to bind just one of the proteins.[26] According to researchers who extensively studied this promoter, these various sites form seven clusters, each of which functions as a regulatory module. They report that during the early stages of development, the module closest to the area where the gene's transcription begins serves as "a central processor" that integrates the output of the other six and causes transcription of the adjacent gene to either start or cease as required to maintain the embryo's proper growth.[27] Then, at a later developmental phase, the module next down the line takes over as the central processor.[28]

In a commentary in *Science* that accompanied a research study on this remarkable promoter, the evolutionary biologist Gregory Wray described it as a "genetic computer." He stated: "The 'program' that runs this tiny computer is directly encoded in DNA as regulatory elements; its inputs are single molecules whose composition varies in time and among various cells of the embryo, and its output is a precise level of transcription."[29]

Yet, as Evelyn Fox Keller points out, there's "tension" between this conception of things and the molecular realities – and it's a stretch to regard the program as " 'directly encoded in the DNA.' "[30] As she explains, the widely "scattered" DNA sequences that give rise to the regulatory proteins that affect the promoter's output merely code for their amino acid sequences; and these sequences do not in themselves fully determine the structural features of the proteins that govern the "dynamics" of their interaction with the promoter's binding sites. She additionally notes that because messenger RNA can be spliced and revised by enzymes independent of the DNA from which it's derived, even the amino acid sequences of the regulatory proteins "cannot be fully predicted" from the sequence of the associated DNA.[31]

Of course, the situation is further complicated by the fact that the genes encoding the 14 regulatory proteins do not contain any instructions for when and at what rate they themselves are to be expressed – and that the factors influencing these variables are in turn influenced by the products of additional far flung genes, which are in turn influenced by another set of genes, and so on. Obviously, this web of relations necessarily extends throughout the entire cell and includes epigenetic factors as well – which confirms Conrad's assertion that genes are ultimately regulated not by individual inputs, but by "global properties" of the system.

Moreover, even though our understanding of *Endo 16's* regulatory program is substantially circumscribed, it's prodigious compared to our knowledge of the programs associated with most other promoters. As Wray noted in his commentary, "In spite of considerable investigation of the function of animal promoters, general principles have remained frustratingly elusive. There is little logic apparent in the organization of regulatory elements. . . . It therefore comes as a surprise to discover a promoter that operates in a logical manner."[32] In further emphasizing the inscrutability of most promoters, he described their operations as "seemingly haphazard."[33]

But the word "seemingly," as well as the word "apparent" that shortly preceded it, should be doubly underscored. That's because the operations of promoters *must* be highly logical. Otherwise, the elegantly integrated organisms that abound in our world could not exist. Therefore, the *actual* lack is on the level of human understanding, and the extent to which these operations appear as logically deficient is a measure of how inadequately we grasp the intricacies of their dynamics. Further, although our understanding has significantly advanced since 1998, when Wray's words were written, it's still quite rudimentary – especially compared to our knowledge of human-made information systems. And the deficiency is not limited to animal promoters. Our knowledge of plant promoters has also remained meager.[34]

So, compared to our knowledge of, and ability to manage, even the most immense and complex human-made computer programs, our capacity to comprehend and control the programs coordinating the processes of living organisms is miniscule.

An Unparalled Level of Parallel Processing

The profound complexity of cellular information systems can be more fully appreciated in light of the fact they perform a profound degree of parallel processing.

The simplest form of computer processing is not parallel but *serial*. In serial processing, instructions are executed sequentially and one at a time. But in parallel processing, distinct operations occur simultaneously. Consequently, parallel programs are more difficult to write because multiple subtasks need to be coordinated. Hence, they're also more prone to problems.

Of these, the most common are caused when separate subtasks do not activate in proper sequence, affecting the system's output in unintended ways.[35] They're called *race conditions*, because it's as if one subtask has raced ahead of the other to capture system resources before it should have.

In order to ensure that subtasks in complex parallel programs remain synchronized and access resources in the proper sequence, the programmers must create barriers to block asynchronous behavior. And this becomes more challenging the more intricately interconnected the system becomes. Accordingly, even the most adept programmers could not create systems that even vaguely approximate the degree of interconnection and harmonization displayed by cellular systems. And even if they could successfully create barriers as needed, they would still face other basic limits. For one thing, if the parallelization within a human-made system increases too greatly, the subtasks must spend so much time communicating with one another that performance is not accelerated but retarded. In contrast, cellular systems sustain intricate communication between components while achieving astounding rates of operation – and avoiding the race conditions that bedevil the complex programs made by man.

Mind-Made vs. Mind Boggling

Thus, in light of what we've considered so far, the claim that bioengineers have adequate understanding of the cellular software they're restructuring is at best naïve – as is the notion that humans could fully comprehend such systems by thoroughly studying their constituent DNA.

As we've seen, DNA does not serve as instructions but as data; and the data occupies a concrete level corresponding to machine code, not the more abstract plane of source code. The correlative of the system's source code would be a set of principles and rules that describes its architecture and governs its operations; and such a code is apparently not inscribed within the physical confines of the organism. Although it's obvious that the system's components are intricately coordinated and highly organized, the rules through which this coordination and organization occurs are not physically expressed like the source code of a computer program, and it's unlikely that they even could be. While there must be some organizing principles, it's doubtful that the human mind could even grasp any but the most general; and the degree of generality would be too great to afford a detailed description of the dynamics.

Accordingly, while computer programs are creations of the human mind, and are therefore well-comprehended by it, the information programs that underlie life not only didn't derive from that mind, the intricacies of their organizing principles and operational dynamics so vastly outstrip its cognitive capacities that they boggle it.

The Stark Contrasts Between Genetic Engineering and Software Engineering: Glaring Gaps in Vision, Precision, and Precaution

Because genetic engineers know so much less about how the programs they alter actually operate than do software engineers, and because the associated risks are so high, one could reasonably expect them to exercise not merely the same degree of caution as do the latter, but substantially *more*. And this expectation is even more justified in light of the fact that their operations are far less precise than those of software engineers. As previous chapters have explained, they cannot control where in the DNA strand the inserted genes end up, nor can they configure them to act in harmony with the myriad doings of the target cell. Instead, these intrusive genetic sequences operate outside the cell's intricate regulatory system in an exceptionally unruly, and potentially disruptive, manner.

Yet, although compared to software engineers, their vision is critically restricted and their acts inexact, the bioengineers have operated with far *less* precaution – which is obvious from surveying the protocols of that other profession.

A Fundamental Facet of Software Development: Testing the Program

Despite the fact that software engineers know so much more about the programs they've fashioned than bioengineers know about cellular information systems, and although their programs have been increasingly designed to reduce unintended effects, the potential for such problems has not been eliminated; and they still arise all too frequently. Therefore, testing is a crucial part of the development process. And because it's so important, it's usually put in the hands of people who had nothing to do with the program's creation – and who therefore are not predisposed to see its reliability confirmed. In fact, those in the testing division are predisposed to find problems, because that's their job; and they attempt to do so by subjecting the program to extensive and intensive trials.

Maintaining and Revising the Program

However, even after several rounds of robust testing, many problems can still go undetected; and software is often released with bugs that are only discovered as it's employed in a large range of conditions. As this occurs, programmers have to make revisions.

Further, maintaining a program comprises more than correcting errors; and the majority of the maintenance costs are incurred because the program must regularly evolve to adapt to changing conditions.[36] The pressure to evolve is so strong that the total maintenance expense typically consumes two-thirds of the life-cycle cost of a successful program.[37] And this expense is so big because revising software is a big process.

Any time a program is revised, whether in correcting errors or adjusting to new circumstances, the very process of making the change could itself disturb the system in some unexpected mode and create another problem elsewhere. In the early days of computing, when a program's parts tended to be highly interconnected, the systems were so susceptible to unwanted interactions that making even a small revision to one section usually caused a disruption elsewhere. Consequently, as a program's errors were corrected, the total number tended to stay constant, because fixing one usually entailed creating another at a different location.

But as programmers learned to insulate the various components from one another and reduce the potential for unwanted effects, the process of error correction began to yield a net benefit. However, although programmers could reduce the potential for problems, they could not eliminate it. That's because, as previously discussed, the components of *any* large, complex information system, even when designed to interrelate more like ravioli than spaghetti, can still interact in ways that their developers not only never intended, but could not even foresee.

Consequently, after any revision, even a small one that's carefully planned and precisely executed, the entire system needs to be thoroughly re-tested. This post-revision scrutiny is called *regression testing*; and it ordinarily entails not only a large portion of the tests that were run prior to the program's first release, but a set of new ones specifically designed to gauge the effects of the novel code that was added.

Imposing Stiffer Standards on Life-Critical Systems

Despite the rigor of the above-described testing procedures, software developers have recognized that if a program's malfunction would pose a substantial degree of risk to human life, it requires a level of testing that's even *more* rigorous. Accordingly, they routinely subject these *life-critical* programs to stricter testing, both before release, and after any revisions are subsequently made. In these tests, the program is put through as many permutations as possible to make sure that it will perform safely under the widest range of conditions, even those that would seldom arise.

Not only has such testing become standard industry practice, it's mandated by regulators. In the United States, the Federal Aviation Administration insists on it in the case of airplane guidance systems, and the FDA requires it in the case of the software governing medical devices such as X-ray machines, radiation therapy equipment, and pacemakers.[38] And the new international standard for life-critical medical devices likewise establishes this stricter level of testing.[39]

Further, it's important to note that even without government mandates, the software industry would be routinely subjecting its high-risk products to thorough testing, because that's standard practice even for software that is *not* life-critical; and although such programs aren't regulated, their developers ordinarily don't release them until they've been carefully tested. Thus, the main difference between the two categories of software is not that the higher risk programs are tested, but that they're tested more strictly; and government regulation did not impel the implementation of a new practice but instead ensured that a common practice would be practiced in a uniformly rigorous manner – commensurate with the related level of risk.

How Bioengineers Fall Deplorably Short in Regard to Testing

In striking contrast to software development, thorough testing for unexpected problems is not a routine feature of genetic engineering – even though the potential for unintended effects is far greater when nature's information programs are being altered. Although the developers of GE crops usually perform nutritional studies to make sure that livestock fed on them will sufficiently grow for commercial needs, such studies don't assess safety, and they're not designed to screen for symptoms of toxicity. So the animals that are stout enough to market may yet harbor a range of undetected, feed-induced infirmities that could also develop in the humans who consume the crops.

Moreover, it's unlikely that actual safety testing would happen at all if it weren't required; and when it *is* required, it's woefully inadequate. Even in the nations that mandate some toxicological testing, the stipulated studies last no more than 90 days, which isn't long enough to detect the many types of problems that develop over an extended time – as was demonstrated by the Séralini study discussed in Chapter 10. Worse, the European Union didn't get around to making even *those* inadequate tests mandatory until 2013. And in the United States, a manufacturer can dump a limitless number of GE crops on the market without performing a speck of testing.[40]

The disparity in how software engineering and genetic engineering are regulated is so vast it's astounding. When dealing with life-critical human-made software, regulators throughout the world are highly sensitive to the potential for unintended consequences, and they won't accept arguments that new programs are substantially equivalent to prior ones, no matter how well-grounded and reasonable those arguments might seem. Instead, they require that the safety of each new program be established through rigorous testing.[41] Moreover, their rejection of the substantial equivalence doctrine is thoroughgoing; and when a program that has been rigorously tested and approved for market is later revised, regulators won't accept assertions that the new version is essentially the same as the old –

no matter how small, well-planned, and precisely executed the alteration, and no matter how well-insulated the system against adverse intercourse between its components. Rather, even when a minuscule change is effected by experts with full knowledge of the system's architecture and designed interactions, they require that the program be treated as a new entity and that its safety be confirmed through another stringent round of testing.

But when faced with radical alterations to the most complex and intricately interconnected information systems on earth, made in a haphazard manner by people who don't understand the system's rules and contours and can barely begin to fathom the full effects of their interventions, regulators have for years allowed the resultant food-yielding organisms to be marketed as long as a superficial case can be made that they're substantially similar to their conventional counterparts – despite extensive evidence that such radical tampering can render the food toxic. And in the United States, such equivalence is automatically presumed, with no requirement for even the most superficial of efforts.[42]

Further, in the US the paradox is more glaring because the same administrative agency is involved in both situations. So while the FDA rigorously regulates the software that drives life-critical medical devices, requiring that even the most minor and well-managed revisions be subjected to extensive testing, it drastically shifts its standards when dealing with GE foods. In that case, it presumes that the unprecedented restructuring of the information programs directing the development and function of living organisms by those who lack capacity to control where the new chunks of code wedge or how they impact the system is nonetheless so innocuous that it need not be regulated at all – despite the fact such restructuring could cause far more extensive damage to human life than a malperforming pacemaker or an errant X-ray machine.[43]

Additionally, not only is there a huge discrepancy in the way regulators treat the two classes of information manipulation, there's a major difference in the diligence displayed by those who do the manipulating.

On the one hand, software engineers have exercised an admirable level of self-regulation and have routinely subjected even low-risk software to careful testing without any government mandate to do so. What's more, they didn't resist government efforts to regulate their high-risk software – and had even recognized and begun to address the need for stricter scrutiny of such programs well before the regulations were imposed.

On the other hand, the behavior of bioengineers has been not merely unadmirable, but reprehensible. As previous chapters have documented, besides failing to responsibly test their creations, they have, from the earliest era of genetic engineering, forcefully resisted regulation and deterred it with dubious and even devious means. And although, despite their deceptions, some meager regulations were finally instituted in many countries, manufacturers have often evaded the imposed obligations by conducting shoddy research, obfuscating adverse data, and inaccurately reporting their findings. Further, even in the cases where a manufacturer has not only obeyed the requirements but exceeded them, the level of testing has still not reached the standard that's voluntarily upheld by software developers when testing programs that aren't close to life-critical – and falls far beneath the one employed for those that are.

The enormity of the gap between the levels of testing performed on life-critical software and GE foods can be better appreciated by recognizing how vastly revamped the biotech industry's current system would have to be in order to approximate the rigor with which the safety of life-critical software is standardly established. At the least, this transformation would entail implementing the minimum set of procedures prescribed by John Fagan and Michael Antoniou that was described in Chapter 10. And such a huge transformation is unlikely to occur.

Yet, even if it did, the testing of GE foods would still lack the reliability of software testing unless another major reform were adopted. The people doing the testing would have to be insulated from pressures to return rosy results – and be even devoid of the desire to do so. Otherwise, the testing could not achieve parity with the procedures governing software. As we've seen, although the individuals testing software are usually employed by the corporations that develop it, their job is to find as many flaws as they can and not to overlook anything suspicious. In contrast, the corporate developers of GE foods have repeatedly demonstrated that their primary concern is not safety but profit; and they've routinely endeavored to relax the rigor of testing in order to cut costs and accelerate the advance toward commercialization. In consequence, because those who test GE foods are either employees of the manufacturers or hired by them, they're likely to reflect the corporate bias toward favorable findings that can hasten the product's release – just as the employees of software firms reflect their employers' desire to detect all problems prior to release, even though the pace of commercialization is thereby retarded.

Of course, it's not known whether this attitudinal discrepancy is due to a difference in rectitude or instead to the fact that programming defects are far more discoverable after release than are most harmful changes to food because, unlike the latter, they usually cause palpable problems that are clearly linked to the product.[44] However, in light of the repeated delinquencies of some of the major GE food manufacturers, it's doubtful the discrepancy solely stems from the fact that flawed food can be more easily passed off than flawed software. But in any event, the disparity does exist; and whatever the reason, the behavior of software companies has in effect been far more responsible than that of the GE food industry.

Further, although GE proponents forcefully contest the charge that industry-controlled testing is less reliable than testing that's independently conducted, it's not only consistent with common sense, but has been confirmed by scientific assessment. A substantial body of research has shown that, for a range of various products, tests conducted by the products' manufacturers, or by researchers in their pay, tend to be biased and are significantly less likely to detect problems than are those performed free of their influence.[45] This bias has also been detected in the case of GE foods and was documented in a published review of the research on these products.[46]

Considering the disparities we've so far discussed, it's not surprising that, besides insisting that testing be rigorous, software developers have devoted a much larger portion of their budget to it and to other measures that enhance the safety and reliability of their products. Thus, testing typically accounts for over 20% of a program's development costs; and according to a standard textbook, the total amount spent on pre- and post-release testing, in combination with the other measures that maintain the program's performance, consumes around 70% of the total expenditure.[47] In contrast, a much smaller fraction of the developmental budget for a GE crop is consigned for safety testing.

Thus, overall, the test-related differential is enormous, because not only are the tests the biotech industry *does* perform unreliably conducted, they're not the kinds that are most needed. None of the various procedures prescribed by Fagan and Antoniou has yet been implemented; and none is likely to be without an arduous struggle. And the prospects for a system of truly independent testing are just as dim. Therefore, the disparity in responsibility shown by bioengineers and software engineers will most probably persist.

Moreover, as stark as this disparity appears when the gap in testing is assessed, it's even starker when we also gauge the variance in the way the two groups have responded to problems.

Discrepancy in the Response to Adverse Incidents: Responsibility vs. Recklessness

Although the majority of software failures have caused only minor to moderate annoyances, a number have entailed serious consequences, including some full-blown catastrophes. And examining how they arose, and how the developers and regulators responded, is highly instructive.

One of the strangest mishaps occurred in the state of New York after AT&T tried to improve a software program that manages telecommunications within a large network of 4ESS switching systems. In mid-December 1989, the company installed the new software in the processors of all 114 of these units with the goal of speeding up the flow of information between them. However, in revising the old program, the programmers had inadvertently omitted a few lines of its code.

Several weeks later, on January 15, 1990, a crisis erupted when one of the switches shut down and rebooted – and then sent a message to neighboring switches signaling that it was again functional. Although this initial crash apparently was not caused by the glitch in the new software, the ensuing problems were. When the former program was running, this type of recovery message was properly sent and processed, but due to the absence of those excluded lines of code, the 'improved' software botched the procedure. According to AT&T's director of technology development, the message "confused the software" in the receiving switches because it "didn't make any sense."[48] He continued that, in effect, the first switch told them, "My CCS7 processor is insane," which induced them to shut themselves down so they wouldn't spread the problem. Ironically, instead of constraining the problem, this preventive measure actually extended it, because when those units rebooted and attempted to signal their neighbors that all was well, the neighbors were in turn confused, whereupon they shut down, rebooted, and then transmitted the same disruptive message to their neighbors. The effect quickly rippled, and soon all 114 switches were crashing and rebooting every six seconds, with the result that for nine hours, an estimated 60 thousand customers had no long distance telephone service – a major predicament in an era when mobile phones were rare.

Six years after the crash of the phone system, another software-induced disaster occurred that was not only much more dramatic, but due to a contrasting cause. Chronicling the incident in the *New York Times Magazine* in December 1996, a half year after it happened, James Gleick encapsulated it as follows:

> It took the European Space Agency 10 years and $7 billion to produce Ariane 5, a giant rocket capable of hurling a pair of three-ton satellites into orbit with each launch and intended to give Europe overwhelming supremacy in the commercial space business.
>
> All it took to explode that rocket less than a minute into its maiden voyage last June, scattering fiery rubble across the mangrove swamps of French Guiana, was a small computer program trying to stuff a 64-bit number into a 16-bit space.[49]

This error was the result of a simple, but stupendous blunder. A subsystem of the software program that had been designed for an earlier version of the rocket, the Ariane 4, had been reused in Ariane 5; and that section of software, which had worked admirably in the older missile, did not mesh with the physical features of its successor. That's because the Ariane 4 was a slower rocket, and all the velocity-related numbers it generated could be successfully handled by the program. But the more powerful and speedier Ariane 5 produced a number that was too big for that program to process, causing the guidance system to shut down – which triggered a series of malfunctions culminating in the mission's explosive end.

Commenting on the calamity, in which a rocket and cargo worth 500 million dollars had been obliterated a mere 39 seconds after launch, the head of the project remarked, "Very tiny details can have terrible consequences. . . . That's not surprising, especially in a complex software system such as this is."[50] Underscoring the degree of complexity, Gleick observed: "Software built up over years from millions of lines of code, branching and unfolding and intertwining, comes to behave more like an organism than a machine."

It's noteworthy that in the phone system crash, the hardware was not altered and the key change was the accidental deletion of a smidgen of code from a prior program, while the rocket exploded because the hardware had changed while an incompatible segment of prior code continued to be employed. And it was not the sole catastrophe caused by an erroneous assumption that a segment of software which behaved beautifully in one physical setting would continue to do so in another that was comparable in most respects. A similar misconception induced a series of medical accidents that not only wrought extensive suffering, but death.

That mistake occurred when the manufacturer of a radiation therapy device developed a new model (the Therac-25) that improved on the previous one (the Therac-20). However, the improvements were primarily on the level of the hardware, and a lot of the older software was retained without revision – on the belief that because it had a long and reliable record, it would still work safely within the new machines. But this belief was unwarranted. Although the Therac-25's did operate smoothly most of the time, there were rare sets of circumstances that caused some segments of older software to misperform, which in turn caused the machinery to seriously malfunction. And this would not have happened in the Therac-20's, because they were capable of safely handling such software slip-ups.

Thus, while the new machines were more versatile and economical than the previous models, because they couldn't accommodate the quirks in the old software they contained, they were significantly less safe. As a result, at least six people were subjected to massive overdoses of radiation, causing them severe pain and injury – and ultimately killing some of them.

Although there have been many other software-created catastrophes, the above three are among the worst; and it's illuminating to consider their implications for genetic engineering. For one thing, virtually all GE food-yielding organisms have one or more pieces of software that derive from one physical system but are being employed in another. And in most cases, there's at least one gene producing a protein within a foreign environment that may not process it properly. As discussed in earlier chapters, in such situations there's a risk that the protein could be misfolded – or dangerously altered through the addition of auxiliary molecules. Further, whereas the Ariane 5 and the Therac-25 were quite similar to the models they replaced, and yet were still perilously incompatible with some of the software those systems utilized, there are *major* differences between the organisms involved in most DNA transfers. Accordingly, there's good reason to suspect increased risk of harmful incongruity between hardware and software.

Moreover, not only are most bioengineered crops vulnerable to the same kind of problem that downed the Ariane and plagued the Therac, they're also open to the type of trouble that crashed AT&T's phone system. That's because, besides their potential hardware/software mismatch, their software is altered in a way that accidentally deletes some information, and scrambles some other.

But the differences in the risks posed by GE foods and those entailed by the failed phone system, the ill-fated rocket, and the misfiring radiation machine don't stop there. For instance, only one of those three (the Therac) significantly endangered human life, and even then the number of individuals directly exposed was miniscule compared to the tens of millions of people world-wide who are ingesting ingredients from one or another bioengineered food every day. Therefore, each of these foods poses a much bigger risk than did any of those entities, because each has the potential to cause a lot more aggregate harm.[51]

There's also a big discrepancy in the way the expert community and governments have responded to the software-related calamities and the way they've reacted to problems associated with bioengineering. In the phone system crash, AT&T promptly determined that the error was in the software and made the necessary fix. When the Ariane 5 exploded, a panel of inquiry convened two weeks later, performed a thorough investigation, identified the source of the problem, and publicized its findings. And although it took longer to determine why several Therac 5's had malfunctioned, a determination was eventually made. Further, in none of these cases did the software industry and programming professionals arise to defend the image of software by declaring it could not have been at fault – nor did they spread disinformation, try to thwart the investigation, or impugn the competence or the integrity of the investigators. Nor did government regulators attempt to protect the involved industries by suppressing evidence or issuing misleading statements. On the contrary, pursuant to its authority to supervise medical devices, the FDA exercised commendable diligence in the Therac case; and even though the manufacturer apparently tried to obfuscate the facts, the agency did not aid the effort but instead endeavored to achieve clarification.[52] Further, the Therac incident motivated the agency to strengthen the regulations governing medical devices and the software that drives them.[53]

At least as important, software engineers have earnestly attempted to learn from their mistakes. Not only have they refrained from portraying their procedures as essentially error-free, they've openly acknowledged their various errors, systematically analyzed them, and steadily improved their methods. And to augment the reliability of their programs, they've routinely subjected them to rigorous testing. Consequently, software engineering is a sounder and safer technology today than it was twenty years ago.

In contrast, as the preceding chapters have demonstrated, the practitioners and proponents of genetic engineering have stubbornly maintained that their technology is precise, reliable, and safe; and they've been averse to even acknowledge its failures let alone to learn from them. Further, they've evaded robust testing and instead have based their claims of safety on obstinately held, albeit thoroughly discredited, beliefs. Consequently, although software engineers have progressively recognized the complexity of the artificial systems they create – and realized that in key respects they're more akin to organisms than machines, bioengineers have grossly discounted the complexity of the living systems they alter and have treated intricate organisms like simple mechanical systems. And they've persisted in the practice despite its dissonance with an ever-growing mass of evidence.

Moreover, (as Chapters 6 and 10 have documented) major governmental regulatory authorities have been equally unwilling to face the facts; and as disquieting data has mounted, they've decreased their diligence and often lowered the standards for testing. This trend is so persistent that, notwithstanding the independent reassessments of several industry-conducted feeding studies that have discovered previously undisclosed evidence of harm, as well as the solid original research that has produced alarming results, in 2012 the European Commission's chief scientific advisor proclaimed that "the precautionary principle is no longer relevant with GMO foods or crops" – which is an artful way of saying that routine pre-market testing is no longer needed.[54]

Drawing the Crucial Contrast: Precise Programming vs. Haphazard Hacking

As the preceding sections have shown, although software engineers and genetic engineers both manipulate complex information systems, the latter possess much less relevant knowledge, exercise far less caution, and are far less willing to learn from (or even acknowledge) their failures. But the differences run much deeper than this – so deep that, notwithstanding the frequent analogies between bioengineering and computer programming, the two are, at basis, not only disparate, but to a large extent opposite.

To appreciate this, we first need to be clear about what the two sets of professionals do – and do not do. And the first fact to note is that bioengineers are *not* engaged in the primary activity of software engineers, nor could they even attempt to be. That activity comprises the design and creation of complex information programs; and the bioengineers who develop new varieties of edible plants and animals are utterly incapable of designing and creating the information systems of the higher organism they deal with – systems that crucially contribute to the generation of living cells, guide their development into complex organisms, and enable those organisms to conduct a multitude of finely tuned and intricately coordinated operations through which they sustain their lives and successfully interact with immensely variegated environments. Instead, they merely make alterations to those systems: systems that were not (and could not have been) created by humans.

As we've seen, software engineers also make alterations to previously developed programs, and it's an important part of programming. But there are major differences between those alterations and the ones effected by biotechnicians on cellular software; and they belong to distinct and contrasting categories.

The changes made to computer programs by the people who've developed them are *revisions*. They're carefully planned improvements that enhance the program's efficiency and effectiveness, enabling it to better accomplish what it was designed to do. And they serve this constructive purpose because they're conceived and executed by individuals with comprehensive knowledge of how the program is structured and how its components are designed to interact. Accordingly, they're performed with precision and prudence; and when old code is edited, or new sections are added, it's in a manner calculated to mesh with the system and minimize the risk of disturbing it in undesirable ways.

In marked contrast, bioengineers are vastly ignorant of the structure and dynamics of the complex systems they alter. They have virtually no understanding of the source code, and even their comprehension of the machine code (the sequence of nucleotide bases in the DNA molecule) is seriously constricted.[55] Consequently, they don't know how an inserted cassette of new code will impact the entire system; and even if they could ascertain the specific location at which it would be least likely to cause broader disruptions, they'd be incapable of putting it there. Instead, their insertions are made in a random manner.

In effect, the bioengineers are adding a new function to a program by acting on the level of the machine code with scant understanding of how the program is put together and how their alterations will affect it – which is a far cry from how programmers make revisions. Not only do the latter possess abundantly greater knowledge and act with far more adroitness and care, they don't operate at the level of the machine code. Working at that level is challenging and problem-prone; and a professional who designed the source code would not even be able to recognize the program on the basis of the machine code, let alone revise it from there in a competent manner. So besides creating programs at the level of the source code, software engineers revise them from that level as well.

Moreover, the doings of bioengineers are further distanced from what could count as revising a program by the fact that they don't improve performance but impede it. As Chapters 4 and 9 have explained, they force the organism to divert energy and assets from essential functions to drive processes that provide it no benefit and impose a net detriment. And the processes they impose behave in an unregulated manner that can induce disruption throughout the system.

Hence, it would be highly illegitimate to liken such actions to the revision of a computer program. But it *would* be appropriate to group them with another class of alterations – ones which, like those of the bioengineers, are conducted by people who don't understand the source code, who act largely on the level of the machine code, and whose insertions impair the program's performance.[56] These are the alterations made by hackers; and in any open-eyed assessment, the manipulations of the bioengineers are far more akin to hacking programs than to revising them.

This kinship is especially striking in light of the fact that in both hacking and bioengineering, the inserted segments of code act like a virus. Not only do they gain entry by breaching the program's defenses against foreign incursions, once inside, they operate independently of, and inimical to, the aims of the invaded system – while commandeering its resources in order to do so.[57] As an article in *Science News* observed, "Computer viruses got their name from . . . 'an obvious but deep biological analogy.' "[58] And many biologists have recognized that the analogy holds for the alien inserts of genetic engineering as well. For instance, Patrick Brown, a professor of plant science at the University of California, Davis, has stated: "Indeed, it can be argued that gene transfer via rDNA techniques resembles the process of viral infection far more closely than it resembles traditional breeding."[59]

However, despite the fact that both bioengineering and hacking degrade the integrity of the invaded system and can compromise its safe function, bioengineering is ultimately less predictable. Although hackers aim to impair the system in some way, they can generally impose the impairments precisely and can usually engender the results that are intended without producing those that are not.[60] In contrast, bioengineers can neither control how their insertions interact with the system nor regularly induce even the intended outcomes; and they have virtually no capacity to restrict, or predict, the unintended ones.

Thus, although proponents of bioengineering portray the alterations it effects as precisely performed, scientifically informed enhancements of genetic programs, in reality, they're tantamount to the hacking of a software system – and in an abnormally haphazard fashion.

Additional Factors that Are Not Merely Astounding, but Insurmountably Confounding

So it's clear that software engineers operate far more safely than do bioengineers; and it should also be obvious that even if the latter earnestly endeavored to act more responsibly and to match the performance of the programmers, they could not come close to succeeding.

Grappling With the Ultimate Spaghetti Code

Moreover, they would still fall substantially short even if they had designed the bioinformation systems and possessed explicitly written copies of the source code. That's because the systems are so intricately intertwined that, from the perspective of software engineering, they resemble extreme instances of wildly tangled spaghetti code. For instance, on a National Public Radio program discussing the similarities of DNA and software, a computer scientist at the University of California referred to DNA as "the worst kind of spaghetti code you could imagine." And he stated that he would have given any student writing such unruly code a bad grade. To help the listeners understand why, the program's moderator explained that with such a code, "even the person who wrote it can't understand it." [61]

Accordingly, biotechnicians who had by some miracle designed an organism's information system could not later alter it in manner remotely approximating the competence and prudence displayed by software engineers when revising life-critical programs – and the risk of inducing accidental, and potentially harmful disruptions would be unacceptably high. Therefore, the fact that, in reality, they have only scant understanding of the system to start with renders their manipulations immeasurably more reckless.

This unsettling truth is more solidly driven home by taking a deeper look at how deep is the deficiency of their knowledge, and how daunting are the differences between altering DNA and revising a computer program.

Disrupting a Finely-Tuned Regulatory Network

As we've seen, genes do not regulate themselves; and the attached promoters that control their capacity to express are modulated by proteins produced by other genes, which are likewise regulated by other genes, which are in turn controlled by others – with the ultimate result that regulation occurs through the cellular system as a whole. Therefore, when a discrete gene is inserted into an organism of a different species, it arrives stripped of the complex set of features that regulate how it expresses. Accordingly, as Chapter 4 explained, such transferred genes generally won't express in their new surroundings because their promoters don't receive the specific input that enables expression to begin – which has forced bioengineers to remove those promoters and replace them with ones from viruses that will continually drive the gene's expression without the need for external input.

In fact, even when the organism is given extra copies of its own genes, the native promoters are often replaced by always-on promoters because the innate ones can't induce a high enough level of expression to satisfy the biotechnicians' aims. As a result, not only do the genes endowed with these high-powered promoters function differently than when possessed of their own, they operate *outside* the cell's regulatory system, which, as Chapters 4 and 9 have discussed, can disrupt things in diverse ways.

For instance, the incessant and uncontrolled production of proteins (whether alien or native) creates potential for two types of problems against which software engineers must be constantly on guard. In the programming context, these problems are called *race conditions* and *buffer overflow*. As previously noted, the former occur when distinct operations compete for system resources in a destabilizing manner; and they've been at the root of many malfunctions and some of the worst disasters (including at least one of the Therac-25 catastrophes and the world's second biggest power blackout).[62]

In the GMO context, such destabilizing competition can arise as a hyper-expressed foreign protein draws so heavily on the cells' chemical resources that some of their own proteins can't be synthesized as necessary. Thus, when a sunflower gene was inserted in rice, the over-expression of its sulfur-rich protein so depleted the plants' sulfur reserves that production of native sulfur-containing proteins slackened.[63] Consequently, although the bioengineers had intended to boost the rice's sulfur content, there was no net gain – and apparently no advance appreciation on the part of the technicians that hyper-consumption of sulfur by one process would proportionately impede others that also required it. Moreover, there seems to have been little appreciation within the biotech community that it's risky to place such extreme demands on one or more resources because it can create imbalances that turn the plant toxic in nonobvious ways.

In the other risky scenario, the disruption doesn't stem from desynchronous competition for the system's resources but from the overwhelming of its buffering mechanisms. Within a software program, a buffer maintains balance between the influx of data and the processing of the data, adjusting the way it's received and arranged so the capacity of the processor components won't be overtaxed. But if an overrun occurs, and those components cannot keep up, the system could crash or otherwise malfunction.

Within a GMO, a similar situation can occur through a cell's inability to accommodate the effects of foreign genes, even if they aren't hyper-expressed. Philip Regal has pointed out that ". . . theory and evidence have suggested that the host's buffering or control systems will often be ineffective for those transgenes that can express well." He explains that because the foreign genes could induce "unusual conditions" that cannot be modulated by the buffering mechanisms, ". . . new factors may be added to the host's biochemical milieu and cause quantitative or qualitative changes in the output of existing biochemical pathways."[64]

Additionally, even a native substance can overburden cellular controls if it's excessively expressed. As we saw in Chapter 3 in the case of L-tryptophan, the hyper-production of one of the amino acids the organism ordinarily makes pushed the buffering mechanisms beyond their limits and led to the formation of at least one unusual toxin.

However, although both software engineers and bioengineers face significant risk of race conditions and buffer overloads, the risks are not evenly apportioned. Software engineers can control their creations, and they assiduously strive to design programs that minimize the risks. On the other hand, bioengineers have little control over the way their insertions impact living systems, and the alterations they effect inevitably induce conditions that significantly *foster* both types of problem. Consequently, the risks inherent in their operations are of much greater magnitude than those ordinarily associated with computer programming.

Multiple Codes, Multiplied Risks

When software engineers develop source code, each line has only one meaning. Accordingly, when that particular line is converted into a segment of binary machine code (the 0's and 1's), it too has but one meaning. Therefore, even though the programmers cannot always precisely predict how various sections of code will interact as the program performs diverse operations, they *can* be confident that what the machine code specifically codes for will remain constant – and that it does not contain additional, unknown meanings embedded within it that might be accidentally activated in surprising ways.

But things are much more complicated in the realm of cellular software; and, in contrast to the information sequences in human-made systems, those in cellular ones can have more than a single signification. Thus, a discrete section of DNA can be read by the transcribing enzymes in alternate ways, with some of its nucleotides involved in the generation of one particular sequence of RNA at some times and another sequence at others – resulting in the production of different proteins.[65]

Moreover, even ostensibly non-coding segments of DNA can contain protein-coding sequences; and such a sequence can escape the recognition of regulators for many years – even if it's present in most of the GE foods on the market and poses a substantial risk (as we've seen in Chapter 6, in the case of the viral gene segment embedded within the promoter from the cauliflower mosaic virus).

Obviously, the fact that DNA contains overlapping coding sequences substantially diminishes the predictability of bioengineering; and this is the case even when only a single code is involved. But other codes exist as well, which shrinks the predictability far more drastically – especially since none of them was even discovered until more than a decade *after* the first GE foods were created, and none is still no more than sketchily understood.

Although for several decades biologists were focused on the three-letter code through which amino acids are specified and proteins are ultimately assembled – and acted as if the genome harbored no others – as DNA was studied more thoroughly, it became increasingly clear that their vision was too constricted. Through the writings of thinkers like Richard Strohman, the realization grew that a huge portion of cellular activity could not be organized through the amino acid code alone – as did the recognition that the basic features of this code enable others to be embedded within it. But substantial time was required before any were discovered.

A major breakthrough occurred in 2006, when a group of scientists reported they had detected a code "superimposed" on the amino acid code that to some degree regulates how the information in that basic code is expressed.[66] This new code appears to specify how the tiny spools of protein around which DNA is looped (the *nucleosomes*) are placed – which in turn influences how genes become accessible to the gene expression machinery.

Soon after this nucleosome code was uncovered, several others came to light; and by the end of 2013, at least seven additional "regulatory codes" had been discerned – and discussed in a paper published in *Science* in December of that year titled, "The Hidden Codes that Shape Protein Evolution."[67] Further, the paper reporting the most recent discovery appeared in that same issue of *Science.*[68] It elucidated how the sites that bind transcription factors, which stimulate gene expression, are specified within the human genome. The authors' research revealed that approximately 15% of the codons in human DNA are "dual-use codons" that "simultaneously specify" both amino acids and transcription factor recognition sites. And they dubbed these double duty codons "duons."

In announcing this discovery, the University of Washington (with which several of the researchers were affiliated) emphasized how it overturned accepted wisdom: "Since the genetic code was deciphered in the 1960s, scientists have assumed that it was used exclusively to write information about proteins. UW scientists were stunned to discover that genomes use the genetic code to write two separate languages. One describes how proteins are made, and the other instructs the cell on how genes are controlled. One language is written on top of the other, which is why the second language remained hidden for so long."[69]

Of course, the revolutionary nature of this discovery was overstated, since previous research had already shown that regulatory codes are embedded within the amino acid code – which provides insight into how long it can take for boundary breaking knowledge to transform the thinking of the life science community.[70] And its relevance for the GE food venture apparently has yet to sink in, despite (or perhaps because of) the fact that, in itself, it invalidates the basic paradigm on which the venture has been based. As the molecular geneticist Ricarda Steinbrecher has noted, the fact that regulatory sequences do not fall exclusively outside the amino acid coding sequences, as had previously been believed, but instead are also woven within them "means that any changes introduced through genetic engineering can potentially result in an altered regulation of any genes affected – and possibly to a much higher degree than previously acknowledged." And she pointed out that any gene could be affected because "the transformation procedures themselves give rise to a multitude of mutations . . . that can occur anywhere in the genome of the plant, not just where a new sequence is inserted or within the gene that is being inserted."[71]

So although the bioengineering venture is based on the premise that a discrete sequence of DNA has but one meaning – and that the meaning is conserved when the sequence is randomly transplanted within the DNA of another species – in reality, the meaning can be radically revised through such an operation. That sequence may have had *multiple* meanings within its native context, and those meanings can be skewed within a foreign context because they to a large extent depend on how the information inside the sequence interacts with information arrayed outside its confines. Moreover, the inserted sequence can disrupt information networks *within* the target organism and jumble the meanings of several of its native DNA sequences as well.

Further, because several regulatory codes could cohabit a single segment of DNA, and because that segment could also harbor overlapping amino acid coding sequences, some codons may play more than two roles – and might serve not merely as duons but as trions, quintons, or something even more multifaceted. This could lead to greater disruption if they were forced into a novel set of interconnections (either by being inserted within a new set or by the alteration of their native set through the insertion of a novel sequence within it).

What's more, even if bioengineers possessed complete knowledge of how the regulatory codes are structured and where all the overlapping amino acid coding sequences are located, their comprehension would still be puny compared to that of programmers because it would only embrace the system's machine code – and fail to encompass its source code: the principles and rules through which its various sub-codes and myriad other components operate as a harmonious and finely tuned whole.

A Compelling Conclusion: Genetic Engineering is Incurably, and Unacceptably, Risky

Thus, when the practice referred to as bioengineering is considered as a technique to alter complex information systems – and carefully compared with software engineering – not only is it shown to be inherently high-risk, but to be incapable of conforming to even the minimal standards of risk management. And so stark are its deficiencies, and so extreme its innate unpredictability, it doesn't deserve to be classed as a form of engineering – and instead of being termed "bioengineering" should actually be called "biohacking."

Further, the ignorance of its practitioners regarding the systems they restructure is not only vast but inevitable, a fact underscored by the assertion of the executive vice president of a pioneering biotechnology corporation that because the genome is so "enormously complex . . . the only thing we can say about it with certainty is how much more we have left to learn."[72] And the reality of just "how much" continues to be driven home by a string of startling discoveries that we can reasonably assume will be an ongoing phenomenon.

For instance, in March 2014, Indiana University announced that a team of its scientists had participated in research that examined the operations of the fruit fly genome "in greater detail than ever before possible" and identified "thousands of new genes, transcripts and proteins." According to this report, the results reveal that the fly's genome "is far more complex than previously suspected and suggests that the same will be true of the genomes of other higher organisms." Further, of the 1,468 newly discovered genes, 536 were found within zones that were previously regarded to be gene-free. Moreover, when the flies were subjected to various stresses, changes were induced in the expression levels of thousands of genes; and four "were expressed altogether differently." This is especially relevant for bioengineering, since, as we've seen, that process imposes multiple stresses on the cells it transfigures.

These astounding results are even more dramatic in view of the fact that the fruit fly genome is one of the most thoroughly studied and comprehensively understood of all genomes, and yet so much basic information about it was still unknown prior to March of 2014 – and so much more still lies beyond the current scope of human knowledge. Further, because biotechnicians know so much less about the genomes they alter than did biologists about the fruit fly genome even prior to the revelations of 2014, it accentuates the meagerness of their knowledge, and the high-risk nature of their genomic incursions.

Software Engineers Are Shocked by the Practices of Genetic Engineers

So deep is the dissonance between bioengineering and software engineering that when those who practice the latter learn what those who pursue the former are actually doing, they're usually aghast. When I started to investigate the technology, I wanted to discover how it stacked up against software engineering, so I began questioning computer professionals. Over the ensuing years, I've talked with many; and once they're apprised of the basic facts, they invariably react with astonishment – usually accompanied by a substantial dose of indignation. One astounded programmer exclaimed, "That's like taking a snippet of code from the program in a toaster oven and splicing it into an airplane guidance system – and yet assuming that nothing will be disturbed."

One of the most powerful and perceptive comments on the defects of bioengineering when assessed in terms of information technology was written by the Australian software engineer and information security expert Stephen Wilson in January 2011.[73]

He began by stating: "As a software engineer, years ago I developed a deep unease about genetic engineering and genetically modified organisms (GMOs). The software experience suggests to me that GM products cannot be verifiable given the state of our knowledge about how genes work." He continued: "Genes are very frequently compared with computer software. I urge that the comparison be examined more closely, so that lessons can be drawn from the long standing 'Software Crisis'."

He then observed:

> It is clear that each genome is an immensely intricate ensemble of interconnected biochemical short stories. We know that genes interact

> with each other, turning each other on and off, and more subtly influencing how each is expressed. In software parlance, genetic codes are executed in a massively parallel manner.
>
> If genomes are like programs then let's remember they have been written achingly slowly over eons, to suit the circumstances of a species. Genomes are revised in a real world laboratory over *billions* of iterations and test cases, to a level of confidence that software engineers can't even dream of. . . . Tinkering with isolated parts of this machinery, as if it were merely some sort of wiki with articles open to anyone to edit, could have consequences we are utterly unable to predict.

He then leveled his most potent critique – the force of which is amplified by the fact that he may have been misled into believing that bioengineered foods have been subjected to rigorous safety testing (as have so many other alert and intelligent individuals). He nonetheless demonstrated the utter unsoundness of the GE food venture by pointing out that in the case of complex life-critical software, even the most rigorously conducted safety testing is not sufficient. As he stated it:

> In software engineering, it is received wisdom that most bugs result from imprudent changes made to existing programs. Furthermore, editing one part of a program can have unpredictable and unbounded impacts on any other part of the code. . . . So mission critical software (like the implantable defibrillator code I used to work on) is always verified by a combination of methods, including unit testing, system testing, design review and painstaking code inspection. Because most problems come from human error, software excellence demands formal design and development processes, and high level programming languages, to preclude subtle errors that no amount of testing could ever hope to find.
>
> How many of these software quality mechanisms are available to genetic engineers? Code inspection is moot when we don't even know how genes *normally* interact with one another; how can we possibly tell by inspection if an artificial gene will interfere with the 'legacy' code?

And those comments were written before extensive research had revealed, to the shock of many biologists, that much, if not most, of the immense regions within DNA previously regarded as "junk" are actually functional parts of the genome. After that revelation, he remarked that it "reinforces" his thesis about the ultimate untestability of the products of bioengineering and increases the genome's "combinatorial complexity enormously."[74] He then added, "If genes are switched on and off by bits of code spread across the genome, then I don't know how genetic engineers are able to predict the effects of gene splicing."[75]

Moreover, the GE venture is even more devastatingly discredited by realizing that even if biotechnicians did possess the comprehensive understanding of cellular information systems that they so sorely lack, the extraordinary interconnectedness of the components, which renders DNA the most mind boggling spaghetti code on earth, would still prevent them from operating with an adequate degree of precision or precaution.

The information technology specialist Roberto Verzola has expressed this idea quite forcefully:

> Let us consider genes as if they are subroutines of a complex piece of software. A plant like corn would then have tens of thousands of these subroutines, combined in a very unstructured program in which each subroutine interacts with hundreds, perhaps thousands, of other subroutines in a kind of very tightly-coupled non-modular spaghetti code that IT [information technology] experts would consider impossible to modify and to maintain. . . .
>
> In a real-world computer program with full documentation and understanding of the instruction set and the functions of every subroutine, making even minor changes in a tightly coupled program will in all probability introduce side-effects ("bugs") which can manifest immediately or only under certain conditions, and which can lead to a major system crash or to subtle changes in the behavior of the program. . . .
>
> And that is for a program that is completely understood. How much more for a tightly coupled genetic system which consists of thousands of subroutines and their interactions which are not even understood?

Genetic Engineering is By Far the Most High-Risk Form of Food Development

Chapter 9 demonstrated that, when risks are rationally assessed, bioengineering tops all other methods for producing new varieties of plants in the potential for causing harm. And Chapter 10 showed that substantial test-based evidence lends support to this assessment. Now this chapter has shown that the high-risk nature of bioengineering is likewise confirmed from the standpoint of computer science – and to a decisive degree.

To appreciate how decisively, it's instructive to gauge how greatly bioengineering differs from radiation breeding when both are examined from the perspective of software engineering. As we saw in Chapter 9, a panel established by the National Academy of Sciences argued that modification via radiation is even more prone to cause unintended disruptions than is bioengineering. But we also saw that, when the panels' various contentions are logically arranged, they actually entail that the latter is the most disruptive. And when we analyze the two processes in light of computer science, bioengineering's status as the most risk-laden technique is clearly confirmed. This analysis appears in Appendix D.

The Irony of the DNA/Software Analogy: Although GE Advocates Use It to Strengthen Their Case, It's Actually the Strongest Argument Against It

As previous chapters have shown, the history of genetic engineering is peppered with ironies; and now we're positioned to savor one of the biggest. Whereas proponents of this practice liken it to software engineering so it will seem more manageable and acceptable, a systematic comparison reveals that it's radically reckless. And software professionals who've examined the contrasts urge that biologists and regulators do so as well in order to learn the important lessons that software engineering can teach.

But, unfortunately, there has been little progress in this direction. Instead, the comparisons remain surprisingly superficial. Consider the case of Richard Dawkins, one of the most prominent and prolific life scientists of recent times. In a 2003 article in *The London Times* lambasting those who protest genetic engineering, he declared that "genetics has become a branch of information technology."[76] In expanding on this theme, he stated:

> The genetic code is truly digital, in exactly the same sense as computer codes. This is not some vague analogy, it is the literal truth. Moreover, unlike computer codes, the genetic code is universal. . . . The consequences are amazing. It means that a software subroutine (that's exactly what a gene is) can be carried over into another species. . . . In the same way, a NASA programmer who wants a neat square-root routine for his rocket guidance system might import one from a financial spreadsheet. A square root is a square root is a square root.

A program to compute it will serve as well in a space rocket as in a financial projection.

However, as we've seen, in reality things are not so simple; and even transferring a segment of code from one rocket into the program of a newer version of the same rocket caused a disastrous malfunction. And although Dawkins' discussion does eventually recognize that the new context in which a subroutine is placed can make a difference, he doesn't go deeply enough. Thus, while he acknowledges that a transferred gene "might not work unless properly tweaked . . . to mesh" with the genes of the target organism, he indicates that such tweaking can be done – but fails to note that enabling the new gene to sufficiently mesh so that its protein will be expressed is not the same as harmonizing its effects with the rest of the system. And, as we've seen, creating a limited mesh can cause a mess.

Nonetheless, Dawkins ultimately agrees that it's justified to make "a rational plea for rigorous safety testing." And he states that "no reputable scientist would oppose such a plea" – apparently unaware that this assertion effectively impugns the integrity of many high-placed members of the scientific establishment. Moreover, he seems oblivious to the fact that in order for the testing of GE foods to come anywhere close to the level of rigor with which life-critical software is scrutinized, the current system would have to undergo a massive revamping that would be strongly resisted by most of the ostensibly reputable scientists who support the products – and that could not be financially borne by their manufacturers.

And the irony is enhanced because not only did Dawkins propound an analogy that actually undercuts his claims, at the time he propounded it, he was Professor for the Public Understanding of Science at Oxford University, a special position endowed by a software developer who had made a fortune during the years he led a key group at Microsoft.[77] Thus, a biologist representing one of the world's top universities in the endeavor to foster public understanding of science was inadvertently misleading the public about the relation between computer science and genetic engineering – and doing it from a platform created by funds derived from computer science.

The Ultra-Irony of the DNA/Software Analogy: Even the World's Most Famous Software Developer Has Not Yet Grasped Its Implications

But the ironies don't stop there. Not only has money earned by a Microsoft executive funded a scientist who's created confusion about how software engineering bears on bioengineering, the man who co-founded the company, served as its chief software architect, and for decades was its CEO has himself failed to appreciate how the lessons of software engineering discredit the GE food venture – and has devoted a substantial part of his vast software-generated fortune toward fostering its growth.

Of course, Bill Gates is highly astute, and he's well aware that the information systems of living cells differ from human-made software in important ways. And in his book, *The Road Ahead*, when he likens DNA to a computer program, he acknowledges that it's "far, far more advanced than any software we have ever created."[78] But on the road he's actually traveling, this reality has apparently slipped from sight. Instead of looming along the way as a regularly posted caution, it seems to have receded beyond the range of even peripheral vision. Consequently, he has pressed ahead to expand the GE food venture, promoting the radical alteration of earth's most intricate software absent even the kind of quality controls employed in developing Microsoft's word processing program, let alone the strict safeguards mandated for life-critical systems.

Accordingly, although he's motivated by altruistic purposes, his efforts are nonetheless misguided. After all, does it ultimately benefit any African nation to develop new varieties of crops that, although they may be salt- or heat-tolerant, might also be intolerably disease dealing? From the perspective of software engineering, this latter risk cannot be discounted and should be taken into serious account, especially since the adverse effects might not manifest for many years. In fact, as we've seen, from such a perspective, one should not even attempt to apply recombinant DNA technology in crop production unless there are absolutely no other reasonable alternatives, which, as Chapter 14 will discuss, is definitely not the case.

It's additionally ironic that, like Dawkins, Gates holds views about testing that aren't attuned with the realities of software development. Thus, although he says that rigorous testing should be conducted on GE foods, he appears to think either that it's regularly being done or that it readily can be – unmindful of the immense reformation that would be required to achieve even partial parity between the safety testing of these products and the testing routinely applied to life-critical software.[79]

Sometimes, when I read another report about the Gates Foundation's granting of many millions to a project that aims to improve nutrition in Africa by inserting genes into one or another species of plant, I wonder if he would have still viewed the grant application favorably had it explicitly framed things within a software revision context. Would he have maintained his confidence in the soundness of the project if the application had stated that the money would be employed to hack into the plant's intricate information system; haphazardly insert a chunk of code cobbled together with pieces of DNA from an unrelated plant, a few bacteria, and a virus; and then release the resultant crop into the food supply of numerous nations on the basis of some simple testing that doesn't come close to meeting Microsoft's standards for a new version of Windows let alone the minimal requirements for certifying the safety of life-critical software?

So, from the standpoint of computer science, the GE food venture is irreparably risky, to an intolerable degree, because there's virtually no way to alter the programs of living organisms with sufficient foresight to avoid harmful unintended effects – and scant practical possibility of successfully screening for them. Further, from such a perspective, even if every test so far performed on GE foods had failed to observe any adverse outcomes, their safety would still not have been demonstrated because the tests that have been employed are deplorably incapable of doing so.

Accordingly, one well may wonder how, despite their crucial importance, the lessons of computer science could have been so consistently ignored by those who promote the GE food venture – even by one who for many years led the world's largest software development company.

As we'll see, this ongoing oversight correlates with the life-science community's chronic embrace of a few fundamental assumptions about the nature of living organisms that were never substantiated and have for years been thoroughly discredited. And it's through such essentially faith-based bias that highly intelligent men and women have been blinded to the manifest risks of tampering with the software of life – and beguiled into believing that these exquisitely integrated systems can be radically altered with far less caution than is employed when making even minor refinements to a human-made computer program.

CHAPTER TWELVE

Unfounded Foundational Assumptions

The Flawed Beliefs that Undergird Agricultural Bioengineering

The first ten chapters developed a comprehensive and thoroughly documented case demonstrating the unsoundness of the GE food venture. They cleared up many misconceptions and established many important points, including the following five:

- The commercialization of genetically engineered foods was enabled by the fraudulent behavior of the US Food and Drug Administration (FDA), could not have happened without it, and continues to be reliant on it.
- The FDA ushered these novel products onto the market in violation of explicit mandates of federal food safety law, and they are still illegally on the market.
- The FDA's falsehoods have been abundantly supplemented with falsehoods disseminated by eminent scientists and scientific institutions, and the entire GE food venture has been chronically and crucially dependent on this disinformation.
- The safety of GE foods has never been established in a scientifically reliable manner, and substantial research has cast doubt upon their safety.
- These foods entail unacceptable risks.

And the book could have easily ended with the end of Chapter 10.

But it continued with Chapter 11 in order to show that besides being unsound from the standpoint of biological science, the GE food venture is additionally unsound from the standpoint of computer science – thereby providing a fuller understanding of just how risky and reckless it really is.

The aim of this chapter is to more fully probe the ideational foundations of this enterprise and to discern the assumptions that initially inspired faith in its soundness – and that continue to sustain the faith despite the dearth of solid supporting evidence and the accumulation of considerable evidence to the contrary.

As we've seen in the preceding chapter, the biotech proponents' seemingly unshakeable faith in the GE food venture has even deluded them into assuming that the largely uncontrollable alterations it imposes upon genomes are somehow equivalent to the precise revisions that are applied to computer software. Further, they have failed to recognize that even these precise revisions entail inevitable risks of unintended consequences – and so they've utterly failed to appreciate how much greater are the risks of imposing imprecise alterations on information systems of living organisms that are far more complex and far less well-comprehended than any human-made information system.

Moreover, as the other chapters clearly revealed, the advocates of GE foods have not only overlooked the lessons of computer science, they've disregarded the aggregate implications of numerous biological facts – and even some of the facts themselves.

Without these crucial and continuing oversights, confidence in the safety of the GE food venture could not have been maintained even by its own advocates, and it almost surely would have withered. However, it was through the initial overconfidence that these oversights actually occurred. Thus, the overconfidence caused the oversights, which in turn sustained the overconfidence.

This overriding overconfidence was rooted in a few notions that, although they provided their adherents a seemingly scientific basis for belief in the safety of genetic engineering, have never been empirically verified and have increasingly clashed with the growing stock of evidence – but have nonetheless been treated as solid scientific facts.

Yet, it took time for this undermining evidence to accumulate, and when recombinant DNA technology emerged in the early 1970's, knowledge of the structure and dynamics of bio-information systems was still quite sparse – which facilitated the formation of significantly oversimplified, and inaccurate, conceptions about them. These erroneous ideas were an essential aspect of the genetic engineering venture, and it was to a large extent grounded on them.

Some Key Presumptions on Which the Bioengineering Venture Was Based

As Chapter 9 pointed out, the safety of genetic engineering was premised on the presumption that the genome is a simple linear system in which the action of a single gene will not significantly impact the others and won't disrupt their normal function. This was emphasized in the 2001 report of the Royal Society of Canada and has been recognized by many other experts.[1] For instance, in a 2007 *New York Times* article, the veteran technology reporter Denise Caruso observed: "The presumption that genes operate independently has been institutionalized since 1976, when the first biotech company was founded. In fact, it is the economic and regulatory foundation on which the entire biotechnology industry is built."[2]

Giorgio Bernardi, a biologist at the University of Rome III who specializes in the study of genome evolution, has pointed out that within such a conceptual framework, the genome's capacity is significantly limited because it "is only endowed with additive and not with cooperative properties."[3] In other words, the genes are viewed as significantly autonomous agents that add to the whole without acting holistically because they don't express their proteins in a closely coordinated manner.

However, this was not the only foundational presumption, and it was linked with some others of the same ilk. One of the most important was the notion that genes aren't arranged in an organized manner and that the sequence in which they occur is essentially unimportant.[4] From such a viewpoint, a gene would function just as satisfactorily if it were relocated to a different chromosome or came in front of a neighboring gene instead of after it. Bernardi refers to this perspective as a "*bean-bag view* of the genome" because it regards the genes as "randomly distributed."[5]

Together, these two presumptions supported the belief that a chunk of recombinant DNA could be put into a plant's genome without inducing disturbance – because if the behavior of the native genes was largely uncoordinated and their arrangement was irrelevant, there would be no important patterns that could be perturbed by such insertions. Accordingly, they engendered confidence in the precision of genetic engineering, because they implied that the outcome of a gene insertion would be exactly what the bioengineers expected. They bolstered the belief that the target organism would continue to function just as before and that change would be limited to the new trait endowed by the inserted gene – which would tidily manifest without altering any of the organism's other qualities.

Despite Being Wrong, the Presumptions Maintained Their Force

These presumptions, which provided an ideological foundation for the bioengineering venture and undergirded a massive endeavor to transform agriculture, turned out to be wrong. As we've seen in previous chapters, abundant evidence has discredited the notion that genes act independently from one another – and has instead demonstrated that their actions are highly coordinated. Commenting on this evidence, Giorgio Bernardi has written that the genome must now be viewed as an "integrated ensemble."[6] Likewise, in her previously mentioned *New York Times* article, Denise Caruso emphasized the fact that "genes appear to operate in a complex network." And she noted the serious implications of this fact for the bioengineering venture, stating: "Evidence of a networked genome shatters the scientific basis for virtually every official risk assessment of today's commercial biotech products, from genetically engineered crops to pharmaceuticals."

Overwhelming evidence has also undermined the tenet that genes are randomly arranged; and by 2004 an article published in *Nature Reviews Genetics* could assert that this tenet was "no longer tenable."[7]

Moreover, besides being discredited, these presumptions share another ignoble feature: they survived discreditation and outlived their plausible legitimacy. Despite the accumulation of undermining evidence, biotech proponents stubbornly clung to them anyway; and they exerted influence long beyond the point they rightfully should have.

For instance, in his testimony to New Zealand's Royal Commission in 2001, the molecular geneticist Michael Antoniou stated that agricultural bioengineering was "based on the understanding of genetics we had 15 years ago, about genes being isolated little units that work independently of each other." And he pointed out that during the ensuing years, sufficient evidence had amassed to refute that view and demonstrate that genes actually "work as an integrated whole of families."[8]

However, the scientists who promote GE foods remained largely unfazed by this evidence, a fact driven home to Antoniou when in 2003 he was selected to represent nongovernmental organizations on the UK's GM Review Panel. Although his arguments for enhanced precaution were based on extensive studies demonstrating that genes are coordinated, most of the 11 other scientists on the panel, who were biotech proponents, dismissed these studies and even argued that it makes no difference how genes are arranged.[9]

The very fact that GE foods are still being marketed and developed, and that their proponents continue to claim their safety has been scientifically certified, attests to the unwarranted staying power of the presumptions, because had they been fully relinquished, confidence in GE foods could not have been sustained.

Thus, the GE food venture was grounded on the belief that, at their deepest level, biological organisms do not display the orderly arrangement and coordination of parts that's commonly denoted by the term "organic." [10] Although its founders and early adherents recognized that on the expressed level of organs and tissues, an organism displays profound interrelatedness, and that the non-chromosomal components of the cells also display it, they believed the situation was quite different *within* the chromosomes, at the level of the DNA. And they regarded this level as distinctly disjointed. Therefore, they believed that the layers of organic wholeness within an organism are underlain by a dimension that's significantly non-holistic – because they regarded the complex information system on which the coordination existing at the other levels depends as an assemblage of units that are substantially uncoordinated.

Further, even though the venture's advocates may have taken note of the contrary evidence that steadily amassed over the years, and even though it may have colored their thinking about specific aspects of biology, to the extent they continued to support the venture and to assert its soundness, they were endorsing an endeavor that still relied on presumptions that this evidence had refuted – which was not the first time intelligent people have compartmentalized their thinking to insulate conflicting ideas.

And such disregard, denial, or avoidance in regard to the evidence was essential for maintaining faith in the venture, because its predictability and safety have always relied on the genome being largely disjointed; and the more the genome instead appears to function as a tightly coordinated system, the more potentially disruptive and unpredictable are the interventions of the bioengineers.

The Venture's Fallback Belief: Natural Breeding as a More Random and Unruly Process than Bioengineering

The GE food venture's endurance despite the discreditation of two of its key presumptions and the steady production of adverse experimental evidence has been enabled not only by resort to disregard and denial, but by reliance on a fallback belief: a belief that's in some ways deeper and more basic than either of the presumptions we've previously examined. It's the belief that, regardless of specific data that may be marshaled by those who express concerns, the biological processes underlying natural reproduction are more random and unpredictable than the mechanics of genetic engineering – and that the latter therefore *must* be more trustworthy.

Although many of the venture's proponents are likely to contest this statement, it's supported by their own repeated assertions. A typical instance appears in the 2000 report of the Institute of Food Technologists that's critiqued in Appendix B. That document declares: "Given the more precise and predictable nature of genetic change accomplished through rDNA techniques as compared to the random genetic changes observed in conventional breeding, such unintended effects would be considered less likely in foods derived from rDNA biotechnology." [11] Such allegations are widely made on behalf of GE foods by their scientist-advocates, and they serve as the ultimate rejoinder whenever an expert mounts a substantial challenge to the technology's safety.

A striking example of such a rejoinder, which attests the centrality of the belief in nature's unruliness, was evoked by a warning about the risks of GE foods that appeared as a comment in *Nature Biotechnology* in 2002. The author was David Schubert, who, as previously mentioned, is a cell biologist at the Salk Institute for Biological Studies. He noted there was mounting evidence that the insertion of even a single gene into a cell's DNA invariably alters the expression pattern of genes throughout the cell; and he explained why other disconcerting facts likewise cast doubt on the soundness of agricultural bioengineering – and entail the conclusion that it "is not a safe option." [12]

In response, 18 biologists at respected universities and institutes published a letter in that journal criticizing Schubert and defending the safety of GE foods. And the way they did so is quite revealing. Faced with a serious challenge written by a professor and laboratory director at one of the world's most prestigious scientific institutes, their response placed primary emphasis on what they described as the "real issue": Dr. Schubert's failure to properly consider "the genetic realities." And the main reality he allegedly failed to recognize is that the natural method of plant breeding is inherently more random than bioengineering. As they put it: "We do not take issue with Schubert's basic contention that unintended genetic and metabolic events can take place. The reality is that 'unintentional consequences' are much more likely to occur in nature than in biotechnology because nature relies on the unintentional consequences of *blind random* genetic mutation and rearrangement to produce adaptive phenotypic results, whereas GM technology employs precise, specific, and rationally designed genetic modification toward a specific engineering goal."[13] [Emphasis added]

This letter thus reveals how strongly the GE food venture relies on the presumption that the natural processes driving biological development are intrinsically more disorderly and risk-bearing than the genetic interventions instigated by the human mind. And it confirms that this belief forms the ideological bedrock on which the venture rests.

But this belief is at odds with the facts, a reality substantiated by the failed attempt of the 2004 NAS report to uphold even the more modest notion that bioengineering and natural breeding pose the *same* risks. As Chapter 9 described, when the panel that produced the report ranked the various modes of plant breeding in terms of their propensity to produce unintended effects, it was compelled to acknowledge that bioengineering has far greater propensity than does pollen-based sexual reproduction – and is far less predictable. Yet, it nonetheless insisted that this disparity in perturbational potential does not entail a difference in risks. However, as the chapter demonstrated, if the disruptive potential is different but the risks are still the same, then the average unpredicted effect of natural breeding has to be much *more* dangerous than the average unplanned alteration caused by genetic engineering – which is a patently ridiculous outcome.

Thus, there's no rational way to reconcile the fact that natural breeding is less disruptive and more predictable than bioengineering with the claim that it poses equal or greater risk, which is why the admission in the 2004 report is a rarity – and why biotech proponents almost always ignore or deny that fact and instead assert that natural breeding is *more* disorderly and unpredictable.

Misrepresenting the Degree of the Randomness

Moreover, not only do biotech proponents routinely impute substantial randomness to the dynamics within living organisms, they overstate the amount and mischaracterize the processes that purportedly display it. According to their version of reality, natural plant breeding is fraught with unruly forces that can wreak havoc within the plants upon which unsuspecting humans feed; and people should feel relieved that these menaces can now be minimized through bioengineering. Previous chapters have provided several specific examples of such misrepresentation, one of the more striking of which is Chapter 9's examination of how the NAS report of 2004 tried to make people jumpy about the mobile elements that are commonly called "jumping genes" by portraying them as more randomly mobile, and more threatening, than they actually are. And it pointed out that in reality, the report got it backwards because, although these entities do *not* pose appreciable risks within natural pollen-based breeding, they *do* when bioengineering is employed because that process tends to stir them up and get them jumping.

An even more egregious case of overstating the randomness in natural processes relative to genetic engineering – and stating things backwards regarding the risks – is the routine attempt of biotech advocates to portray sexual reproduction as a disturbingly random and messy phenomenon. For instance, in their letter critiquing Schubert, those 18 life scientists did not stop at calling genetic mutation a "random" phenomenon, they said the same about natural "genetic rearrangement." [14] And the most frequent and important form of such rearrangement is an essential phase in the reproductive process in plants and animals that provides a large part of the genetic diversity required for a species to remain robust. Like genetic engineering, it involves the recombination of DNA; but unlike that artificial technique, it does so in a manner that doesn't disrupt or imbalance the organism.

This natural form of recombination occurs during the formation of gametes (the sperm and egg cells). It includes a step called *crossover* in which two partner chromosomes break at corresponding points and then exchange complementary sections of DNA; and every time a gamete is produced, every set of paired chromosomes engages in it. In this way, all the chromosomes end up with genes from both parents instead of from only one. However, all the genes are preserved, as is the sequence in which they're positioned. The only changes are in the relationships between alleles. As was discussed in Chapters 2 and 4, alleles are alternative versions of a gene, and when chromosomes recombine, alleles that were formerly on separate chromosomes can occupy the same one, and an array of new assemblages can arise while the integrity of the genome stays intact. So this natural recombination augments diversity while maintaining stability. And without it, except for the occasional favorable mutation, the composition of chromosomes would stay the same from generation to generation, and genetic diversity would grow at far too sluggish a pace.

Moreover, not only does natural recombination preserve the order of the genes, it's predictable in how it cuts DNA. In preparation for crossover, enzymes cleave the DNA at specific sites; and these sites are not randomly located. So the entire process displays a high degree of orderliness, and, in the words of Phil Regal, "is amazingly organized."

Further, the steps of reproduction that follow it are also highly ordered. Consequently, when a sperm and egg unite to form the cell from which a new organism develops, each contributes the same number of chromosomes, each chromosome is matched with a similar partner, and the total number of chromosomes is maintained.

However, although sexual reproduction is, from start to finish, an exquisitely well-coordinated and coherent process, from what the GE proponents say, you would never know it. Instead, you would think it was far from orderly; and the two examples already cited are not in the least atypical. For instance, an article published in 2012 by six scientists who advocate GE foods contrasts the allegedly precise modifications made through bioengineering with the "random genetic modifications that occur in conventional breeding."[15] And the 2004 report by the National Academy of Sciences (NAS) tries to make the contrast more vivid. It asserts: "Genetic engineering methods are considered by some to be more precise than conventional breeding methods because only known and precisely characterized genes are transferred. In contrast, conventional breeding involves transferring thousands of unknown genes with unknown function along with the desired genes."[16]

But besides being inaccurate, the arguments about the randomness and unpredictability of natural breeding are deceptive, because they shift the focus away from the issue of whether the plant is safe to eat and place it on an unrelated one – while pretending that *it* is the key safety-related question. This misleading tactic fixates on the predictability of the plant's specific agronomic traits; and it portrays traditional breeding as less predictable than bioengineering because undesired attributes are often transferred along with the one that is desired. However, those who employ this ploy don't acknowledge that (as Chapter 9 has shown) if both parents are safe to eat, the unwanted traits hardly ever pose risk to human health. Rather, they're undesirable for reasons irrelevant to risk (such as aesthetic appearance or seed size), and breeders must then perform back-crossing to eliminate them while retaining the trait they want. However, although the inclusion of unwanted traits entails more work, it does not increase attendant risks. Therefore, while breeders can't fully predict what traits will appear, they can confidently predict that the resulting plant will be safe to eat.[17]

Thus, the pro-GE portrayal of nature is shamefully misleading. Although it describes the sexual reproduction of food-yielding plants as a messy and risky affair that involves the transfer of "thousands of unknown genes with unknown function," we actually know quite a lot about those genes. And what we know is far more important than what we don't know. We know that they're all where they're supposed to be, and that they're arranged in an orderly fashion. And we know that during the essential process in which some of them are traded between partnered chromosomes in order to promote the diversity that strengthens the species, their orderly arrangement is marvelously maintained. Most important, we know that their functions mesh to form an exquisitely efficient system that generates and sustains a plant that regularly provides us wholesome food.

This sharply contrasts with genetic engineering. Although the gene that's transferred via this technology is known, not only is it impossible to predict all its unintended effects, there's no sound basis for assuming they'll be safe. In fact, as several previous chapters have demonstrated, there's good reason to presume they pose significant risk.

Therefore, when the two processes are fairly compared, it's bioengineering that is more random and risky. The inserted cassettes are haphazardly wedged into the cell's DNA, they create unpredictable disruptions at the site of insertion, the overall process induces hundreds of mutations throughout the DNA molecule, the activity of the inserted cassettes can create multiple imbalances, and the resultant plant cannot be deemed safe without undergoing a battery of rigorous tests that has yet to be applied to any engineered crop.[18]

Nonetheless, despite its gross inaccuracy, the GE proponents' depiction of natural processes as disorderly and untrustworthy has, through persistent repetition by scientists who should know better, been adopted by the media and ingrained in general awareness. Thus, in discussing the conclusions of the NAS's 2004 report, *The New York Times* contrasted bioengineering and natural breeding in the same misleading manner as had the report. Apparently, neither the reporter nor editors critically assessed the report's version of reality or considered whether employing its language would convey a false impression. Instead, they told their readers that "[g]enetic engineering involves the transfer of a specific gene from one organism to another" but that "[c]ross-breeding, by contrast, involves the mixing of thousands of genes, most unknown."[19]

A Dubious Distinction: Ag-Biotech is the First Essentially Faith-Based Technology

Many scientists have pointed out that science itself requires faith in some basic ideas, and every technology requires it too. But in these fields, faith does not function to substantiate theories; and in the case of technology, it does not serve as the primary basis for confidence in the safety of products and procedures. Instead, safety is initially assessed via testing, and thereafter it's continuously gauged through real-world performance. If the buildings constructed through a particular technique are generally stable over time, the technique is considered safe; and if there are instead a significant number of failures, it's deemed too risky. And in most fields of engineering, although some defects may go undetected during testing, they will eventually manifest in an obvious manner. Bridges collapse, airplanes crash, software malfunctions. So, flawed techniques will eventually be exposed.

But agricultural bioengineering is different. As previous chapters have demonstrated, not only has the scope of the testing never been adequate,[20] adverse results have been routinely ignored. Moreover, after the products have been brought to market, there's been no reliable way to monitor their performance. Although acute toxicity has been essentially ruled out in most cases (except the toxic tryptophan supplement that caused an EMS epidemic in 1989), due to the lack of proper monitoring, it's been impossible to determine whether any GE foods are creating long-term health problems. And, as we've seen, although biotech proponents commonly claim that the lack of observed calamities serves as proof of safety, significant damage might be occurring that even epidemiological testing couldn't detect; and in the utter absence of meaningful monitoring, a GE food could be causing cancer at a greater rate than cigarettes yet still appear benign.

Consequently, because GE foods have not been demonstrated safe via testing nor shown to be safe through experience, their safety is solely a matter of belief. The conviction that they're safe is not based on reliable clinical or practical evidence but instead is fed by faith that natural breeding processes are more unpredictable and risky than producing foods through genetic engineering. And this faith is founded on erroneous assumptions.

The proponents of GE foods have unwarrantedly presupposed that although the structures of living organisms are elegantly coordinated at most levels, this coordination does not prevail at their deepest dimension – and that their seminal components are a random assemblage of partially attuned parts. Accordingly, they regard the information systems that undergird living beings as less soundly constructed, less comprehensively coordinated, and hence less susceptible to disruption through alteration than are the systems of software produced by the human mind. And they therefore presume that the supposedly well-managed interventions wrought by genetic engineering will be largely innocuous. At the same time, they wrongly presume that the operations of natural breeding are more haphazard than those of bioengineering, and hence more dangerous.

Thus, agricultural bioengineering stands apart from all other technologies because it's the only one that is so crucially reliant on faith. And even if there were another technology in which faith plays as critical a role, ag-biotech would still be unique because it's the only one based on belief in the disorderly nature of some vital biological processes – and the only one for which such a belief serves as the basis for decisions about predictability and safety.

So, notwithstanding its claim to be based on solid science and hard evidence, the GE food venture is ultimately grounded on faith; and, from an objective standpoint, the faith is misplaced because its suppositions are false.

Yet, even though these presumptions have been discredited, and even though adverse test results have repeatedly emerged, this vast enterprise has managed to continue. And, as previous chapters have amply demonstrated, its continuation has been crucially dependent on incessant disinformation and deception.

As we shall see, when the mass of falsehood and fraud that has sustained the venture is assessed against the backdrop of history, its magnitude is clearly unprecedented – and it stands as a unique phenomenon in the annals of science.

CHAPTER THIRTEEN

The Devolution of Scientists into Spin Doctors

Genetic Engineering's Most Malignant Mutation

> "The scientist . . . must conform to the facts. The sanction of truth is an exact boundary which encloses him."
>
> Jacob Bronowski, *Science and Human Values*[1]

In his influential book, *Science and Human Values*, the renowned scientist Jacob Bronowski emphasizes the need for scientists to maintain a rigorous relationship with the facts, and he states it's essential that they cultivate "the habit of truth."[2] Moreover, he highlights the importance of this practice by titling the middle section of the book "The Habit of Truth."[3]

However, despite the necessity of this habit for the proper practice of science, since the advent of genetic engineering it has so steadily and substantially eroded within the life science community that many members have instead grown habituated to twisting the truth in order to promote that controversial technology – and their penchant for false pronouncement has become quite pronounced.

Blurring the Boundary between Scientists and Spin Specialists

As the previous chapters have documented, the scientist-proponents of genetic engineering have not merely failed to adhere to the facts as assiduously as scientists are supposed to, in several respects, they haven't even maintained higher standards of truthfulness than people engaged in politics. And we've seen that as science became more politicized in regard to genetic engineering, and as its scientist-promoters strove to augment their influence over the political process, the media, and public opinion, they increasingly adopted the techniques of spin specialists advancing political campaigns.

The transformation of doctors of science into spin doctors began at least as early as 1977, when GE proponents mounted a major effort to quell the growing concerns about the risks of genetic engineering. As we saw in Chapter 1, the historian Susan Wright, in chronicling this endeavor, documented several examples of the proponents' propensity to spin the facts in a misrepresentative manner. This tendency toward spin has also been noted by Diana B. Dutton, a Senior Research Associate of the Stanford School of Medicine, who observed that as the proponents strained to project a positive image of their technology, "[e]ven accumulating evidence that there were, indeed, risks was interpreted in a positive light." [4]

As Chapter 1 recounted, this endeavor soon evolved into an enormous political lobbying campaign to quash proposed legislation that would have regulated genetic engineering. Exemplifying the zeal with which scientists took up their new lobbyist role, Norton Zinder, an eminent microbiologist at Rockefeller University and a member of the National Academy of Sciences, "urged his colleagues to 'lobby like crazy' with the Congressmen from their states." [5]

Dutton reports that as this campaign gained strength, the numerous scientists who did have concerns "began to see the burden of proof concerning risks shift from the proponents' camp to their own." [6] Accordingly, many signed a statement forcefully critiquing the " 'misrepresentation and exaggeration of recent data purporting to show the safety of recombinant DNA research' " and alleging that the scientist-proponents of genetic engineering were using scientific data for their own political purposes.[7] But because there were so many prestigious scientists and scientific institutions engaged in the "full-blown lobbying effort," [8] including the American Society for Microbiology, the National Academy of Sciences, and major universities, its force could not be blunted by the scientists who protested its excesses.[9]

However, in achieving victory, and blocking regulatory legislation, the scientists incurred substantial costs, the most serious of which were not financial. As Dutton observes, "Scientists had won their political battle, but in the process lost some of their innocence." [10] What's more, they lost some of their integrity as well. And the significance of this loss was not lost on Norton Zinder, the microbiologist who had exhorted his colleagues to "lobby like crazy." Dutton reports that in reflecting on his own lobbying efforts, he confided to Paul Berg, " 'I've been busy so long calculating the results of moves – did I push too soon? too late? were the right people contacted? . . . how far can I stretch the 'truth' without lying? – that I may have lost all perspective.' " [11]

Regrettably, in order to keep the GE venture rolling, the kinds of excesses displayed during that lobbying campaign had to continue, and as the enterprise expanded into food production, too many of its scientist-promoters definitely did lose their perspective – and stretched the truth beyond the bounds of mere spin into the realm of clear-cut falsehood.

As we've seen, due to their truth-twisting, the burden of proof not only shifted as a practical matter, but was eventually instituted as a formal governmental policy – starting within the US National Institutes of Health and eventually spreading throughout the federal executive branch and its regulatory agencies. And this shift was even implemented by the Food and Drug Administration, despite the fact that in order to relieve the manufacturers of GE foods from the obligation of establishing their products' safety, this agency had to perpetrate a major fraud and also violate one of the nations' most important and long-standing consumer protection laws.

As previous chapters have demonstrated, without this shift in the burden of proof, which scientists induced by shedding their burden of truth, the GE food venture could not have progressed and none of its products would have been commercialized. Moreover, as the chapters have also demonstrated, not only would enforcing the law (and keeping the burden properly imposed) have blocked the marketing of GE foods, the maintenance of integrity within the scientific establishment would independently have done so, because if scientists had honestly described the facts about these novel foods, and acknowledged their deep differences with naturally produced ones, the public would never have accepted them.

Recognizing the Centrality of the Scientists' Duplicity

The Key Deceptions Have Come from the Scientific Establishment, Not the Biotech Industry

Thus, the deceptions of the mainstream scientific establishment have not merely played a crucial role in enabling the advance of agricultural bioengineering, but the *key* role, and the history of the enterprise cannot be properly comprehended without recognizing it. Accordingly, this group of individuals and institutions must ultimately be held responsible for all of the enterprise's delinquencies and associated problems – which would not have arisen if they had spoken honestly.

This assertion will probably come as a surprise to most people with concerns about GE foods, because they tend to focus on the transgressions of Monsanto and the other multi-national corporations that sell them and portray these entities as solely responsible for the problems their products pose. But in doing so, they overlook the reality that these corporations could not have commercialized any GE foods if the scientific establishment (and especially the molecular biologists) had not prepared the way by systematically deluding the government and the public about the basic facts. And had this deception not been achieved, and widespread concerns not been substantially mollified, it's doubtful such profit-seeking entities would have invested the vast sums required to develop GE foods in the first place.[12]

Further, it's important to realize that the endeavor to avoid regulation of genetic engineering pre-dated the modern biotechnology industry. When more than a hundred biologists convened at Asilomar in February 1975 in an effort to maintain control over how their research with recombinant DNA technology would be supervised, and to deter the involvement of outside regulatory agencies, no companies employing that technology even existed. And when the first one eventually appeared, it was founded by a scientist who was one of the technology's inventors (and co-founded not by a big corporation but by a lone venture capitalist).[13] Moreover, most of the early biotech companies were likewise launched by molecular biologists and venture capitalists, and major chemical companies like Monsanto and DuPont did not significantly enter the picture until much later. Thus, when the first political lobbying campaign was mounted by GE proponents in 1977, whatever biotech industry existed was not only small, but essentially an extension of the scientific research community – not an arm of the major corporations that purvey pesticides. Nor was it yet involved in the production of bioengineered plants but was instead focused on the technology's medical applications. Moreover, that initial lobbying endeavor was primarily conducted by university scientists, universities, and other scientific institutions.[14]

Furthermore, even after bioengineering had expanded to agriculture, the GE food venture had kicked into high gear, and Monsanto and other multi-nationals had become heavily engaged, the scientific establishment continued to play the chief role in dispensing the disinformation on which the venture's survival depended.

The scientists' pivotal position was due to their perceived authority. Because the public trusts scientists at universities and non-profit institutes much more than private corporations, the latter try to get potentially controversial products endorsed by scientists who are ostensibly independent from them. But in the case of genetic engineering, scientists took the initiative *before* there was a related industry, and they still needed no prodding after the industry had developed. Further, their influence in regard to this technology has been exceptionally strong. Research has shown that the American public regards university scientists, along with the federal regulatory agencies, as the most reliable sources of information on GE foods;[15] and it was through the misrepresentations spread by such scientists that the FDA was enabled to distort the facts as well, because if they had upheld the truth, the agency could not have gotten away with twisting it. After all, if the majority of university-based biologists had forthrightly acknowledged the big differences between genetic engineering and traditional breeding, and had not obfuscated and distorted basic facts, the FDA could not have issued claims about the essential similarity between the two. And if they had spoken as frankly and responsibly about risks and the need to test for them as did the expert panel of the Royal Society of Canada and the scientist-plaintiffs in the Alliance for Bio-Integrity lawsuit, the agency could not have plausibly asserted that GE foods are generally recognized as safe within the community of experts.

Moreover, when making public pronouncements, the scientist-proponents of biotech have tended to be even *more* biased than industry representatives. For instance, Researchers at the Center for Biotechnology Policy and Ethics at Texas A&M University analyzed 132 newspaper articles relating to bioengineering published in the US during 1991 and 1992 and found that the bulk of quoted information came from industry and university sources – and that the university scientists generally presented a more one-sided picture than did the industry. While industry representatives were as likely to comment on potential dangers of biotechnology as were its critics, academic scientists made much less mention of them and overwhelmingly argued for the projected benefits.[16]

The AMA as a Clinical Case of Chronic Irresponsibility

One of the most remarkable examples of unabashed and biased promotion of GE foods by a major member of the scientific establishment has been provided by the American Medical Association (AMA). As we saw in Chapter 9, this organization has been at the forefront of the promotional endeavor, declaring in a 1990 policy statement on agricultural biotechnology its intent "to endorse or implement programs that will convince the public and government officials that genetic manipulation is not inherently hazardous. . . ."[17] And this pronouncement about the inherent safety of agricultural biotechnology was released two years before the FDA issued its own policy statement on this technology's products and long before any genuine safety testing had even been conducted. Further, besides announcing a commitment to belittle the risks of agricultural bioengineering, the

AMA's statement pledged "to actively participate in the development of national programs to educate the public about the benefits. . . ."

This blatant promotional policy stands in stark contrast to the earlier stance of the medical community in regard to tobacco. During the 1950's, although a substantial percentage of American doctors smoked, they were barred by their ethical code from appearing in cigarette advertisements. So tobacco companies had to pay actors to pose as doctors in their ads in order to project the impression their cigarettes were endorsed by this esteemed class of health professionals. But when GE foods were developed a few decades later, the AMA *itself* unreservedly championed them – and even encouraged individual members to endorse them as agents of its official policy. Further, its promotional endeavor has been unstintingly maintained since 1990; and although its successive pronouncements about GE foods were imbued with an authoritative aura, several have been significantly misleading.[18]

The National Academy of Sciences: Another Example of Protracted Irresponsibility

As prior chapters have revealed, the case of the AMA is far from unique, and many other respected scientific institutions have likewise sullied themselves in order to promote the bioengineering venture. One whose misbehavior has been as deplorable as it has been effectual is the US National Academy of Sciences (NAS). And Chapters 1, 2, 4, 9 and 10 have solidly documented the following delinquencies.

- In 1977, this esteemed organization not only abetted the dissemination of disinformation about the risks of genetic engineering, it increased the degree of distortion.
- Moreover, it failed to conduct a proper examination of the risks of releasing GMOs because the molecular biologists who influenced its agenda were concerned they would lose control of the issue.
- And when Phil Regal finally convinced the Environmental Protection Agency to sponsor a workshop at which a meaningful examination could be conducted, the Academy tried to commandeer and cripple it.
- Then, in 1987, the Academy released a report that minimized the environmental risks of GMOs by mishandling scientific issues – a report one of the authors privately admitted was essentially political rather than scientific.
- Although the mid-section of its next report on the topic (in 1989) did contain science-based acknowledgements of risk, the NAS staff

affixed opening and closing chapters with unwarranted claims about safety that enabled the document to be passed off as an affirmation that there was no cause for concern.

- In 1997 and 2004, the Academy published influential books that downplayed the risks of genetic engineering – but contained significant inaccuracies.
- In 2004, the organization issued a report on GE foods that, although it has had a major impact, is substantially illogical and seriously misleading. Not only does it disregard or misrepresent several important facts, its arguments are in key respects incoherent and even self-contradictory.

How Scientists Have Deliberately Thwarted Labeling

The deceptions of the scientific establishment have even played an essential part in blocking attempts to achieve the labeling of GE foods in America that would have adversely affected their marketing. This became especially clear in 2012 during the intense public debate in connection with a ballot initiative in California that would have required the labeling of GE foods within its borders. That controversy attracted substantial media attention, and eminent scientists eagerly fed reporters statements that supported the safety of GE foods. But not only were many of them couched in misleading language, far too many were flagrant falsehoods.

Some of the most influential of these were dispensed by a scientist who wielded considerable authority: Bob Goldberg, a professor in the Department of Molecular, Cell, and Developmental Biology at the University of California Los Angeles (UCLA) who is also a member of the National Academy of Sciences. For instance, in defending the safety of GE foods, he asserted to a *Los Angeles Times* reporter, "When you put a gene into a plant . . . it behaves exactly like any other gene."[19] But this bold statement, so prominently featured in an article in California's largest newspaper, is false. As discussed in previous chapters, most inserted genes won't even function unless they're artificially boosted by alien promoters, and the powerful promoters from viruses that are usually employed radically alter the genes' behavior and cause them to act very differently than the other genes in the target plant.[20] Moreover, we've seen there are other respects in which the inserted genes don't always behave "exactly" like the native genes either – and that the various ways in which their behavior differs entails unusual risks to the health of the consumer.[21]

But Goldberg would not desist, and he freely delivered other quotable but grossly erroneous pronouncements as well. Thus, an article in the *San*

Francisco Chronicle contained his declaration: "Bioengineered crops are the safest crops in the world. . . . We've been testing them for 40 years. They're like the Model T Ford."[22] But this claim about the 40-year span of testing is astounding in light of the fact that in 1972, 40 years prior to the date of his statement, the first genetically engineered bacteria had not even been created.[23] Moreover, not only were no GMO's in existence when he alleged that testing on GE crops had begun, bioengineers weren't able to produce even one functional GE plant until 1982; and the studies on the Flavr Savr™ tomato in the early 1990's (discussed in Chapter 10) probably marked the first time a GE crop had undergone any meaningful safety testing. So his claim about the duration of testing exceeded reality by around 20 years.

Nevertheless, this false claim commanded a lot of attention, especially because the section in which it appeared was introduced by the bold-type heading: **40 years of tests**. Further, his claim that GE crops are the safest crops in the world entails that those produced through sexual reproduction are somehow riskier, a highly dubious proposition for which there's no supporting evidence – and which is contrary even to the 2004 report by the NAS.[24]

But Goldberg's store of bogus statements was not exhausted. When asked about the studies that have reported negative effects of GE foods, he dismissed them as never having been peer reviewed.[25] However, as we saw in Chapters 6 and 10, several have undergone peer review. So Goldberg either had kept himself insulated from any evidence that could shake his certitude or he was lying – or perhaps had himself been deceived by lies that were circulated by other GE proponents. In any case, whether he was carelessly speaking from ignorance or shamefully attempting to deceive, his falsehoods must surely have misled a large number of people, most of whom were probably unaware that he had co-founded an agricultural biotechnology corporation and stood to profit handsomely from public acceptance of GE foods.[26]

Yet, although the misrepresentations by Goldberg and other individual scientists no doubt dissuaded many voters from casting their ballots in favor of the labeling initiative, a far greater dissuasive effect was generated by the unexpected intervention of a major scientific organization. For many weeks, commercial entities like Monsanto, Du Pont, and the other manufacturers of GE foods – joined by major corporations like PepsiCo and General Mills, whose products contain ingredients derived from them – had been pumping tens of millions of dollars into a mammoth advertising campaign to defeat the labeling initiative. This massive corporate opposition came as no surprise.

Nor was it surprising that the ads these corporations funded contained several distortions and were highly deceptive. But it *was* surprising when, on October 20th, the American Association for the Advancement of Science (AAAS), one of the world's most esteemed scientific bodies and the publisher of the prestigious journal *Science*, decided to enter the fray and officially lend its weight to the anti-labeling campaign. On that date, a few weeks before voters went to the polls, the association's Board of Directors issued a statement that purported to demonstrate why mandatory labeling of GE foods is not only unnecessary but unwise. And it ended with the assertion that such a measure "can only serve to mislead and falsely alarm consumers."[27]

But what was truly misleading was the statement itself, which, like the advertisements of the big commercial corporations, contained several false allegations.

Its main contentions were:

1. That "every respected organization" that has examined the evidence has determined GE foods to be "no riskier" than conventional ones.
2. That these determinations are backed by solid scientific evidence.
3. That in order to receive regulatory approval in the United States, each new GE crop "must be subjected to rigorous analysis and testing."

And it's amazing these assertions were issued under the auspices of the AAAS, because each is clearly untrue. Chapter 9 has demonstrated the invalidity of the first, Chapter 10 has done the same for the second (revealing how the studies that were cited to support it fall far short of doing so), and Chapter 5 has established the flagrant falseness of the third.

Regrettably, while many savvy Californians were wary of the advertisements thrown at them by Monsanto and its cohorts, they naturally presumed that the widely-publicized assertions of such a renowned scientific organization were accurate. And it's almost certain that these apparently authoritative yet erroneous assertions, in combination with those issued by Bob Goldberg and other scientist-proponents of GE foods, provided the margin of victory to the forces that opposed labeling – especially since the margin was so slim.

The Disinformation Is Even More Dominant Today

Thus, the scientists promoting the GE food venture have been much better at skillfully manipulating perceptions than at safely manipulating genomes, and the perceptions they've inculcated are to a large degree false. Indeed, if the clarity that scientists are supposed to facilitate had instead been fostered, the venture could not have survived.

But not only has it survived, through the ongoing dissemination of seemingly science-based disinformation, it's been steadily gaining the support of respected journals and journalists. For example, an article posted in April 2014 on *The New Yorker's* online blog observed that "there's been a shift toward G.M.O.s among editorial boards and science writers" to such an extent that Michael Pollan, a famous author, Berkeley professor, and prominent critic of GE foods, has confided, "I feel pretty lonely among my science-writing colleagues in being critical of this technology, at this point." [28]

Moreover, even *his* criticism is now substantially limited. He told the reporter that although he's concerned that the technology's current applications are harming the environment and entrenching undesirable farming practices, he's not significantly concerned about the effects of its products on human health. And the 700 students in the course he was teaching on how to create a more healthful and sustainable food system didn't seem concerned about potential health risks of GE foods either. According to the journalist, they were primarily bothered by what they perceived as growing corporate control of the food supply.

The fact that neither the professor nor the students in a course on sustainable agriculture at UC Berkeley in the Spring of 2014 regarded GE foods as riskier to eat than naturally produced ones, and instead believed they can be safely employed in global agriculture as long as the specific crops and methods are aligned with proper principles, reveals how thoroughly the deceptions of their scientist-supporters have succeeded.

Flawed Thinking Has Accompanied the False Talking

Further, not only are these scientist-promoters unable to consistently talk straight about GE foods, they can't always think straight about them either. This debility is well-exemplified within the reports released by the National Academy of Sciences. As we saw in Chapter 9, none of its first four reports on GE crops properly employed the concept of risk, and each blurred the distinction between risks and hazards. Further, Chapter 4 revealed that, in attempting to assert the equivalence of genetic engineering and traditional breeding, the 1989 report served up an absurd statement; and Chapter 9 showed that when the 2004 report tried to demonstrate the parity of their risks, it became logically dysfunctional.

An incisive assessment of the extent to which the arguments of the proponents are both deceptive *and* rationally defective is provided by Guy Cook in his book, *Genetically Modified Language*, which he began researching when he held the Chair of Applied Linguistics at the University of Reading in the UK. In the introduction, he says the book demonstrates that many of the arguments for genetic engineering "... exemplify disturbing trends in the public use of contemporary English by powerful individuals and organizations, in which language, while purporting to be rational, honest, informative, democratic and clear, is in fact none of these things, but, on the contrary, often illogical, obscure, patronizing and one-sided, populated with false analogies, misleading metaphors, and impenetrable ambiguities."[29]

The Most Immediate Damage Has Been to Science

Besides enabling the imposition of great potential harm on consumers and the environment, the delinquencies of the scientific establishment have inflicted concrete harm on science – and the harm has been major. What's more, this damage has been inflicted in the name of protecting science. Starting in Chapter 1, we've seen how the scientist-promoters of the genetic engineering venture have routinely branded any opposition to it as an attack on science itself – and how in mounting their defense, they've increasingly employed deceptive practices that are foreign to science and injurious to its spirit. As Patrick Brown, a professor of plant sciences at the University of California, Davis, has observed: "To date many in the scientific community have been unwilling to rationally consider the concerns surrounding the current GMOs and have wrongly considered that a defense of GMOs is a prerequisite to protect the science of plant biotechnology. Nothing could be further from the truth. . . ."[30]

But instead of facing the truth about the wrong-headedness of trying to protect science by protecting the image of GE crops, their scientist-proponents have significantly *effaced* the truth about how these crops are created, how they differ from traditionally bred crops, and how extensive evidence has cast doubt on their safety. And in striving to manage (and censor) the flow of information to the public, they've suppressed the free flow of ideas within the scientific community, which is the life-blood of scientific progress.

Stifling Free Discussion, Inculcating Fear

Chapter 10 has documented how experts who've dared to publish research showing problems with GE food have suffered vicious and unjust attacks that have disgraced them and discredited their research. Further, the repressive climate within the scientific community has been so intense that its members have not only been inhibited from performing research that could cast doubt on GE foods, but even from accurately describing inconvenient facts. In commenting on this deplorable phenomenon, Philip Regal has noted: "Traditionally, scientists regarded intellectual honesty as part of collegiality, and there was accountability if one was caught telling lies. Accordingly, liars were blackballed. But since the rise of genetic engineering, the situation in molecular biology has to a significant degree become inverted, and, when it comes to that technology, one gets blackballed for telling the truth."

This suppression began well before the advent of any GE crops, when the molecular biology establishment strove to deter regulation during the 1970's. Diana Dutton reports how, even then, ". . . dissident scientists had to endure increasingly overt professional ostracism" and were subjected to harsh criticism. And she notes that, in a widely quoted interview, James Watson referred to those who expressed concerns as "'kooks, shits, and incompetents.'" Accordingly, she states: "It was especially difficult for younger faculty members without tenure to withstand hostility and intimidation from senior colleagues, and many withdrew from the controversy, fearing for their careers."[31]

Further, because the scientific establishment intensified its defense of bioengineering after GE foods arrived, has stubbornly insisted that they're safe, and has harshly denigrated opposing viewpoints, scientists who do have reservations are reluctant to express them. And there's abundant evidence that those who are bold enough to voice doubts are routinely censured by superiors, denied tenure at universities, refused choice employment in the private sector, or otherwise degraded in the scientific community.

For instance, when Ann Clark, a scientist at the University of Guelph in Canada, publicly criticized the deficient safety testing for GE foods, adverse consequences came swiftly. "Within two hours of the press conference releasing the report, my dean had called me unethical," Clark said. "It became quite ugly, because the national media picked it up, and people whose views aren't parallel to mine have used [the dean's remarks] extensively."[32] Because Clark had tenure, she continued to speak out without fear of losing her job. But she says her treatment has deterred others: "There aren't many academics who will say something if they know their administrators – the people who sit in judgment on their performance – are going to publicly lambaste them."[33]

Similarly, agricultural economist John Ikerd's refusal to get on the biotech bandwagon brought him problems at the University of Missouri. "You become labeled as not a team player, as not one of the trusted members of the faculty," he says. "You are not on committees you used to be on, you're not involved in the leadership of the department, and you don't get write-ups in the university publications. You have to decide before you speak out that you don't care about these repercussions. It's like being a whistleblower."[34]

And at least one American university has tried to purge its ranks of scientists who won't espouse the official position on GE foods. For instance, in 2001 Oregon State University sent a letter to its bioscience faculty informing them that if they didn't support genetic engineering, they did not belong there.[35] Interestingly, it was signed by a former president of the institution, perhaps because the presiding president, who was an ecologist, couldn't bring himself to putting his signature on such a document.

Besides its chilling effect on the faculty who remained, the letter drove at least one professor away: Elaine Ingham, the soil scientist who had performed such a valuable service in helping to discover the serious risk posed by the bioengineered *Klebsiella planticola* bacteria that was discussed in Chapter 7 – and who dissuaded the EPA from approving their release. Ingham had already suffered a scolding from her department head after speaking about the risks of that bioengineered bacterial strain at an international conference, and when she read the ominous letter, she realized that because she was not prepared to stay silent about what she perceived to be the potential problems of genetic engineering, her life at the university would become even more unpleasant. So she decided to resign and start her own research institute where she could freely function as a scientist.

Due to the various pressures, numerous experts profess the safety of GE foods in public yet privately confess concerns, as Phil Regal noted in the declaration he submitted to the court in the Alliance for Bio-Integrity lawsuit.[36] His observation was based on extensive participation on expert panels and in scientific conferences; and it's confirmed by a study that found significant repression of opinion among Cornell University agricultural and nutrition-science faculty and extension staff. Although 63% had reservations about the safety of GE crops, they said they felt uncomfortable about expressing their views, in contrast to the minority who were strong backers of biotech.[37] Not only is such suppression a blight on the scientific spirit, it sustains the illusion that GE foods are far more supported within the society of scientists than is actually the case.

Thus, it's a sad fact that during the bioengineering era, the ethical standards of the scientific establishment have been so substantially perverted that (a) speaking deceptively about GE foods has become the expected norm, (b) speaking truthfully on that topic is not merely discouraged but punished, and (c) those who attempt to uphold the traditional ethics of science by openly communicating the facts are branded as unethical by scientists who wield authority.

A Widespread Failure to Critically Assess the Promotional Claims

Further, not only have a large number of life scientists been inhibited from forthrightly communicating facts about GE foods due to fear of reprisals, a large number have failed to seriously assess the facts at all and have instead unthinkingly repeated the standard claims propagated by the establishment. Phil Regal has commented on the prevalence of this practice and noted how strikingly it manifested when scientists routinely repeated the "party line" in regard to the EMS epidemic induced by a bioengineered food supplement (discussed in Chapter 3) – and unequivocally asserted that the bioengineering had in no way caused the catastrophe. As he observed:

> University scientists who had not studied the documentation itself began parroting the arguments that the public relations persons for the industry had developed. As though any proof was necessary, it became crystal clear that ideas within the community of molecular biologists were largely being generated and spread from the top down. This was clear because opinions that were being stated with authority were not being based on studies of the actual facts by the individual scientists who were speaking as 'scientific' authorities, but only on what was being said by those who were in positions of power. It was clear that gossip had become as good as scientific evidence in the profession, even on matters where human lives could be at stake.

In reflecting on the persistence of this behavior, Regal has stated: "Of course, because it's in the self-interest of many molecular biologists to trust their leaders, I cannot assume that they've all been lying. But arguing that you know something to be true when you have not studied it, and instead parroting the party line, is certainly a form of intellectual dishonesty even if it is not outright lying." Nonetheless, although he's being fair in refraining from condemning all the scientists who've routinely repeated dubious claims as liars, it's obvious that far too many have been so careless in the claims they make, and so intent to conceal unfavorable facts, that their statements can be justly judged fraudulent.

Moreover, many scientists who have not actively disseminated the promotional claims themselves have yet been badly misled by them. As the molecular biologist David Schubert (of the Salk Institute) has noted:

> I have spoken with many molecular and even plant biologists who are not directly involved in producing genetically engineered foods, and it's clear that most have been misled about the basic facts. For instance, they assume these foods have undergone rigorous safety testing (as in the case of drugs) and are surprised to learn that they haven't. Nor are they aware that the insertions of the recombinant cassettes are not precise but random. And they don't understand the potential effects on the plants' secondary metabolism that could generate harmful substances. Instead, due to the disinformation dispensed by the life scientists who practice and promote agricultural bioengineering, they have the impression that there are no unusual risks and that everything is under control. In most cases, after I've explained the key facts, they change their position.[38]

Many Scientists Have Been Deliberately Lying

But Schubert has also pointed out that too many of the scientists who've been making the statements that have misled their colleagues have not been innocently mistaken themselves. In his words, "Some plant biologists are making statements about GE foods that they almost certainly know are not true."[39] And this book has presented extensive evidence that supports this observation. Thus, while some biotech proponents dismiss scientists who critique GE foods as "outliers," it's clear that a substantial number of those who champion them are *outright* liars.[40]

Scientific Standards Have Been Eroded

As the integrity of scientists has steadily eroded in the endeavor to promote GE foods, so have several of the standards, in addition to truthfulness, that scientists are supposed to uphold. For instance, as Chapter 6 has described, regulators (and the scientific community) have allowed the manufacturers of GE foods to re-write basic rules on how experiments should be conducted, permitting them to dilute statistically significant differences between GE crops and their parental variety grown under the same conditions by instead comparing the former with a wide range of varieties grown under substantially different conditions.

Further, as that chapter and Chapter 10 have documented, in several instances regulators like the European Food Safety Authority (EFSA) have ignored statistically significant differences between animals fed GE foods and the control animals that ate the non-altered counterparts by upholding the manufacturers' claims that the differences are not "biologically meaningful." Moreover, the EFSA repeatedly dismissed such significant differences as not biologically relevant without providing clear criteria for what counts as relevant. And when, in response to repeated criticism by independent scientists, it finally attempted to furnish a definition, the result was markedly deficient.[41] As the scientists John Fagan and Michael Antoniou have noted, the EFSA's attempt "fails to give a rigorous scientific or legal definition of what makes a statistically significant finding 'biologically relevant' or not." They point out that it instead "allows industry to come to its own conclusion on whether changes found in an experiment are 'important', 'meaningful', or 'may have consequences for human health.' " And they note that because "[t]hese are vague concepts for which no measurable or objectively verifiable endpoints are defined . . . they are a matter of opinion, not science."[42]

And, as Chapter 10 has shown, the scientist-proponents of GE foods have been inconsistent in the way they've relaxed scientific protocols, doing so in a biased manner that's created a double standard under which any study reporting problems with these products is subjected to far stricter requirements than those purporting not to find any. Through such a duplicitous set up, rigorous studies published in peer-reviewed journals are pilloried or even forced into retraction if they've detected ill effects while shoddy ones that couldn't qualify for publication in such journals have been treated as authoritative as long as they claim reassuring results.

This gross degradation of scientific standards is an important part of the GE food fraud, because proponents have created the illusion that science has been assiduously upheld when, in reality, it has been systematically subverted in order to make the products of bioengineering appear safe.

The Very Nature of Science Has Been Misunderstood and Misrepresented

Not only have many scientist-proponents of GE foods misrepresented scientific facts and subverted scientific standards, they've misrepresented the very nature of science itself. And it seems that a significant number have done so because they themselves have grown confused about what it is and isn't. Thus, as several observers have pointed out, many proponents tend to erroneously equate the technology of genetic engineering with science and treat any criticism of the technology as an attack on science.

But in reality, the technology of agricultural bioengineering is not in itself science but merely an attempt to apply science to achieve practical results, and the safety of its products is not automatically assured. Rather, their safety must be assessed in a scientific manner. One way is through the theoretical approach, which involves applying our best biological knowledge and making an earnest evaluation of risks. The other is by performing scientifically rigorous tests on each product.

As we've seen, when the theoretical approach has been employed by scientists who were sincerely endeavoring to consider all the relevant evidence and render an objective assessment (as was the case with the experts on the FDA's biotechnology task force and those on the Royal Society of Canada's expert panel), genetic engineering has been recognized as deeply different than conventional breeding, and the foods it produces have been deemed to entail greater risk than those produced via traditional means. On the other hand, when scientists who apparently wanted to uphold the image of GE foods have performed risk assessments (as in the case of those that produced the 2004 NAS report discussed in Chapter 9), they tend to ignore a lot of evidence and employ a lot of loose logic, even to the point of altering the concept of risk. And only in this way have they been able to conclude that GE foods are not inherently riskier than conventionally produced ones.

Additionally, as Chapter 12 demonstrated, the belief in the safety of GE foods is ultimately grounded on a set of assumptions that have been thoroughly discredited, which renders the venture that produces them bereft of any sound theoretical support.

Moreover, as we've seen in Chapters 6 and 10, the venture does not have a sound empirical foundation either, and when actual tests have been performed, they've yielded a crop of disturbing data. Further, if the data are examined in an honest and scientifically rigorous manner, there's ample evidence to support the view that no GE food has yet been proven safe, that the safety of several is in serious doubt, and that none of them should be on the market.

Nonetheless, despite the lack of solid theoretical or test-based support, the scientist-proponents of GE foods have persistently proclaimed that these products are safe. And in doing so, they've gone beyond falsely conflating the technology of genetic engineering with science and have also mistaken their own unsubstantiated opinions for science. Therefore, their claims are ultimately based on their own purported authority; and Chapter 9 has furnished a prime example, demonstrating how the arguments of the panel that produced the 2004 NAS report essentially boiled down to the assertion, "GE foods are safe because we say they are."

Such parading of unfounded opinion in the guise of solid science has been a constant feature of the GE food venture – and one of its deepest and most enduring deceptions. As Phil Regal has noted, although the scientist-supporters of GE foods have been unable to provide adequate scientific backup for their claims, they nevertheless present them as science-based – thus "trying to wave the flag of science without a staff to support it."

Putting Things in Perspective: The Biggest Fraud in the History of Science

To properly gauge the size and severity of the aggregate fraud that has enabled the advance and survival of the GE food venture, we need to view it within the context of history and assess it against major science-related frauds that have already happened. As we'll see, although some of the earlier frauds have been both sizable and harmful, the one that's been perpetrated on behalf of GE foods surpasses them all.

The Notorious Piltdown Hoax is Dwarfed by the GE Food Fraud

In December 1912, the scientific community was electrified by the unveiling of a seemingly momentous discovery. At the Piltdown quarry in Sussex, England, fossil remains had been unearthed over several years that, when combined, appeared to form a skull that exhibited both human and pre-human features. The brow was distinctly human while the jawbone was far more primitive and ape-like. In the eyes of many, the "missing link" in the evolutionary chain from apes to *Homo sapiens* had finally been found and the Darwinian theory of human descent confirmed. However, although numerous experts regarded the specimens as the authentic remains of an early hominid, many had doubts. For one thing, when the jaw was discovered, it was not attached to the skull, so it could have come from a different animal.

In 1953 the issue was decisively settled. Sophisticated analysis revealed that the bones were not sufficiently ancient but had been doctored to look so. And the jawbone appeared ape-like because it in fact had belonged to an ape. What's still unresolved is the identity of the person or persons who doctored the fossils and planted them in the quarry, although it's evident that at least one individual with scientific expertise must have been involved.[43]

The Piltdown forgery is one of the greatest frauds inflicted on science. It purported to confirm one of the most revolutionary and vehemently contested scientific theories of the modern era; it duped a large number of people, including many experts; and the confusion continued for decades. Yet, compared to that fraud, the deception that underlies the genetic engineering of our food is far bigger – and far more insidious. Bigger in terms of the number of scientists perpetrating it, bigger in terms of the number of people misled, and bigger in terms of the damage to science and the potential harm to society and the natural environment.

The UK's Mad Cow Disease Deception Doesn't Come Close Either

Further, although there have been frauds that, unlike the one at Piltdown, did involve risks to public health, they're not on a par with the GE food-related fraud either. One of the biggest of these deceptions involved the UK government's attempts to dispel fears about "mad cow" disease (BSE). As documented in a report by researchers at the University of Sussex, officials kept insisting in public that British beef was completely safe when they knew this claim was unjustified. In the words of the report: "Policy-makers were repeatedly told, both by the scientific experts on whom they claimed to rely, and by the wider scientific community, that it was impossible to be certain that consuming meat, milk and dairy products from animals with BSE posed no risk."[44] Nonetheless, government officials continued to claim solid grounds for certainty, as when the Agricultural Minister declared to the House of Commons there was ". . . clear scientific evidence that British beef is perfectly safe."[45]

While there are parallels between the deceptions on behalf of BSE and those that enabled GE foods, there are major differences; and the fraud in the latter case is of far greater magnitude. Most of the deceit about BSE was perpetrated by the UK government, and only a few other governments also employed deception to cloud its risks. In contrast, many governmental bodies around the globe have engaged in significant misrepresentation regarding GE foods, and they've done so over a longer period of time. Moreover, while it appears that the majority of the scientific community acknowledged the potential risks of BSE and emphasized the lack of scientific certainty, in the case of GE foods, the majority of scientists have failed to be forthright about the risks and uncertainties and a large number have instead dealt in systematic duplicity. Thus, while in the case of BSE, the fraudulent behavior of government officials was not abetted by the bulk of the scientific establishment, in the case of GE foods, it has been.

The Worst Scientific Fraud of the Stalinist Era Was Far Less Egregious

Even an enormous fraud that was abetted by the Stalinist Soviet government and seriously damaged science and agriculture in the USSR for decades pales in comparison to the GE food fraud. This ugly episode was due to the endeavors of the biologist and agronomist Trofim Lysenko, who promoted theories for boosting agricultural production that were attractive to the Communist party bosses but misaligned with reality. And because he was backed by a totalitarian state, he was able to impose his ideas in a thorough and repressive manner for at least 30 years. However, when scientists finally had sufficient freedom to speak out against him, it sparked an official investigation that produced a "devastating" report accusing him of misrepresenting facts and deliberately falsifying data.[46] The findings indicated that the methods he propounded were unsound and were causing significant losses.[47]

Yet, despite the duration of Lysenko's influence and the extent of its harmful effects, the aggregate fraud that has accompanied and enabled agricultural bioengineering is in most respects of greater magnitude than the fraud that he wrought. He was the main scientist driving it, and although the majority of the other scientists were cowed into silence by fear of the Soviet regime, only a minority became ardent promoters of his views. For instance, merely four out of the thirty-five geneticists in the Academy of Sciences' Institute of Genetics became Lysenkoites when Lysenko became the director in 1940.[48] Thus, his agenda gained the support of the government, not because it was pushed by a large number of scientists or alleged to represent a scientific consensus, but because it was appealing on both economic and ideological grounds – and because the government had the power to ignore and manipulate scientific opinion.[49] In contrast, a large segment of the scientific community has been avidly engaged in twisting the truth on behalf of genetic engineering, and the biotech agenda has gained government support through the intensive efforts of the scientific establishment. Further, whereas Lysenko's fraud was facilitated by only a single government, which was rigidly totalitarian, the GE food fraud has been actively abetted by several governments in societies that are supposed to be open and democratic. And although the Soviet government supported Lysenko, it apparently did not participate in the misrepresentation of research as has the US Food and Drug Administration, which has disseminated disinformation about GE foods just as vigorously as have their scientist and industry promoters.

Moreover, although the repression suffered by Soviet scientists during the Lysenko era was in several cases more severe than that inflicted on scientists whose speech or research has been deemed threatening to the GE food venture, far more scientists have been negatively impacted in the latter case; and the repression has occurred on a global rather than regional scale. Worse, their mistreatment is in important respects more egregious because it has occurred in open societies that are supposed to protect and nurture freedom of thought and speech.

In addition, there's a major discrepancy in the degree of risk imposed on human and environmental health. According to two *New York Times* journalists who analyzed the Lysenko episode in their book on scientific frauds, *Betrayers of the Truth*, although his projects yielded no benefits, in general they didn't cost very much nor did they produce significant harm.[50] On the other hand, the GE food venture has required massive expenditure, has produced extensive environmental harm (as documented in Chapter 7), has imposed excessive risks on human health, and (as far as we know, given the lack of proper monitoring) may have actually been harming millions of consumers.

Finally, not only has the GE food fraud been perpetrated by more scientists in more countries, been abetted by more governments, imposed more extensive repression, and entailed far greater risk, its influence has also lasted longer. Lysenko's influence was dominant for 30 years, and was significant for no more than 35,[51] while the scientist-proponents of genetic engineering have already succeeded in misleading the scientific community, governments, and the media for more than 37.[52]

The GE Food Deception Encompasses All Previous Forms of Fraud – and Has Even Introduced a New One

Further, although *Betrayers of the Truth* presents a comprehensive study of scientific frauds, none of the other delinquencies it describes comes close to the magnitude of the cumulative fraud that has empowered the GE food venture either. For one thing, most of them were pulled off by a single scientist. And even the worst didn't impose risks on human or environmental health anywhere near the enormity of those entailed by the GE-related fraud.

What's more, while most of the transgressions discussed in *Betrayers of the Truth* exemplify only one of the types of fraud that it describes, the GE food fraud encompasses *all* of them – including the one the authors classify as the most serious (and rarest) form: inventing an experiment "out of thin air."[53] As Chapter 1 pointed out, during the early era of genetic engineering its proponents allayed public and Congressional concerns about its risks by claiming that research had produced important "new evidence" demonstrating its safety, despite the fact that no such research had been performed and no such evidence existed. And that chapter, along with Chapter 4, also revealed that during the same era, Stanley Cohen helped defeat Congressional attempts to regulate genetic engineering by so grossly misrepresenting an experiment he had conducted that people were duped into thinking it was performed under natural conditions and achieved results it had fallen far short of attaining. Thus, he in effect invented a fantasy experiment, because the version he instilled in people's minds was very different from the one he'd actually carried out. And this chapter has shown how Bob Goldberg boasted about a multi-year span of safety tests that was pure fiction.[54]

Moreover, the promoters of GE foods have effected a type of fraud that *Betrayers of the Truth* doesn't even describe. They have not only fabricated research studies that don't exist but have essentially expunged real ones that they don't like. Through their fervid attacks, they forced retraction of Séralini's and heaped so much derision on Pusztai's that the scientific community and the media largely ignore it, not only as if it had never been published in a major peer-reviewed journal, but as if it never even happened. Moreover, as we saw in Chapter 3, they've even managed to effectively disappear an entire epidemic that was associated with a GE food supplement, along with the published research related to it.

Entirely Unprecedented and Uniquely Unsavory

Consequently, when viewed in the context of history, the aggregate fraud that has fostered the growth of the bioengineering venture, has allowed it to avoid the regulation it should have received, has enabled the foods it produces to be commercialized, and has kept those foods on the market despite the accumulation of evidence that should have forcefully driven them from it is the biggest and most pernicious ever connected with science. Many more scientists and respected scientific institutions have in one way or another abetted the spread of misinformation about GE foods than have been involved in any other deception, and hundreds (if not thousands) around the world have been complicit. Also unprecedented are the number of people who've been fooled and the degree to which they've been deluded. Hundreds of millions on all continents have been given a distorted picture of key biological processes, false accounts of the research and testing, and misleading reports about risk. In addition, regulators have often been misled by the misreporting of data, and in several instances, the regulators themselves have colluded in the misrepresentation of facts.

None of the other frauds significantly jeopardized the integrity of science, nor did any rely on misrepresenting basic processes of biology.[55] Further, those that have posed some threat to health have usually involved a single drug or food additive. In contrast, the potential negative impacts of the GE food deception are wide-ranging and deeply endangering. On the dimension of public health alone, this fraud has already permitted several foods of questionable safety to be broadly marketed and extensively consumed throughout the US, Canada, and several other nations; and its ultimate goal is to substantially transform the genetic core of virtually every food-yielding organism.

Thus, the evidence has borne out an assertion initially made in this book's introduction: that the delinquencies of the scientific establishment in promoting GE foods have not only been unsavory but unprecedented. And it has revealed that the cumulative fraud perpetrated to enable and sustain their commercialization is by far the biggest, most brazen, and most dangerous in the history of science.

The Psychology That Drives the Deception

What has motivated so many scientists to betray the standards they were taught to honor and engage in persistent deception? Obviously, as several chapters have pointed out, many have been at least partly motivated by the prospect of financial gain, either because they own equity in a biotechnology company, receive substantial income from consulting for such companies, or receive generous grants to pursue research utilizing recombinant DNA technology. But many scientists who promote the GE food venture have no such conflict of interest, and even most of those who stand to profit from the venture are probably not entirely driven to support it by the desire for personal wealth. There are deeper factors at play, and they have also been a powerful force.

For 40 years life scientists have been imbued with the idea that genetic engineering is not only efficacious and safe but integral to applied biology and essential for the progress of society. Further, many – perhaps most – have developed the belief that it's crucial for feeding the burgeoning population in the Third World as well as instrumental for protecting the environment and enhancing the quality of food within the industrialized nations. They also believe it will enable profound advances in medicine and many fields of manufacturing. So certain are they that the benefits of this technology will be profound and the risks negligible that many feel not only justified to employ it, but deeply obligated to do so. As one microbiologist declared, he and his colleagues have a "moral and ethical responsibility" to pursue the promise of genetic engineering.[56] And several prominent British scientists have argued that developing GE crops in the Third World is a "moral imperative."[57]

Concomitant with this belief in the necessity of genetic engineering is the felt need to vigorously defend it, and a large number of life scientists have come to regard any critique against a particular application of rDNA technology as a threat to all its other applications as well. Accordingly, the bioscience community tends to behave as if public rejection of genetic engineering in agriculture will endanger its wide-spread acceptance in other areas and to react to criticism of GE foods as if it were an assault on biotechnology as a whole – and to a significant extent, an attack on the life sciences as currently structured and practiced. It also increasingly regards all concerns about the safety of GE foods as based in ignorance. This notion colors a World Bank report on GE crops that says those who are "technically competent" have no qualms about safety and that the public perceives risks only because its members lack such competence.[58]

Within this mind set, things tend to be viewed in the extreme, and an attitude has arisen that the bioscience community needs to close ranks and defend itself against the public's ignorance and irrational fears which, left unchecked, could impede if not thwart the full development and deployment of biotechnology. For instance, the eminent botanist Norman Borlaug warned that when it comes to new modes of agriculture such as bioengineering, "science and technology are under growing attack" by "misinformed environmentalists" who "seem to be doing everything they can to stop scientific progress in its tracks."[59]

Because so much progress is felt to be in jeopardy, the conviction has steadily grown that preserving the vast benefits of biotechnology from the ravages of irrationality is of such importance that those with genuine knowledge are justified in resorting to whatever techniques will favorably sway public perceptions. And as many scientists progressively indulged in the projection of alluring images and the obfuscation of facts, they eventually crossed the vague boundary between obfuscation and misrepresentation – and seem to have felt legitimized in doing so. Ironically, this *ends justifies the means* approach, in which misrepresentation in the service of scientific progress is viewed as essentially benign, is reminiscent of the "pious frauds" through which religious officials sometimes manipulated information so as to bolster faith among the flock.

The Importance of Assigning Culpability

Whatever the underlying motivations, the misleading pronouncements of the scientific community have already done great harm – and imposed much greater potential harm. Moreover, it's important to recognize that most of these pronouncements have been genuinely fraudulent and that the scientists who have made them are guilty of fraud, even when they have not technically told lies.

Our legal system recognizes that fraud can exist without overt falsehood and that its defining feature is deception. As one court stated: "Acts constituting fraud are as broad and as varied as the human mind can invent. Deception and deceit in any form universally connote fraud."[60] Because the essence of deception is to cause a false impression in the minds of others, one can be guilty of it not only by employing misleading words, but also by withholding words. Therefore, according to the law, failing to reveal pertinent facts is a form of fraud, as is the attempt to hinder others from gaining or understanding them.

So from the perspective of the legal system, a large number of scientists have clearly engaged in fraudulent behavior in order to promote genetically engineered foods. Whether or not they have intentionally lied, they have generated widespread confusion, and often delusion, about the facts; and they are therefore guilty of fraud.[61]

The misrepresentations that have surrounded GE foods are varied, ranging from blatant lies issued by FDA officials to nuanced distortions dispensed by university professors. But while the forms vary, they are all in some significant way deceptive – and have all been effective. And the individuals who have dispensed them should be held accountable.

Thus, the advent of genetic engineering has indeed induced a serious attack on science; but the attack has come from within, as hundreds of scientists have systematically subverted the standards they were trained to uphold in order to uphold that enterprise. Consequently, not only is agricultural bioengineering the sole faith-based technology, it's the only technology that's been chronically and crucially reliant on the persistent dissemination of disinformation – with the scientific establishment the disseminator-in-chief. And if that esteemed group of individuals and institutions had maintained its integrity and spoken honestly, the venture could never have gained traction.

Furthermore, the myriad distortions, deceptions, and downright lies issued by scientists and scientific institutions on behalf of genetic engineering since it emerged in the 1970's constitute the most colossal and pernicious scientific fraud ever perpetrated; and besides the serious damage they've done to the integrity of science, they've imposed unacceptable risks on human and environmental health.

Accordingly, in light of the enormity of this fraud, the complicity of so many government agencies and officials, and the magnitude of the risks it has inflicted, major remedial action is required within both the public and private sectors. Fortunately, sensible solutions are available.

CHAPTER FOURTEEN

New Directions and Expanded Horizons

Abandoning Genetic Engineering and Advancing to Safe, Sustainable, and Sensible Modes of Farming

Facing Up to the Critical Facts

The influential bioethicist Gary Comstock has argued that it is ethically justified to develop genetically engineered foods "assuming we proceed responsibly and with appropriate caution."[1] But the preceding chapters have demonstrated that the GE food venture has abjectly failed to meet this criterion and has instead routinely proceeded in an irresponsible and reckless manner. As we've seen, it has advanced not by honoring the principles and protocols of science but by evading them, not by following the food safety laws but by violating them, and not by openly and fairly communicating the facts but by systematically clouding and frequently distorting them. Further, we've seen that because of the nature of the genetic engineering process – and the economic realities inherent in applying it to commercial agriculture – these abuses are not avoidable aberrations that can ultimately be eliminated from the enterprise but rather are intrinsic features necessary for its survival. And it's become quite clear that if the corporations that produce genetically engineered foods and the government agencies that are supposed to regulate them were actually to follow sound science, to uphold the law, and to consistently communicate the truth, the entire venture would quickly collapse.

Moreover, Chapter 11 has demonstrated that from the perspective of computer science, the technique referred to as bioengineering is actually biohacking – and that it's inherently unsafe because, due to the vast complexity, extreme interconnectivity, and substantial inscrutability of bio-information systems, scientists are incapable of altering them according to the standards by which software engineers revise life-critical computer programs and can only exercise a mere modicum of the caution required in that far more manageable enterprise.

The GE Food Venture's Defining (and Debilitating) Attribute: Ethical Unsustainability

Consequently, although the proponents of the GE food venture consistently claim that it's essential for establishing sustainable agriculture, it has in fact introduced a new dimension of unsustainability. Regardless of the farming practices associated with the crops it produces, the enterprise is *ethically* unsustainable because it cannot continue without consistent evasion of sound scientific practices, violation of the law, and misrepresentation of the facts. Just consider what would happen if all the information in this book became widely known and most people (including most US, Canadian, and European legislators and other government officials) learned how radically genetic engineering differs from traditional breeding, how methodically the risks have been misrepresented, how frequently the tests have returned unsettling results, how routinely the protocols of science have been violated, how badly the food safety laws of the United States have been broken, and how thoroughly they themselves have been deceived.

It's obvious that the venture could not survive. And by now it should be obvious that it doesn't deserve to.

The Biotechnicians' Lack of Necessary Knowledge Is Far More Evident Today than When Earlier Warnings Were Issued

Anyone who may still be unconvinced about the unacceptability of the GE food venture should again consider how deeply deficient is the knowledge on which it rests. For instance, in 2000 Patrick Brown, a professor of plant science at the University of California, Davis, wrote a cautionary article about agricultural bioengineering asserting: "As scientists it is our responsibility to recognize that we do not yet have sufficient knowledge of the process to use it safely."[2] As he explained: "We must recognize that our knowledge of the processes that regulate gene incorporation and expression are in their infancy and that our capacity to manipulate the plant genome is crude. . . ." He then pointed out that most of what we *do* know indicates the "profound manner" in which this artificial process differs from traditional techniques – and that it's "well known to cause unexpected metabolic perturbations."

As well-founded as Professor Brown's warnings were in 2000, they're even more compelling today because (as Chapter 11 has shown) startling discoveries have revealed that bioinformation systems are far more complex, and far more poorly comprehended, than was recognized when he wrote. Accordingly, we now know that biotechnicians have even *less* capacity to safely manage the alterations they impose upon those intricate systems than he assumed they possess. This also entails that the similar precautions issued by the Royal Society of Canada a year after Dr. Brown issued his are likewise even more pertinent today than when first written.

Underscoring how vast is the ongoing ignorance about the intricacies of biological systems, and how utterly incapable biotechnicians are of reconfiguring them in a predictably safe manner, scientists still have not learned how to alter one of the most rudimentary of these systems without unexpected outcomes. Thus, after several years of attempts to computationally model a virus that's "one of the simplest and most well-studied biological systems," a biologist at the Massachusetts Institute of Technology was still unable to predict how mutations would affect its development, with his simulations regularly failing to match actual results.[3] In commenting on this failure, an article in *Harvard Magazine* observed, "Evolution may have been responsible for the diversity of biological functions, but to a human scientist, those functions could appear byzantine and impossible to comprehend, let alone engineer."[4] Consequently, in order to better "understand and manipulate" the virus, the biologist and his team re-built it in a way that was much simpler. However, although this restructuring made the new version of the virus easier for them to alter in a predictable manner, it substantially impaired the entity's integrity, and "its fitness was considerably reduced."

This incident undercuts the idea that agricultural bioengineering can be performed safely. Viruses are not even living cells, and the virus involved was one of the simplest among even that class of simple systems. Further, it had been studied for 60 years.[5] Yet, its genome was still not comprehended well enough to manipulate in a predictable fashion. Therefore, it would be outlandish to suppose that biotechnicians could artificially alter the far more complex, far less well-studied genomes of higher organisms with greater foresight and enhanced reliability – and far more realistic to conclude that they'll probably never be able to.

Another Compelling Reality Check

In case you still shy away from the idea of completely curtailing GE foods, ask yourself whether you would be willing to daily fly on an airplane that's dependent on a complex computer-run guidance system that had been radically revised without undergoing the safety testing necessary to ensure that even minor revisions to such software programs have not disrupted them in dangerous ways. Further, even if you'd personally be willing to take this gamble, would you want to subject your children or grandchildren to such a repetitive risk?

Of course, this kind of situation would not arise in the case of a commercial airliner because the federal regulators wouldn't allow the deployment of such an altered but improperly tested guidance system, and it's only because the regulators that should have been applying parallel safeguards to GE foods have been grossly delinquent that the latter products are on the market at all. Moreover, if you wouldn't want your loved ones subjected to the risk of regularly flying on an airplane with a radically revised but deficiently tested guidance system, how could you countenance subjecting hundreds of millions of people to the risks of consuming food-yielding organisms whose complex information systems have also been radically revised but inadequately tested?

Exposure to Genuine Facts Can Prompt Dramatic Turnarounds

As compellingly as the software analogy can function in solidifying opposition to GE foods, there's ample evidence to do so without bringing that analogy into play at all. A striking example of how powerfully even a partial set of the pertinent facts can reverse opinions about these products is provided by Dr. Thierry Vrain, who was for many years the Head of Biotechnology at Agriculture Canada's Summerland Research Station. Following are some illuminating remarks he made in 2014.

> I retired 10 years ago after a long career as a research scientist for Agriculture Canada. . . . I was the designated scientist of my institute to address public groups and reassure them that genetically engineered crops and foods were safe. . . .
>
> I have in the last 10 years changed my position. I started paying attention to the flow of published studies coming from Europe, some from prestigious labs and published in prestigious scientific journals, that questioned the impact and safety of engineered food.
>
> I refute the claims of the biotechnology companies that their engineered crops yield more, that they require less pesticide applications, that they have no impact on the environment and of course that they are safe to eat. . . .
>
> The whole paradigm of the genetic engineering technology is based on a misunderstanding. . . .

> I think there is cause for alarm and it is my duty to educate the public.
>
> One argument I hear repeatedly is that nobody has been sick or died after a meal . . . of GM food. Nobody gets ill from smoking a pack of cigarettes either. But it sure adds up, and we did not know that in the 1950s before we started our wave of epidemics of cancer. Except this time it is not about a bit of smoke, it's the whole food system that is of concern. The corporate interest must be subordinated to the public interest, and the policy of substantial equivalence must be scrapped as it is clearly untrue.[6]

Confidence in GE Foods Is Substantially Based on Misinformation

Although there have also been cases in which individuals who previously objected to GE foods have reversed their position, I'm not aware of any in which the reversal was, like Dr. Vrain's, based on an accurate understanding of the facts. Instead, the shifts seem to have significantly stemmed from confusion. For instance, in an article describing why he changed his position on GE foods and has come to think that it's legitimate to develop them, the bioethicist Gary Comstock indicates he has trusted the pronouncements of their scientist-promoters – and has not realized they are significantly inaccurate.[7] He also appears to be mistaken about the quality of the testing that's being conducted, because he stipulates that the safety of GE foods must be assured "through a rigorous and well-funded risk assessment procedure" – and evidently believes that GE crops have been subjected to one.

Moreover, he additionally bases his position on the belief that it's legitimate to permit the potential benefits of GE foods to outweigh their harms, even though (as Chapter 9 has explained) when it comes to food safety, US law strictly forbids such a practice and demands demonstration of a reasonable certainty that novel products will not be harmful. But not only is he unaware of this critical fact, he seems unaware that the touted benefits have been overblown and that several rigorous tests have detected harm to laboratory animals. Thus, his realignment of attitude has been unduly influenced by erroneous ideas.[8]

Furthermore, as we've seen in Chapter 13, it appears that most scientists who themselves don't actively promote GE foods and yet believe they're safe, have (like the bioethicist Comstock) formed their beliefs due to disinformation dispensed by scientists who do promote them. And it also appears that when provided accurate information by a respected source, they change their minds. We've additionally seen (in Chapter 11) that Bill Gates' confidence in GE foods is significantly based on the mistaken belief that they're being adequately tested (or readily can be), despite the fact that from the perspective of software engineering, the testing has been ludicrously deficient – and could not vaguely approach the rigor with which life-critical software is tested without a tremendous revamping of the current system.

GE Foods Are Unacceptably Risky from Every Significant Angle of Analysis

As revealed in previous chapters, the risks of GE foods have been shown to be unacceptable through each of three distinct lines of investigation: (1) a genuinely scientific, biological-based risk assessment, (2) an assessment based on the principles of computer science, and (3) an assessment of the aggregate evidence of adverse outcomes they've induced. But there's yet another angle of analysis through which the extraordinary riskiness of these products can be demonstrated: a formal statistical approach based in probability theory and the properties of complex systems. Such an analysis was published in 2014 by a team of five experts in risk assessment, headed by Nassim N. Taleb, a Distinguished Professor of Risk Engineering at the New York University School of Engineering and widely renowned for his book, *The Black Swan.*

These experts assert the importance of distinguishing two basic forms of potential harm: (a) "localized nonspreading impacts" and (b) "propagating impacts resulting in irreversible and widespread damage." They state that the first type is more common and more easily dealt with because it can be calculated through past data and managed through cost-benefit analyses and mitigation techniques. Moreover, they point out that even when miscalculations are made in regard to such risks, the resultant damage "is bounded."[9]

In contrast, they emphasize that the second type of potential harm, which entails the possibility of nonlocalized irreversible damage, requires a much more precautionary approach. They state that if an activity poses such a risk, unless there is "scientific near-certainty" about its safety, its proponents must bear the burden of proving that it's safe before implementation is permitted.

But they're quite conservative in regard to this highly conservative approach, and they think it should only be applied in extraordinary cases.

Thus, they don't even consider it to be warranted in many operations that employ nuclear energy, because they view the potential harm as essentially local and nonsystemic – and thus capable of being managed through conventional risk assessment and cost benefit analysis.

Yet, although in the case of nuclear energy these experts caution against extreme caution, and advise against applying the strict precautionary approach as a general rule, they categorically prescribe it in the case of GE crops. That's because they deem the attendant risks to be systemic – and to entail potential widespread harm to the ecosystem as well as to human health. And they provide extensive analysis to back their judgment up.

Moreover, they note the lack of sound analysis or evidence to support the permissive approach urged by biotech proponents. As they observe: "Rather than recognizing the limitations of current understanding, poorly grounded perspectives about the potential damage with unjustified assumptions are being made. Limited empirical validation of both essential aspects of the conceptual framework as well as specific conclusions are being used because testing is recognized to be difficult."

Their conclusion about GE crops is unequivocal: that strict precaution should be exerted to avoid the risk of "considerable and irreversible environmental and health damage" – which, in their approach, entails that no new GE crops should be approved and all those currently on the market should be withdrawn.

An Inescapable Conclusion: GE Foods Must Be Promptly Banned

Thus, from whatever angle we consider the relevant evidence, it's clear that the GE food venture has not been, nor can be, conducted responsibly or safely and that it must therefore be terminated as rapidly as possible. And the nation that could play the key role in halting it is the United States. Although the US has been the chief driver of the GE food venture, and although the venture could not have advanced or even survived without the fraud of the US government, because the nation's food safety laws are actually inimical to this venture as currently conducted, it could be most quickly and thoroughly arrested there. And the sooner this fact becomes more widely realized, the quicker the change can happen.

The Focus Should Expand from Mere Labeling to Full Elimination

The time has come for American consumers with concerns about GE foods to broaden their focus. Instead of exclusively endeavoring to get these foods labeled, they should also concentrate on getting them banned. Not only are the products illegally on the US market, it's clear that the process by which they're produced is inherently high-risk and could never adequately conform to the requirements of the food safety laws, the standards of science, or the protocols of information technology.

Although the right to know what's in our food is an important one, it's underlain by one that is more fundamental and (in the case of the United States) more explicitly granted by statute: the right not to be exposed to inadequately tested genetically engineered foods in the first place. After all, labeling is technically appropriate for foods that are legitimately on the market, and if a group of foods are instead being marketed illegally, the proper remedy is not to label them but to remove them. In fact, placing the emphasis on labeling implies that the foods are on the market legally and obscures the reality that they're actually being sold in violation of the law.

The campaigns to obtain labeling that have been undertaken in so many states within the US have performed a highly valuable function in informing citizens about the presence of GMOs in most of the foods they've been buying, in educating them about the downsides, and in highlighting the lack of federal regulation. But now that people are generally more aware and better informed, it's important they recognize the key issue is not that GE foods are on the market without labeling but that they're on the market at all – and that federal law would have kept them off the market if the FDA had not fraudulently broken it.

Moreover, the issue of labeling is, as both a practical and technical matter, ultimately intertwined with the question of whether the FDA's presumption that they're *Generally Recognized as Safe* (GRAS) is valid. As a technical matter, the judge in the Alliance for Bio-Integrity lawsuit linked the two issues and ruled that if the FDA's presumption is legitimate, its determination that bioengineering is not a material fact that must be disclosed through labeling is likewise legitimate. (This has been discussed in Chapter 5.) And as a practical matter, any law that a state passes (either through direct legislative action or ballot initiative) that requires labeling will be challenged in court; and the strongest defense will be to demonstrate that the FDA's rebuttable GRAS presumption not only has always been illegitimate but has been repeatedly rebutted – and cannot therefore legitimize the agency's failure to require labeling. Thus, although the court adjudicating such a lawsuit would not be empowered to order GE foods off the market (as would a court adjudicating a direct action challenging the FDA's GRAS presumption), the presumption's invalidity provides grounds for ruling that states have a legitimate right to require that these foods be labeled.[10]

It's also important to bear in mind that even if labeling becomes required in the United States, it would probably not completely curtail the GE food venture. While it might deter Monsanto and major multinational corporations from continuing to produce herbicide resistant and pest-resistant GE crops, which don't provide consumers any direct benefit, it might not hinder them from developing the so-called second generation of GE crops designed to provide nutritional enhancement (such as the addition of Omega 3 fatty acids) or to eliminate undesirable features (such as an allergen). Manufacturers would want to label such foods anyway, because they would need to inform consumers of the purported beneficial differences between them and their conventional counterparts.

Further, it's doubtful that a labeling requirement would stop the Bill and Melinda Gates Foundation or other foundations and institutions from developing GE crops to assist the Third World. The narrative that would predominate is that although affluent US consumers had been driven by irrational fears to demand labeling for themselves, the impoverished peoples of the developing world should not thereby be denied the benefits of GE crops; and there would no doubt be an increased effort to "educate" these populations, and the world at large, about the purported science-based safety of these products. Moreover, because the labeling campaigns have focused on the right to know and have largely ceded the scientific high ground to the proponents of GE foods, they've made it easier for such a misleading narrative to prevail.

In contrast, by instead pressing for a complete ban, and focusing on the illegitimacy of the FDA's GRAS presumption, the covered up warnings of its own scientists, and the other reasons for regarding GE foods as unacceptably risky, the standard fictional narrative will be less likely to stand. And it's important that it does not, and that the prevailing narrative be fact-based, no matter how unpleasant those facts by now have become. The truth has been twisted for far too long, and the record must finally be set straight.

Effecting a Ban Can Be Accomplished Quite Simply

a. There's No Need to Pass a New Law, but Merely to Enforce One that's Already on the Books

Whereas achieving labeling at the state level requires not only the passage of new laws but the successful defense of those laws in court, removing GE foods from the market does not require any additional law or regulation. The statute that should have forced them to be adequately tested (and that would have effectively kept them off the market had it been honored) was passed in 1958, and the FDA regulations that gave it additional strength have also been on the books for many years. The only novelty that's needed is for the FDA to start enforcing the law rather than to continue breaking it.

And that change could be easily effected. It only requires one person to take the definitive step, and that person is the president of the United States, who at the time of this writing is Barack Obama. If President Obama were to learn, either from this book or another source, how badly he and his predecessors have been misled about GE foods and how flagrantly the law has been broken for more than twenty years, it's quite likely he would be moved to take remedial action. And all he would need to do is issue an executive order to the commissioner of the FDA instructing her that the agency (1) must openly acknowledge that its rebuttable GRAS presumption regarding genetically engineered foods has been solidly rebutted and (2) must take steps to remove each GE food from the market until it has at minimum been demonstrated safe (according to the *reasonable certainty of no harm* standard) through rigorous long-term, multi-generational toxicological feeding tests. Such an order would be well within the law and, moreover, would have the force of law.

Although presidential executive orders are often criticized by members of the opposing party in Congress on the grounds the action should have first received Congressional approval, and although the ones issued by President Obama have tended to incur such criticism, there would be no plausible grounds for such objections in the case of an order to reform FDA policy on GE foods. That's because the president would be correcting a longstanding dereliction in which the executive branch has subverted the will of Congress by violating an important law that Congress had already passed. So such an executive order would not be circumventing Congress but remedying an illegal circumvention of that body's will. Thus, the president would be upholding and implementing the expressed intention of the legislative branch, an act for which none of its members could legitimately criticize him.

b. Quick Removal Is Agriculturally Feasible Too

But in terms of economic and agricultural feasibility, could GE crops be rapidly removed from the market? According to the agricultural economist John Ikerd, an emeritus professor with the University of Missouri, it would be difficult to do it within one year due to the fact there wouldn't be an adequate supply of non-GE seeds with which to plant the next crop of corn, soybeans, and canola. In his view, the ban could apply to all new GE varieties, with existing ones phased out over perhaps as few as two growing seasons to allow for an orderly adjustment. He says, "Although these changes couldn't be made immediately, they could still be achieved rather quickly." [11]

Of course, the mere announcement by the FDA that GE foods could no longer be presumed GRAS, that each must undergo the formal food additive petition process and be demonstrated safe via rigorous safety testing, and that each would be banned until this had happened would have a powerful effect. And even if the agency specified that the ban would be implemented over a two year period, it would substantially depress consumer demand and could create significant complications. But such considerations should not be permitted to delay the announcement, because otherwise it would be continually postponed.

Once we accept the reality that GE foods are unacceptably risky, decisive action must be taken, with the understanding that it's far better to weather whatever short-term economic difficulties may be entailed by promptly banning them than to suffer the potential long-term health and environmental damage that could result from inaction.

In Addition to the President, Other Key Individuals Could Play a Major Role

Even if the president of the US did not take the initiative, several other individuals are in a position to do so. For instance, what if Bill or Melinda Gates were to read this book? They are astute individuals with a deep understanding of computer science, and it's difficult to believe they would not be affected by the presentation of evidence and by the analysis demonstrating the unsoundness of genetic engineering from the perspective of software engineering. If Mr. Gates concluded that he's been significantly misinformed about the facts, decided that it's unwise for his foundation to continue to invest in the development of GE crops, and then spoke out about why he had changed his mind, the GE food venture would nosedive.

Another person who could generate an equally profound effect is Bill Clinton, who has been one of the GE food venture's strongest boosters. In 2006 he declared, "I did everything I could as President to support the development of biotechnology and its practical applications in American life." [12] He also expressed his ongoing support for GE crops, but with the proviso, "If anybody could give me any evidence why I shouldn't do it, I'd be happy to change my position." [13] And I'd be happy if someone could put this book in front of him, which would provide such evidence in abundance. If he realized how seriously the actual facts clash with what he was led to believe by people he had good reason to trust, and how, under pressure from his predecessor's administration, the FDA had violated the law in order to put GE foods on the market, it's likely that he would not only change his position, but feel obliged to try to make amends for the irresponsible policy he has for so long mistakenly supported. And because he's so skilled at communicating ideas and continues to wield great influence, he could pull the rug out from under the GE food venture if he wanted to.

Further, there are several current and former heads of state in other nations who could put the brakes on the venture, especially when armed with the pertinent facts.

But even if it takes considerable time before highly influential individuals speak out, as citizens learn the facts and exert their collective influence, progress can occur. Because the GE food venture is grounded on disinformation, it's deeply vulnerable. So, in one way or another, it will inevitably be stopped.

This Is Not an Extreme Position

Although many may brand this stance an extreme position, it's not. After all, is it extreme to insist that the food safety laws be followed? Is it extreme to demand that humans refrain from rewriting the most complex and least understood information systems on earth unless they can do so with at least the same degree of care that's exercised by the technicians who revise man-made systems that are far simpler and far better comprehended? Is it extreme to ask that major decisions about the wholesomeness of our food and the future of food production be based on the best scientific knowledge and solid empirical evidence rather than on discredited presumptions and deficient testing? Is it extreme to reject the pronouncements of scientists and scientific institutions that have so consistently displayed dishonesty that no responsible jury would accept their testimony – even in a case where the stakes are, by comparison, trifling?

Of course not. In reality, it's quite conservative.

What's extreme is the GE food venture itself and the claim that its products are essentially as reliable and safe as crops produced via the processes of nature. What's extreme is the notion that this radical venture should not only be continued but be extended into organic agriculture. What's extreme is the idea that although this venture has consistently depended on deception and imposed extraordinary risks, because it has become so entrenched we should just put up with it.

Curtailing GE Foods Is Just as Much a Conservative Cause as a Progressive One

Although many proponents of the GE food venture portray any opposition to it as the product of a left-leaning agenda, and imply that conservatives should as a matter of principle support it, this is inaccurate. In reality, not only is the venture contrary to basic principles shared by liberals and conservatives alike, it's especially offensive to some that are specifically upheld by the latter.

For instance, Friedrich Hayek, one of the most influential conservative thinkers of the last hundred years, whose theories have been lauded by free market economists and politicians like Ronald Reagan and Margaret Thatcher, emphasized the importance of respecting complex, spontaneously ordered systems and the inability of human intelligence to impose purportedly rational plans onto such systems without inducing unintended consequences. And although this thinking is just as applicable to bioinformation systems as to economic systems, too many conservatives fail to realize that the bioengineers are doing precisely what Hayek objected to and are attempting to manipulate complex natural systems in a top-down, interventional manner with scant knowledge of the intricate dynamics through which these systems function.

Unfortunately, as the conservative columnist David Brooks has observed, many of today's conservatives have become so narrowly focused on reducing government regulation that they tend to overlook other traditional conservative aims.[14] And, from a Hayek-inspired perspective, it would seem that preserving the integrity of the bioinformation systems upon which humanity relies for nourishment clearly qualifies as one of those goals. Accordingly, it could be cogently argued that conservatives should not be willing to tolerate the heavy-handed and radical restructuring of these most complex naturally formed systems on earth merely because some governmental regulation would be needed to prevent it.

Furthermore, the numerous conservatives who are devoutly religious have even deeper reasons for objecting to such heavy-handed interventions.

There Are Also Strong Religiously-Based Reasons for Rejecting the GE Food Venture

As the introduction to this book noted, the plaintiffs in the Alliance for Bio-Integrity lawsuit not only included nine scientists but a group of individuals and organizations from diverse religious traditions that objected to GE foods on the basis of religious principle. Seven of these plaintiffs were ordained Christian priests and ministers (including a Roman Catholic, an Episcopalian, a Lutheran, and a Baptist) and three were rabbis (Orthodox, Conservative, and Reform). Further, the positions taken by most of the religiously-motivated plaintiffs (whether Christian, Jewish, or Hindu) were based in traditional theism, a system of belief in which divine intelligence is viewed as having in some significant way been directly and purposively involved in the development of the various forms of life.[15] Therefore, since a large proportion of humankind also embrace traditionally theistic beliefs, it's important to examine some basic reasons why people who hold such beliefs could view the GE food venture as spiritually offensive.

Because from a traditional theistic perspective the natural cross-breeding barriers can be seen as basic features of the divine plan, this logically engenders the idea that limited human intelligence should refrain from artificially altering such an intricate system – especially when it involves reconfiguring the genetic structure of numerous organisms.

Moreover, such an attitude is not exclusively religious, and some scientists have developed it based on entirely secular considerations. As Chapter 4 noted, the Nobel Laureate biologist George Wald repeatedly underscored how radically genetic engineering differs from all previous methods of manipulating nature and how ominous are the alterations it brings about. He warned that it presents "problems unprecedented not only in the history of science, but of life on the Earth," and he emphasized that "such intervention must not be confused with previous intrusions upon the natural order of living organisms."[16] Consequently, he referred to this new level of intervention as "the biggest break in nature that has occurred in human history."[17]

A similar outlook was evidenced by one of the pioneers of molecular biology, Erwin Chargaff, whom *The Guardian* described as "one of the giants of the world of biochemistry."[18] In expressing his concern about the sundering of natural boundaries that recombinant DNA technology had wrought, he stated: "I have the feeling that science has transgressed a barrier that should have remained inviolate."[19]

Accordingly, if from a purely secular standpoint the incursions of genetic engineering can be regarded as serious transgressions of the natural order, there's far more reason to regard them as such from a traditionally theistic perspective. Thus, one of the plaintiffs in the Alliance for Bio-Integrity lawsuit, Rabbi Alan Green, stated in a filing with the court that he regards the cross-breeding boundaries as divinely set barriers that humanity should not sunder. He then asserted: "I believe that genetic engineering greatly exceeds all other methods of creating new varieties of food-producing organisms in its distortion and disruption of natural boundaries and structures." And he declared: "Therefore, as a matter of religious principle, I feel obliged to avoid consuming the products of this radical technology."[20] The Christian clergy who were his co-plaintiffs expressed similar positions. For instance, in a statement typical of those entered by this set of individuals, an Episcopalian minister said he believed that "the forcible transfer of genetic material across nature's cross-breeding barriers for the purpose of redesigning food is a disruption of the divine plan."[21]

Further, even short of regarding the cross-breeding barriers as "inviolate" boundaries and genetic engineering as a disruption of the divine plan, there are still strong religiously-based reasons for rejecting the venture. That's because any religious individual could legitimately insist that limited human intelligence should at minimum treat the cross-breeding barriers and the complex web of life with substantial respect and exercise great care in attempting to artificially alter such an intricate system about which so little is yet comprehended.

Moreover, because even the more reverential, and stricter, attitude toward these natural boundaries discussed above is legitimate from a secular as well as a religious standpoint, it's obvious that this more moderate one must be too. Such an attitude has been expressed by the American public interest organization Consumers Union. In contemplating the unprecedented powers of genetic engineering to reconfigure fundamental facets of nature, CU urged policy makers to recognize this technology ". . . represents something that is fundamentally new and, as such, should be approached with caution, care and some humility."[22]

Although CU's plea for humility was based on secular grounds and is well-justified on such grounds, it's far more compelling from a perspective in which the structures of living organisms and the barriers between them are seen as features of a divinely instated system; and the virtual absence of humility from the thinking of the bioengineers and their governmental promoters is a glaring defect that rightly undermines confidence in their actions. Furthermore, from such a perspective the routinely reckless approach that has characterized the GE food venture can justly be viewed as displaying not merely a lack of humility, but a high degree of arrogance – and a disrespect for God.

In particular, theists could regard the stark discrepancy between genetic engineers and software engineers in terms of the precaution exercised by each as a clear case of such disrespect. As Chapter 11 pointed out, when software engineers make even minor, well-planned revisions to life-critical information programs that they themselves have designed, they proceed under the presumption that, despite their best efforts, unintended disruptions have probably occurred that could unduly endanger human life. They therefore follow strict procedures and subject the revised programs to rigorous rounds of testing. But when bioengineers make radical revisions to the information programs of food-yielding organisms that they have not created, that are far more complex and interconnected than any man-made system, and about which they have minimal comprehension, they and their supporters nonetheless presume either that no harmful disruptions have occurred at all (as in the case of the FDA) or that any which may have happened can be detected by tests that, relative to those employed in software engineering, are grossly superficial. Accordingly, from a traditionally theistic perspective, they fail to honor the fact that God has designed the systems they're altering, and fail to appreciate that His intricately exquisite software should be treated with *at least* the same degree of respect as is accorded the software fashioned by the limited human mind.

Moreover, from such a religious perspective, the disrespect is amplified by the fact that the lack of proper precaution stems from one of the GE food venture's foundational presumptions: the erroneous notion (discussed in Chapter 12) that the processes of natural reproduction are more disorderly and unpredictable than those of human-conducted genetic engineering – and are therefore riskier.

It's also noteworthy that there are religiously based reasons for rejecting GE foods that don't rely on any theistic beliefs. Thus, one of the plaintiffs in the lawsuit was the Chancellor of a Buddhist university, and although Buddhism is classified as a nontheistic religion, he stated that he felt religiously obliged to avoid GE foods because he regards the artificial genetic restructuring that occurs through recombinant DNA technology as contrary to Buddhist principles.

GE Foods Are Not Sufficiently Beneficial to Outweigh Their Risks – and They Are Not Needed to Solve the Problems of Agriculture

Because the advocates of agricultural bioengineering routinely tout its purported benefits and argue that they must be given great weight, and because so many influential people have been led to believe that GE crops can solve major problems in the developing world, it's important to explain why this is not the case.

However, any discussion of potential benefits must begin by re-emphasizing the fact that, as a technical matter, they're irrelevant. According to US food safety law, they should play no role in assessing the risks; so within that key nation, not only is it irrelevant to factor them in, it's illegal. Moreover, as discussed earlier in this chapter, the probability-based risk analysis performed by Nassim Taleb and his colleagues also determined that it's illegitimate to consider benefits. According to these experts, the risks of GE foods are so great they fall outside of the bounds of those that can be properly offset by potential benefits, and they must instead be contained by the products' strict prohibition.[23]

But even if it *were* legitimate to consider potential benefits, GE crops don't possess enough to justify their use. Chapters 7 and 9 have shown not only how the benefits of those that are currently commercialized have been highly exaggerated, but how those crops have actually been causing significant agricultural and ecological problems. And when the plausible risks to the health of human consumers are also factored in, the balance tilts so heavily against the products that their use is clearly unwarranted.

Nor can the projected benefits of a long-promised new generation of GE crops, which are supposed to provide genuine boons to consumers and tangible boosts to production, outweigh the risks either. For one thing, there would still be substantial risk to human health – even in cases where the degree of genetic alteration is significantly less than in the current forms of GMOs. (This point is more fully elaborated in Appendix D.) For another, in many cases naturally bred alternatives would be available that don't entail the downsides of their GE counterparts. For instance, although biotech proponents repeatedly tell us that genetic engineering is necessary for producing crops that are drought-tolerant, not only can such crops be produced via conventional breeding, conventional breeding has been more successful in doing so.[24] That's because drought resistance is a complex attribute and is not based in a single gene. Instead, it arises from many genes operating in a coordinated manner. Therefore, it's very difficult for bioengineers to endow crops with this trait.

Due to this difficulty, genetic engineering has rarely played a part in creating drought tolerance in the GE crops that display it. Instead, this capacity was developed through conventional, non-GE techniques and the drought-tolerant plant was then bioengineered to additionally render it herbicide- or pest-resistant.[25]

Moreover, even when genetic engineering can endow plants with other desirable traits, it's no better at it than conventional breeding. As two experts with the Union of Concerned Scientists have noted: "Genetic engineering might be worth the extra cost if classical breeding were unable to impart such desirable traits as drought-, flood- and pest-resistance, and fertilizer efficiency. But in case after case, classical breeding is delivering the goods."[26] Comprehensive lists of non-GE crops that possess beneficial traits such as high yield, drought resistance, salt tolerance, pest resistance, disease resistance, and nutritional fortification (without posing the inherent risks of genetic engineering) are provided at the free resource, GMO Myths and Truths (2nd Edition) on pages 285 and 318 – 321, http://earthopensource.org/files/pdfs/GMO_Myths_and_Truths/GMO-Myths-and-Truths-edition2.pdf.

The simple fact is that we don't need GE crops, a reality that's been recognized by numerous independent experts who have analyzed the extensive evidence. For example, in 2008 the World Bank and four United Nations agencies completed a four-year study on the future of farming: the International Assessment of Agricultural Knowledge, Science and Technology for Development (IAASTD).[27] This massive study was conducted by more than 400 experts from 80 countries, and 58 governments have endorsed it. And its assessment of genetic engineering flew in the face of the promotional claims by concluding that this technology is not essential for solving the problem of hunger. What's more, it noted that yields of GE crops were "highly variable" and that in some cases there had been "yield declines." It also noted that there were continuing concerns about the safety of the crops.

To the further consternation of the technology's proponents, the scientists in charge of the IAASTD study have not minced words about its inability to deliver on the promises that they've been making about what it will do. When the project's director (Bob Watson) was asked at a press conference whether GE crops were the answer to world hunger, he replied, "The simple answer is no."[28] And when the co-chair of the study (Hans Herren) was interviewed, he provided an answer that, while not as simple, was no less explicit:

> [GMOs] haven't actually proven anything yet in terms of increased yields, as far as any of the major food crops are concerned. . . . I don't really see any proper use for GMOs, now or even in the future. I think that the solutions for problems with agricultural food security lie elsewhere – not in the seed or GMO seeds in particular. . . . The fact of life is that right now, we produce enough food for 14 billion people. . . . In the developed countries in particular, we produce more food than is required. In developing countries, we under-produce and that's not because we need GMOs, that's because those countries have bad agronomic practices, farmers don't have the right information on when to plant and how to best manage their farms. It's an issue of more and better information to farmers in the developing countries."[29]

Thus, as a legal, a theoretical, and a practical matter, it's futile to assert the benefits of GE crops; and it's necessary to recognize that they're unnecessary. Moreover, not only are these crops unneeded, they're impeding progress because they're diverting attention and resources from the approaches that *are* necessary.

The Paramount Need Is for Fuller Development of Agroecological/Sustainable Methods

While the IAASTD study did not support genetic engineering, it clearly endorsed a different approach. It called for the development of "agroecological" methods of production (which include those that are classified "organic"), methods that require fewer inputs, conserve resources, and preserve the soil. Such methods can build pest protection through natural means and also induce both hardier crops and greater soil fertility without reliance on synthetic additives.

Moreover, although the agroecological approach employs many traditional, time-honored practices, it's not limited to them. For instance, it also makes use of modern techniques such as Marker Assisted Selection (MAS), which enables the development of plants with important complex traits that bioengineering cannot produce – but without the risks that it *does* engender.

Most important, not only are agroecological methods better suited to the developing world than are the high-input practices of industrial agriculture, they've been highly successful at producing higher yields. For instance, a recent UN report that surveyed 114 farming projects in 24 African countries determined that through the adoption of organic or near-organic practices, yields increased on average by over 100%.[30] Further evidencing the widespread successes of such methods in Africa, the UN Special Rapporteur on the Right to Food has reported: "Yields went up 214% in 44 projects in 20 countries in sub-Saharan Africa using agroecological farming techniques over a period of 3 to 10 years. . . ." And he pointed out that this accomplishment is "far more than any GM crop has ever done."[31]

He has also stated:

> To feed 9 billion people in 2050, we urgently need to adopt the most efficient farming techniques available. Today's scientific evidence demonstrates that agroecological methods outperform the use of chemical fertilizers in boosting food production where the hungry live – especially in unfavorable environments. To date, agroecological

> projects have shown an average crop yield increase of 80% in 57 developing countries, with an average increase of 116% for all African projects. Recent projects conducted in 20 African countries demonstrated a doubling of crop yields over a period of 3–10 years. Conventional farming relies on expensive inputs, fuels climate change and is not resilient to climatic shocks. It simply is not the best choice anymore today. Agriculture should be fundamentally redirected towards modes of production that are more environmentally sustainable and socially just."[32]

Furthermore, agroecological methods can also succeed in industrialized nations. Long-term studies in the United States have demonstrated that well-managed organic farming systems can produce yields that are comparable to conventional systems.[33, 34] And small farms employing agroecological practices are on balance substantially more productive than large industrialized farms. As Miguel Altieri, a Professor of Agroecology at University of California at Berkeley, explains: "A large farm may produce more corn per hectare than a small farm in which the corn is grown as part of a polyculture that also includes beans, squash, potatoes, and fodder. But, productivity in terms of harvestable products per unit area of polycultures developed by smallholders is higher than under a single crop with the same level of management. Yield advantages can range from 20 percent to 60 percent, because polycultures reduce losses due to weeds (by occupying space that weeds might otherwise occupy), insects, and diseases (because of the presence of multiple species), and make more efficient use of the available resources of water, light, and nutrients."[35]

But even if agroecology could not yield as well as industrialized monocultures, it would still offer better benefits for the industrialized nations. As the University of Missouri agricultural economist John Ikerd has pointed out, because global food security "does not depend on continued increases in productivity in industrial agricultural countries such as the U.S. and Canada," the most important goal in these countries should be "to increase the sustainability of agriculture rather than to increase agricultural yields or productivity."[36]

Thus, it's clear that besides being unsustainable, conventional agriculture is not maximally productive – and that agroecological agriculture can outperform it, especially in the less developed regions of the world. It's also clear that genetic engineering is not the answer – and not even a sound option, whatever the level of economic development or set of environmental conditions. This high-tech but low-foresight approach is the most unsustainable form of agriculture, because not only is it inherently and unacceptably risky in regard to both human and environmental health (no matter what genes are employed in the reconfigurations), it depends on consistent contortion of the truth – and cannot survive an accurate airing of the facts.

Nonetheless, it has been far more lavishly funded over the last 30 years than the agroecological methods that can outperform it and outlast it. And the lopsided emphasis on genetic engineering has restricted the more sustainable and valuable approaches from achieving as much as they could if they were better supported.

The experience of a professor of soil science at the University of Hawaii is indicative of this restriction. After I'd given a lecture there that (among other things) noted the gross imbalance in funding and its ill effects, he expressed full agreement with what I'd said. He told me that he had submitted several grant applications for sustainable projects in developing nations that had been turned down because, as he was informed, they didn't involve molecular bioengineering. He said that this had happened enough times that he had given up on writing more proposals because he did not want to employ that technology – and did not want to waste more of his time.

It's Time to Decisively Move Forward

We've seen how a mass of disinformation, much of it spread by eminent scientists who abused their positions of authority, has kept most people confused about the products of genetic engineering for nearly 40 years. And we've seen how even highly astute individuals with training in science or engineering have been taken in.

But we've also observed that, when the cloud of disinformation is systematically dispelled, the actual facts become not only clear but compelling. And the conclusion they compel is that GE foods should never have been commercialized in the first place and must be curtailed as quickly as possible. They were never generally recognized as safe within the scientific community, there has never been genuine evidence of their safety, and substantial evidence indicates that several are most likely unsafe – and that the process through which they're *all* created is inherently risky.

Further, they were initially commercialized through the fraud of the United States Food and Drug Administration (the FDA) and the flagrant violation of that nation's food safety laws, and their continued marketing in the US continues to be illegal. Yet, the US government plays the leading role in promoting them world-wide and impelling their use in other nations.

However, things can be turned around quickly – even more quickly than in 1906, when America's food industry underwent its first dramatic reform. On February 26th of that year, a book titled *The Jungle* was published that graphically exposed the unsanitary and gruesome conditions in the nation's meat-packing plants. Its revelations created a shockwave, and one of the individuals most shocked was the President, Theodore Roosevelt, who declared that "radical action must be taken."[37] So he dispatched his Labor Commissioner and a social worker to investigate the Chicago factories, and he submitted their unsettling report to Congress on June 4th.[38] Before the end of the month, due to that report and strong public pressure, Congress had passed both the Meat Inspection Act and the Pure Food and Drug Act, the first federal statutes to protect the public from dangerous practices in the related industries.

Now, more than a hundred years later, the time is ripe for another dramatic food-related reform; and it can be achieved more simply. President Obama doesn't have to convince Congress to act; he merely has to order the FDA to get its act together – and to stop violating the nation's central food safety law and start applying it to GE foods. And that simple step would quickly send the GE food venture into a death spiral.

But even if he doesn't take that step, the venture cannot survive for long because it rests on a perilously flimsy foundation. It has always been supported by distortions of truth – and it will topple when the distortions are exposed and expunged.

We don't need new laws, we don't need new research, we just need new awareness. All that's required is for a few key people, or enough people in general, to learn the basic facts. And when that happens, as it is bound to, it will bring the fact-averse GE food venture to an end.

Abandoning GE foods will not be a sacrifice but a liberation. It will free us from the debilitating influence of disinformation and delusion. It will free us from unacceptable risks to our health, the health of future generations, and the health of our environment. It will free up immense resources and enable them to be re-directed to the development of safe, sustainable, and sensible modes of farming.

For far too long the genes of the food-yielding organisms on which we depend have been altered in radical and risky ways, and for far too long the truth about what's been happening has been badly twisted.

The time has come to set things straight. The time has come to restore the integrity of genomes, the integrity of science, and the sustainability of agriculture. The time has come to transcend the mistakes of the past and advance to a brighter and more fruitful future.

<u>APPENDIX A</u>

EXTENDED EXAMINATION OF THE JUDGE'S DECISION IN *ALLIANCE FOR BIO-INTEGRITY V. SHALALA*

This Appendix provides additional analysis of the flaws in Judge Kollar-Kotelly's opinion by examining some prior decisions of federal courts that are relevant to the issues but that were not discussed in Chapter 5 due to space limitations. Some of these decisions were relied on by the judge, and some were relied on by the Plaintiffs in their written arguments.

I. *The Illogic of Allowing FDA Administrators to Disregard the Input of Their Scientific Staff*

Judge Kollar-Kotelly ruled that the FDA's administrators could presume that GE foods are generally recognized as safe despite the fact that their own experts had repeatedly warned about their unusual risks. She justified her ruling by arguing that an agency's interpretation of its own regulations is not invalidated by contrary opinions of "lower-level" officials; and she cited a 1986 decision of the D.C. Circuit Court of Appeals (the court that reviews decisions by the judges in her district) to back her up.[1]

But that case does not support her argument. It involved interpretation of a Nuclear Regulatory Commission (NRC) regulation that prevented nuclear power plants from being licensed unless there was a finding that adequate protective measures could be taken in event of an emergency. The Circuit Court upheld the Commission's decision that the potential effects of earthquakes need not be considered, despite the opinions of two of its staff that they should.

However, the issue in that case substantially differed from the issue in our suit, since it centered on whether the general language of a regulation should be interpreted to pertain to a specific type of risk – which to a large extent is a policy decision. In contrast, the issue in our case was not about how the language of an agency regulation should be interpreted but about whether, in light of well-established regulations, there were rational grounds on which an agency could make a particular presumption. Further, the opinion of the agency's scientific staff was a highly relevant factor in determining whether such grounds existed. The question was whether upper level administrators may rightfully presume that there is an overwhelming consensus among scientists that GE foods are safe despite the obvious fact that most of their own experts did not regard them to be. It was not a question of allowing one opinion about interpreting regulatory language to prevail over another. It was a question of allowing administrators to disregard (and contradict) a crucial fact in making what was supposed to be a fact-based determination.[2] By adopting a presumption that GE foods are generally recognized as safe, despite direct knowledge that they were not, the administrators were acting in an arbitrary and capricious manner forbidden by law.

II. *Ignoring the Issue of Inconsistency*

There's another significant respect in which the action of the NRC discussed above is distinguishable from the FDA's actions in adopting its 1992 policy. In that earlier case, the court emphasized that the NRC's decision was not inconsistent with its prior practices. In contrast, the FDA's application of its GRAS-related regulations to GE foods is clearly inconsistent with its previously established interpretations and applications of them.

In our written submissions, we noted that courts are wary of agency inconsistency. We called the judge's attention to a warning the D.C. Circuit Court had issued in 1982 that "sharp changes of agency course constitute 'danger signals' to which a reviewing court must be alert;" and we cited the court's statement that such changes are grounds for denying deference to the agency's actions.[3] Further, we demonstrated that the FDA's interpretation of the GRAS standards in regard to GE foods constitutes a "sharp change" of course, pointing out several legal actions the agency had previously pursued in which it sought to bar unapproved additives by arguing for a strict interpretation of the GRAS requirements.[4] We also noted the change in course was so severe that the director of the FDA's Biological and Organic Chemistry Section reproved the agency for turning the prior understanding of the term *food additive* "on its head."[5] Moreover, one FDA document observed that a "disadvantage" of the type of policy that the agency eventually did adopt was the fact it would be "At odds with emerging FDA legal interpretations of what is required to achieve GRAS status..."[6]

We additionally discredited the notion that the FDA's actions deserved deference by citing a ruling of the US Supreme Court that judges should not defer to an agency when its decision fails to consider an important aspect of the problem, or when it gives an explanation for its decision that runs counter to the evidence. Because the FDA administrators failed to consider the unusual aspects of GE foods, and because their justification for their policy clashed with the evidence, this ruling was right on point.[7]

Surprisingly, Judge Kollar-Kotelly did not see fit to mention this case, or the case in which the D.C. Circuit had warned about inconsistency, or any of the cases (or other evidence) demonstrating that the FDA's 1992 policy was at odds with its prior practice. Instead, she chose to disregard the issue of inconsistency altogether.

III. *Misplaced Emphasis on Agency Expertise*

Rather than acknowledging that inconsistency deserves no deference, Judge Kollar-Kotelly sought to justify her exercise of deference by stressing the import of agency expertise. To do so, she cited a decision of the D.C. Circuit Court which, in upholding an action of the Environmental Protection Agency (EPA), remarked that "the rationale for deference is particularly strong" when an agency "is evaluating scientific data within its technical expertise."[8]

However, like the NRC case, this one provided no support for deferring to the FDA's presumption about GE foods. For one thing, as in the NRC case, the court emphasized that the agency had acted consistently with its prior practices.[9] Moreover, while the EPA had applied its expertise in thoroughly analyzing the data, the FDA staff members who possessed the scientific expertise (and evaluated the technical issues) had argued *against* the presumption of GRAS – and it was adopted in spite of their input.

Thus, fuller examination of the relevant decisions of federal courts more fully demonstrates the unsoundness of Judge Kollar-Kotelly's reasoning.

APPENDIX B

Two Reports by Other Respected Organizations that Misrepresent the Risks of GE Foods

The Royal Society

One widely cited report on the safety of GE foods was released in 2002 by the UK's Royal Society. And, like its counterpart issued by the National Academy of Sciences in 2004, not only did it fail to address the arguments of the 2001 Canadian Royal Society report, it didn't even acknowledge them. Although it did at least mention that report (in contrast to the NAS's complete disregard), the notice it took was minimal. It merely remarked that the report had criticized regulatory reliance on the concept of substantial equivalence.[1] But it neglected to describe the criticisms, and certainly never refuted them. Nor did it mention any of the other concerns the Canadian experts had raised; and it avoided discussion of many of the important issues they examined. For instance, while it did note the routine use of viral promoters in GE crops, it gave no indication that they impel hyper expression of the gene to which they're attached – nor any hint that the Canadian experts, along with many others, have regarded this as risky.[2]

Further, like the NAS report, that of the Royal Society grossly overstated the risks of conventional breeding. For instance, it alleged that such breeding methods could give rise to "unknown toxins, anti-nutrients or allergens."[3] But because there's no evidence this has ever happened, it had to prop its claim with inapt examples involving the same species (celery and potatoes) that were later employed by the NAS for the same purpose – examples in which toxins that were already present became elevated, but in which not a single "unknown" toxin was produced. Moreover, not only did the authors employ these invalid examples to bolster their false assertion, they also unfairly used them to suggest that the risks of conventional foods are on a par with those produced through rDNA technology, stating that this purported evidence "raises the question" of whether both sets of foods should be required to meet the same safety assessment criteria.

But, in reality, the main question raised by a careful reading of the report is whether its authors were more committed to promoting GE foods than to upholding the standards of science. And in light of the additional, and more egregious, derelictions documented in Chapter 10, it's clear that the commitment to the latter has indeed been eclipsed by the former.

The Institute of Food Technologists

Remarkably, other groups have even exceeded the excesses of the Royal Society. Consider the case of the 29,000-member Institute of Food Technologists (IFT), which published a report in 2000 because (in its words) it was "eager to contribute to a meaningful dialogue on scientific issues and consumer concerns. . . ."[4] But, as have most other such "meaningful" contributions, it mangled the meaning of risk.[5] It likewise baselessly raised the specter of unknown and potentially harmful ingredients infiltrating the market through conventional breeding.[6] And it claimed that because bioengineering is "more precise and predictable," this trusty technology is "less likely" to induce such unintended effects.[7] Of course, it was easier for the panel that wrote the food safety section to make such a claim because they didn't mention several aspects of bioengineering that set it apart from conventional breeding (and entail a higher risk of unintended effects), such as the use of viral promoters to over-express gene products – or the genome-scrambling effects of cassette insertion.[8] Nor did they note that, in contrast to most conventional crops, almost all of those produced through bioengineering are subjected to the mutative process of tissue culture.

Moreover, their case for the harmlessness of GE foods to human health was substantially augmented by the indiscriminate inclusion of, and reliance on, statements from other reports that had nothing to do with food safety. Thus, they featured several pronouncements from the NAS reports of 1987 and 1989, despite the fact those documents dealt only with environmental effects of field trials within the continental US.[9] And they gave prominence to the assertion in the 1989 report that ". . . no conceptual distinction exists between genetic modification of plants and microorganisms by classical methods or by molecular methods that modify DNA and transfer genes." As Chapter 4 explained, this assertion is logically absurd. Nonetheless, the IFT panel treated this absurdity as a verity – and proffered it as support for the safety of GE foods.

Thus, careful scrutiny reveals that, like the 2004 report of the NAS, the two discussed above are deeply flawed – and contribute to the methodical misrepresentation of risk.

NOTES

Introduction

1 The FDA acknowledges that it has been operating under a policy "to foster" the US biotechnology industry. See, e.g., "Genetically Engineered Foods," *FDA Consumer* (Jan. - Feb., 1993), 14.

2 Keller, Evelyn F. *The Century of the Gene*. (Cambridge: Harvard University Press, 2002), 142-43.

3 Ibid. 143.

4 Ibid. 144.

5 Ibid. 148.

1. The Politicization of Science

1 Among the Harvard professors who regarded Mayr as the greatest 20th century biologist were E.O.Wilson and Stephen Jay Gould. See Meyer, A., "On the importance of being Ernst Mayr," *PLoS Biol* 3(5):e152 (2005): 0100.

2 When used in this way, the term "gene-splicing" refers to manipulations of biotechnicians. As will be discussed in subsequent chapters, although segments of DNA are also spliced into DNA molecules through natural processes, the details of these processes significantly differ from those of recombinant DNA technology.

3 Regal recorded these words in a set of recollections about his endeavors to set the genetic engineering venture on a more scientific track – recollections that he sent to me for use in this book. The statements from him that follow in this and other chapters are largely drawn from these recollections and from my extensive conversations and email correspondence with him. Accordingly, except for quotes excerpted from his published articles, I will not provide specific references for his various statements.

4 Crichton completed the first draft of *The Andromeda Strain* in 1967.

5 Morrow, J.F., Cohen, S.N., Chang, A.C.Y., Boyer, H.W., Goodman, H.M., Helling, R.B., "Replication and transcription of eukaryotic DNA in Escherichia coli," *Proceedings of the National Academy of Sciences* 71 (1974): 1743- 47. Prior to that accomplishment, other researchers had learned how to join two pieces of DNA together. The initial fusion was achieved by a team in Paul Berg's lab at Stanford University; and Berg subsequently received a Nobel Prize in recognition of this and other groundbreaking research in recombinant DNA technology. Jackson, D. A., Symons, R. H., and Berg, P., "Biochemical methods for inserting new genetic information into DNA of Simian Virus 40: Circular SV40 DNA molecules containing lambda phage genes and the galactose operon of Escherichia coli," *Proceedings of the National Academy of Sciences* [PNAS] 69 (1972): 2904.

6 As discussed in the previous note, Berg had been able to create some recombinant DNA even before scientists had discovered how to isolate individual genes from one species, copy them, and then splice them into the DNA of other species. But his technique was relatively complicated and could not be widely employed. The tumor-inducing virus that he planned to work with is referred to as SV40.

7 For a discussion of this incident see the preface to the 2013 paperback edition of: Pollack, R., *The Faith of Biology and the Biology of Faith* (New York: Columbia University Press, 2013).

8 Berg, P., "A Stanford Professor's Career in Biochemistry, Science Politics, and the Biotechnology Industry," an oral history conducted in 1997 by Sally Smith Hughes, Regional Oral History Office, The Bancroft Library, University of California, Berkeley (2000), 92; http://texts.cdlib.org/view?docId=kt1c6001df&doc.view=entire_text.

9 Ibid., 93.

10 Singer, Maxine and Soll, Dieter, "Guidelines for DNA Hybrid Molecules," *Science* 181 (September 21, 1973): 1114.

11 Wright, Susan, *Molecular Politics: Developing American and British Policy for Genetic Engineering 1972-1982* (Chicago: University of Chicago Press, 1994), 136.

12 Ibid.

13 Berg, Paul et al., "Potential Biohazards of Recombinant DNA Molecules," *Science* 185 (July 26, 1974): 303.

14 Ibid.

15 Barinaga, Marcia, "Asilomar Revisited: Lessons for Today?" *Science* 28, no. 5458 (March 3, 2000): 1584–85.

16 Watson, J. and Tooze, J., *The DNA Story* (San Francisco: W.H. Freeman, 1981), 49.

17 Wright (1994), op. cit. note 11, 135.

18 Ibid., 26.

19 Cited in Goodell, Rae, "How to Kill a Controversy: The Case of Recombinant DNA" in *Scientists and Journalists: Reporting Science as News*, Friedman, S.M., Dunwoody, S. and Rogers, C., eds. (New York: The Free Press/ Macmillan, 1986), 172.

20 Bennett, William and Gurin, Joel, "Science that Frightens Scientists: The debate over DNA," *The Atlantic Monthly* 239 (February, 1977): 43-62.

21 Lewin, Roger, "The Asilomar Conference: Was the Asilomar Conference a Justified Response to the Advent of Recombinant DNA Technology, and Should It Serve as a Model for Whistle-Blowing in the Future?" in *Bioscience Society; Report of Schering Workshop*, Roy, D.J. et al., eds., (Chichester, New York: John Wiley & Sons, 1991), 206.

22 Ibid.

23 Wright, Susan, "Molecular Biology or Molecular Politics? The Production of Scientific Consensus on the Hazards of Recombinant DNA Technology," *Social Studies of Science* 16, no. 4 (Nov. 1986): 593-620, 595.

24 Ibid.

25 Watson, James D., "An Imaginary Monster," *Bulletin of the Atomic Scientists* 33 (May 1977): 12.

26 Watson expressed his regret in a speech quoted in McAuliffe, Sharon and McAuliffe, Kathleen, *Life For Sale* (New York: Coward, McCann & Geoghegan, 1981), 176.

27 Kay, Lily, *The Molecular Vision of Life: Caltech, The Rockefeller Foundation, and the Rise of the New Biology* (New York: Oxford University Press, 1993); Pnina Abir-Am, "The biotheoretical gathering, transdisciplinary authority and the incipient legitimation of molecular biology in the 1930s: new perspectives on the historical sociology of science," *Hist Sci* 25 (1987):1-70.

28 Weaver, Warren, *Scene of Change: A Lifetime in American Science*, (New York: Scribner's, 1970), 56.

29 Ibid., 57.

30 Regal, Phil, "Metaphysics in Genetic Engineering: Cryptic Philosophy and Ideology in the 'Science' of Risk Assessment." In *Coping with Deliberate Release: The Limits of Risk Assessment,* Van Dommelen, Ad, (ed.), International Centre for Human and Public Affairs, Tilburg/Buenos Aires (1996).

31 Weaver, op. cit. note 28, 183.

32 Regal, "Metaphysics in Genetic Engineering," op. cit. note 30.

33 Alibek, Ken, *Biohazard: The Chilling True Story of the Largest Covert Biological Weapons Program in the World – Told from Inside by the Man who Ran It* (New York: Random House, 1999), xi.

34 Chargaff, Erwin, "On the Dangers of Genetic Meddling," *Science* 192 (1976), 940.

35 Ibid., 938.

36 King, Jonathan, quoted in McAuliffe and McAuliffe, op. cit. note 26, 174; See also Bennett and Gurin, op. cit. note 20, 56-57.

37 Wald, George, "The Case Against Genetic Engineering," in *The Recombinant DNA Debate*, Jackson, D. and Stich, S., Eds. (Prentice-Hall, 1979), 127-28.

38 Ibid.

39 Wald, George, speaking at a press conference in Washington, D.C. March 1977. Quoted in Kimbrell, A., *The Human Body Shop: The Engineering and Marketing of Life,* (New York: Harper Collins, 1994), 159. Although Wald's statement that genetic engineering is the "biggest break in nature" occurred at a press conference, I think it's appropriate to include it along with statements he wrote in an earlier article because doing so does not in any way misrepresent his thinking – and enables it to be expressed in a compact manner.

40 Wald, George, "The Case Against Genetic Engineering," op. cit. note 39.

41 Lewin, Roger, (1991), op. cit. note 21, 206.

42 Wright (1986), op. cit. note 23, 593.

43 Ibid., 601.

44 Ibid., 600.

45 Ibid., 600-01.
46 Ibid., 601.
47 Ibid.
48 Ibid.
49 Ibid., 602.
50 Thomas, Gavin; http://www.microbiologyonline.org.uk/ecoli.htm.
51 Wright (1986), op. cit. note 23, 603.
52 Ibid., 602, no. 20.
53 Ibid., 604.
54 Ibid.
55 Ibid., 604-5
56 Ibid., 605.
57 Ibid., *emphasis in original.*
58 Ibid., 606. The full quote that appears on p. 45 of the transcript of the meeting is: "I think that is what you have to deal with. It may not mean a thing, but that is very easy to do. Its molecular politics, not molecular biology and I think we have to consider both, because a lot of science is at stake." In an email to me (in answer to my questions) Wright explained that from the prior discussion, it is clear the word "that" refers to the problem of convincing the public. Accordingly, in her article, Wright renders the first sentence as: "I think (the problem of convincing the public) is what you have to deal with."
59 Ibid.
60 Ibid., 600. In an email to me, Wright confirmed that the media were not invited to any of the conferences or even informed of them – and so were not present.
61 Ibid., 607.
62 Quoted in Wright (1986), op. cit. note 23, 607.
63 Ibid., 608.
64 Ibid.
65 Quoted in Dutton, Barbara, *Worse than the Disease: Pitfalls of Medical Progress* (New York: Cambridge University Press, 1992), 193.
66 Ibid.
67 Ibid.
68 Wright (1986), op. cit. note 23, 613.
69 Ibid.
70 Ibid.
71 Wright (1994), op. cit. note 11, 275. The Academy's misrepresentation appeared in a report issued by its Assembly of Life Sciences.
72 Ibid., 269.
73 Dutton, op. cit. note 65, 193.
74 Ibid.
75 Ibid., 193-94.
76 Wright (1994), op. cit. note 11, 269.

77 Ibid., 270.

78 Ibid., 271.

79 Edward M. Kennedy, speech to the Association of Medical Writers, September 27, 1977, New York, quoted in Wright (1994), 272.

80 Wright (1994), op. cit. note 11, 272.

81 Ibid., 245.

82 Chang, Shing and Cohen, Stanley N., "*In Vivo* Site-Specific Genetic Recombination Promoted by Eco RI Restriction Endonuclease," *Proceedings of the National Academy of Sciences* 74 (November 1977): 4811-15. The fragments of mouse DNA were not integrated within the central area of the bacterial DNA (its *chromosome*) but within a small ring of DNA outside of it (called a *plasmid*). Chromosomes and plasmids will be discussed in Chapter 4, which will also more thoroughly examine Cohen's experiment and the deceptive claims that were made about it.

83 Stanley Cohen quoted in Wright (1994), op. cit. note 11, 272.

84 Stanley Cohen to Donald Fredrickson, September 6, 1977, ORDAR, quoted in Wright (1994), op. cit. note 11, 246.

85 Dutton, op. cit. note 65, 194.

86 Wright (1994), op. cit. note 11, 272. Wright says Kennedy used Cohen's claim as an "escape hatch."

87 Roy Curtiss to Donald Fredrickson, April 12, 1977, *ORDAR* 8, quoted in Wright (1994), op. cit. note 11, 244. I learned of the conflict between the letters through Wright's observations.

88 Wright (1994), op. cit. note 11, 246.

89 Ibid., 291.

90 In light of the extensive information I've read, it seems reasonable to assume that most legislators, including Kennedy, were never adequately informed about the illegitimacy of Cohen's claim. However, I have seen no explicit evidence to that effect. Further, although several legislators were sent copies of the Curtiss letter in April 1977, it seems that when Cohen's claim was issued six months later, they did not realize that it was undercut by the earlier document. For one thing, Cohen's letter did not mention the unusual conditions under which the research was performed; and the fact that he'd employed them was not well-publicized.

91 Wright (1986), op. cit. note 23, 609.

92 Ibid., 610.

93 Transcript quoted in Wright (1986), op. cit. note 23, 611.

94 Ibid., 612.

95 Ibid.

96 Ibid.

97 Ibid.

98 Ibid., 615 and Wright, op. cit. note 11, (1994), 513, n. 56. Wright obtained the quote in an interview she conducted.

99 Wright (1994), 256.

100 Wright (1986), op. cit. note 23, 614.

101 Ibid.

102 Ibid., 596, 615

103 Wright (1994), op. cit. note 11, 351.

104 Newmark, Peter, "WHO Looks for Benefits from Genetic Engineering," *Nature* 272 (20 April 1978): 663-64, quoted in Wright (1986), op. cit. note 23, 614.

105 Wright (1994), op. cit. note 11, 366.

106 Wright, Susan, email communication.

107 Ibid.

108 Wright (1994), op. cit. note 11, 64. The molecular biologist who proposed a dangerous experiment was Sydney Brenner, of the University of Cambridge.

109 Wright (1994), op. cit. note 11, 250.

110 Ibid., 248-50.

111 Ibid., 249, 463.

112 Rowe quoted in Wright (1994), op. cit. note 11, 372.

113 The study was published as: Israel, M.A., Chan, H.W., Hourihan, S.L., Rowe, W.P. and Martin, M.A., "Biological activity of polyoma viral DNA in mice and hamsters," *J. Virol* 29 (1979): 990–96. The specific type of polyoma virus that was employed is referred to as PY. It's in the same viral group as the tumor-producing SV40 that Paul Berg had, several years previously, intended to insert within an E. coli-infecting virus – which roused the concern of Robert Pollack and ultimately spurred the development of the precautionary measures that the Rowe-Martin experiment was intended to relax.

114 Israel, Mark A. et al., "Interrupting the Early Region of Polyoma Virus DNA Enhances Tumorigenicity," *Proceedings of the National Academy of Sciences* 76 (August 1979): 3714.

115 Wright (1994), op. cit. note 11, 373. For a discussion of the various results, see 368-74.

116 Ibid., 368.

117 Ibid., 366-67.

118 Ibid., 375-66. At an RAC meeting in May 1979, Jonathan King of MIT noted that an experiment confined to *E. coli* (such as Rowe-Martin) could not confirm the safety of rDNA research with other organisms.

119 On November 10, 2010, I accessed the false claim at: http://www.niaid.nih.gov/labsandresources/labs/aboutlabs/lmm/viralpathogenesisvaccinesection/Pages/martin.aspx.

In January 2014, I discovered that this URL is no longer functional and that Martin's current biographical information omits the misrepresentation that was present in 2010 – and had presumably been posted for many years. In fact, the current page does not specifically mention the Rowe-Martin experiment at all. The new URL is: http://www.niaid.nih.gov/LabsAndResources/labs/aboutlabs/lmm/viralPathogenesisVaccineSection/Pages/martin.aspx#niaid_inlineNav_Anchor.

It's quite plausible that the falsehood was removed as a result of being exposed by information I circulated describing it. A supporter of the bioengineering venture may well have read it and alerted Dr. Martin about the need for revision.

120 Wright (1986), op. cit. note 23, 616.

2. Expansion of the Biotech Agenda

1 Because commercialization of a GE food was still more than a decade away, attention at that time was primarily focused on whether a gene-altered crop could damage the environment during field testing, not on whether it might eventually bring new risks to the dinner table. Food safety did not become a salient issue until much later – and it will be discussed in subsequent chapters.

2 Interview with Arnold Foudin, Ph.D., Deputy Director, Biotechnology Permits, PPQ, APHIS, USDA, Washington, DC (October 6, 1997), cited in Jones, Mary Ellen, "Politically Corrected Science: The Early Negotiation of U.S. Agricultural Biotechnology Policy," a Doctoral Dissertation in Science and Technology Studies at Virginia Polytechnic Institute (1999), 63.

3 Ibid., 88.

4 Interview with David MacKenzie cited in Jones, (1999), op. cit. note 2, p. 89, n. 231. The word "disbelief" is the term that Jones uses in describing his reaction as related to her.

5 Ibid., 101.

6 Ibid.

7 Ibid., 105-06.

8 Ibid., 108.

9 Ibid.

10 Berg, Paul et al., "Potential Biohazards of Recombinant DNA Molecules," *Science* 185 (July 26, 1974): 303.

11 At the April 22, 1981 meeting of an RAC working group, concern was raised that risk assessment data was still limited to *E. Coli* K-12; and I have seen no indication that by December of that year the situation had changed. *See* Minutes of Large-Scale Review Working Group of the RAC, April 22, 1981, in US Department of Health and Human Services, (1982); Documents Relating to "NIH Guidelines for Research Involving Recombinant DNA Molecules," November 1980-August 1982, Office of Recombinant DNA Activities, NIH Publication No. 83-2604, 78.

12 Jones (1999), op. cit. note 2, 109. Jones reports that Jonathan King (of MIT) and Ruth Hubbard and George Wald (both of Harvard) "reproached" Baltimore ". . . for conflict of interest, accusing him of promoting the deregulation of an industry in which he had a considerable economic interest." In referencing their allegations (in footnote 287), she cites documents they filed with the NIH in 1982 as: Documents relating to "NIH Guidelines for Research Involving Recombinant DNA Molecules" November 1980-August 1982, Office of Recombinant DNA Activities, NIH Publication No. 83-2604; Hubbard (p.717), Jonathan King (p.

719), and George Wald (p.701). These documents are no longer available on the NIH website.

To my knowledge, there were no allegations that Paul Berg had any conflict of interest; and the evidence indicates that he endeavored to avoid such conflicts – and to maintain proper boundaries between academia and industry.

13 Jones, op. cit. note 2, 113.

14 Ibid., 113-14. Due to these lacks, the director of NIH deferred action pending receipt of more information about containment – information that was not provided until the following year. Eventually, the proposal was given final approval on August 7, 1981.

15 Ibid., 146-47.

16 Ibid.

17 Ibid., 156.

18 As noted in Chapter 1, because the various statements from Regal in this book are largely drawn from his written recollections and from my extensive conversations and email correspondence with him, except for quotes excerpted from his published articles, I will not provide specific references for his various statements.

19 The generic arguments did exclude at least one class of GMO: those derived from organisms known to be pathogenic.

20 Regal, Philip, "Models of Genetically Engineered Organisms and Their Ecological Impact," in *Ecology of Biological Invasions of North America and Hawaii* , Mooney, H. and J. Drake., eds. (New York: Springer-Verlag, 1986), 117.

21 Cited in Ibid., 118.

22 Ibid. (emphasis in original).

23 Because the wrinkle-inducing allele is recessive, it doesn't get expressed in as many peas as does the allele that confers smoothness, which is dominant. So the majority of peas are smooth.

24 Regal further notes the illegitimacy of the claim that gene-splicing is akin to crossing two distantly related (yet inter-breedable) species – a process that generally reduces the fitness of the resultant organisms. He points out that this infirmity results because the offspring of sexual reproduction receive one set of chromosomes from each parent, and that when the parents are distantly related, these sets are not *co-adapted.* He contrasts this with genetic engineering, where a few foreign genes are added to a *stable* genome in which the sets of chromosomes *are* co-adapted.

Moreover, Regal points out another significant respect in which genetic engineering radically differs from traditional breeding: it can alter parts of the genome that the latter cannot touch. In his words:

"Traditional breeding is limited to rearranging only a fraction of the DNA molecule – the fraction that varies within a population. Yet much of the DNA molecule does not vary among individuals in a population. For example, dogs have only two eyes. Even though eyes are under genetic control, they do not vary in number and thus we cannot select for dogs with three or more eyes.

This part of their DNA code is 'locked up' in various ways beyond the reach of natural or artificial selection. Mutagenesis can induce random changes in parts of the code that are locked up, but mutagenesis cannot systematically rewrite this part of the code in a biologically coherent way. With rDNA techniques one can, in principle, go into the DNA code books that are 'locked up' and rewrite them."

25 Raven's current views on GMOs are discussed in endnote 8 of Chapter 14.

26 A member of Reagan's White House staff recounts: "We wanted it to be an American technology," Jones (1999), op. cit. note 2, 231.

27 For instance, this was the view of Warren Weaver and Max Mason, two of molecular biology's prime promoters. See: Schwartz, J. "The Soul of Soulless Conditions? Accounting for genetic fundamentalism," *Radical Philosophy* 86 (November/December 1997): 4.

28 The NAS website devotes many pages to its building; and over several years, the main page has displayed the prominent heading: "The NAS Building . . . a Temple of Science." (The heading was present when I first visited that web page in 2007, and as of January 2014, it's still in use. Further, it's reasonable to presume that it was there long before I first accessed the page.) http://www.nasonline.org/about-nas/visiting-nas/nas-building/

29 Einstein was elected a foreign associate of the Academy in 1922 and became a member in 1942, two years after he became a naturalized citizen.

30 Jones (1999), op. cit. note 2, 231.

31 Ibid., 227-28.

32 Ibid., 164.

33 Ibid., 69, 217, 239.

34 Ibid., 217.

35 Ibid., ii.

36 Ibid., 330.

37 National Academy of Sciences, "Introduction of Recombinant DNA-Engineered Organisms into the Environment: Key Issues" (1987): 6.

38 Ibid., 22.

39 Ibid., 20.

40 Marchant, Gary, "Modified Rules for Modified Bugs," *Harvard Journal of Law and Technology* 1 (Spring and Summer, 1988): 165.

41 Krimsky, S. and R. Wrubel, *Agricultural Biotechnology and the Environment: Science, Policy, and Social Issues* (Champaign, IL: University of Illinois Press, 1996), 219.

42 Dumanoski, "Academy Report Challenged," *Boston Globe*, August 24, 1987, 44 (Cited in Marchant, G., op. cit. note 40, 166.)

43 Krimsky and Wrubel (1996), op. cit. note 41, 219.

44 Regal relates that after the task force carefully drafted a set of regulations, the state's department of agriculture (which was boosting biotech) managed to

gain control and fashioned so many loopholes that the regulations were rendered virtually meaningless.

45 *Reporters' Guide: Genetic Engineering in Agriculture* (Washington, D.C.: Environmental Media Service, 2000), 29.

46 Ecological Society of America, "The Planned Release of Genetically Engineered Organisms: Ecological Considerations and Recommendations," *Ecology* 70, no. 2 (April 1989).

47 http://www.ourdocuments.gov/doc.php?flash=true&doc=90&page=transcript

48 Ibid.

3. Disappearing a Disaster

1 Janet O'Brien, National EMS Network Newsletter, Fall 1997.

2 Story on Harry Schulte, WCPO-TV, 11 PM News, Cincinnati, Ohio, February 26, 1998.

3 Personal Communication from Betty Hoffing to Jeffrey Smith as reported in Smith, J., *Seeds of Deception: Exposing Industry and Government Lies About the Safety of the Genetically Engineered Foods You're Eating* (Fairfield, IA: Yes! Books, 2003), 107.

4 Personal communication from Gerald Gleich, M.D.

5 The website of a medical doctor who is an authority on tryptophan states that the 2% figure came from the *New England Journal of Medicine*: http://craighudsonmd.com/tryptophan.html (accessed 1-2-12).

6 Hertzman, P. et al., "The Eosinophilia-Myalgia Syndrome: The Los Alamos Conference," *Journal of Rheumatology* 18, no. 6 (1991): 867-73.

7 The CDC's final estimate (which was made almost two decades ago) put the number of deaths at between 80 and 100. The figure of 1,500 permanent disabilities is from Wikipedia.

8 Although the causal link between Showa Denko's LT and the epidemic has been questioned by some researchers (e.g., Shapiro, S., "Epidemiologic studies of the association of L-tryptophan with the eosinophilia-myalgia syndrome: a critique," *J Rheumatol Suppl* [Oct. 1996]: 44-59), the link has been firmly established by published research conducted not only at the Center for Disease Control (CDC), but at the Mayo clinic and other respected institutions. The strength of the connection is attested by CDC scientists who have written: "These studies constitute overwhelming evidence that the cause of the EMS epidemic was ingestion of L-tryptophan produced by Showa Denko," (emphasis added), *The Lancet* 343 (April 23, 1994): 1037.

9 Philen, R.M. et al., "3-(Phenylamino)alanine—a Link Between Eosinophilia-Myalgia Syndrome and Toxic Oil Syndrome?," *Mayo Clinic Proceedings* 68 (1993): 197-200.

10 Ibid.

11 Slutsker, L. et al., *Journal of the American Medical Association* 264, no. 2 (July 11, 1990): 213-17.

12 Garrett L., "Genetic engineering flaw blamed for toxic deaths," *Newsday*, August 14, 1990: C-1.

13 Roberts, Leslie, "L-tryptophan puzzle takes new twist," *Science* 249 (August 31, 1990): 988.

14 Ibid.

15 Ibid.

16 Belongia, E.A. et al., "An Investigation of the Cause of the Eosinophilia-Myalgia Syndrome Associated with Tryptophan Use," *New England Journal of Medicine* 323, no. 6 (August 9, 1990): 357-65.

17 Scientists also sometimes referred to it as "Peak E." Subsequent to the initial publication discussing its structure, a more precise determination was made by a group at the Mayo Clinic, Mayeno, A.N., et al., Characterization of "peak E," a novel amino acid associated with eosinophilia-myalgia syndrome, *Science* 21 (December 1990) 250, no. 4988: 1707-08.

18 Belongia et al., (1990), op. cit. note 16.

19 Charles, Dan, *Lords of the Harvest: Biotech, Big Money, and the Future of Food* (Cambridge: Perseus, 2002), 224.

20 Raphals, P., "Does medical mystery threaten biotech?," *Science* 249, no. 619 (1990).

21 Unpublished study by the FDA and Showa Denko K.K., cited in Toyoda, M. et al., "Formation of a 3-(Phenylamino)alanine Contaminant in EMS-associated L-Tryptophan," *Bioscience, Biotechnology, Biochemistry* 58 (1994): 1318.

22 Ibid., 1318-1320.

23 Belongia et al., (1990), op. cit. note 16.

24 Philen RM, Hill, RH Jr, Flanders WD, et al., "Tryptophan contaminants associated with eosinophilia-myalgia syndrome. The Eosinophilia-Myalgia Studies of Oregon, New York and New Mexico," *Am J Epidemiol* 138 (1993): 154-59; Hill, R.H. et al., "Contaminants in L-tryptophan Associated with Eosinophilia-Myalgia Syndrome," Archives of Environmental Contamination and Toxicology 25 (1993): 134-42.

25 Love, L.A., Rader, J.I., et al., "Pathological and Immunological Effects of Ingesting L-Tryptophan and 1, 1'-Ethylidenebis (L-Tryptophan) in Lewis Rats," *Journal of Clinical Investigation* 91 (March 1993): 804-11.

26 Hill, R.H. et al., "Contaminants in L-tryptophan Associated with Eosinophilia-Myalgia Syndrome," Archives of Environmental Contamination and Toxicology 25 (1993): 134-42.

27 Ibid.

28 Ibid.

29 Love et al. (1993), op. cit. note 25, stated that the research indicates that several factors were probably involved. Further, EBT may have sometimes played a subordinate role. Although it was not significantly related to a lot's harmful status, there was a positive association, which suggests it may have sometimes acted in combination with other factors to facilitate the occurrence of EMS (even though the disease was often induced in its absence).

30 Yanofsky, C., email to William Crist, June 2, 1998.

31 Yanofsky, C., quoted in Raphals, P., "EMS deaths: Is recombinant DNA technology involved?," *The Medical Post*, November 6, 1990.

32 Yanofsky, C., email to William Crist, June 2, 1998.

33 Personal communication from Gerald Gleich, M.D. to Jeffrey Smith, as reported in Smith, J., *Seeds of Deception: Exposing Industry and Government Lies About the Safety of the Genetically Engineered Foods You're Eating* (Fairfield, IA: Yes!Books, 2003), 275.

34 Email from William Crist.

35 Edwin M. Kilbourne et al., "Tryptophan Produced by Showa Denko and Epidemic Eosinophilia-Myalgia Syndrome," *Journal of Rheumatology Supplement* 23, no. 46 (October 1996): 81-92.

36 National Eosinophilia-Myalgia Syndrome Network, position statement, approved quote by Gerald J. Gleich, M.D., Mayo Clinic and Foundation, May 25, 2000.

37 Mayeno, A.N. and Gleich,G.J., "Eosinophilia-myalgia syndrome and tryptophan production: a cautionary tale," *Trends in Biotechnology* (TIBTECH) (September 1994), Vol. 12, pp. 346-352.

38 Email from Dennis Mackin of January 31, 2012. Mr. Mackin and other plaintiffs' attorneys eventually received this information, which included the dates on which the various strains had been used. Years later, Bill Crist obtained a chart listing the strains and dates from Mackin, and I received a copy from Crist. Although this latter document was not the copy of the fax the FDA had sent in September 1990, it contained the critical information.

39 Hill, Robert H. et al., "Contaminants in L-tryptophan Associated with Eosinophilia-Myalgia Syndrome," *Archives of Environmental Contamination and Toxicology* 25 (1993): 134-42. (This study indicates that Strain IV was more toxic than III.)

40 Email from William Crist.

41 "Bitter Pill," Dateline NBC, NBC News, August 22, 1995; NHK Special in Japan, "Product Liability Litigation in America," August 5, 1995.

42 Torigoe memo of Aug 23, 1988, Exhibit 90 in specific lawsuits against Showa Denko, 3.

43 Email from Adrian Gibbs, Emeritus Professor of Virology, Australian National University,

44 Although there's currently insufficient basis for determining whether the viral problem arose within Strain III or IV, the erratic employment of IV suggests that it was the affected strain.

45 SD records obtained by attorneys during the lawsuits indicate such a shutdown; and Paul Rheingold (an attorney who represented hundreds of victims) confirmed in a phone conversation in January 2012 that a significant shutdown did occur.

46 Within SD records that follow Torigoe memo, op. cit., in what appears to be part of Exhibit 90 at official reference number 333755 0885.

47 Email from William Crist that included a copy of the Maryanski letter.

48 Emails from William Crist regarding information received from Don Morgan (of Cleary, Gottlieb, Hamilton & Steen) via phone and emails of March 5 and April 19, 2001. According to Morgan's March 5 email: "SDK invited FDA to send someone to Japan to receive the bacteria and learn how to care for them and run jar fermentations correctly. FDA never followed up on this offer, or expressed any criticism of SDK's reluctance to send live bacteria by mail."

49 Crist conveyed this information to me via email.

50 Ibid.

51 Bains, William, *Biotechnology from A to Z* (Oxford: Oxford University Press, 1993), 10. This false statement is repeated in the second edition, published in 1998.

52 Report of the Royal Commission on Genetic Modification (New Zealand, 2001), 43.

53 Ibid.

54 Ibid.

55 The Royal Commission's report gave no specific reference for any assertion it made in the tryptophan section. Rather, it listed all fifteen sources for the entire section in one large reference note. These sources include submissions to the commission by interested persons as well as several journal articles and even a special report prepared for the commission. But none of the assertions in the tryptophan section was specifically linked with any of these sources, nor were any page numbers provided to indicate which parts of these sources contained the relevant information. Consequently, it is extremely difficult to discover from which source each assertion was supposedly derived and to ascertain its reliability. Moreover, as of the time of this writing, neither the submissions by interested persons nor the special report on which the Commission relied are provided on the government site at which the report can be downloaded – and it's not clear for how long, if ever, they were readily available. http://www.mfe.govt.nz/publications/hazards/report-royal-commission-genetic-modification.

In effect, this strange way of reporting references has essentially barred people from checking the sources for the Commission's assertions about the tryptophan incident.

56 Douglas L. Archer, Deputy Director Center for Food Safety and Applied Nutrition (CFSAN), FDA, Testimony before the Subcommittee on Human Resources and Intergovernmental Relations, US House of Representatives, July 18, 1991. As for the agency's knowledge that the bacteria had been engineered, Jeffrey Smith reports that a former FDA employee was astounded when he informed her that Archer had failed to mention that fact, declaring that by then "everyone in the agency" knew about the bioengineering, Smith, *Seeds of Deception* (cited in note 33), 121.

57 Cited in Beisler, Joshua, H., "Dietary Supplements and Their Discontents: FDA Regulation and the Dietary Supplement Health and Education Act of 1994,"

Rutgers Law Journal (Winter 2000): 531. Although this particular report was issued in 1993, a few years after Archer's testimony, it indicates that the agency's desire to regulate supplements had continued unabated for decades, and had not been satiated by the ban imposed on LT.

58 Ibid. In the law journal, the entire piece is termed a *note*, but that does not indicate it is ·either short or otherwise minor. Law journals typically reserve the term *article* for pieces authored by professors, judges, and practicing attorneys and refer to those written by members of their staff, who are law students, as *notes*. But notes receive the same editorial scrutiny as articles, they are often of similar length, and they're frequently cited in judicial opinions and articles in other journals. Because Mr. Beisler's piece runs forty pages and is carefully reasoned and thoroughly researched, I decided to refer to it as an article rather than a note to avoid conveying the false impression that it's a short statement of opinion that's also short on supportive evidence.

59 Manders, Dean, "The FDA Ban of L-Tryptophan: Politics, Profits and Prozac," *Social Policy*, vol. 26, no. 2, Winter 1995: http://www.ceri.com/trypto.htm (accessed February16, 2012).

60 FDA Public Affairs Office, Press Release of May 18, 1994. The enzyme was introduced in 1990, and Showa Denko first marketed GE-derived LT in 1984.

61 Stephen Naylor, personal communication.

62 "Genetically Modified Foods," Australia New Zealand Food Authority, November 2001. (downloaded from the agency's website on March 3, 2002.) The agency's name was later changed to Food Standards Australia/New Zealand.

63 Fedoroff, N. and Brown, N.M., *Mendel in the Kitchen: A Scientist Looks at Genetically Modified Foods* (Washington, DC: Joseph Henry Press, 2004); Ronald, P. and Adamchak, R., *Tomorrow's Table: Organic Farming, Genetics, and the Future of Food* (New York: Oxford University Press, 2008).

64 Charles, *Lords of the Harvest*, op. cit. note 19, 224.

65 Aldridge, Susan, *The Thread of Life: The Story of Genes and Genetic Engineering* (Cambridge: Cambridge University Press, 1996), 185-86.

66 http://www.bastyrcenter.org/content/view/1828/ (accessed 1-24-12)

67 http://www.bastyrcenter.org/content/view/590/#top (accessed 1-24-12)

68 Lambrecht, B., *Dinner at the New Gene Café: How Genetic Engineering Is Changing What We Eat, How We Live, and the Global Politics of Food* (Thomas Dunne Books: St. Martin's, 2001); Hart, K., *Eating In the Dark: America's Experiment With Genetically Engineered Food* (Pantheon: Random House, 2002).

69 Mann, L.R.B., Straton, D., and Crist, W. E., "The Thalidomide of Genetic 'Engineering.'" http://www.gmfoodnews.com/trypto.html

70 The FDA rescinded its ban on LT in 2006. Of course, if the fatal batches of LT had not been marketed in 1989, there would not have been a ban in the first place, and no barriers for a new engineered line of LT to face. Regarding Europe, a report by the BBC indicated that the toxic LT would gain entry there too. (*Seeds of Deception* [op. cit. note 33 above], p. 275, n. 20.)

71 Schubert, David, *Journal of Medicinal Food* 11(4) (December 2008): 601-05

72 Hill, R.H., Caudill, S.P., et al, "Contaminants in L-Tryptophan Associated with Eosinophilia Myalgia Syndrome," *Archives of Environmental Contamination and Toxicology* 25 (1993): 134-42.

73 In technical terms, AAA is "an unsaturated fatty acid conjugate of tryptophan," email from Stephen Naylor.

74 Email from Stephen Naylor.

75 So there is no question as to the accuracy of my account of Dr. Naylor's research, he executed a notarized affidavit confirming the facts as conveyed in this chapter. It's reproduced at www.alteredgenestwistedtruth.com/stephen-naylor-affidavit/

4. Genes, Ingenuity, and Disingenuousness

1 Miller, Henry, I., "Happy Earth Day, Mr. Rifkin," *Washington Times*, April 22, 1997.

2 Wald, George, "The Case Against Genetic Engineering," in *The Recombinant DNA Debate*, Jackson, D. and S. Stich, eds. (Prentice-Hall, 1979), 127-28.

3 Wald, George, speaking at a press conference in Washington, D.C. March 1977; Quoted in Kimbrell, A., *The Human Body Shop: The Engineering and Marketing of Life* (New York: Harper Collins, 1994), 159. Although Wald's statement that genetic engineering is the "biggest break in nature" occurred at a press conference, I think it's appropriate to include it along with statements he wrote in an earlier article because doing so does not in any way misrepresent his thinking – and enables it to be expressed in a compact manner.

4 Bains, William, *Biotechnology from A to Z* (Oxford: Oxford University Press, 1993). The introduction was written by G. Kirk Raab, President and CEO of Genentech, Inc., vi-viii.

5 Ibid., 224.

6 For example, in 2004 a sociologist who has conducted many surveys of consumer attitudes in the US observed that although two-thirds of the respondents voice support for GE foods, they know very little about the issue and that the minority who oppose them tend to be most educated about the facts. "Change of Heart By Thomas Hoban," September 23, 2004: http://www.lobbywatch.org/archive2.asp?arcid=4387 (accessed Aug. 15, 2012)

7 Communications Programmes for EuropaBio, January 1997; Prepared by Burston Marsteller, Government and Public Affairs: http://home.intekom.com/tm_info/geleak1.htm (accessed Aug. 15, 2012).

8 Miller, Henry, I. "Happy Earth Day, Mr. Rifkin," *Washington Times*, April 22, 1997. In subsequent years, Dr. Miller has continued to sound the "seamless continuum" theme. See e.g., Miller, H.I., Point of View, Dec 1, 2007 (27, no. 21): http://www.genengnews.com/keywordsandtools/print/1/12239/ (accessed Aug. 15, 2012).

9 May, Sir Robert, BBC Interview, March 9, 2000.

10 Viruses are an interesting case. They are not full cells, and most biologists do not consider them to be living organisms in their own right. They cannot

reproduce on their own and can only do so when they have invaded a living cell of another species, commandeered its genetic apparatus, and re-directed it to generate their components.

11 While overall, there is still a net loss of the capacity of energy to perform work, the dissipation is far more gradual within living systems, more of the energy contributes to the formation of organized structure, and the degree of organization is much higher than in the processes that typify nonliving nature.

12 Grace, Eric., *Biotechnology Unzipped* (Washington, DC: Joseph Henry Press, 1977), 22.

13 While the mature red blood cells of mammals have shed their nucleus (and the DNA within it), they did possess a DNA- packed nucleus in their earlier stages and so functioned as information processing machines as they developed.

14 Most bacteria possess one main molecule of DNA (referred to as a chromosome) and several smaller, auxiliary molecules (called plasmids). This will subsequently be discussed in more detail.

15 Gitt, W., *In the Beginning Was Information* (Green Forest, AR: Master Books, 2007), 98.

16 In plants and animals, most genes can generate multiple proteins because the transcriptional machinery can arrange the information they contain in various ways.

17 Sex chromosomes are an exception. The sex chromosome of the human male (the Y chromosome) does not contain all the genes in the female chromosome (the X chromosome).

18 Although some plants fertilize themselves, they usually also send some of their pollen outward to fertilize the female gametes of other plants while some of their female gametes receive pollen from other plants. Some plants also have four or more sets of chromosomes instead of two. But their gametes will still contain only one half of the number possessed by their parental cells, so if the plant's cells contain four different sets of chromosomes, its gametes will have only two.

19 http://link.springer.com/chapter/10.1007/978-94-011-5794-0_3

20 Thomas, C. and Nielsen, K., "Mechanisms of, and Barriers to, Horizontal Gene Transfer between Bacteria," *Nature Reviews Microbiology* 3 (September 2005): 712.

21 Griffiths et al., *An Introduction to Genetic Analysis*, Sixth Edition (New York: W.H. Freeman, 1996), 424.

22 Ibid.

23 Not all restriction enzymes make staggered cuts, but biotechnicians are able to draw on a wide variety of those that do. Further, they've learned how to add sticky ends to fragments that are cut with enzymes that leave sheer edges. But this ability could not have developed without the application of the enzymes that do make staggered cuts.

24 Although evidence indicates that restriction enzymes probably have additional roles, and that they at times may have facilitated the integration of

foreign DNA into bacterial genomes, it's clear that their primary function is to inhibit such integration. For instance, Werner Arber, who received a Nobel Prize for discovering these enzymes, has postulated that there are two basic types of what he terms "evolution gene products." These products act either to generate genetic variation or to modulate it "to low levels that are tolerable for the long-term maintenance of a given strain or species." He classes restriction enzymes as modulators because they "seriously reduce both the chance of DNA acquisition and the size of a DNA segment that may eventually be acquired by the recipient cell." And he does so even though he thinks that they stimulate "occasional DNA acquisition to occur in small steps." (Arber, W., "Molecular Evolution: Comparisons of Natural and Engineered Systems, The Challenges of Sciences, A Tribute to Carlos Chagas," Pontifical Academy of Sciences, *Scripta Varia* 103 [Vatican City 2002]): www.pas.va/content/dam/accademia/pdf/sv103/sv103-arber.pdf; Arber, W., "Genetic Variation and Molecular Evolution," *Encyclopedia of Molecular Cell Biology and Molecular Medicine* (2006).

In contrast, bioengineers employ restriction enzymes not only to radically increase the rate at which genomes are altered by alien DNA, but to vastly expand the range of species that can interact – to the point where none of the natural species barriers remains intact.

25 One experiment did find that when *E. coli* are within river or spring water that naturally contains a sufficient level of calcium ion, they can take up plasmid DNA without additional calcium or abnormal shifts in temperature (Baur, B. et al., "Genetic Transformation in Fresh Water: *Escherichia coli* is Able to Develop Natural Competence," *Applied and Environmental Microbiology* 62 (Oct. 1996): 3673-78). And increasing evidence suggests that many species of bacteria that aren't ordinarily able to take up external DNA can do so under some naturally occurring conditions. But such instances appear to be relatively rare. In any case, the fact remains that biotechnicians have routinely resorted to unnatural means to induce *E. coli* to receive external DNA.

26 The only bacteria in which these particular elements have been found belong to the Archaebacteria kingdom, bacteria that exist in extreme environments, such as boiling water. These bacteria are rarely used in commercial applications of bioengineering.

27 For a good discussion of reverse transcriptase, see: http://www.nature.com/scitable/topicpage/the-biotechnology-revolution-pcr-and-the-use-553.

28 *Biotechnology from A to Z*, First Edition, (cited in note 4), 131.

29 Chang, Shing and Cohen, Stanley N., "*In Vivo* Site-Specific Genetic Recombination Promoted by Eco RI Restriction Endonuclease," *Proceedings of the National Academy of Sciences* 74 (November 1977): 4811-15.

30 Although genes are sometimes inserted to block the expression of another gene rather than to express any of their own proteins, this was seldom the aim of commercial applications when Cohen's experiment was performed; and even today, the predominant goal of commercial bioengineering is to produce a functional protein from the inserted gene.

31 Chang, A. C. Y., Lansman, R. A., Clayton, D. A. & Cohen, S. N., "Studies of Mouse Mitochondrial DNA in Escherichia coli," Cell 6 (1975): 231-44, at 241. This study was performed two years prior to the one on which Cohen based his influential claim. It's discussed more fully in the following note.

32 Actually, one cannot know to what extent (if any) mitochondrial proteins were produced in the 1977 experiment conducted by Chang and Cohen because their technical report makes no mention of that topic. But in 1975 Cohen and three other colleagues published a paper that did discuss the protein production that was observed after mouse mitochondrial genes were inserted into *E. coli*. And the degree of production was disappointing. Those are the results I reported in describing the 1977 study. In the absence of data directly bearing on the level of production in that study, I think it's reasonable to assume that the results were no better than in the 1975 experiment, because if the authors had determined that the production was more impressive, they would have made that fact known. However, it appears that in 1977, they didn't even try to make such a determination. Performing one in the earlier experiment had been a complicated, time consuming endeavor, so because there was little likelihood it would reveal better results, they probably decided it wasn't worth the effort. Of course, the results in 1977 might even have been worse; but I decided to give Cohen the benefit of the doubt by reporting the 1975 data as if they applied to the 1977 experiment.

One might wonder why Cohen even performed that experiment if he had already discovered that the proteins encoded by the genes of mouse mitochondria will not be adequately produced within *E. coli*. The reason, it seems, was to provide the basis for arguing that bacteria can integrate plant and animal genes in a natural manner. The 1975 experiment fell short of providing such a platform. In that research, rDNA technology was used to splice the mouse DNA into plasmids that were *outside* the *E. coli*, and only thereafter did the fully engineered plasmids enter the bacteria. But in his subsequent study, Cohen wanted to show that the splicing could occur *within* the bacteria. He aimed to demonstrate that fragments of mouse DNA and fragments of plasmid DNA could be taken up by the bacteria and then joined together *inside* the bacterium. However, as we've seen, he failed to demonstrate that this process can occur through purely natural means because he induced the uptake through means that were highly unnatural. (These were the same means he'd employed in 1975 in order to get the fully engineered plasmids into the bacteria.) Moreover, he compounded the artificiality by cutting both the plasmid and mouse mitochondrial DNA with the same restriction enzyme to make it easier for them to fuse together after they entered the bacteria.

33 Quoted in Wright, Susan, *Molecular Politics: Developing American and British Policy for Genetic Engineering 1972-1982* (Chicago: University of Chicago Press, 1994), 82.

34 Itakura, K. et al., "Expression in Escherichia coli of a chemically synthesized gene for the hormone somatostatin," *Science* 198(4321) (Dec 9, 1977): 1056-63. When human insulin was produced from bacteria the following year, the same

set of techniques was employed (Crea, R. et al., "Chemical Synthesis of Genes for Human Insulin," *PNAS* 75, no. 12, 5765-69).

35 Fedoroff, N. and Brown, N.M., *Mendel in the Kitchen: A Scientist Looks at Genetically Modified Foods* (Washington, DC: Joseph Henry Press, 2004), 4-5.

36 One other species of soil bacteria, *Agrobacterium rhizogenes*, is known to have capacity for inducing plants to express some of its genes. The result is a malady termed "hairy root disease."

37 Ibid., 129.

38 In recent years, biotechnicians have found other promoters that are suited for some of their purposes; and several GE plants have been developed (or are in the developmental pipeline) that do not rely exclusively on the 35S, while some don't contain it at all. Further, some of these promoters derive from plants as well as viruses. Yet, as will be discussed in subsequent chapters, in almost every case, the inserted genes are still impelled to express in unnatural ways; and the regulatory integrity of the engineered organisms continues to be compromised.

39 Charles, Dan, *Lords of the Harvest: Biotech, Big Money, and the Future of Food* (Cambridge: Perseus, 2002), 75.

40 Ibid., 78.

41 Dr. Sherri Brown, quoted in *Milling and Baking News*, August 4, 1997, as cited in Kneen, B., *Farmageddon: Food and the Culture of Biotechnology* (British Columbia: New Society Publishers, 1999), 26.

42 Latham, J. R. et al., "The Mutational Consequences of Plant Transformation," *Journal of Biomedicine and Biotechnology* Article ID: 25376, (2006), 1-7.

43 Ibid., 3.

44 Quoted in Reese, A., *Genetically Modified Food: A Short Guide for the Confused* (London: Pluto Press, 2006), 46-47.

45 Srivastava, M., Eidelman, O. & Pollard, H.B., "Pharmacogenomics of the Cystic Fibrosis Transmembrane Conductance Regulator (CFTR) and the Cystic Fibrosis Drug CPX Using Genome Microarray Analysis," *Mol. Med.* 5 (1999): 753-67. The researchers assessed a sample of 588 genes and found significant changes in 5% of them.

46 Schubert, D., "A different perspective on GM food," *Nat Biotechnol* 20 (2002): 969.

47 Personal communication from David Schubert.

48 See discussion in: Clark, E. Ann, "Parliamentarians and Technology: Meeting the Challenges for the New Millennium, Workshop on Ensuring Food Safety," uoguelph.ca, May 9, 2000: http://www.uoguelph.ca/plant/research/homepages/eclark/Hc.htm. See also Document #4 at: http://biointegrity.org/24-fda-documents

49 Ordinarily, only a relatively small number of an organism's genes are continuously active; and they are the ones whose products are constantly required. It's unnatural for a gene to be expressed during times when there's no need for its product.

50 Elmore et al., "Glyphosate-Resistant Soybean Cultivar Yields Compared with Sister Lines," *Agron J* 93 (2001): 408-12.

51 Wilson, A., J. Latham, and R. Steinbrecher, "Genome Scrambling -Myth or Reality? Transformation-Induced Mutations in Transgenic Crop Plants," *Technical Report* (October 2004): 1, http://www.econexus.info/taxonomy/term/12.

52 *Lords of the Harvest* (cited in note 39), 85.

53 See e.g., Kaeppler et al., "Epigenetic aspects of somaclonal variation in plants," *Plant Molecular Biology* 43 (2000): 179–88; 181.

54 Although tissue culture is used in embryo rescue too, because a seed is involved, there tend to be less perturbations than when genetically engineered cells are cultured. Chapter 9 provides further discussion of the risks of tissue culture in producing GE crops.

55 Finnegan, H. and McElroy, "Transgene inactivation: plants fight back!" *Biotechnology*, 12 (1994): 883-88.

56 Hansen, Michael, "Genetic Engineering is Not an Extension of Conventional Plant Breeding," Consumer Policy Institute/Consumers Union, Jan. 2000, http://www.mindfully.org/GE/GE-Not-Ext-Michael-Hansen.htm.

57 Ibid.

58 Lambrecht, Bill, *Dinner at the New Gene Café: How Genetic Engineering Is Changing What We Eat, How We Live, and the Global Politics of Food,*(Thomas Dunne Books: St. Martin's Press, New York, 2001), ix.

59 Fox, M. W., *Superpigs and Wondercorn* (New York: Lyons and Burford, 1992), 102 and 117; Commoner, Barry, "Unraveling the DNA Myth: The spurious foundation of genetic engineering," *Harper's* (February 2002).

60 *Superpigs and Wondercorn* (cited above), 102.

61 Ibid.

62 Jabed, A. et al., "Targeted microRNA expression in dairy cattle directs production of β-lactoglobulin-free, high-casein milk," *PNAS* 109, no. 42 (2012): 16811-16, http://www.pnas.org/content/109/42/16811.full.

63 Reddy, A.S., Thomas, T.L., "Modification of Plant Lipid Composition: Expression of a Cyanobacterial D6 – desaturase Gene in Transgenic Plants," *Nature Biotechnology* 14 (1996): 639-42.

64 Inose, T. and Murata, K., "Enhanced Accumulation of Toxic Compound in Yeast Cells Having High Glycolytic Activity: A Case Study on the Safety of Genetically Engineered Yeast," *International Journal of Food Science and Technology* 30 (1995): 141-46.

65 "Making Crops Make More Starch," BBSRC Business, UK Biotechnology and Biological Sciences Research Council, January 1998, 6-8.

66 Yamada, Ken, "Genetic Vegomatics Splice and Dice With Weird Results," *Wall Street Journal* 18 (April 1992).

67 *Mendel in the Kitchen,* op. cit. note 35, x; also back cover.

68 Nina Fedoroff quoted at http://sciblogs.co.nz/guestwork/2010/02/15/is-genetic-engineering-just-like-breeding/ (accessed April 11, 2012).

69 Within recent years, evidence has accumulated indicating that, although still rare compared to point mutations, larger mutations that lead to adaptive

change occur more frequently than biologists had recognized. *See* Shapiro, James, *Evolution: A View from the 21st Century* (FT Press, 2011). However, there are significant differences between such natural changes and those induced by the artificial modes of bioengineering.

70 Even when breeders intentionally induce mutations with radiation or chemicals, the selected changes are usually small and could have occurred through natural means. No alien promoters are introduced, nor are several other hazards imposed that are inherent to rDNA technology. Such purposely induced mutations are more thoroughly discussed in Chapter 9 and Appendix D.

71 Dreifus, C., "A Conversation With Nina V. Fedoroff," *New York Times*, August 19, 2008.

72 Although she does say that "a gene from a bacterium can, with the proper switches added, work in a plant cell", she is evidently referring to the promoters and terminators that must be added, not to the codons that require revision. Moreover, she goes on to make a misleading statement about the results of such cross-kingdom transfers. She asserts: "The plant cell will make the *very same* protein from that gene that the bacterium made." (*Mendel in the Kitchen,* op. cit., note 35, 91. [*emphasis added*]) However, as will be discussed in Chapter 6, plants can add sugar chains to these proteins that are never added by bacteria; and these changes could render the protein toxic or allergenic.

73 Although an experiment that combined different species of tobacco via grafting resulted in transfer of chloroplast DNA from cells in one section of the grafted plant to cells in the other, this phenomenon, too, significantly differs from what occurs in genetic engineering. (Stegemann, S. and Bock, R., "Exchange of Genetic Material Between Cells in Plant Tissue Grafts," *Science*, vol. 324, no. 5927 [May 1, 2009]: 649-51.)

Chloroplasts are small units within plant cells that perform *photosynthesis*, the process through which the energy in sunlight is transformed into the plant's food. Like mitochondria (the cell's power houses), chloroplasts possess their own genomes; and their genomes are substantially different from those that reside in the cellular nucleus.

In this grafting experiment, none of the nuclear DNA transferred between the genetically distinct cells. It all stayed at home. Nor did any of the mitochondrial genes relocate. The only genes that moved across the graft junction were those of chloroplasts. And they did not move piecemeal. A large section of chloroplast DNA (perhaps all of it) moved as a unit and functioned as a unit in its new cellular location, performing essentially the same useful function as the cell's native chloroplasts. And these genes operated within chloroplasts, not within the nucleus. This phenomenon contrasts starkly with genetic engineering, where isolated segments of nuclear DNA from one species are wedged into the nuclear DNA of another species – and then act independently of their neighbors in performing unregulated activities that are alien to the host cell.

Further, for transfer of chloroplast DNA to occur via grafting, not only must the biological distance between the species be small (again in contrast to GE), so must the physical distance between the cells that are involved. Only cells in close proximity to the splice junction between the two parts of the plant are involved in the transfer. Thus, the phenomenon is restricted to a tiny fraction of the total cells, whereas in most engineered plants, all the cells contain some DNA that was transferred from another species.

Moreover, a search I performed did not detect any published reports of such transfers occurring in grafts involving other species, so it's possible the phenomenon is restricted to tobacco plants.

74 When I refer to genes in their "natural" state, I'm denoting genes whose internal chemical structure has not been directly reconfigured by human artifice in a way that could never otherwise occur. According to this denotation, the genes restructured via recombinant DNA technology are unnatural whereas most of those mutated via intentional application of ionizing radiation are not, because the same types of radiation-induced mutations could have occurred via natural sources of radiation, without any human intervention. Of course, if the dose of radiation is much greater than would ordinarily be received absent human intervention, then unusual mutations that result would be considered unnatural. Further, even in most of the cases where the individual mutations are not deemed unnatural, the overall pattern of mutations would be, because the entire seed would have been subjected to an extraordinary application of radiation. Even so, as Chapter 9 and Appendix D discuss, the use of radiation in plant breeding significantly differs from genetic engineering – and entails a lower level of risk.

75 *Mendel in the Kitchen*, op. cit. note 35, 127. In her exact words: "It is as natural – or as artificial – as an apple tree." Although the specific reference is to the recombinant DNA molecule, it logically extends to the organism in which that molecule functions.

76 Ibid.

77 Even when branches from several different species are grafted onto the same root stock, the various fruits remain on separate branches in the mature tree, and their respective genes do not intermingle. Further, although it may be possible for chloroplast DNA to move between the sections of a grafted tree (as can occur in grafted tobacco plants, described in note 73 above), it appears that no such cases have been observed. Moreover, even if such transfers do occur in trees, they wouldn't "combine" genes from different species. According to the American Heritage Dictionary of the English Language (4th Edition), to "combine" is "to bring into a state of unity," to "merge," "to join (two or more substances) to make a single substance." But, although foreign genes *are* joined to form a single molecule in bioengineering, when chloroplast genes move to chloroplasts in a neighboring foreign cell, they do not suddenly enter the nucleus and merge with its DNA. And even though, over the lifetime of the tree, it's possible that a few foreign chloroplast genes would migrate to the nucleus that would hardly make grafting a

significant combiner of foreign genes. This is especially so considering that only a miniscule fraction of the tree's cells (those along the splice junction) would even be candidates for such combination. Further, there would be no migration of foreign genes into the leaves or fruit.

There's another important point. When Fedoroff wrote her book, there was no evidence to suggest grafting might enable even the trivial level of foreign gene transfer just discussed. Her book was published in 2004, and the experiment showing the transfer of chloroplast DNA in grafted tobacco plants wasn't published until 2009. Thus, at the time she made it, Fedoroff's claim that grafting combines foreign genes was not even *remotely* supported by the known evidence; and it conflicted with the prevailing scientific consensus – which indicates her willingness to ignore (and even contradict) the best available biological knowledge in order to promote GE foods.

78 *Mendel in the Kitchen,* op. cit. note 35, 123-27.

79 Ibid., 127.

80 For one thing, the bacteria don't ordinarily attack the region where the gametes form. Moreover, if somehow a gamete did become infected and then survived to participate in forming an embryo, if the bacterial tumor-inducing genes were active, the embryo would probably become tumorous – as would any cells derived from it. So it's highly unlikely that a mature organism would ever emerge with a functional gene from *A. tumefaciens.*

Further, although there's another species of soil bacteria, *Agrobacterium rhizogenes,* that also inserts some of its genes into plants (inducing a malady known as "hairy root disease"), and although there's evidence implying that in the distant past it transferred four of these genes into the germ line of a tobacco species, there's no evidence that any of the proteins they code for is being expressed (Aoki, S. and Syono, K., "Horizontal gene transfer and mutation: Ngrol genes in the genome of Nicotiana glauca," PNAS 96, no. 23 [Nov. 9, 1999]: 13229–34).

Recent research confirms that transfer of *Agrobacteria* genes to the germ line of plants is indeed extremely rare. In a study published in December 2012 to determine if any plant species besides tobacco contained such genes, 127 were screened and *none* was found to contain any genes from *A. tumefaciens.* Further, only one of them was observed to contain some DNA from *A. rhizogenes* – and that DNA is apparently not being expressed. (Matveeva, T. et al., "Horizontal Gene Transfer from Genus Agrobacterium to the Plant Linaria in Nature," *Mol Plant Microbe Interact* 25, no. 12 [December 2012]: 1542-51).

81 Although *Agrobacteria* have been used to create more varieties of GE food than the gene gun, because the latter has been used to create commercialized GE corn and soy, it has produced the greatest volume of GE organisms actually consumed.

82 ANZFA Occasional Paper Series No. 1, *GM foods and the consumer* (June 2000): 28. When the guide was released, the agency was called the Australia/New

Zealand Food Authority (ANZFA). In July, 2002, it was renamed Food Standards Australia New Zealand (FSANZ).

83 On February 15, 2001 I met with ANZFA's chief scientist, biotechnology manager, and general standards manager for almost two hours and informed them about the false statement regarding promoters, as well as some other errors the document contained. They invited me to submit further comments to them at their personal email addresses. On July 25, 2001 I sent them formal comments more fully explaining these falsehoods. It is evident that they read my comments, and I also sent them to several other officials within the agency.

Nonetheless, the agency continued to distribute the document in pamphlet form for many months after being informed of the errors; and as of September 12, 2002, more than a year later, it was still offering the document on its website for downloading with the false statements intact. This indicates a severe lack of integrity, since making corrections to the digital version would have been quite simple. (The uncorrected document may well have been available long thereafter, but I stopped checking at that point. I don't know how many more months or years elapsed before it was finally removed.)

84 Neither the first edition of Baines' book, published in 1993, nor the second edition, published in 1998, even mentioned the term "promoter," although they did indicate that a foreign gene would generally not express within the host cell without being given a suitable "genetic context" (as in p. 132 of the first edition). And, while the third edition (in 2004) does enlarge this discussion by noting that an "appropriate" promoter sequence must be part of the context, there's still no indication that, in a transgenic plant, the promoter is derived from a virus – or that as a result, the expression of the foreign gene is completely deregulated.

85 Ronald, P. and Adamchak, R., *Tomorrow's Table: Organic Farming, Genetics, and the Future of Food* (New York: Oxford University Press, 2008), xiii. The quotes from the journals appear on the book's front and back cover.

86 The index has no entry for "promoter" or "cauliflower mosaic virus," and in my reading of the book, I found no reference to either of them in the main text. The only appearance of the word "promoter" that I discovered is within the definition of "transgene" in the glossary at the back of the book. According to the pertinent sentence, "Along with the genes of interest, may contain promoter, other regulatory, and marker genetic material." (p. 177) However, because the term "promoter" has (as far as I could ascertain) not been previously defined or discussed (there's not even an entry for it in the glossary), most readers would not even know what a promoter is. And even if someone did already understand a promoter's general role in gene expression, this sentence implies that the gene's native promoter is transferred along with it. There's no indication that virtually all the foreign genes inserted in commercialized GE crops are attached to a powerful promoter derived from a plant virus.

87 The terms "particle bombardment," "bioballistics," and "gene gun" are not in the index and, to my knowledge, they don't appear in the rest of the book either.

88 *Field Testing Genetically Modified Organisms: Framework for Decisions*, The National Academy of Sciences (1989), 14.

89 Antoniou, Michael, quoted in Smith, Jeffrey, *Genetic Roulette* (Fairfield, IA: Yes!Books, 2007), 8.

90 The bizarre statement appears in the report's introduction. As Chapter 2 noted, the introduction was written by NRC staff, not by the university scientists who wrote most of the rest of the report.

91 Grace, Eric., *Biotechnology Unzipped* (Washington, DC: Joseph Henry Press, 1977), xiii. The Joseph Henry Press is an imprint of the National Academies Press.

92 Ibid., xiii-xiv.

93 Ibid., 45.

94 Ibid., 29.

95 Ibid., xv. While Fedoroff was not on the review team that interacted with Grace, she was sent a review copy so that she could write a comment about the book (email from Eric Grace).

96 The email to him was sent on March 1, 2012; he replied on March 5.

97 The Dr. Oz Show, December 7, 2010: http://www.doctoroz.com/videos/genetically-modified-foods-pt-3.

98 *Mendel in the Kitchen*, op. cit. note 35, xii.

99 There were only two other speakers that morning: a Monsanto official to represent the views of the biotech industry and an executive from a public interest organization to represent the viewpoint of consumers.

100 Alliance for Bio-Integrity/ICTA Press Conference, National Press Club, Washington D.C., May 27, 1998.

5. Illegal Entry

1 Regal, Philip, "Are Genetically Engineered Foods Safe? A Scientist's Quest for Biosafety": http://www.iatp.org/files/Are_Genetically_Engineered_Foods_Safe_A_Scient.htm.

2 Ibid.

3 Lloyd, T., "Monsanto's new gambit: Fruits and veggies," *Harvest Public Media*, April 8, 2011: http://bit.ly/LQTNxp

4 Goodman, M.M., "New sources of germplasm: Lines, transgenes, and breeders." Paper presented at: Memoria Congresso Nacional de Fitogenetica; 2002; Univ. Autonimo Agr. Antonio Narro, Saltillo, Coah., Mexico.

5 Millstone et al., "Beyond 'substantial equivalence,'" *Nature* (Oct. 7, 1999) estimated that a combination of short and medium-term tests would cost at least an additional $25 million; and Dr. Millstone informed me that full long-term testing would be significantly more expensive – and that if multi-generational testing were included, the cost would be even higher. Because Major Goodman put the cost of developing a GE crop at around $60 million (absent any toxicological testing), adding $25 million to the cost would amount to an increase of over 40% (See Goodman, 30).

6 Regal, op. cit., note 1.

7 For a basic history of the regulation of food, see http://www.fsis.usda.gov/Fact sheets/Additives_in_Meat_&_Poultry_Products/index.asp (accessed April 26, 2012).

8 "Profiles in Toxicology," *Toxicological Sciences* 70, (2002): 159-60: http://toxsci.oxfordjournals.org/content/70/2/159.full.pdf; accessed April 26, 2012.

9 S. Rep. 2422, 1958 U.S.C.C.A.N. 5301- 2.

10 21 U.S.C. § 321

11 Although one of the provisions of the 1958 reforms (the so-called Delaney Act) was repealed by Congress in 1996 that did not change the fundamentals of the food additive requirements.

12 Document #4 at: http://biointegrity.org/24-fda-documents

13 Ibid. All the quotes in this paragraph are from #4.

14 Ibid.

15 Document #5 at: http://biointegrity.org/24-fda-documents

16 Document #2 at: http://biointegrity.org/24-fda-documents

17 Comments from the Division of Food Chemistry and Technology and the Division of Contaminants Chemistry, "Points to Consider for Safety Evaluation of Genetically Modified Foods. Supplemental Information" (November 1, 1991). Document #6 at: http://biointegrity.org/24-fda-documents

18 Ibid. This statement is in a section that is not reproduced on the Alliance for Bio-Integrity website. Its page number in the FDA administrative record (A.R.) for the lawsuit is 18613, and this is the way it was referenced in the briefs that were filed with the Court.

19 The statement cited is at page 18744 of the administrative record.

20 Document #9 at: http://biointegrity.org/24-fda-documents

21 Document #10 at: http://biointegrity.org/24-fda-documents

22 Ibid., at A.R. 18990-91.

23 Document #7 at: http://biointegrity.org/24-fda-documents

24 Document #1 at: http://biointegrity.org/24-fda-documents

25 Eichenwald et al., "Biotechnology Food: From Lab to Debacle," *New York Times*, January 25, 2001. In this article, executives who directed Monsanto's policy during the 1980's purported that they had pushed for meaningful regulation from 1986 until the early 1990's, when "a new management team" took over, decided to reverse course, and convinced the White House that it should scrap plans for such regulation. However, this account is suspect. It does not square with Regal's observations of the realities during the late 1980's, and it's also dubious in light of the industry's persistent anti-regulatory efforts during the previous years. As earlier chapters have shown, the industry has consistently endeavored to avoid regulation while projecting the illusion that responsible regulations were in place. Accordingly, one could reasonably surmise that the Monsanto executives interviewed for the article were attempting to shift blame for the company's unpopular policies to their successors – and that whatever differences existed between their agenda and the one pursued in the early 90's were not about whether meaningful regulations

should be implemented but about how best to create the impression that they had been.

26 "What Would You Do with a Fluorescent Green Pig?," Ecology Law Quarterly 201, no. 34 (2007): 238.

27 Document #1 at: http://biointegrity.org/24-fda-documents. In her memo, Dr. Kahl stated that this forced conclusion "is because of the mandate to regulate the product, not the process," p. 2.

28 Document #4 at: http://biointegrity.org/24-fda-documents

29 Document #21 at: http://biointegrity.org/24-fda-documents

30 Document #23 at: http://biointegrity.org/24-fda-documents

31 Document #24 at: http://biointegrity.org/24-fda-documents

32 The letterhead on his stationery says: Executive Office of the President
Office of Management and Budget

33 "From Lab to Debacle," op. cit. note 25.

34 "Statement of Policy: Foods Derived From New Plant Varieties," May 29, 1992, *Federal Register* 57, no. 104, sec. VI.

35 Ibid., 22991.

36 21 U.S.C. § 321(s)

37 The text of the statute states that the substance must be ". . . generally recognized, among experts qualified by scientific training and experience to evaluate its safety, as having been adequately shown through scientific procedures . . . to be safe under the conditions of its intended use . . . . ," 21 U.S.C. § 321(s).

38 The agency did not extend the presumption to genes that are known to produce a toxin or an allergen. But manufacturers would want to avoid such genes anyway.

39 Document #8 at: http://biointegrity.org/24-fda-documents.

40 21 CFR Sec. 170.30(b); 21 CFR 170.3(i)

41 21 CFR Sec. 170.30 (b)

42 Document #1 at: http://biointegrity.org/24-fda-documents

43 Document #19 at: http://biointegrity.org/24-fda-documents. "FDA Regulation of Food Products Derived from Genetically-Altered Plants: Points to Consider." The authorship of this document is not indicated, and it is also undated. However, its page numbers in the administrative record support an inference that it was written in July or August of 1991.

44 Ibid., at A.R. 18195.

45 Ibid., at A.R. 18196.

46 Maryanski, J., "Safety Assurance of Foods Derived by Modern Biotechnology in the United States," July 1996. In January 1999, the FDA affirmed that it still was not conducting scientific reviews, stating: "FDA has not found it necessary to conduct comprehensive scientific reviews of foods derived from bioengineered plants . . . consistent with its 1992 policy" (Reported in *The Lancet* 353 (May 29, 1999): 1811.)

For a detailed discussion of the unreliability of the FDA's voluntary program, see Freese, W. and D. Schubert, "Safety Testing and Regulation of Genetically Engineered Foods," Biotechnol Genet Eng Rev. 21 (2004): 299-324.

47 Lacey Declaration, *Alliance for Bio-Integrity v Shalala.*

48 Regal Declaration, *Alliance for Bio-Integrity v Shalala.* http://alteredgenes twistedtruth.com/declaration-of-philip-regal/.

49 Defendants' Opposition to Plaintiffs' Cross Motion for Summary Judgment, 1.

50 *Alliance for Bio-Integrity v. Shalala.,* 116 F. Supp. 2d 166 (D.D.C. 2000) at 172-73.

51 116 F. Supp. 2d at 174, 175.

52 The exception is for foods containing genetic material transferred from one of the most commonly allergenic species. To date, it does not appear that any such food has come to market.

53 If the judge had ruled that the FDA had violated the Administrative Procedure Act by not holding formal notice and comment, or had violated NEPA by not performing an environmental assessment, it would also have meant that the FDA's policy had been implemented in violation of the law. But neither of these rulings would have cast doubt on the safety of GE foods for human consumption

54 116 F. Supp. 2d at 177; Citing 21 *C.F.R.* Sec. 170.30 (a-b); 62 Fed. Reg. 18940.

55 116 F. Supp. 2d at 177; Citing 62 *Fed. Reg.* At 18939.

56 116 F. Supp. 2d at 177

57 *United States v. Seven Cartons . . . Ferro-Lac,* 293 F. Supp. 660, 664 (S.D. Il. 1968*),* modified on other grounds, 424 F. 2d 136 (7th Cir. 1970).

58 *Statement of Policy: Foods Derived From New Plant Varieties,* May 29, 1992, Federal Register 57, no. 104 at 22991.

59 A.R. at 11723-24.

60 A.R. at 37744-45.

61 Document #8 at: http://biointegrity.org/24-fda-documents

62 Document #1 at: http://biointegrity.org/24-fda-documents

63 Ferro-Lac, 293 F. Supp. 660 (S.D. Il. 1968). Cited in note 57.

64 Ibid., at 664.

65 Ibid., at 665.

66 While some biotech proponents claim that a proposed rule the FDA introduced in 1997 independently legitimizes the presence of GE foods on the market, in actuality, it does not. That proposal aimed to simplify procedures through which manufacturers can interact with the agency in establishing the GRAS status of additives. (*Federal Register* 62 (1997): 18938-964) But it has never formally moved beyond the stage of a "proposed" rule. Although the FDA has been treating it as an "interim policy," it has not been finalized and lacks the force of law that officially enacted rules possess. So, despite what many people have been led to believe, it has no legal authority or effect. Accordingly, when the Alliance

for Bio-Integrity lawsuit was litigated in 1998-99, the FDA did not try to assert the relevance of that proposed rule, and it is not mentioned in the court's opinion.

Moreover, that proposed rule makes no attempt to alter the two basic criteria that must be satisfied in order for an additive to validly possess GRAS status (technical evidence of safety and general recognition within the scientific community that such evidence in fact exists.) And, as the analysis in this chapter has demonstrated, neither of those criteria has ever been met by any GE food. So even if the rule *had* been finalized, it would not have rendered GE foods GRAS.

67 21 U.S.C. § 321(n)

68 116 F. Supp. 2d at 179.

69 Defendants' Opposition to Plaintiffs' Cross Motion for Summary Judgment, pp. 2 &4.

70 "Proposed Rules," *Federal Register* 66, no. 12 (January 18, 2001): 4709.

71 Ibid., 4710.

72 Ibid.

73 Ibid.

74 Ibid., 4710-11.

75 Ibid., 4711.

77 *Temp. Envtl. L. & Tech. J.* 225 (2000-2001): 19. Although this piece is 17 pages in length, in law review parlance, it's referred to as a "note" and not an article, because law review articles are not authored by law students and are much longer. But I'm referring to it as an article, since most readers would be misled by the term "note" into thinking that the piece is much shorter than it is.

77 Ibid., 229-30. It's noteworthy that although the author uncritically accepted the judge's erroneous rulings regarding the GRAS issue, she critiqued her handling of the labeling issue. This indicates that her acquiescence in the former was not due to excess deference but to deep confusion.

78 E.g., 22 *Rev. Litig.* 669 at 681 (2003); 22 Berkeley Tech, *L.J.* 671 at 698 (2007) (discussed at note 80).

79 34 Ecology L.Q. 201 (2007) at 224

80 22 Berkeley Tech., *L.J.* 671 at 695. The article stated that in contrast to the EU, "[t]he U.S. policy . . . is that GMOs should be permitted to flourish in the absence of proven hazards." Although this statement referred to food safety as well as the environment, there was not even a hint that the policy violates explicit mandates of the law – and illegally shifts the burden of proof to those who question the safety of GE foods.

This article also implied that we failed to present evidence of expert conflict. Its superficial (and somewhat misleading) statements about our lawsuit appear on p. 698 in the main text and also in note 156.

81 Lambrecht, Bill, *Dinner at the New Gene Café: How Genetic Engineering Is Changing What We Eat, How We Live, and the Global Politics of Food* (Thomas Dunne Books: St. Martin's Press, New York (2001), 51.

82 Glickman, Dan, quoted in Simon, Stephanie, "Biotech Soybeans Plant Seed of a Risky Revolution," *Los Angeles Times*, July 1, 2001.

83 Glickman, Dan, quoted at http://eap.mcgill.ca/MagRack/RH/RH_E_97_02.htm (accessed 6-14-12)

84 Glickman, Dan. March 1999. Quoted at: http://www.ces.ncsu.edu/depts/foodsci/ext/pubs/biotech.html (accessed: 7-17-12)

85 Pickett, M., "Ag Official Defends Rules for Biotech Crops," *Billings Gazette*, November 23, 2004.

86 Shacinda, S., "U.S. Needs Good Plan to Give AIDS Funds-health Chief," *Reuters*, December 1, 2003.

87 Lambrecht, B., "Outgoing Secretary Says Agency's Top Priority is Genetically Modified Food," *St. Louis Post- Dispatch*, January 25, 2001.

88 Clinton, Bill, Conference call with farm radio broadcasters from Hermitage, Arkansas (as reported by *Reuters*, November 5, 1999).

89 *Statement of Policy: Foods Derived From New Plant Varieties*, May 29, 1992, Federal Register 57, no. 104. See the discussion in Section V.

90 US Department of Health and Human Services. Press Release. May 3, 2000.

91 FDA quoted in: "Health Risks of Genetically Modified Foods," *The Lancet* 353, no. 9167 (May 29, 1999): 1811.

92 Weise, E., "FDA Tries to Remove Genetic Label Before it Sticks," *USA Today*, October 9, 2002.

93 Cole, M., "FDA Objects to Food Labeling Initiative," *Crop Choice News*, October 9, 2002, http://www.cropchoice.com/leadstry6df6.html?recid=1028.

94 Statement of Robert E. Brackett, Ph.D., Director, Center for Food Safety and Applied Nutrition before the Senate Committee on Agriculture, Nutrition and Forestry, June 14, 2005: http://www.fda.gov/NewsEvents/Testimony/ucm112927.htm (accessed: 7-17-12).

95 Leahy, Stephen, "Crop Testing Rules Menace Food Supply, Say Critics," IPS News Agency, November 25, 2004.

96 Ibid.

97 Philpott, Tom. Monsanto's man Taylor returns to FDA in food-czar role, *GRIST*, July 8, 2009. Taylor's most recent return to the agency was not made directly from Monsanto. After serving as one of the corporation's vice presidents, he worked for a few other organizations prior to rejoining the FDA.

98 http://www.fda.gov/AboutFDA/CentersOffices/OfficeofFoods/ucm196721.htm

99 Personal communication from Dr. Marion Healy, ANZFA Offices, Canberra, AU, February 15, 2001.

6. Globalization of Regulatory Irregularity

1 Millstone, E. et al., *"Beyond 'substantial equivalence',"* *Nature*, October 7, 1999.

2 Ibid.

3 Ibid.

4 Clark, E. Ann, "Food Safety of GM Crops in Canada: toxicity and allergenicity," GE Alert, 2000.

5 Faust, Marjorie, "Biotech Crops for the Dairy and Livestock Industries," *Proceedings of the 2001 California Animal Nutrition Conference*, 76-86.

Although by 2001, there had been tests in which animals were fed the whole GE food, they were not the kinds of tests the FDA experts had called for. Either they aimed to discover whether the foreign DNA and the resultant foreign protein that get implanted in the crops are later found in the animals, or they were designed as nutritional feeding studies, which gauge how well the animals grow but do not assess toxicological effects in a thorough manner. Chapter 10 discusses how biotech promoters have misrepresented the relevance of such studies.

6 "Elements of Precaution: Recommendations for the Regulation of Food Biotechnology in Canada; An Expert Panel Report on the Future of Food Biotechnology prepared by The Royal Society of Canada at the request of Health Canada Canadian Food Inspection Agency and Environment Canada," The Royal Society of Canada, January 2001.

7 Calamai, P., "Ottawa Rapped, Expert Study Considered Major Setback for Biotech Industry," *Toronto Star*, February 5, 2001.

8 "Elements of Precaution," op. cit. note 6.

9 Ibid.

10 "Ottawa Rapped," op. cit. note 7.

11 Personal communication from Lucy Sharratt, Coordinator, Canadian Biotechnology Action Network, June 28, 2012.

12 Smith, Jeffrey, *Seeds of Deception: Exposing Industry and Government Lies About the Safety of the Genetically Engineered Foods You're Eating* (Fairfield, IA: Yes! Books, 2003), 7-9.

13 Ibid., 9.

14 The organization's name was subsequently changed to Food Standards Australia New Zealand (FSANZ).

15 Comments to ANZFA about Applications A346, A362 and A363 from the Food Legislation and Regulation Advisory Group (FLRAG) of the Public Health Association of Australia (PHAA) on behalf of the PHAA (October, 2000).

16 GE proponents might argue that changes between the GE plant and the control line might not necessarily be the results of the gene insertion but of the somaclonal variation that usually accompanies the generation of plants via tissue culture. However, even if the changes were due to tissue culture, that process is an essential part of plant genetic engineering. Therefore, unless rigorous testing eventually demonstrated that the changes were harmless, it would be proper to regard them as potential risks of the bioengineering process. Tissue culture and somaclonal variation will be more thoroughly discussed in Chapter 9.

17 Comments to ANZFA about Applications A372, A375, A378 and A379 from the Food Legislation and Regulation Advisory Group (FLRAG) of the Public Health Association of Australia (PHAA) on behalf of the PHAA (April 2001).

18 PHAA Report of 2000 (emphasis in original).

19 Millstone et al., op. cit. note 1.

20 Ibid.

21 Kawata, Masaharu ,"Inspection of the Safety Assessment of the Roundup Tolerant Genetically Modified Soybean: Monsanto's Dangerous Logic as seen in the Application Document submitted to Japan." The report was originally published in the Japanese journal *Technology and Human Beings*, vol.11 (Nov. 2000): 24-33. As of November 2014, it was available at: http://www.mindfully.org/GE/GE2/Monsanto-Safety-Japan-Inspection.htm

22 http://www.biosafety-info.net/article.php?aid=22 (accessed: July 5, 2012). Although the differences were not statistically significant, the 8% mortality rate was double the usual rate in the UK broiler chicken industry.

23 Ibid.

24 Ibid.

25 Lean, Geoffrey, "Europe Split Over Safety of GM Corn," *The Independent*, December 21, 2003.

26 Smith, Jeremy, "EU Lifts Biotech Ban," *Reuters*, May 19, 2004. See also: http://www.gmo-safety.eu/archive/218.moratorium-ends.html (accessed: June 22, 2012).

27 Ibid.

28 It's probable that the commissioner (David Byrne) was not intentionally deceiving the public but believed what he said because he himself had been deceived by pro-biotech forces. In fact, in light of his prior history, it's most reasonable to assume that this was the case.

29 Schubert, David, "A Different Perspective on GM Food," *Nature Biotechnology*, vol. 20, no.10, October 2002, 969: http://www.biotech-info.net/different_perspective.html

30 The transfer of the protein from bean to pea is described in: Shade, R. E. et al., "Transgenic pea seeds expressing the alpha-amylase inhibitor of the common bean are resistant to bruchid beetles," *Biotechnology* 12 (1994): 793-796.

The research that discovered the adverse effects occurred several years later and is described in: Prescott, V.E. et al, "Transgenic expression of bean alpha-amylase inhibitor in peas results in altered structure and immunogenicity," *Journal of Agricultural Food Chemistry* (2005) Nov 16;53(23): 9023-9030.

31 The test that's commonly used is the SDS gel test. The more sensitive test is the MALDI-TOF test.

32 Commoner, Barry, "Unraveling the DNA Myth: The spurious foundation of genetic engineering," *Harper's Magazine*, Feb 2002.

33 However, Mad Cow Disease has not resulted from genetic engineering; and although it can be induced by feeding cattle diseased tissue from sheep (thereby transgressing a natural boundary by crossfeeding), there is no known link between it (or any of its related diseases) and crossbreeding.

34 Hagan, N. et al., "The redistribution of protein sulfur in transgenic rice expressing a gene for a foreign, sulfur-rich protein," *Plant J.* 34 (2003): 1–11. This study is more extensively discussed in Chapter 9, note 111.

35 Gurian-Sherman, Doug, "Holes in the Biotech Safety Net," Center for Science in the Public Interest (2003), 14: cspinet.org/new/pdf/fda_report__final.pdf. *Note:* The report provides citations to the scientific studies that support Gurian-Sherman's assertions. For a major review published since his report, see: Latham, J., Wilson A. and Steinbrecher, R., "The Mutational Consequences of Plant Transformation," *Journal of Biomedicine and Biotechnology*, vol. 2006, Article ID 25376, 3. Chapter 9 discusses the messiness of the insertional process in more detail.

36 Ibid.

37 The first GE crop (a tomato) was commercialized in 1994.

38 Podevin, N. and du Jardin, P., "Possible consequences of the overlap between the CaMV 35S promoter regions in plant transformation vectors used and the viral gene VI in transgenic plants," *GM Crops Food* 3 (2012): 296–300; doi:10.4161/gmcr.21406.

39 De Tapia, M. et al., "Molecular dissection of the cauliflower mosaic virus translation transactivator," *EMBO J* 12 (1993): 3305-14.

40 Takahashi, H., Shimamato, K., Ehara, Y., "Cauliflower mosaic virus gene VI causes growth suppression, development of necrotic spots and expression of defence-related genes in transgenic tobacco plants," *Mol Gen Genet.* 216 (1989): 188–94.

41 Park, H.S., Himmelbach, A., Browning, K.S., Hohn, T., Ryabova, L.A., "A plant viral 'reinitiation' factor interacts with the host translational machinery," *Cell.* 106 (2001): 723-33.

42 Haas, G., Azevedo, J., Moissiard, G., Geldreich, A., Himber, C., Bureau, M., et al., "Nuclear import of CaMV P6 is required for infection and suppression of the RNA silencing factor DRB4," *EMBO J* 27 (2008): 2102-12.

43 http://www.independentsciencenews.org/commentaries/regulators-discover-a-hidden-viral-gene-in-commercial-gmo-crops/

44 http://www.efsa.europa.eu/en/faqs/faqinsertedfragmentofviralgeneingmplants.htm

http://archive.foodstandards.gov.au/consumerinformation/gmfoods/gmfactsheetsandpublications/gmfoodsandtheuseofdn5796.cfm

45 http://independentsciencenews.org/commentaries/gmo-regulators-hidden-viral-gene-vi-regulators-fail/

46 The Science and Environmental Health Network made the observation. Quoted at: http://www.precaution.org/lib/pp_def.htm)

47 The statute exempts specific classes of substances from being defined as "additives," such as pesticidal chemicals, which are regulated under the provisions of a different statute. See: 21 U.S.C. § 321(s).

48 Communication of April 30, 1997 on consumer health and food safety (COM(97) 183 final).

49 Resolution of March 10, 1998 on the Green Paper: General Principles of Food Law in the EU.

50 EC Communication on the Precautionary Principle, Feb. 2, 2000, Reference 11.

51 EC Communication on the Precautionary Principle, Feb. 2, 2000.

52 Kok, E.J. and Kuiper, H.A.. "Comparative safety assessment for biotech crops," *Trends in Biotechnology* 21 (2003): 439–44.

53 Hilbeck, A., Meier, M., Römbke, J., Jänsch, S., Teichmann, H. and Tappeser, B., "Environmental risk assessment of genetically modified plants - concepts and controversies," *Environmental Sciences Europe* 23, no. 13 (2011).

54 Dalli, John., "GMOs: Toward a Better, More Informed Decision-Making Process," March 17, 2011.

55 Séralini et al., "Genetically modified crops safety assessments: present limits and possible improvements," *Environmental Sciences Europe*, 23, no. 10 (2011): http://www.enveurope.com/content/23/1/10/ (accessed July 12, 2012).

56 Ibid.

57 Ibid.

58 Fleming, J., "No risk with GMO food, says EU chief scientific advisor." EurActiv.com, July 24, 2012: http://www.euractiv.com/innovation-enterprise/commission-science-supremo-endor-news-514072 (accessed August 2, 2012).

59 The meeting took place in the ANZFA offices in Canberra. I subsequently emailed the participants a letter that summarized what they had said and critiqued their policy. I also requested that they inform me of any factual errors I might have made in my recounting of the meeting. Dr. Healy, the chief scientist, sent an email acknowledging receipt of my letter and did not object to any of my statements about what she had said. On September 6, 2001 I sent her an email stating: "When I sent my comments, I asked that you inform me of any factual misstatements you might find in them. Over a month has passed and you have not pointed out any such misstatements. Therefore, I assume you assent to the correctness of my statements of fact." Copies of my initial letter and my follow-up are available at: www. alteredgenestwistedtruth.com

60 ANZFA Occasional Paper, Series No. 1, *GM foods and the consumer* (June 2000). As previously noted, the agency's name was subsequently changed to Food Standards Australia New Zealand (FSANZ).

61 The degree to which the agency's officials have evaded the implications of adverse evidence, misrepresented the facts, and persisted in derelict practices is driven home in the two letters I emailed them subsequent to our meeting that are referred to in note 59.

62 Séralini et al. (2011), op. cit. note 55.

63 Ibid; the authors state that it was necessary to obtain court orders, but they don't provide the details. Some of the relevant ones are as follows. The three GE plants involved were varieties of maize (corn) produced by Monsanto: MON863, MON810 and NK603. At the time approval was first sought for these plants, a manufacturer initiated the process by applying to the appropriate regulatory agency in an EU member state. Because Monsanto claimed that the raw data was confidential, the regulatory agencies that possessed the data refused to release it to

the public, or to the researchers. But the courts ruled that the public had a right to see the data and ordered the regulators to release it.

64 Quoted in Smith, Jeffrey, "An FDA-Created Health Crisis Circles the Globe," 3: http://www.seedsofdeception.com/utility/showArticle/?objectID=1477

65 "Elements of Precaution," op. cit. note 6, 214.

66 "Throwing Caution to the Wind: A review of the European Food Safety Authority and its work on genetically modified foods and crops," *Friends of the Earth Europe* (November 2004): 3.

67 Ibid., 13.

68 Ibid.

69 Ibid.

70 Ibid., 13 & 14. These two pages are the source for the various assertions made in the paragraph.

71 Seralini et al., "New Analysis of a Rat Feeding Study with a Genetically Modified Maize Reveals Signs of Hepatorenal Toxicity," *Archives of Environmental Contamination and Toxicology* 52, no. 4 (May 2007): 596-602.

72 *Friends of the Earth Europe and Greenpeace*, "Hidden Uncertainties: What the European Commission doesn't want us to know about the risks of GMOs," April 2006.

73 Ibid.

7. Erosion of Environmental Protection

1 Ingham, Elaine, "Ecological Balance and Biological Integrity," posted at http://www.purefood.org/ge/klebsiella.cfm.

2 Doyle, J.D. et al., "Effects of genetically engineered microorganisms on microbial populations and processes in natural environments;" in, Neidleman, S., Laskin, A.J. (eds.), *Advances in Applied Microbiology*, vol. 40 (Academic Press, San Diego, CA, 1995), 237-87; see also, Short et al., "Effects of 2,4 dichlorophenol, a metabolite of a genetic engineered bacterium and 2,4 dichlorophenoxyacetate on some microorganism-mediated ecological processes in soil," Appl. Environ. Microbiol. 57 (1991): 412-18.

3 Jones, R.P., "Biological principles for the effects of ethanol," *Enzyme Microbiol. Technol.* 11 (1989): 130-53.

4 Ingham, E.R., Doyle, J.D., and Hendricks, C.W., "Effects of Klebsiella planticola SDF20 on soil biota and wheat growth in sandy soil," *Applied Soil Ecology* 11 (1999): 67-78.

5 In fact, when Michael Holmes, the graduate student who initiated the study, continued the research, he discovered that in some circumstances the GE bacteria could out-compete the parent strain. (Elaine Ingham, the professor who supervised and participated in the original research, informed me of this in a personal conversation. She said that these findings were described in Holmes' doctoral dissertation, which has not been published.)

6 Ingham, E., "Ecological Balance," op. cit. note 1; Ingham, E., Letter to the Editor: "Engineered Bacterium Could Have Serious Implications for Human Life on Earth," *Agribusiness Examiner*, Issue 119, June 11, 2001.

7 Suzuki, D. and Dressel, H., *From Naked Ape to Superspecies: Humanity and the Global Eco-Crisis* (Vancouver: Greystone Books, 2004), 121.

8 Ingham, E., quoted in Luke Anderson, *Genetic Engineering, Food and Our Environment* (White River Jct., VT: Chelsea Green, 1999), 39-40.

9 Ingham, E., Letter to the Editor, op. cit. note 6.

10 Ingham, "Ecological Balance" op. cit. note 1.

11 In this context, "regulators" refers to those with the authority and capacity to set policy. It is not meant to include all employees of the regulatory agencies. From 1983 to the present, there have been many members of these institutions who endeavored to pursue a genuinely science-based and responsible policy on GMOs. However, their collective influence has been insufficient to shape outcomes.

12 Anderson, Luke, op. cit. note 8, 40.

13 US General Accounting Office, *Biotechnology: Managing Risks of Genetically Engineered Organisms* (Government Printing Office [GAO/RCED-88-27, Washington D.C., 1988, 108 pp.])

14 PEER White Paper, "Genetic Genie: The Premature Commercial Release of Genetically Engineered Bacteria," September 21, 1995; re-issued, January 25, 2000.

15 Roslin, Alex, "Germs gone wild," *Georgia Straight*, July 21, 2005: http://www.ibiblio.org/london/SoilWiki/message-archives/JoeCummins/msg00517.html

16 PEER White Paper, op. cit. note 14, v.

17 Ibid., reporting on the comments of Suzanne Wuerthele.

18 PEER News Release, Jan. 26, 2000.

19 Pollack, Andrew, "Lax in Tests of Gene-Altered Crops," *New York Times*, January 3, 2006.

20 Brasher, Philip, "Report Blasts Oversight of Field Tests," *Des Moines Register*, Dec. 30, 2005.

21 Weiss, Rick, "U.S. Rice Supply Contaminated, Genetically Altered Variety Is Found in Long-Grain Rice," *Washington Post*, August 19, 2006; Weiss, Rick, "Firm Blames Farmers 'Act of God' for Rice Contamination," *Washington Post*, November 22, 2006; A05.

22 See for example, Doering, Christopher, "ProdiGene to spend millions on bio-corn tainting," *Reuters News Service, USA*, December 9, 2002.

23 Ibid. The corn got mixed with the soybeans, but it did not cross-pollinate them (an outcome that is biologically barred).

24 Smith, Jeffrey, *Institute for Responsible Technology Newsletter*, August 2006.

25 Press Release, Center For Food Safety, February 6, 2007: http://www.centerforfoodsafety.org/GTBC_DecisionPR_2_7_07.cfm

26 Séralini, G.E. et al., "Genetically modified crops safety assessments: Present limits and possible improvements," *Environmental Sciences Europe* 23(10) (2011);

Freese, W. & Schubert, D., "Safety testing and regulation of genetically engineered foods. *Biotechnol Genet Eng.* (rev. 2004): 299-324.

27 Castaldini, M. et al., "Impact of Bt corn on rhizospheric and soil eubacterial communities and on beneficial mycorrhizal symbiosis in experimental microcosms," *Appl Environ Microbiol.* 71(11) (Nov. 2005): 6719-29; Zwahlen, C. et al., "Degradation of the Cry1Ab protein within transgenic Bacillus thuringiensis corn tissue in the field," *Mol Ecol.* 12(3) (Mar 2003): 765-75.

28 Ibid.

29 Cheeke, T.E., Pace, B.A., Rosenstiel, T.N., Cruzan, M.B., "The influence of fertilizer level and spore density on arbuscular mycorrhizal colonization of transgenic Bt 11 maize (Zea mays) in experimental microcosms," *FEMS Microbiol Ecol.* 75(2)(Feb. 2011): 304-12; Cheeke, T.E., Rosenstiel, T.N., Cruzan, M.B., "Evidence of reduced arbuscular mycorrhizal fungal colonization in multiple lines of Bt maize," *American Journal of Botany.* 99(4) (2012): 700–07.

30 Tank, J.L., Rosi-Marshall, E.J., Royer, T.V., et al., "Occurrence of maize detritus and a transgenic insecticidal protein (Cry1Ab) within the stream network of an agricultural landscape," *PNAS* 27 (September 2010).

31 Rosi-Marshall, E.J., Tank, J.L., Royer, T.V., et al., "Toxins in transgenic crop byproducts may affect headwater stream ecosystems," *Proc Natl Acad Sci USA* 104(41) (Oct 9, 2007): 16204-08.

32 Bohn, T., Traavik, T., Primicerio, R., "Demographic responses of Daphnia magna fed transgenic Bt-maize," *Ecotoxicology* 19(2) (February 2010): 419-30.

33 Marvier, M. et al. "A meta-analysis of effects of Bt cotton and maize on nontarget invertebrates," *Science* 316(5830) (June 8, 2007): 1475-77; Losey, J.E., Rayor, L.S., Carter, M.E., "Transgenic pollen harms monarch larvae," *Nature* 399(6733) (May 20, 1999): 214; Jesse, L.C.H. and Obrycki, J.J., "Field deposition of Bt transgenic corn pollen: Lethal effects on the monarch butterfly," *J. Oecologia* 125 (2000): 241–48; Lang, A. and Vojtech, E., "The effects of pollen consumption of transgenic Bt maize on the common swallowtail, Papilio machaon L. (Lepidoptera, Papilionidae)," *Basic and Applied Ecology* 7 (2006): 296–306; Ramirez-Romero et al., "Does Cry1Ab protein affect learning performances of the honey bee Apis mellifera L. (Hymenoptera, Apidae)?," *Ecotoxicology and Environmental Safety* 70 (2008): 327–33.

34 Lövei, G.L., Arpaia, S., "The impact of transgenic plants on natural enemies: A critical review of laboratory studies," *Entomologia Experimentalis et Applicata* 114 (January 2005): 1–14. This paper systematically reviewed the studies in peer-reviewed journals that examined how Bt crops affect insects that prey on plant pests. The authors determined that 57% of the parameters measured showed "significant negative impacts," (p. 7). Even though the authors noted that several studies had methodological shortcomings, they concluded: "Nevertheless, the overall skew towards negative impacts . . . is a signal that we ought to consider seriously. The negative impacts are too numerous to just explain them [sic] away as non-significant or non-relevant" (p. 11).

35 Mellon, M., "Introduction," *Now or Never: Serious New Plans to Save a Natural Pest Control* (Union of Concerned Scientists, 1998), 2.

36 Gassmann AJ, Petzold-Maxwell JL, Keweshan RS, Dunbar MW. "Field-evolved resistance to Bt maize by Western corn rootworm." *PLoS ONE.* 2011; 6(7): e22629.

37 Ibid; Associated Press, "Monsanto shares slip on bug-resistant corn woes," August 29, 2011; Gray M., "Severe root damage to Bt corn confirmed in northwestern Illinois," *Aces News*, August 24, 2011.

38 Fagan, J., Antoniou, M.C., and Robinson,C., *GMO Myths and Truths: An Evidence-Based Examination of the Claims Made for the Safety and Efficacy of Genetically Modified Crops*, 2nd Edition, version 1.0, (London: Earth Open Source, 2014), 249.

39 Freese, Bill, "Going Backwards: Dow's 2,4-D-Resistant Crops and a More Toxic Future," *Food Safety Review* (Spring 2012), 1: http://www.centerforfoodsafety.org/wp-content/uploads/2012/04/CFS_FSR_spring_2012.pdf

40 Benbrook, C.M., "Impacts of genetically engineered crops on pesticide use in the United States: The first thirteen years," *The Organic Center* (November 2009): http://www.organic-center.org/reportfiles/13Years20091126_FullReport.pdf

41 Ibid.

42 Stanley, T., "The Superweed Invasion," National Public Radio, October 4, 2010; Neuman and Pollack, "Farmers Cope with Roundup-Resistant Weeds," *New York Times*, May 3, 2010: http://www.nytimes.com/2010/05/04/business/energy-environment/04weed.html?_r=0 (accessed: 7-15-12).

43 Neuman and Pollack (2010), op. cit. note 42.

44 Benbrook (2009), op. cit. note 40.

45 Breeze, V.G. and C.J. West "Effects of 2,4-D butyl vapor on the growth of six crop species," *Ann. Appl. Biol.* 111(1987): 185-91.

46 AAPCO (1999 & 2005), "1999/2005 Pesticide Drift Enforcement Survey," Association of American Pesticide Control Officials, survey periods 1996-1998 and 2002-2004, respectively.

47 Freese (2012), op. cit. note 39, 2.

48 Ibid.

49 Paganelli, A., Gnazzo,V., Acosta, H., López, S.L., Carrasco, A.E., "Glyphosatebased herbicides produce teratogenic effects on vertebrates by impairing retinoic acid signaling," *Chem Res Toxicol.* 23(10) (2010): 1586–95.

50 Gasnier, C., Dumont, C., Benachour, N., Clair, E., Chagnon, M.C., Seralini, G.E., "Glyphosate-based herbicides are toxic and endocrine disruptors in human cell lines," *Toxicology* 262(3) (August 21, 2009): 184-91.

51 Kremer RJ, Means, N.E., Kim, S, "Glyphosate affects soybean root exudation and rhizosphere microorganisms." *Int J of Analytical Environmental Chemistry* 85(15) (2005): 1165–1174; Sanogo S, Yang XB, Scherm H, "Effects of herbicides on Fusarium solani f. sp. glycines and development of sudden death syndrome in glyphosate-tolerant soybean." *Phytopathology* 90(1) (Jan 2000): 57-66.

52 Food Standards Agency, *About mycotoxins*, undated: http://www.food.gov.uk/safereating/chemsafe/mycotoxins/about/

53 Alm, H. et al. "Influence of Fusarium-toxin contaminated feed on initial quality and meiotic competence of gilt oocytes," *Reprod Toxicol.* 22(1) (July 2006): 44-50; Diaz-Llano, G. and Smith, T.K., "Effects of feeding grains naturally contaminated with Fusarium mycotoxins with and without a polymeric glucomannan mycotoxin adsorbent on reproductive performance and serum chemistry of pregnant gilts," *J Anim Sci.* 84(9) (September 2006): 2361-66.

54 In September 2014 the US Department of Agriculture deregulated Dow's 2,4-D-resistant soybeans and corn

55 Freese (2012), op. cit. note 39, 2.

56 Press Release, Center for Food Safety, May 3, 2007: http://www.centerforfoodsafety.org/2007/05/03/federal-judge-orders-first-ever-halt-to-planting-of-a-commercialized-genetically-altered-crop/

57 Waltz, E., "Industry exhales as USDA okays glyphosate resistant alfalfa," *Nature Biotechnology* 29(3) (March 2011): 179–81.

58 Leslie TW, Biddinger DJ, Mullin CA, Fleischer SJ., "Carabidae population dynamics and temporal partitioning: Response to coupled neonicotinoid-transgenic technologies in maize." *Environ Entomol.* 38(3) (Jun 2009): 935-943.

59 Tennekes, H.A., "The significance of the Druckrey-Kupfmuller equation for risk assessment--the toxicity of neonicotinoid insecticides to arthropods is reinforced by exposure time," *Toxicology* 276(1) (September 30, 2010): 1-4.

60 Pettis, J.S., Vanengelsdorp, D., Johnson, J., Dively, G., "Pesticide exposure in honey bees results in increased levels of the gut pathogen Nosema," *Die Naturwissenschaften* 99(2) (February 2012): 153-58; Krupke, C.H., Hunt, G.J., Eitzer, B.D., Andino, G., Given, K., "Multiple routes of pesticide exposure for honey bees living near agricultural fields," *PLoS ONE* 7(1) (2012), e29268.

61 Bindraban, P.S., Franke, A.C., Ferrar, D.O., et al., "GM-related sustainability: Agro-ecological impacts, risks and opportunities of soy production in Argentina and Brazil," *Plant Research International* (Wageningen, the Netherlands, 2009).

62 Neuman and Pollack (2010), op. cit. note 42.

8. Malfunction of the American Media

1 I can't recollect his name or the network he worked for.

2 Alliance for Bio-Integrity/ICTA Press Conference, National Press Club, Washington D.C., May 27, 1998.

3 Press Release, Grocery Manufacturers of America, May 27, 1998.

4 E.g., "FDA Sued Over Genetically Altered Food," *Omaha World-Herald*, May 28, 1998, at 9. (The statement was made by Eric Flamm, a senior policy adviser at the FDA).

5 Krimsky, S. and Wrubel, R., *Agricultural Biotechnology and the Environment: Science, Policy, and Social Issues* (Champaign, IL: University of Illinois Press, 1996), 163-64.

6 Ibid.

7 To refer to these scientists as not among the pro-biotech mainstream is not to imply that they were opposed to all forms of genetic engineering. Many believed that some of its applications might prove to be safe and beneficial. However, they were not partisan promoters of the technology but were willing to provide objective scrutiny and to critique aspects of the enterprise that they perceived to be problematic.

8 Lewenstein, B. et al., "Historical survey of media coverage of biotechnology in the United States, 1970 to 1996." Paper presented to AEJMC Annual Meeting, Baltimore, MD, August 8, 1998.

9 Priest, S. H., and Talbert, J., "Mass Media and the Ultimate Technological Fix: Newspaper Coverage of Biotechnology," *Southwestern Mass Communication Journal* 10(1) (1994): 76-85.

10 Susanna Priest quoted in "The Odd Couple: Biotechnology and the Media," *AgBiotech Buzz* 2 (11) (December 20, 2002).

11 http://www.commondreams.org/news2002/0429-06.htm (accessed 7-23-12)

12 Ibid.

13 Hencke, D. and Evans, R., "How US put pressure on Blair over GM food," *The Guardian*, February 28, 2000.

14 Hankinson, S.E. et al., "Circulating Concentrations of Insulin-Like Growth Factor 1 and Risk of Breast Cancer," *Lancet*, vol. 351, no. 9113 (1998): 1393-96; Chan, J. et al., "Plasma Insulin-Like Growth Factor-1 [IGF-1] and Prostate Cancer Risk: A Prospective Study," *Science* 279 (January 23, 1998): 563-66.

15 Smith, Jeffrey, *Seeds of Deception: Exposing Industry and Government Lies About the Safety of the Genetically Engineered Foods You're Eating* (Fairfield, IA: Yes! Books, 2003), 188.

16 "Can two reporters take on Murdoch and win?," *The Independent*, London, Sept. 14, 1999.

17 Ibid.

18 Quoted in *Seeds of Deception*, op. cit. note 15, 189.

19 *The Independent* (1999), op. cit. note 16.

20 Personal communication from Jane Akre.

21 *The Independent* (1999), op. cit. note 16.

22 Quoted in *Seeds of Deception*, op. cit. note 15, 190-92.

23 *The Independent* (1999), op. cit. note 16.

24 Ibid.

25 Oddly, although Wilson also sued Fox for the same reason, his claim was not successful.

26 http://www.foxbghsuit.com/2D01-529.pdf

27 At that point, the documents had not yet been posted to the Alliance for Bio-Integrity website. After they were, I no longer needed to fax them to interested individuals.

28 Weiss, Rick, "Biotech Food Raises a Crop of Questions," *Washington Post*, August 15, 1999. Although the article did note that some experts were concerned

that some of the inserted genes might be allergenic, greater space was devoted to the experts who stated that no unusual risk was posed. Further, the risk of toxicity was never even mentioned.

29 Eichenwald et al., "Biotechnology Food: From Lab to a Debacle," *New York Times*, January 25, 2001.

30 Licthblau, E. and Shane, S., "Vast F.D.A. Effort Tracked E-Mails of its Scientists," *New York Times*, July 15, 2012.

31 That headline was even more dramatic: "Vast Effort by F.D.A. Spied on E-Mails of its Own Scientists."

32 Brody, Jane, "Facing Biotech Foods Without the Fear Factor," *New York Times*, January 11, 2005.

33 In a special report marking the 40th anniversary of Watergate, the *Washington Post* noted how the White House had "continued to denounce" its coverage as "biased and misleading" and had also dispensed "unveiled threats and harassment": http://www.washingtonpost.com/wp-srv/politics/special/watergate/part1.html

34 Leonard Downie, Jr., who was the *Post's* deputy metro editor during that period and helped supervise the Watergate coverage, has recently recounted the strain that he and his colleagues endured: "We were ignored and doubted by the rest of the news media and most of the country, and under heavy fire from the Nixon administration and its supporters. It was a tense time . . . , with our credibility and our newspaper's future on the line;" Downie, Leonard, Jr., "Forty years after Watergate, investigative journalism is at risk," *Washington Post*, June 7, 2012: http://www.washingtonpost.com/opinions/forty-years-after-watergate-investigative-journalism-is-at-risk/2012/06/07/gJQArTzlLV_story.html.

35 See note 34.

36 While the *Post* and other members of the media may have sometimes refrained from revealing questionable government actions in matters of foreign policy, in the interest of national security, I'm not aware of any other instances in which it has suppressed facts about government fraud on the domestic front – especially fraud that compromises public safety.

37 Apple, R.W., "Lessons from the Pentagon Papers," *New York Times*, June 23, 1996: http://www.nytimes.com/books/97/04/13/reviews/papers-lessons.html.

38 Correll, John T., "The Pentagon Papers," *Air Force Magazine*, February 2007.

39 Ibid.

40 Ibid.

41 Ibid.

42 Ibid.

43 Although the government ultimately decided not to bring criminal charges against the newspapers, it did bring them against Ellsberg. However, due to gross irregularities in the behavior of the FBI and some other government employees in relation to his case, the judge eventually declared a mistrial and dismissed the charges against him. But he was never formally acquitted of violating the Espionage

Act, and had it not been for the government's bungling of the case, he would probably have been convicted (See Correll's article, cited above).

44 Downie, 2012, op. cit., n. 34. Downie was the *Post's* executive editor from 1991 to 2008. Although I don't know if he was directly involved in the decision to remove the revelations about the FDA from Weiss's article, it's difficult to believe that the editor with whom Weiss was interacting would have made such an important policy decision on his own – and it seems likely that he was acting within an editorial framework that had already been established at higher levels of authority. Thus, there's good reason to assume that Downie had in some significant way been involved in the formulation and implementation of a policy restricting what would be written about the risks of GE foods. After all, he held the same position at the *Post* during the first 15 years of the GE food era as had Ben Bradlee during the Watergate era; and those familiar with the book (or movie), *All the President's Men*, know how actively engaged Bradlee was in the supervision of the Watergate reporting.

Further, regardless of the degree to which Downie may have been involved, I think it's hypocritical for executives at the *Post* to sustain their chest-thumping about the paper's courageous actions regarding Watergate while they cling to their cowardly policy about GE foods. If Downie, who is currently a professor of journalism at Arizona State University's Walter Cronkite School of Journalism and Mass Communication, and is also a vice president at large of the *Post*, is sincerely committed to "accountability journalism," he will openly assume responsibility for whatever role he may have played in the *Post's* irresponsible behavior in regard to GE foods – or, if he played no role at all, he will identify those who should be held accountable. Moreover, he should use his influence to rescind the restrictive policy and replace it with one that allows full reporting of the facts. What's more, I think that he (and/or others at the *Post*) should start making amends by publishing a series of articles that communicate not only the facts that were removed from Weiss's report, but many of the other key facts that are documented in this book – facts that the American people have a right to know. Only then can their boasts about the paper's Watergate triumphs be free from hypocrisy.

45 *New York Times, Co. v. United States*, 403 U.S. 713 (1971).

9. Methodical Misrepresentation of Risk

1 http://www.aaas.org/news/releases/2012/media/AAAS_GM_statement.pdf

2 http://www.who.int/foodsafety/publications/biotech/20questions/en/

3 "Elements of Precaution: Recommendations for the Regulation of Food Biotechnology in Canada; An Expert Panel Report on the Future of Food Biotechnology prepared by The Royal Society of Canada at the request of Health Canada Canadian Food Inspection Agency and Environment Canada," The Royal Society of Canada, January 2001.

4 Calamai, P., "Ottawa Rapped, Expert Study Considered Major Setback for Biotech Industry," *Toronto Star*, February 5, 2001.

5 "Genetically modified food and health: A second interim statement," British Medical Association Board of Science and Education, March, 2004.

6 Kmietowicz, Z., "GM Foods Should Be Submitted to Further Studies, says BMA," *British Medical Journal* 328(7440) (March 13, 2004): 602.

7 Public Health Association of Australia Letter to Government Officials, November 2, 2000.

8 *The Lancet* 353 (May 29, 1999): 1811.

9 Fedoroff, N., and Brown, N.M., *Mendel in the Kitchen: A Scientist Looks at Genetically Modified Foods* (Washington, DC: Joseph Henry Press, 2004), xii.

10 National Research Council and Institute of Medicine of the National Academies (NAS), "Safety of Genetically Engineered Foods: Approaches to Assessing Unintended Health Effects" (Washington D.C.: The National Academies Press, 2004). (As the report's title indicates, it was prepared by two of the NAS's divisions: the National Research Council and the Institute of Medicine.) It can be read online or downloaded at: http://www.nap.edu/openbook.php?record_id=10977&page=R1.

11 "Elements of Precaution," op. cit. note 3, 184.

12 Ibid., 185.

13 Ibid., 184.

14 Ibid., 22.

15 Ibid., 185.

16 Ibid., 186.

17 Ibid., ix.

18 Ibid., 48.

19 In previous parts of this book, where the discussion of risks was not technical, the terms have also sometimes been employed synonymously, according to customary usage.

20 For simplicity, this example assumes that every bite from a venomous snake in Arizona entails the same potential for death as every bite incurred in Ohio. Under this assumption, the degree of harm in each case – the death of one person – is the same, enabling the difference in risk to be determined merely by assessing the different probabilities of being bitten. However, in reality, some species of venomous snakes are more deadly than others. So to accurately calculate the risk differential, we would need to factor in the difference between the toxicity of the average venomous snake bite in Arizona and the average bite incurred in Ohio. This would render the Arizona walk even riskier.

21 According to a recent *New York Times* article, runway collisions are the biggest threat in aviation. It contained the following quote from the chairman of the National Transportation Safety Board: "Where we are most vulnerable at this moment is on the ground . . . . To me, this is the most dangerous aspect of flying," Wald, M. "For Airlines, Runways Are the Danger Zone," *NY Times*, April 25, 2008.

22 In 2008, the National Safety Council compiled an odds-of-dying table comparing the risks of flying and driving. The odds of dying in a motor vehicle were calculated to be 1 in 98 over a lifetime. In contrast, for air travel the lifetime

odds were only 1 in 7,178. http://traveltips.usatoday.com/air-travel-safercar-travel-1581.html.

23 *Introduction of Recombinant DNA-Engineered Organisms into the Environment: Key Issues* (National Academy of Sciences, 1987), 6.

24 Ibid., 22.

25 *Field Testing Genetically Modified Organisms: Framework for Decisions* (The National Academy of Sciences, 1989), 14.

26 E.g., 40 and 43; although the report does cite the standard definition of risk (p. 41), and although it at times does speak about the "magnitude" of risk, its approach is inconsistent; and its language is loose. So is its reasoning. However, because the 2004 report makes similar mistakes that are also more varied, and because it's the broader and more important of the two, I'll focus on it and will not expend space examining the defects of the 2000 report in greater depth.

27 NAS 2004 Report, op. cit. note 10, 2.

28 Because the 2004 report was focused solely on human health effects, the 2000 report was broader in respect of issues addressed, since it also dealt with environmental ones. But it only dealt with those issues in regard to a limited class of GMOs, and so was narrower in that respect; and its conclusions weren't technically applicable to all engineered food crops.

29 The release was issued on July 27, 2004 by the National Research Council and the Institute of Medicine, the divisions of the NAS that had produced the report. It was titled: "Composition of Altered Food Products, Not Method Used to Create Them, Should Be Basis for Federal Safety Assessment."

30 NAS 2004, op. cit. note 10. This finding is stated on p. 180 (and also on p. 9 in the Executive Summary). The report states that "the policy to assess products based exclusively on their method of breeding is scientifically unjustified." Since this statement follows reference to the fact that the EU subjects GE foods to a higher level of assessment than other foods, it's clear that it intends to convey the idea that this practice is unjustified and that GE foods, as a class, should not be treated differently than conventional ones.

31 Jennifer Hillard quoted in: Pollack, A., "Panel Sees No Unique Risk from Genetic Engineering," *New York Times*, July 28, 2004.

32 Document #1 at: http://biointegrity.org/24-fda-documents

33 The NAS 1989 report states (on p. 2) that the maxim was a "fundamental principle" of the 1987 report; and it notes that this principle was then "adopted" by those responsible for preparing it (the 1989 report) and "reemphasized" in Chapter 2.

34 Ibid: In the PDF format, the chart appears on p. 64 and the explanatory text on p. 65. See: http://www.nap.edu/openbook.php?record_id=10977&page=64; The chart also appears in the Executive Summary on p. 4.

35 The committee specified that selection from a homogeneous population has less potential for unintended effects than selection from one that is heterogeneous.

36 Ibid., 65. In the chart, the category comprising both radiation and chemical-based processes is referred to as "mutation breeding." However, I prefer not to use this term, since it implies these are the only techniques that induce mutations, even though (as will be discussed) tissue culture-based breeding also generates new traits by inducing mutations. And genetic engineering causes mutations as well. Therefore, instead of employing the term "mutation breeding" to refer to the techniques that employ radiation and chemicals, I'll refer to them jointly as "radiation and chemical-based breeding."

37 As Chapter Four explained, although an actual rifle firing a .22-calibre bullet was initially employed in such transfers, as the gun evolved, macroscopic bullets were no longer used, and the microscopic particles were propelled by a blast of air. But the device providing the blast is still a type of gun. The process is often referred to as *particle bombardment*, *bioballistics*, or *biolistics*. The 2004 report uses the latter term.

38 NAS 2004, op. cit. note 10, 63.

39 For stylistic purposes, I'm substituting "dangerous" for "risky," with the intent that it conveys the same technical meaning. Further, the number 10 was arbitrarily selected for the sake of argument, not because bioengineering has been determined to induce that many side effects for every effect induced by pollen-based breeding. However, in this regard the hypothetical value may be on the low side, because in the report's comparative chart, the bar depicting the unintended effects of bioballistic gene transfer between distantly related species is around 14 times longer than the bar adjoining the least disruptive form of pollen-based breeding.

40 NAS 2004, op. cit. note 10, 63.

41 Ibid.

42 Ibid.

43 Nor do we "know" that any of the foods created through tissue culture is actually safe.

44 In this context, the words "proven safe" do not denote the certainty involved in a mathematical proof. They denote the standard of proof instituted by the FDA for purposes of evaluating food: a demonstration that there is "reasonable certainty" of no harm.

It's also important to note that the NAS report does not primarily attempt to establish its claim about the safety of GE foods by citing actual safety tests. Instead, it relies on the specious arguments that are critiqued in this chapter's analysis. Moreover, as Chapter 6 revealed (and as Chapter 10 will more fully elucidate), several of the tests on GE foods raise reasonable doubts. This research cannot be lightly dismissed, and it further undercuts the committee's claim about what we presently know.

45 While some tests have been conducted on whole foods that were irradiated for the purpose of reducing microbes, safety testing has not been performed on foods grown from irradiated seeds, which present a different set of hazards.

46 NAS 2004, op. cit. note 10, 27.

47 Ibid., 45.

48 The committee acknowledged the lack of records regarding radiation (on p. 28 of their report); and they also noted that it has not been feasible to track for effects of GE foods, while urging the FDA to institute practices that would facilitate it.

49 Schubert, David, "Pharmed Food: Consume with Caution," *The Ecologist*, November 2008.

50 Although the evidence doesn't prove that GE was the cause, it strongly points to that conclusion; and, as Chapter 3 explains, it's more likely than not that the process was to blame for the toxic contamination – which would be sufficient to hold the process liable in a court of law.

51 NAS 2004, op. cit. note 10, 47.

52 Ronald, P. and Adamchak, R., *Tomorrow's Table: Organic Farming, Genetics, and the Future of Food* (New York: Oxford University Press, 2008), 102. Although she doesn't explicitly cite the report as the basis for this particular assertion, her foregoing discussion demonstrates that the document is the primary source on which it relies. For instance, on p. 69 she states the report indicates that the GE crops currently on the market "are safe to eat."

53 While it is not in principle impossible that she could know they are safe, given the present state of the evidence, it's not currently possible.

54 NAS 2004, op. cit. note 10, 63.

55 NAS 2004, op cit. note10, 131-32. Although the committee's language lacks precision, they appear to include GE foods among those that need not be proven safe prior to marketing and can only be removed if obvious problems emerge later. And if they in fact were not confused, they should have avoided confusing language that imparts the impression they were unaware of what the law requires.

Further, besides accepting the basics of the FDA's hands-off policy in regard to GE foods, the committee also defended the lax regulatory policy in most of the rest of the world, where the concept of *substantial equivalence* reigns (Ibid., 129-30). But in doing so, they relied on the type of simplistic linear model that the Canadian experts had discredited. They indicated that putting primary attention on the protein the inserted gene expresses is a sound approach, while failing to acknowledge that, even if the protein is safe to consume, its unregulated expression (as well as the insertion process itself) could disrupt cellular function in deleterious ways. Thus, in explaining (approvingly) how the *substantial equivalence* approach is applied to almost all the GE foods then on the market (including *Bt* corn and Roundup Ready soybeans), the report noted that the assessment focuses "primarily on the introduced trait or gene product" (Ibid., 130). But, as their Canadian counterparts had demonstrated, the presumed sufficiency of this narrow approach is itself based on a presumption that's significantly flawed: the presumption that whatever unintended side effects are induced by the transformation process will be adequately detected by superficial compositional comparisons. And such constricted thinking is based on the notion that attention should mainly rest on the product – and that the process has no significant bearing on the risk that the product will harbor harmful side effects that are difficult to discover.

Moreover, to the extent the NAS report faulted the regulatory policy of the European Union and other regions embracing the *substantial equivalence* approach, it was not for applying this approach to GE foods, but for requiring that it be applied to all of them while exempting all conventional products. In rejecting this aspect of the policy, the report emphasized that it's "scientifically unjustified" to set assessment criteria based exclusively on the manner of production (Ibid., 180).

56 Ibid., 29. Here's how the committee described the way GE plants are developed via the use of reconfigured bacteria: "By substituting the DNA of interest for the crown gall disease-causing DNA, scientists derived new strains of *Agrobacterium* that deliver and stably integrate specific new genetic material into the cells of target plant species. If the transformed cell then is regenerated into a whole fertile plant, all cells in the progeny also carry and may express the inserted genes."

57 The term "genomic shock" is used in connection with tissue culture by several biologists. One example is: Kaeppler et al., "Epigenetic aspects of somaclonal variation in plants," *Plant Molecular Biology* 43 (2000): 179–88; 181.

58 When genetically identical cells go through tissue culture, they tend to mutate in different ways. This differential in mutations is referred to as *somaclonal variation*. The NAS report generally employs this term in referring to the process of tissue culture, and it's used as the heading of the relevant section on page 26. Because I think it's simpler and more straightforward to speak of tissue culture instead, since it is the name of the technique through which somaclonal variation occurs, I don't employ the latter term.

59 NAS 2004, op. cit. note 10, 27.

60 Ibid., 28-29. It took twenty-seven more pages before they finally acknowledged that tissue culture is an aspect of the bioengineering process. In describing a few of the potential unintended effects of GE, they said: ". . . spontaneous mutation may occur in the tissue culture phase of the transformation regeneration processes" (p. 56). But, unless one already knew how widely the technique is relied on in producing GE plants, this sentence would be unlikely to induce such understanding. Further, as will be seen, when the committee subsequently presented a chart depicting differences in disruptive potential between the various modes of plant breeding, it treated tissue culture as distinct from bioengineering.

61 Although in the case of a few species, there are ways in which isolated cells can be regenerated without resort to tissue culture, it's the standard method through which engineered cells are transformed into mature plants.

62 Neelakandan, A. and Wang, K., "Recent progress in the understanding of tissue culture-induced genome level changes in plants and potential applications," *Plant Cell Rep* 31 (2012): 597-620; 611. I emailed Dr. Wang, the director of the Center for Plant Transformation at Iowa State University, inquiring if the statement about "high probability" of changes referred not merely to regenerated plants before they've been crossed, but to the final, commercialized products as well – even though

the total number of changes would have been reduced in those products via crossing. She emailed to confirm that the statement applied to the final products too.

63 It's logical to presume that all the bars represent the potential for unintended effects to remain in the final products of the respective methods, because if they instead are meant to pertain to plants that have not undergone crossing, then they couldn't reflect differences in the potential for unintended effects that remain after crossing has occurred and the product is ready for marketing – the phase at which the differences are most important. And it's reasonable to think that such differences exist, as will be explained shortly.

64 Skirvin et al., "Sources and frequency of somaclonal variation," *Hort Sci* 29 (1994):1232-1237.

65 It's also more likely that the more intensive culturing processes would generate a higher percentage of dramatic mutations; but it's also likely that most of these would not remain in the final product, since they would either prevent plant development or result in observable (and more readily removable) abnormalities.

66 As we shall see, even without registering the effects of tissue culture, the bar associated with that mode of bioengineering should be substantially longer and darker.

67 Its bar, when adjusted, would extend 1.3 centimeters beyond the right vertical axis of the chart (the point at which the bar for radiation ends). In the context of the chart, this is a significant difference. Moreover, even if the GE bars are adjusted by adding only one-fourth the length of the tissue culture bar, the one associated with the most disruptive mode is longer than that of radiation; and the other is almost as long.

Note: In order to take measurements, I first reproduced the chart that's in the PDF version of the report on an 8.5 x 11 inch piece of paper. I then used a ruler to ascertain the lengths of the bars. It was difficult to be precise because of the way the gray tails shade toward the ends. Some of the values I obtained are: tissue culture (SCV): 5.6 cm; bacterial transfer of rDNA between distantly related species: 9.0 cm; biolistic transfer of rDNA between distantly related species: 10.2 cm; radiation breeding: 10.8 cm. Other people may get slightly different values; but the overall result will most likely be similar. Further, it's important to keep in mind that the lengths of the bars only reflect the committee's rough estimates.

68 While this analysis is illuminating, it's important to note that in neither the committee's chart nor the adjusted versions of it do the ratios between the lengths of the bars precisely reflect reality. The committee's calculations are not based on evidence that enables exact determinations; and the available data don't provide a basis for anything more than reasonable estimates – although the estimates the committee made did not always express this attribute.

69 Ronald and Adamchak, op. cit. note 52, 88. Of course, in stating that radiation is riskier, she's at odds with the committees' claim that there's no correspondence between placement on the chart and degree of risk. However, as

we'll see in Chapter 14, when discussing risks, Ronald not only contradicts the NAS, she even contradicts herself.

70 At the close of a section arguing that genes inserted via bioengineering are not drawn to "hotspots" in the DNA that foster genetic instability, they stated: "Similarly, there is no evidence to suggest the CaMV 35S promoter in GE plants is any more unstable than the CaMV 35S promoter in ordinary plants infected with CaMV" (NAS 2004, 61). The question of whether the 35S promoter inserted in plants is, itself, genetically unstable is separate from the other issues that I noted. Although there's still room for scientific debate about this additional issue, because properly presenting it would add a significant amount of text to an already long chapter, I've decided to forgo it.

71 Hohn, T. and Rothnie, H., "Plant pararetroviruses: replication and expression," *Current Opinion in Virology* 3 (2013): 621–28.

72 NAS 2004, op. cit. note 10, 60 & 62.

73 Ibid., 60.

74 E.g., Fedoroff, *Mendel in the Kitchen*, op. cit. note 9, 103; where it's stated that "neither genes nor transposons normally move."

75 E.g., Wu, R., Guo, W.L., Wang, X.R., Wang, X.L., Zhuang, T.T., Clarke, J.L., Liu, B., "Unintended consequence of plant transformation: biolistic transformation caused transpositional activation of an endogenous retrotransposon Tos17 in rice," ssp. japonica cv., Matsumae, *Plant Cell Rep* 28 (2009): 1043–51.

76 *Mendel in the Kitchen*, op. cit. note 9, 105.

77 David Schubert, personal communication. Mutation breeding via radiation and chemicals also stirs up transposons. But, as will be discussed, there's good reason to think bioengineering entails at least as great a transposon-related risk.

78 *Mendel in the Kitchen*, op. cit. note 9, 104-05. However, as the book points out, wide crosses between "very distantly related plants" can activate transposons.

79 As in several other sections of the report, the committees' discussion is not as coherent as one would expect, and it's difficult to discern the structure of their argument. But their words do create the impression that transposon mobilization is somehow separate from the GE process. Leaving aside the issue of whether this obfuscation was deliberate, it seems they may have been trying to advance the following argument:

(a) Plant genomes contain numerous transposons; (b) many of the associated insertion events either created, or could have created, disruptions; (c) any disruptions caused by insertions of rDNA would be no riskier than those associated with transposons; (d) therefore, such insertions present no cause for concern.

But such an argument is flawed. Not only does it disregard the fact that genetic engineering can induce transposon movement (through three distinct modes) and thereby impose additional transposon-related risks, it mistakenly equates whatever risks may linger from ancient transposon insertions with the risks entailed by present-day insertions of rDNA cassettes. Scientists recognize that transposons and their movements have played a significant role in the evolution of plants and have contributed to important features that are found in contemporary varieties. And it's known that over great expanses of biological time, while positive effects of the transpositional events have been conserved, most deleterious effects have not been maintained. But the situation is otherwise with rDNA insertions. Instead of a long process of screening by natural selection, the screening for harmful effects in whatever plants survive the transformation process is performed by human inspection; and, as the 2001 Canadian report repeatedly warned, the current monitoring process is unable to detect all the subtle changes that could harm consumer health.

Moreover, even if effects of transpositional events in the distant past remain that don't impair the function of the plant but do impair the health of those that consume them, the insertional effects of bioengineering add to this baseline of risk to a more significant degree than does pollen-based breeding – and, as will be demonstrated, more greatly than do all other forms of breeding as well.

80 Forsbach A., Schubert, D., Lechtenberg, B., Gils, M., Schmidt, R., "A comprehensive characterization of single-copy T-DNA insertions in the *Arabidopsis thaliana* genome," *Plant Molecular Biology* 52(1) (2003): 161–76. The researchers selected only plants that contained a single insertion site.

81 Latham, J., Wilson, A. and Steinbrecher, R., "The Mutational Consequences of Plant Transformation," *Journal of Biomedicine and Biotechnology*, vol. 2006, article ID 25376, 3.

82 Ibid. (emphasis added). They noted that conclusions about particle bombardment had to be provisional, because very few of its insertion events were well-studied at that time.

83 NAS 2004, 66.

84 Regarding gene loss, e.g., Kaya, H., Sato, S., Tabata, S., Kobayashi, Y., Iwabuchi, M., Araki, T., "*hosoba toge toge*, a syndrome caused by a large chromosomal deletion associated with a T-DNA insertion in *Arabidopsis*," *Plant & Cell Physiology* 41(9) (2000): 1055–66. Re: deletion-related disruption of gene function, see e.g., Amedeo, P., Habu, Y., Afsar, K., Mittelsten Scheid, O., Paszkowski, J., "Disruption of the plant gene *MOM* releases transcriptional silencing of methylated genes," *Nature* 405(6783) (2000): 203–06. Re: potential disturbance of native genes through the influence of the inserted DNA; E.g., Hannon, G.J., "RNA interference," *Nature* 418(6894) (2002): 244–51; Bartel, B. and Bartel, D.P., "MicroRNAs: at the root of plant development?" *Plant Physiology* 132(2) (2003): 709–17."

85 Amedeo et al., op. cit. note 84; Ichikawa, T., Nakazawa, M., Kawashima, M., et al., "Sequence database of 1172 T-DNA insertion sites in *Arabidopsis* activationtagging lines that showed phenotypes in T1 generation," *The Plant Journal* 36(3) (2003): 421–29; Weigel, D., Ahn, J.H., Bl´azquez, M.A., et al., "Activation tagging in Arabidopsis," *Plant Physiology* 122(4) (2000): 1003–13.

86 Freese, W. and Schubert, D., "Safety testing and regulation of genetically engineered foods," *Biotechnology and Genetic Engineering Reviews* 21 (2004): 314 (emphasis added).

87 Latham et al., op. cit. note 81, 4.

88 Ibid., 3.

89 Ibid., 4.

90 Ibid.

91 Further, if plants are not grown from seed but are propagated clonally (as is usual with potato and banana), none of the genome-wide mutations are removed, and they'll be present in every future clone of the original GE plant.

92 Latham et al., op. cit. note 81, 4.

93 Ibid., 5. According to the molecular biologist Allison Wilson, who has extensively examined the data submitted to regulators, although the standard Southern analyses submitted in applications to the USDA are claimed to detect whether additional copies of all or parts of the cassette have been deposited in distant sites, the plants' developers do not submit sequence data for the entire genome. Therefore, subsequent whole genome sequence analysis will likely reveal transgenic inserts missed by Southern analysis – as was the case with the commercialized transgenic papaya. Comparison of the transgenic genome with the genome of the parent plant would also be necessary to determine the presence (and extent) of any additional genome-wide differences between the transgenic plant and its parent (e.g. movement of native transposons, rearrangements or deletions of plant DNA) [personal communication].

94 Latham et al., op. cit. note 81, 5.

95 E.g., Windels, P. et al., "Characterisation of the Roundup Ready soybean insert," *European Food Research and Technology* 213(2) (200):107–12; Hernandez, M. et al., "A specific real-time quantitative PCR detection system for event MON810 in maize YieldGard based on the 3-transgene integration sequence," *Transgenic Research* 12(2) (2003): 179–89.

96 Wilson, A. et al., "Genome Scrambling – Myth or Reality? Transformation-Induced Mutations in Transgenic Crop Plants," *EcoNexus Technical Report* (October 2004). The report presented such scrambling as a reality.

97 Numerous studies cited in the review were published in 2003 or earlier, so the committee could have taken account of them, since its report was not released until the summer of 2004. And although they did refer to four papers mentioned in that review, none of these were the ones that examined (or even expressly discussed) deletions and rearrangements in the regions surrounding the insertion site. Nor did any mention the insertion of superfluous DNA.

98 E.g., Latham et al., op. cit. note 81; Freese, W., Schubert, D., "Safety testing and regulation of genetically engineered foods," *Biotechnology and Genetic Engineering Reviews* 21 (2004): 299–324; Spok, A. et al., "Risk Assessment of GMO-Products in the European Union," *Bundesministerium f¨ur Gesundheit und Frauen*, 2004.

99 NAS 2004, 27.

100 NAS 2004; their main discussion runs from p. 41 through p. 45. Although they also noted that tomatoes could contain problematic levels of a naturally occurring toxin, they acknowledged this was due to environmental factors rather than to the breeding process.

101 Steiner, H.Y. et al., "Evaluating the Potential for Adverse Interactions within Genetically Engineered Breeding Stacks," *Plant Physiology*, April 2013, vol. 161 no. 4: 1588.

102 NAS 2004, op. cit. note 10, 43.

103 Ibid., 56.

104 http://wildflowerfinder.org.uk/Flowers/P/Potato/Potato.htm; http://www.sigmaaldrich.com/catalog/product/sigma/d5649?lang=en&region=US. Further, the fact that demissidine is present in potatoes was reported in the scientific literature as early as 1981: Jadhav, S.J., R.P. Sharma, D.K. Salunkhe, "Naturally occurring toxic alkaloids in foods," *Crit Rev Toxicol* 9 (1981): 21–104.

105 Ibid., 43.

106 Steiner, H.Y. et al., op. cit. note 101, 1588. It's noteworthy that these authors are proponents of GE foods, and one is employed by Pioneer Hi-Bred, a major biotech corporation. Yet, they admit there's no evidence that conventional breeding has produced novel toxins (or is even likely to), despite the NAS committees' contention that this has actually happened – and that it could well be happening in several cases of which we're unaware.

Further, although many people are under the impression that through the process of pollination, modern varieties of hybridized wheat have become endowed with one or another novel proteins, there appears to be no sound evidence this has actually happened. Instead, the relative concentrations of native proteins have changed. See, e.g., van den Broeck, H.C. et al., "Presence of celiac disease epitopes in modern and old hexaploid wheat varieties: wheat breeding may have contributed to increased prevalence of celiac disease," Theor Appl Genet (2010) 121:1527–1539. DOI 10.1007/s00122-010-1408-4.

107 Schubert, David, quoted in Smith, Jeffrey, *Genetic Roulette* (Fairfield, IA: Yes! Books, 2007), 56.

108 Hagan, N. et al., "The redistribution of protein sulfur in transgenic rice expressing a gene for a foreign, sulfur-rich protein," *Plant J.* 34 (2003): 1–11. For a discussion, see note 111.

109 Although the study on the allergenic effects of the protein produced by the GE peas was not published until after the committee had released its report, the study revealing that the foreign gene expressed in GE rice might have been misfolded was published the year prior to the report's release. Yet, the report makes no mention of it. Moreover, the potential for a protein synthesized within a foreign species to be adversely altered via either add-ons or misfolding was recognized well before the report was written.

110 It's clear that serious problems could result from either class of insertions. In regard to adverse outcomes induced by foreign genes, Philip Regal has pointed out that ". . . theory and evidence have suggested that the host's buffering or control systems will often be ineffective for those transgenes that can express well." He explains that because the foreign genes could induce "unusual conditions" that cannot be modulated by the buffering mechanisms, ". . . new factors may be added

to the host's biochemical milieu and cause quantitative or qualitative changes in the output of existing biochemical pathways." Regal, P., "Scientific Principles for Ecologically Based Risk Assessment of Transgenic Organisms," *Molecular Ecology* 3 (1994): 5-13 (The sentences cited above were from a section relevant to food safety as well as ecological safety).

On the other hand, when the inserted gene comes from a closely related species, the organism's control system could be stressed in trying to cope with the hyper-expression of a native substance, resulting in the formation of unusual toxins – as happened with Showa Denko's tryptophan-producing bacteria.

111 For instance, in an attempt to increase the sulfur content of rice, a transgenic variety was created containing a sunflower gene that expresses a protein rich in sulfur. However, the amount of sulfur in the rice did not increase, apparently because the high demand for sulfur imposed by the over-expression of the sunflower gene drew heavily upon the plants' sulfur pools and decreased the production of some of their own sulfur-containing proteins.

Further, there were at least two other types of change that were not directly related to the competition for sulfur production but were induced by some other mechanics. In one, the level of two native proteins significantly *increased*; and because they chaperone the correct folding of proteins, and because their levels tend to increase in plants in response to stresses that impair proper protein formation, the researchers regarded this result as a warning sign. They stated it ". . . raises the possibility that at least some of the SSA [the foreign protein] is misfolded."

The second unusual outcome that was not directly related to the competition for sulfur involved a failure to process another native protein [glutelin B] in a normal manner, which led to elevated levels of its unprocessed form.

Although the researchers didn't determine whether the various alterations could exert negative impacts on consumer health, or whether other potentially hazardous changes had also occurred, a process that can induce such significant shifts in the way a plant operates clearly has the potential for doing so. Hagan, N. et al., "The redistribution of protein sulfur in transgenic rice expressing a gene for a foreign, sulfur-rich protein," *Plant J.* 34 (2003): 1–11. See also, Islam, N. et al., "Decreased accumulation of glutelin types in rice grains constitutively expressing a sunflower seed albumin gene," *Phytochemistry* 66 (2005): 2534–39.

112 Nestle, Marion, "The AMA's Strange Position on GM Foods": http://www.theatlantic.com/health/archive/2012/06/the-amas-strange-position-on-gmfoods-test-but-dont-label/258968/ While her allegation was specifically directed at a particular, and glaring, instance of inconsistency between two of the report's main assertions, even when the document *is* self-consistent, it is *not* consistent with good science – as will be demonstrated.

113 American Medical Association, Policy Statement on Biotechnology and the American Agricultural Industry, 1990.

114 American Medical Association, Report 2 of the Council on Science and Public Health (A-12) (2012)2.

115 The AMA's opposition to labeling is clearly stated in the 2012 report, which was heavily relied on by the opponents of a California ballot initiative that would have required labeling. The initiative was narrowly defeated.

116 AMA (1990), op. cit. note 113. The first GE whole food that came to market was the *Flavr Savr* tomato, introduced in May 1994. It is discussed in Chapter 10.

117 Document #1 at: http://biointegrity.org/24-fda-documents

118 In one indication of deficient benefits, the *Des Moines Register* reported that studies of Iowa farmers conducted for 1998 and 2000 by Iowa State University economist Dr. Michael Duffy showed: "Farmers who plant genetically modified corn and soybeans fare no better financially than those who grow traditional crops. . . ." And it noted Duffy's statement that seed companies and chemical companies have reaped the primary benefits of biotechnology so far (Perkins, Jerry, "Biotech Crops Fail to Reap More Cash," *Des Moines Register,* January13, 2002). Dr. Duffy also found that, in both years, yields for the GE soybean (Monsanto's Roundup Ready variety) were lower than for the non-GE beans.

119 The yield drag of the Roundup Ready soybean was confirmed by researchers at the University of Nebraska. In controlled studies comparing RR soy with non-engineered sister lines, they found consistent yield decreases with the GE beans of 5%. They concluded that the study "demonstrates that a 5% yield suppression was related to the [foreign] gene or its insertion process. . . ." And they made it clear that the reduction was not due to the application of the herbicide, because they determined that it had exerted no effect on yields (Elmore et al., "Glyphosate-Resistant Soybean Cultivar Yields Compared with Sister Lines", *Agron J* 93 [2001]: 408-12). The other problems have been discussed in Chapter 7.

120 21 CFR 170.3(I)

121 American Medical Association, Policy Statement on Biotechnology and the American Agricultural Industry, 1990.

10. A Crop of Disturbing Data

1 As it turned out, this hypothesis was incorrect; and the translation of the PG gene was inhibited via a different mechanism.

2 As will be discussed later in this chapter, the tomato also contained a marker gene; and that gene had a bacterial origin.

3 Martineau, Belinda, *First Fruit: The Creation of the Flavr Savr™ Tomato and the Birth of Biotech Foods*, 2001 (McGraw-Hill: New York), 146.

4 Calgene also conducted acute oral toxicity tests of eight other lines of the Flavr Savr and five corresponding control lines. (Martineau, personal communication)

5 *First Fruit*, op. cit. note 3, 150. The term "gross lesions" was used by FDA pathologists who reviewed the data. They also referred to the lesions as "gastric erosions" (Document #14, p. 1 at: http://biointegrity.org/24-fda-documents).

6 Pusztai, A., "Can Science Give Us the Tools for Recognizing Possible Health Risks of GM Food?," *Nutrition and Health* 16 (2002): 73-84.

7 *First Fruit,* op. cit. note 3, 152.

8 Document #14, p. 2 at: http://biointegrity.org/24-fda-documents.

9 Document #17, pp. 2-3 at: http://biointegrity.org/24-fda-documents.

10 Document #16 at: http://biointegrity.org/24-fda-documents.

11 Document #15, p. 3 at: http://biointegrity.org/24-fda-documents.

12 *First Fruit*, op. cit. note 3, 181.

13 Agency Summary Memorandum, Re: Consultation with Calgene, Inc., Concerning FLAVR SAVR™ Tomatoes, May 17, 1994, U.S. Food and Drug Administration.

Because the Commissioner selects the FAC members, and because Kessler had had a long tenure, it's reasonable to assume he had appointed a substantial number of the standing members.

14 *First Fruit*, op. cit. note 3, 182.

15 Cony, A., "FDA Scientists find Flavr Savr Safe," *Sacramento Bee*, April 6, 1994. (The FDA's statement was released several days before this article was published; and the article referred to it as a past event.)

16 *First Fruit*, 182. Belinda Martineau, the author of the book, was a member of the Flavr Savr development team and made this observation first-hand.

17 Cony, A., "FDA Scientists find Flavr Savr Safe," *Sacramento Bee*, April 6, 1994; As cited in *First Fruit*, 252, n. 1. The FDA's subsequent public releases regarding the Flavr Savr continued to falsely assert that its scientists had determined that the tomato was safe, *e.g.* "First Biotech Tomato Marketed," U. S. Food and Drug Administration, Center for Food Safety and Applied Nutrition, *FDA Consumer*, September 1994.

18 FDA Food Advisory Committee Meeting, April 6-8, 1994, Transcript: vol. 2, 153.

19 Ibid., 159-61.

20 Ibid., 162; 167-68.

21 *First Fruit*, op. cit. note 3, 186. The words "love fest" were in quotation marks in Martineau's report.

22 FDA Public Affairs Office, HHS News, May 18, 1994.

23 "Biotechnology of Food," U. S. Food and Drug Administration, Center for Food Safety and Applied Nutrition, *FDA Backgrounder*, May 18, 1994.

24 "FDA'S Policy for Foods Developed by Biotechnology," U. S. Food and Drug Administration, Center for Food Safety and Applied Nutrition, CFSAN Handout: 1995. The document was also published as a chapter in American Chemical Society Symposium Series no. 605, 1995.

25 *First Fruit*, 146. This page discusses the initial consensus, and the words in quotation marks are from Martineau's take on it.

26 Document #15, p. 3 at: http://biointegrity.org/24-fda-documents.

27 Ibid., 4.

28 Additionally, some substances have been granted a specific statutory exemption from meeting the test requirements.

29 According to the FDA's own regulations, foods claimed to be GRAS ". . . require the same quantity and quality of scientific evidence as is required to obtain approval of the substance as a food additive." (21 CFR Sec. 170.30[b])

30 Agency Summary Memorandum Re: Consultation with Calgene, Inc., Concerning FLAVR SAVR™ Tomatoes, May 17, 1994, US Food and Drug Administration.

31 Pusztai, A. et al., "Genetically modified foods: potential human health effects," in *Food Safety: Contaminants and Toxins*, in D'Mello, J.P.F., ed., Scottish Agricultural College, Edinburgh, UK, April 2003, 351.

32 Belinda Martineau's files attest that the deaths occurred in a different line than did the lesions.

33 Because the relevant files were not with me at the time but at the offices of the International Center for Technology Assessment (the attorneys of record in our lawsuit) in Washington, D.C., I conveyed the request to them; and they sent him the information.

34 Pusztai, A. et al., (2003), op. cit. note 31, 351.

35 http://www.responsibletechnology.org/posts/throwing-biotech-lies-at-tomatoes-part-1-killer-tomatoes/

36 Pusztai, A. et al. (2003) op. cit., note 31, 350. Although the quoted words were specifically made in discussing the acute toxicity study, they also applied to the 28-day studies, in which the variation in starting weights was even greater.

37 Ibid., 351.

38 Ibid., 351-52.

39 Although the administrators were apparently abetted by a few agency scientists, their scientific standards seem to have been significantly lower than were those of the experts who wrote the critical memos. Further, it's not clear if they scrutinized the data as carefully as did that set of experts.

40 When the cassette already contains a gene conferring resistance to an herbicide, that herbicide can be used to kill off the non-transformed cells, eliminating the need to add an antibiotic resistance marker gene.

41 Although biotech proponents have tried to discredit concerns by arguing that kanamycin has largely fallen into disuse and is no longer medically significant, the facts show otherwise. Not only is it used prior to endoscopy of the colon and rectum, it's used to treat ocular infections, and also in blunt trauma emergency treatment. It's additionally applied in veterinary medicine.

Perhaps even more significant, the effectiveness of other antibiotics could also be compromised. That's due to a phenomenon called *cross resistance*, wherein bacteria that become resistant to a particular antibiotic subsequently develop resistance to others within its family. And kanamycin belongs to an important family. Its relatives include antibiotics that are substantially relied on today. So it's a matter of concern that this family has displayed appreciable cross resistance (Onaolapo, J., *Afr. J. MedSci* 23 [1994]: 215-9). Moreover, according to the medical doctor Jaan Suurkula, two of kanamycin's cousins are of "great value" in treating serious infections because they entail gentler side effects than their alternatives. See: http://www.psrast.org/antibiot.htm. Therefore, it's ominous that a strain of bioengineered bacteria in which the kanamycin resistance marker gene had been used was found to be cross resistant to these two valuable drugs (Smirnov, V.V. et al., *Antibiot-Khimiorec* 39(4) [Apr 1994]: 23-28). As Dr. Suurkula has observed, "it would be an important drawback" if resistance to these antibiotics were to increase.

42 *First Fruit*, op. cit. note 3, 161. Belinda Martineau has informed me that despite the fact people in both the FDA and the biotech industry routinely claim that Calgene decided to submit the application on its own volition, she was (as she reports in her book) in the office of the company's CEO when he received a call from an FDA official (which she heard via his speaker phone) informing him that the agency preferred to have Calgene's request for an advisory opinion converted into, and submitted as, a formal food additive petition.

43 Document #11 at: http://biointegrity.org/24-fda-documents

44 Document #12, p. 6 (at AR # 013136) at: http://biointegrity.org/24-fda-documents

45 Ibid. (at AR #013130)

46 Document #13 (at AR # 013139) at: http://biointegrity.org/24-fda-documents

47 The scientist was Nega Beru. FAC Meeting Transcript, op. cit. note 18, vol. 2, 178.

48 Dr. Beru's assertion that the FDA had concluded that the use of the gene is safe appears on page 187 of the above transcript.

49 As Appendix C explains, although the FDA did expressly approve the marker gene, it refrained from doing so in the case of the tomato itself. And in the agency's letter to Calgene regarding the latter, it cleverly chose its words to give the illusion that an approval was being granted while, in actuality, there was no express approval or certification of safety. But due to the artful illusion, Calgene declared, and the media reported, that formal approval had been received and safety certified.

50 Belinda Martineau's files indicate that Calgene had decided to market that particular line; and she informed me that, as far as she knows, it's the one that was commercialized.

51 *United States v. Seven Cartons . . . Ferro-Lac*, 293 F. Supp. 660, 664 (S.D. Il. 1968). That case is discussed in Chapter 5. In it, the FDA was challenging the GRAS status of an additive, the experts testified on behalf of its challenge, and they asserted that they were not aware of any studies in the standard literature demonstrating the substance was safe. Obviously, the outcome would almost surely have been different if those scientists had questionable credentials and were countering a large number of well-qualified experts whose opinion was based on solid evidence of safety published in standard peer-reviewed journals. But if even a few experts can show that a widely-held opinion is not based on such evidence and instead rests on assumptions and hypotheses that are open to reasonable doubt, then the substance they challenge cannot be legitimately deemed GRAS. (This ruling was subsequently

modified on other grounds by an appellate court, 424 F.2d 136 (7th Cir. 1970) – grounds that did not affect the holding about the sufficiency of two experts.)

52 Document #12, pp. 5 & 7 at: http://biointegrity.org/24-fda-documents

53 Ibid., 7.

54 Document #13, p. 3 at: http://biointegrity.org/24-fda-documents

55 Ibid., 2.

56 I asked Belinda Martineau, who was intimately involved in Calgene's endeavors regarding the tomato, whether any additional data about the spread of resistance to gut flora had been submitted in response to the concerns of the FDA experts. She graciously searched her files and reported that none had. (The only additional data that was submitted was unrelated to those specific concerns.)

57 21 CFR 170.3 (3) (I).

58 Because this fact was more than sufficient to negate any claim the gene was GRAS, it certainly could not be circumvented merely by converting Calgene's submission into a food additive petition, in which the same standard of safety is in effect.

Moreover, even if Calgene had decided to commercialize a Flavr Savr line that had not been linked with either lesions or deaths, that line would still have entered the market in contravention of the law. For one thing, it would have contained the illegally-approved marker gene. For another, its safety would still have been subject to reasonable doubt. Because the FDA experts had identified an unresolved safety issue, a shadow was cast on *every* line of the Flavr Savr. And until that issue was resolved, the shadow would remain. After all, if the lesions (or the deaths) were in some way caused by one or another aspect of the bioengineering, there would be a possibility that deleterious effects could be induced by other lines of tomato altered with the same cassette – and that these effects might not be evident unless more thorough studies were conducted.

59 *First Fruit*, op. cit. note 3, 203, 223.

60 Ibid., 221-22.

61 Ibid., 222.

62 Ibid., 221 (for the information that there was no public opposition).

63 As Appendix C explains, this profession about safety having been demonstrated went far beyond what the agency was willing to state in the actual letter it had sent Calgene in regard to the tomato.

64 "Elements of Precaution: Recommendations for the Regulation of Food Biotechnology in Canada; An Expert Panel Report on the Future of Food Biotechnology," The Royal Society of Canada, January 2001, 48.

65 Pusztai, A., Submission to the New Zealand Royal Commission on Genetic Modification (2001).

66 Pusztai has stated that "our task was to establish novel testing methods." Pusztai, A., "Responses to the Royal Society's (RS) six referees' reviews on the Audit and Alternative Report." (placed on the internet by the Rowett Research Institute on February 16, 1999, but no longer accessible).

67 Pusztai, A, Transcript of testimony to the New Zealand Royal Commission on Genetic Modification, February 7, 2001, 3406.

68 Smith, Jeffrey, *Genetic Roulette* (Fairfield, IA: Yes!Books, 2007), 23.

69 The differences could also have been due to variable effects of tissue culture, as will be discussed subsequently.

70 Pusztai, A., SOAEFD flexible Fund Project RO 818: Report of Project Coordinator on data produced at the Rowett Research Institute (RRI): http://www.worldcat.org/title/soaefd-flexible-fund-project-ro-818-report-of-projectcoordinator-on-data-produced-at-the-rowett-research-institute-rri/oclc/041214388

71 Pusztai testimony, op. cit. note 67, 3430.

72 Ibid.

73 Ibid.

74 Ewen, S.W.B. and Pusztai, A., "Effects of diets containing genetically modified potatoes expressing *Galanthus nivalis* lectin on rat small intestine," *The Lancet* 354 (1999): 1353-54.

75 However, no tumors were observed.

76 Email from Susan Bardocz, one of the scientists on the research team (and Pusztai's wife).

77 Smith, Jeffrey, *Seeds of Deception: Exposing Industry and Government Lies About the Safety of the Genetically Engineered Foods You're Eating* (Fairfield, IA: Yes!Books, 2003), 13. Smith based his account of Pusztai's experiences and subjective reactions on extensive interviews. Unless otherwise specified, the following statements about Pusztai's subjective outlook are based on Smith's account, which provides a much more detailed and dramatic exposition of events than is provided in this chapter.

78 *GM-FREE Magazine*, vol. 1, no. 3, August/September 1999.

79 Transcript from, "World in Action," sent by Arpad Pusztai to Jeffrey Smith. Quoted by Smith in *Seeds of Deception*, op. cit. note 77, 15.

80 Ibid., op. cit. note 77, 16.

81 Ibid., 18.

82 Rowell, A., "The sinister sacking of the world's leading GM expert – and the trail that leads to Tony Blair and the White House," *Daily Mail*, July 7, 2003. http://www.gmwatch.org/latest-listing/42-2003/4305

83 Ibid.

84 Rowell, A., *Don't Worry, It's Safe to Eat* (London, UK: Earthscan, Ltd, 2003).

85 https://royalsociety.org/

86 Up until the 1960's, every issue of its journal *Philosophical Transactions* bore a notice that "It is an established rule of the Royal Society . . . never to give their opinion, as a Body, upon any subject."

87 The proactive nature of the policy was acknowledged in the President's Address in The Royal Society Annual Review 1998-99, which declared that "We have contributed early and proactively to public debate about genetically modified plants."

88 Flynn, L. and Gillard, M., "Pro-GM food scientist 'threatened editor'," *The Guardian*, October 31, 1999.

89 Jeffrey Smith reported (in the *Huffington Post*) that Pusztai informed him about having offered the data to the Society and being refused: http://www.huffingtonpost.com/jeffrey-smith/biotech-propaganda-cooks_b_675957.html

90 In his response to the review (op. cit. note 66) Pusztai pointed out that the Royal Society bore "a great deal of the blame" because it gave the reviewers internal documents that were "manifestly inappropriate for peer-review."

91 Ibid.

92 Ibid.

93 Ibid.

94 Ibid.

95 Ibid.

96 Editorial: "Health risks of genetically modified foods," *The Lancet* 353 (May 29, 1999): 1811. The editorial said that by May 22, 1999, the Society had completed its review; but a document issued by the Society in 2002 stated the report was published in June. Apparently, the editor of the *Lancet* had seen a copy of the report prior to formal publication.

97 Horton, R., "GM Food Debate," *The Lancet* 353, issue 9191 (November 13, 1999): 1729.

98 *The Guardian* stated it had "established that the Royal Society was involved in trying to prevent publication." And it noted that these efforts began *before* the Society learned that the *Lancet* was reviewing the research. Flynn and Gillard (1999), op. cit. note 88. (While this article did not employ the term "unsavory," it provided a comprehensive, and unflattering, report on the Society's actions to discredit Pusztai and his research that clearly revealed their unsavory character.)

99 Ibid.

100 Flynn and Gillard, op. cit. note 88.

101 Ibid.

102 Ibid. All words in quotation marks in this paragraph were in quotes within *The Guardian* article.

103 Bateson, P., "Mavericks are not always right," *Science and Public Affairs*, June 2002. Bateson's allegation distorts the truth by ignoring the fact that five out of the six referees voted for publication. Instead, he imparts the impression that more than one objected (which is false), and that no one with statistical competence voted for publication (which is almost surely false as well).

104 Royal Society, "Genetically modified plants for food use and human health – an update," February 2002, 5.

105 The review was published in June 1999, and the Lancet paper was not published until October 15th.

106 Royal Society (2002), op. cit. note 104, 5. The crucial sentences were: "In June 1999, the Royal Society published a report, *Review of data on possible toxicity*

of GM potatoes, in response to claims made by Dr. Pusztai (Ewen & Pusztai, 1999). The report found that Dr Pusztai had produced no convincing evidence of adverse effects from GM potatoes on the growth of rats or their immune function." Thus, the first sentence clearly implies that the review centered on the paper authored by Pusztai and Stanley Ewen that was published in the *Lancet*, not the incomplete summary that had been prepared for scientists at the Rowett Institute (which was the *only* submission the participants in the 1999 report examined). An extensive examination of the Royal Society's misbehavior in regard to Pusztai is contained in: Rowell A. *Don't Worry, It's Safe to Eat.* (London, UK: Earthscan Ltd, 2003).

107 Royal Society, *Review of data on possible toxicity of GM potatoes,* June 1999, 1 & 2.

108 http://www.publications.parliament.uk/pa/ld200001/ldhansrd/vo010216/text/10216-02.htm (accessed: June 2, 2014).

109 Ibid.

110 Arthur, Charles, "Scientists blame media and fraud for fall in public trust," *The Independent*, January 31, 2003: http://www.independent.co.uk/news/science/scientists-blame-media-and-fraud-for-fall-in-public-trust-609014.html

111 Report of the Royal Commission on Genetic Modification (New Zealand, 2001), 209.

112 For instance, the report alleges that differences detected in the feeding studies were due to the fact that raw potatoes were used – and that because rats don't like them, they were starving and the 110-day trial had to be abandoned after 67days. It then asserted: "Starvation affects gut histology, and the lining of the gut of control rats eating unmodified potatoes was shown to be abnormal."

But these assertions are erroneous. First, there were four major studies, and only one was designed to last 110 days. The others were completed in 10 days, and even though the longer one that used raw potatoes ended earlier than planned, it still yielded significant results. (Pusztai Witness Brief, p. 3) Second, all four tests showed significant differences in several physiological indices between rats fed GE potatoes and those fed on the non-GE ones. In all, 39 statistically significant differences were found (by independent multivariate statistical analysis), of which no more than five could have been the result of random error. (Pusztai Testimony, transcript of February 7, 2001, p. 3430.) Further, whatever negative effects the raw potato diet had on the control group were of significantly less magnitude than the effects observed in the rats eating the GE potatoes, which indicates that something unique to the GE potatoes was also a causative factor. Third, the rats were not "starving." They were continuing to put on weight, but not at the rate required by UK government regulations on animal feeding studies. (Ibid., pp. 3435, 3441.) Fourth, even starvation does not produce abnormal gut histology. It merely contracts the gut. (Private communication from Dr. Pusztai.) Fifth, trials were also conducted using *boiled* potatoes. On this diet, the longer study did run for a full 110 days. As in the case of the raw potato diet, there were statistically significant differences between the rats eating GE and non-GE potatoes. (Ewen Witness Brief;

Pusztai transcript, p. 3442.) Moreover, the commission was fully informed of these facts (except for the fourth) but nonetheless misrepresented them.

113 Fedoroff, N. and N.M. Brown, *Mendel in the Kitchen: A Scientist Looks at Genetically Modified Foods* (Washington, DC: Joseph Henry Press, 2004), 181-83.

114 Although the substance of this charge was made in her book, the precise wording was extracted from an article she posted on a pro-GE website on February 25, 2006, around 18 months after her book had been published. See: http://www.agbioworld.org/biotech-info/articles/biotech-art/pusztai-potatoes.html.

That article was for the most part the same as the section on Pusztai's research in her book, except that it elaborated more fully in some places and contained a few revisions.

115 In her words: "But oddly enough, in the entire poisoned rat debate no one seems to have seen the central flaw in Pusztai's experiments: the absence of appropriate controls," *Mendel in Kitchen*, op. cit. note 113, 182.

116 Royal Society, *Review of data on possible toxicity of GM potatoes,* June 1999.

117 Royal Society Report (2002), op. cit. note 104, 5. The relevant section discussed the Society's 1999 critique of Pusztai and then stated: "It concluded that the only way to clarify Dr Pusztai's claims would be to refine his experimental design and carry out further studies to test clearly defined hypotheses focused on the specific effects reported by him. Such studies, on the results of feeding GM sweet peppers and GM tomatoes to rats, and GM soya to mice and rats, have now been completed and no adverse effects have been found (Gasson and Burke, 2001)."

118 Although it eventually did get reviewed and published, that didn't occur until well *after* it was relied on to refute Pusztai's findings. And, according to David Schubert, the study was nonetheless deficient in several respects, and not nearly as strong as was Pusztai's. The paper was published as: Chen Z. et al., "Safety Assessment for Genetically Modified Sweet Pepper and Tomato," *Toxicology* 188 (2003): 297-307.

119 David Schubert pointed this out in an email to me.

120 Pusztai, A., Letter to Royal Society, February 6, 2002: http://ngin.tripod.com/300103f.htm

121 Ibid.

122 Email from David Schubert.

123 Gasson, M. and Burke, D., "Scientific perspectives on regulating the safety of genetically modified foods," *Nature Reviews Genetics* 2 (2001): 217-22.

124 Royal Society Report (2002), op. cit. note 104, 209.

125 In their paper, Ewen and Pusztai only discussed intestinal abnormalities. So the only tests that could cast serious doubt on the studies would have to demonstrate that the particular lectin involved actually causes the problems they detected in the rats. But it's extremely unlikely that such findings could be legitimately registered, since several published studies have demonstrated that that type of lectin is harmless to mammals at levels hundreds of times higher than was produced within the GE potatoes – and Pusztai's experiment showed that the

control potatoes that were spiked with the lectin used in his studies did not induce the problems that the GE potatoes did.

126 Burke, Derek, "GM crops: time to counter the scare stories and relax barriers," March 27, 2014: https://theconversation.com/gm-crops-time-tocounter-the-scare-stories-and-relax-barriers-24678 (accessed May 2014).

127 Burke specifically (and falsely) alleged that Pusztai made the claims in 1998. While it is true that in 2002 Stanley Ewen cautioned (in a submission to a committee of the Scottish Parliament) that the 35S viral promoter used in most GE foods could increase the risk of stomach and colon cancer by over-stimulating cellular growth, he did *not* claim that any GE foods had caused cancer. http://www.sundayherald.com/29821.

128 Fedoroff's article was posted February 25, 2006 at: http://www.agbioworld.org/biotech-info/articles/biotech-art/pusztai-potatoes.html. As of March 14, 2006, Pusztai had not received a reply from Fedoroff to his message informing her of her error, so he authorized his critique of her statements to be posted at: http://gmwatch.org/index.php/news/archive/2006/1937-pusztai-replies-to-fedoroff

As of Aug. 25, 2013, Fedoroff's article was still on the AgBioWorld site in its original, erroneous form.

129 Although the EMS epidemic caused by the toxic tryptophan supplement broke out during 1989, it extended into 1990, and the research linking it to Showa Denko's genetically engineered bacteria didn't occur until that year.

130 Gab-Alla, A.A. et al., "Morphological and biochemical changes in male rats fed on genetically modified corn" (Ajeeb, Y.G), *J Am Sci.* 8 (9)(2012): 1117–23.

131 Gab-Alla, A.A. et al., "Histopathological changes in some organs of male rats fed on genetically modified corn" (Ajeeb, Y.G.), *J Am Sci.* 8 (10)(2012): 684–96.

132 Finamore, A., Rosell, M., Britti, S., et al., "Intestinal and peripheral immune response to MON810 maize ingestion in weaning and old mice," *J Agric Food Chem.* 56 (2008): 11533–39; doi:10.1021/jf802059w.

133 Krzyzowska, M., Wincenciak, M., Winnicka, A., et al., "The effect of multigenerational diet containing genetically modified triticale on immune system in mice," *Pol J Vet Sci.*13 (2010): 423-30.

134 Tudisco, R., Lombardi, P., Bovera, F., et al., "Genetically modified soya bean in rabbit feeding: Detection of DNA fragments and evaluation of metabolic effects by enzymatic analysis," *Anim Sci.* 82 (2006): 193–99. doi:10.1079/ASC200530.

135 Malatesta, M., Boraldi, F., Annovi, G., et al., "A long-term study on female mice fed on a genetically modified soybean: effects on liver ageing," *Histochem Cell Biol.* 130 (2008): 967–77.

136 Malatesta, M., Caporaloni, C., Gavaudan , S., et al., "Ultrastructural morphometrical and immunocytochemical analyses of hepatocyte nuclei from mice fed on genetically modified soybean," *Cell Struct Funct.* 27 (2002): 173–80; Malatesta, M., Caporaloni, C., Rossi, L., et al., "Ultrastructural analysis of pancreatic acinar cells from mice fed on genetically modified soybean," *J Anat.* 201 (2002): 409–15; Malatesta, M., Biggiogera, M., Manuali, E., Rocchi, M.B.L., Baldelli, B.,

Gazzanelli, G., "Fine structural analyses of pancreatic acinar cell nuclei from mice fed on genetically modified soybean," *Eur J Histochem.* 47 (2003): 385–388.

137 Interview in documentary film: Robin, M.M., "The World According to Monsanto [film]," 2008.

138 Séralini, G.E., Clair, E., Mesnage, R., et al., [RETRACTED:] "Long term toxicity of a Roundup herbicide and a Roundup-tolerant genetically modified maize," *Food Chem Toxicol.* 50 (2012): 4221-31.

139 Hammond, B., Dudek, R., Lemen, J., Nemeth, M., "Results of a 13 week safety assurance study with rats fed grain from glyphosate tolerant corn," *Food Chem Toxicol.* 42 (2004): 1003-14. doi:10.1016/j.fct.2004.02.013.

140 European Food Safety Authority (EFSA); Opinion of the Scientific Panel on Genetically Modified Organisms on a request from the Commission related to the safety of foods and food ingredients derived from herbicide-tolerant genetically modified maize NK603, for which a request for placing on the market was submitted under Article 4 of the Novel Food Regulation (EC) No 258/97 by Monsanto (QUESTION NO EFSA-Q-2003-002): Opinion adopted on November 25, 2003, *EFSA J.* 2003(9) (2003): 1–14.

141 De Vendomois, J.S., F. Roullier, D. Cellier, G.E. Séralini, "A comparison of the effects of three GM corn varieties on mammalian health," *Int J Biol Sci.* 5 (2009): 706–26.

142 Even if the substance being tested has a tendency to induce tumors at a higher than normal rate, if each group of rats contains only 10 of each sex (as was the case in Séralini's study), no tumors might be observed, whereas if 50 per sex per group are used (as is ordinarily done in cancer studies), the study has a much better chance of detecting some tumors. However, that doesn't entail that using a lower number of rats somehow invalidates any statistically significant increase in tumors that *is* observed. As long as a sufficient number was used to reliably register such differences (which was enabled by the number Séralini employed), those differences are valid.

143 Thus, Peter Saunders, emeritus professor of mathematics at King's College London, has stated that the smaller number of rats "makes the results if anything *more* convincing, not less." He explained: "Using a smaller number of rats actually made it *less* likely to observe any effect. The fact that an effect was observed despite the small number of animals made the result all the more serious." Saunders, P., "Excess cancers and deaths with GM feed: The stats stand up," *Sci Soc.* (2012). Available at: http://www.i-sis.org.uk/Excess_cancers_and_deaths_from_GM_feed_stats_stand_up.php.

144 Only one in ten control rats developed a tumor, and even then it occurred late in their lives.

145 Committee on Publication Ethics (COPE), Retraction guidelines, 2009. Available at: http://publicationethics.org/files/retraction%20guidelines.pdf

146 Hayes, A.W., "Response to Letters to the Editors," December 2013. Available at: http://www.elsevier.com/about/press-releases/research-and-journals/food-and-chemical-toxicology-editor-in-chief,-a.-wallace-hayes,-publishes-response-to-letters-to-the-editors#sthash.tTW2LCGq.dpuf.

147 In his response to the retraction, Seralini pointed out: "It should be noted that tumorigenesis is not synonymous with cancer. Tumors can be in some cases more rapidly lethal than cancers because their size can cause hemorrhages and possible impairments of vital organs, as well as secretion of toxins." Séralini, G.E. et al., "Conclusiveness of toxicity data and double standards, Food and Chemical Toxicology" (2014), doi: http://dx.doi.org/10.1016/j.fct.2014.04.018

148 OECD guideline no. 452 for the testing of chemicals: Chronic toxicity studies: Adopted September 7, 2009.

149 He stated: "While the number of animals used may have been sufficient to reach conclusions regarding oral toxicity, it proved insufficient for conclusions related to the carcinogenicity of the test substances." Hayes Response (2013), op. cit. note 146.

150 As reported in the *New York Times:* "The editor of the journal, Food and Chemical Toxicology, said in a letter to the paper's main author that the study's results, while not incorrect or fraudulent, were 'inconclusive, and therefore do not reach the threshold of publication. " Pollack, A., "Paper Tying Rat Cancer to Herbicide is Retracted," *New York Times*, November 28, 2013. Seralini quoted that letter as having stated that the raw data were "'not incorrect'". Séralini et al., op. cit. note 147.

151 Schubert, D., "Science study controversy impacts world health," *U-T San Diego*: http://www.utsandiego.com/news/2014/jan/08/science-food-health/, published January 8, 2014.

152 The words "could have been caused" have been used because the results provided reasonable grounds for thinking that this had happened but did not decisively demonstrate it.

153 The research demonstrates that the level of herbicide residue on marketed Roundup-ready plants can induce substantial damage to animals that eat them, which entails that all such plants are dangerous; and it also indicates that harmful effects might be induced by the expression of the gene that's inserted to confer the Roundup resistance, which casts additional doubt on the safety of all the plants that contain it.

154 Pollack, A., "Paper Tying Rat Cancer to Herbicide is Retracted," *New York Times*, November 28, 2013.

155 Environmental Sciences Europe (ESEU, 2014, 26:14).

156 Fagan, J., Antoniou, M.C., and Robinson,C., *GMO Myths and Truths: An Evidence-Based Examination of the Claims Made for the Safety and Efficacy of Genetically Modified Crops*, 2nd Edition, version 1.0, (London: Earth Open Source, 2014), 147.

157 It's become so standard, and so obvious, that it's been noted by several other commentators.

158 Statement by the AAAS Board of Directors On Labeling of Genetically Modified Foods, October 20, 2012: http://www.aaas.org/sites/default/files/migrate/uploads/AAAS_GM_statement.pdf

159 European Commission, A decade of EU-funded GMO research (2001–2010), 2010.

160 http://earthopensource.org/index.php/3-health-hazards-of-gm-foods/3-2-myth-eu-research-shows-gm-foods-are-safe

161 European Commission, A decade of EU-funded GMO research (2001–2010), 2010.

162 Snell, C., Aude, B., Bergé, J., et al., "Assessment of the health impact of GM plant diets in long-term and multigenerational animal feeding trials: A literature review," *Food Chem Toxicol.* 50 (2012): 1134-48.

163 *Myths and Truths*, op. cit. note 156, 138, 140, 161.

164 Ibid., 162.

165 E.g., Fares, N.H., El-Sayed, A.K., "Fine structural changes in the ileum of mice fed on delta-endotoxin-treated potatoes and transgenic potatoes," *Nat Toxins* 6(6) (1998): 219-33.

166 *Myths and Truths*, op. cit. note 156, 105. I'm indebted to the authors of this document for this and the following insights about the weaknesses of the Nicolia review.

167 Ibid.

168 Ibid., 107-08. The authors of this critique point out that Nicolia and colleagues do not base their conclusion "on empirical data, reasoned scientific argument, or even peer-reviewed papers" but on four non-peer-reviewed opinion pieces "containing inaccuracies and unsubstantiated personal views."

169 Ibid., 106. Although another long-term study on a GE glyphosate-tolerant crop (a soybean) has been done, Claire Robinson has noted that it's doubtful the beans were sprayed with glyphosate – at least not to a degree even close to the amount that's typically applied by farmers. That's because the herbicide was detected at an extremely low level: far lower than when beans are sprayed in actual farming operations. So the test was limited to assessing the effects of the insertional event itself, and it was ill-designed to do so because the control bean was not the isogenic variety. (Sakamoto Y. et al., "A 104-week feeding study of genetically modified soybeans in f344 rats." *Shokuhin Eiseigaku Zasshi* 49 (2008):272-282. (An English translation of the study is available at: net.gedal.fr/knowledgebase/docs/A593742.pdf)

170 Ibid., Section 2.3. For example, the main review paper also failed to discuss a multigenerational study in which rats that consumed GE Bt maize over three generations not only displayed alterations in blood chemistry but suffered liver and kidney damage. (Kilic A, Akay MT., "A three generation study with genetically modified Bt corn in rats: Biochemical and histopathological investigation," *Food Chem Toxicol.* 46 (2008):1164–70. doi:10.1016/j.fct.2007.11.016.)

The other review papers that purport to demonstrate the safety of GE foods have likewise failed due to multiple deficiencies. One that's frequently cited is a survey of the data on livestock that were fed GE crops. (Van Eenennaam, A. and Young, A.E., "Prevalence of impacts of genetically engineered feedstuffs on livestock populations." J Anim Sci, 92 (2014):4255-4278.) But besides relying on uncontrolled studies of animals with dissimilar digestive systems than humans to try to refute the adverse findings of well-controlled studies on standard laboratory animals that are much better models for our physiologies, this paper contains several other serious flaws that undercut its claims, as is incisively demonstrated at: http://www.gmwatch.org/index.php/news/archive/2014/15717-junk-science-and-gmo-toxicity and http://www.gmwatch.org/index.php/news/archive/2014/15669

In contrast, around the same time that defective review appeared, another review was published in a peer-reviewed journal providing further confirmation that the safety of the GE crops on the market has *not* been established. It focused on those containing at least one of the three most prevalent genes used in creating GE plants, and it investigated the extent to which these crops had been subjected to reliable tests employing standard laboratory animals and histopathological study of the digestive tract – a far more sensitive type of study than the superficial inspections relied on by the Van Eenennaam review. The researchers discovered that although there were 47 crop varieties with one or more of these genes that had been approved by regulators for animal or human consumption, only 9 (19%) had been tested via such histopathological study. Further, 76% of those tests were done *after* the crop had been approved for marketing, and half were published at least 9 years after approval. Worse, the researchers could not find a single study that was properly conducted or reported. Zdziarski I.M. et al., "GM crops and the digestive tract: A critical review," *Environment International* 73 (2014): 423-433. http://gmojudycarman.org/wp-content/uploads/2014/10/Zdziarski-et-al-14-GM-crops-and-rat-digestive-tract-review.pdf

171 For example, during the meeting of the Food Advisory Committee in regard to the Flavr Savr tomato, a participant asked the FDA to respond to a charge that had been made by Rebecca Goldburg (who represented a public interest organization) that its policy on GE foods had illegitimately shifted the burden of proof. But instead of admitting this obvious truth, the biotechnology manager, James Maryanski, engaged in denial. He argued: "The standards by which a substance must be determined to be GRAS are based on the case history and the law, and so we did not change that standard whatsoever." But Goldburg would not let him get away with it; and she pointed out how the FDA had, in practical effect, changed the standard. FAC transcript, op. cit. note 18, vol. 3, 138-41.

172 Document #15, p. 3 at: http://biointegrity.org/24-fda-documents.

173 Ibid.

174 Ibid.

175 Parrott, W., Chassy, B.M., "Is this study believable? Examples from animal studies with GM foods," 2009, 8. Available at: http://agribiotech.info/more-details-on-specific-issues.

176 Ibid., 6.

177 Folta, K., quoted in Johnson, N., "Food for bots: Distinguishing the novel from the knee-jerk in the GMO debate," *Grist*, August 22, 2013: http://grist.org/food/dodging-argument-bot-crossfire-to-revisit-some-gm-research-controversies/

Further, it's clear Folta was arguing that trying to discredit the belief that GE foods are as safe as naturally produced ones is on a par with trying to disprove that gravity exists – and not with merely attempting to replace an accepted theory about what gravity is with another that's alleged to explain the phenomenon better. That's because of how he illustrated his point that "pro- and anti-GM science" have different verificational thresholds. He stated: "For example: To test the hypothesis that gravity does not exist on earth I need some elaborate mechanisms, many replicates, tons of math and new models of thinking that change our understanding of basic fundamentals of natural science. To test the hypothesis that gravity exists, I have to push a pencil off of my desk. Two very different evidence thresholds."

178 21 CFR Sec. 170.30(b); 21 CFR 170.3(I). While I'm aware that it's impossible to prove a food is safe as conclusively as one can prove a mathematical proposition is true, I'm employing the terms "prove" and "burden of proof" as understood within the context of food safety regulation. And within the context of US law, the manufacturer bears the burden of proof; and the standard of proof is clearly defined: a demonstration that there's a "reasonable certainty" the product won't be harmful under its intended conditions of use.

179 *Myths and Truths*, op. cit. note 156, 142-144.

180 Ibid., 143.

181 Ibid.

182 Antoniou, Michael, email communication.

183 In *United States v. An Article of Food, etc.* 678 F. 2d 735 (5th Cir. 1982), the court upheld the FDA's charge that a substance was not GRAS based on the testimony of five doctors. And in *United States v. Seven Cartons . . . Ferro-Lac*, 293 F. Supp. 660, 664 (S.D. Il. 1968), the court denied GRAS status to a substance based on the affidavits of two scientists who said that they were not aware of any studies in the scientific literature showing it was safe. (This ruling was subsequently modified on other grounds by an appellate court, 424 F.2d 136 (7th Cir. 1970) – grounds that did not affect the holding about the sufficiency of two experts.)

184 http://www.ensser.org/increasing-public-information/no-scientific consensus-on-gmo-safety/. Although some of the signatories do not have graduate degrees in one of the directly relevant life sciences, a large number do; and their number is more than sufficient to defeat the claim that consensus about safety exists within the expert community.

185 Schubert, D., Letter to the *Los Angeles Times*, October 28, 2012.

186 This statement assumes there won't be any hasty, and successful, attempts to alter the laws.

11. Overlooked Lessons from Computer Science

1 Writing in 2002, Evelyn Fox Keller observed: "Computer metaphors have been commonplace in biology for almost half a century," Keller, E.F., *Making Sense of Life*, Harvard University Press, Cambridge, MA (paperback edition, 2003), 247.

2 http://scienceblogs.com/tomorrowstable/2012/09/24/rachel-carsons-dream-of-a-science-based-agriculture-may-come-as-a-surprise-to-those-who-believe-that-sustainability-and-technology-are-incompatible/

3 I'm employing the word "misrepresent" merely to denote that an inaccurate representation has been conveyed, not to imply that the inaccuracy was part of an intentional effort to mislead people. In regard to software, I think the erroneous statements of the GE proponents have stemmed from a failure to fully comprehend the facts, not from a desire to obfuscate them.

4 This example comes from: http://en.wikipedia.org/wiki/Data_(computing)

5 http://en.wikipedia.org/wiki/Computer_programming

6 A prevalent form of such programming is termed *object-oriented design.*

7 Although some programs are not command-based *imperative codes* and instead specify outcomes without dictating the discrete steps through which they're to be attained (and are thus called *declarative codes*), the programmers still aim to achieve predictable outcomes that follow linearly from the initial specifications.

8 For one discussion of ravioli code, see: http://www.techopedia.com/definition/26876/ravioli-code. Another alternative, in which the program is somewhat layered, is called *lasagna code.*

9 Strohman, Richard, "The Coming Kuhnian Revolution in Biology," *Nature Biotechnology* 15, March 1997.

10 Ibid., 197.

11 Ibid., 199.

12 Ibid., 197.

13 Strohman, Richard, "Beyond Genetic Determinism: Toward a New Paradigm of Life," *Pressing Times*, Spring 2002: http://www.mindfully.org/GE/GE4/Beyond-Genetic-Determinism-Apr02.htm

14 Keller, E.F., *The Century of the Gene* (Harvard University Press, Cambridge, MA, 2000), 100.

15 Ibid.

16 Ibid., 100-01.

17 Ibid., 101.

18 Ibid., 162, n. 52.

19 Strohman, "The Coming Kuhnian Revolution," op. cit. note 9, 194.

20 Even when one segment of code sometimes acts as an instruction and sometimes serves as data, the role it's playing is clear in any given instance.

21 Conrad, Michael, "The Importance of Molecular Hierarchy in Information Processing," in *Towards a Theoretical Biology* 4: *ESSAYS*, Waddington, C.H., ed. (Chicago: Aldine-Atherton, Inc., 1972), 222-28. The quoted text is on p. 224.

22 Ibid., 225.

23 Ibid., 226.

24 Whitehouse, D., "Scientists Hail New 'Map of Life'": http://news.bbc.co.uk/go/pr/fr/-/2/hi/science/nature/3223318.stm

25 A minority of the promoters are always in an open, receptive state. They will be discussed a bit later.

26 Yuh, Chiou-Hwa and Davidson, E.H., "Modular cis-Regulatory Organization of *Endo 16*," *Development* 12 (1996): 1069-82.

27 Yuh, Chiou-Hwa et al., "Genomic cis-Regulatory Logic: Experimental and Computational Analysis of a Sea Urchin Gene," *Science* 279 (1998): 1896-1902.

28 Yuh, Chiou-Hwa et al., "Cis-Regulatory Logic in the Endo 16 Gene: Switching from a Specification to a Differentiation Mode of Control," *Development* 128 (2001): 617-29.

29 Wray, G.A., "Promoter Logic," *Science* 279 (1998): 1872.

30 Keller, *Making Sense of Life*, op. cit. note 1 above, 338-9, n. 12.

31 Ibid., 339.

32 Wray, op. cit. note 29, 1871.

33 Ibid., 1872.

34 In fact, the knowledge regarding plant promoters may be even more deficient than in the case of animal promoters. According to molecular biologist Allison Wilson, greater resources have been expended on the study of the latter.

35 https://www.owasp.org/index.php/Race_Conditions

36 Lehman, M. M., "Program, Life-Cycles and the Laws of Software Evolution," *In Proceedings of IEEE* 68(9) (1980): 1060-76; Lientz, B., E. Swanson., *Software Maintenance Management* (Addison Wesley, Reading, MA, 1980).

37 Schach, R., *Software Engineering*, Fourth Edition (McGraw-Hill, Boston, MA, 1999), 11.

38 21 CFR Parts 807, 814

39 IEC 62304

40 This statement pertains to safety testing; and it describes the situation according to the standpoint of the FDA. If the inserted cassette produces a pesticidal protein, then the EPA has authority to regulate it, but it has rarely required meaningful safety testing of such substances either. See, e.g., Marden, E., "Risk and Regulation: U.S. Regulatory Policy on Genetically Modified Food and Agriculture," 44 B.C. L. Rev. 733.

41 For instance, FDA regulations state that in the case of "high risk devices that pose a significant risk of illness or injury" (Class III devices) safety claims must be supported by "the submission of clinical data" (21 CFR Part 814). Extensive non-clinical data is also required. In contrast, devices that entail lower risk (Class II) can gain approval by establishing that they're substantially equivalent with specific types of legally marketed devices (CFR Part 807 Subpart E). And some of the least risky (Class I) don't even require pre-market notification: http://www.millerassociates.net/files/SW_Risk_Mgmt_Arch.gif.

However, according to Tom Miller, whose company (Miller Associates, Inc.) performs quality assurance for software used in medical devices, the 'substantial equivalence' route doesn't eliminate the need for testing, and to satisfy FDA standards, the software in Class II devices must undergo "significant testing," albeit not as stringent as for Class III (email communication). Mark Rainbow, a software engineer who works for a medical device company, concurs, stating that when substantial equivalence is claimed, "the FDA still requires full testing of the device operations and complete reviews of the design and an analysis of potential hazard" (email communication).

Further, the international standard for the testing of life-critical software, which the FDA recommends but does not formally require, is even stricter than the standard imposed by the latter. A flow chart on the Miller Associates website provides insight into the thoroughness with which software is assessed and tested in order to satisfy this standard enforced in the EU nations and other countries – and provides a glaring contrast between the toughness these nations exhibit in the case of life-critical software and the laxness they display when it comes to GE foods: http://www.millerassociates.net/files/SW_Risk_Mgmt_Arch.gif.

42 This presumption is made by the FDA in regard to the safety of GE crops. The agency sets GE animals in a different category.

43 Because a much greater percentage of the population consume a particular variety of GE food than rely on a particular type of pacemaker or are exposed to a specific kind of X-ray machine, that food could cause much more harm than either.

44 As discussed in previous chapters, most harmful alterations to food don't cause immediate problems but do damage incrementally; and so the majority of such changes in GE foods would likely go unnoticed absent epidemiological testing, especially if they cause common ailments such as cancer.

45 E.g., Michaels, D., *Doubt is Their Product: How Industry's Assault on Science Threatens Your Health* (Oxford University Press, 2008); Barnes, D.E. and Bero, L.A., "Why review articles on the health effects of passive smoking reach different conclusions," *JAMA* 279 (1998): 1566-70; Lexchin, J., Bero, L.A., Djulbegovic, B., Clark, O., "Pharmaceutical industry sponsorship and research outcome and quality: systematic review," *Br Med J.* 326 (2003): 1167; doi:10.1136/bmj.326.7400.1167; Lexchin, J., "Those who have the gold make the evidence: How the pharmaceutical industry biases the outcomes of clinical trials of medications," *Sci Eng Ethics* (2011); doi:10.1007/s11948-011-9265-3; Bekelman, J.E., Li, Y., Gross, C.P., "Scope and impact of financial conflicts of interest in biomedical research: a systematic review," *JAMA* 289 (2003): 454-65.

46 Diels, J. et al., "Association of financial or professional conflict of interest to research outcomes on health risks or nutritional assessment studies of genetically modified products," *Food Policy* 36 (2011): 197–203.

47 Schach, op. cit. note 37.

48 Larry Seese, quoted in *The Risks Digest* 9(62), February 26, 1990.

49 Gleick, James, "A Bug and a Crash," *New York Times Magazine*, December 1, 1996.

50 Quoted in Gleick, above.

51 A GE food would only pose less risk if there was scant likelihood that its associated hazards would manifest – which, in light of the analysis in Chapter 9,

does not appear to be the case. The issue of risk will be examined more thoroughly as the chapter continues.

52 Leveson, N., and Turner, C., "An Investigation of the Therac-25 Accidents," *IEEE Computer* 26(7), July 1993, 18-41. This article describes both the "impressive" behavior of the FDA and the less than impressive behavior of the manufacturer, including statements from FDA officials critical of how it withheld evidence.

53 Personal communication from Tom Miller, whose qualifications are described in note 41 above.

54 Glover, Ann, "Interview" by *Eur Activ*, July 24, 2012: http://www.euractiv.com/innovation-enterprise/chief-scientifc-adviser-policy-p-interview-514074

55 For instance, there's still debate about how much of the DNA has a function, and even the functions of many sections that are known to be functional are not well understood. And the genomes of some of the plants that are engineered have not been fully sequenced. More importantly, most commercial lines are poorly characterized on the molecular level and only one has been fully sequenced: papaya. Further, even though the examination of that GMO was not rigorous enough to detect unintended mutations (because it was not compared to the isogenic parent line), it yet revealed that the insertion of new DNA was not restricted to the site from which the foreign gene functioned and that many small fragments of that gene were lodged in other locations.

56 Although hackers sometimes know the source code of the program they're invading (as in the case of open source code), they usually don't; and hacking is generally performed in the absence of such knowledge.

57 Although hackers insert several types of sequences, which have distinct functions, each almost always harms the software system or injures the interests of the user. Accordingly, they're generally referred to as "malware," because they serve malicious purposes. And although a specific class of malware are referred to as "viruses," all malware exhibits some basic features of biological viruses; and those are the features described in the main text.

58 Christensen, D., "Beyond Virtual Vaccinations," *Science News*, July 31, 1999 (The words the article quoted were those of an expert at IBM, Steve R. White).

59 Brown, Patrick, "The Promise of Plant Biotechnology – the Threat of Genetically Modified Organisms," July 2000. Available at: http://www.campaignforrealfarming.org/2012/01/the-promise-of-plant-biotechnology-the-threat-of-genetically-modified-organisms/

60 Although there have been exceptions, such as the unintended consequences of the "Morris worm" released in 1988.

61 "DNA as Software," *All Things Considered, National Public Radio*, April 25, 2003: http://www.npr.org/templates/story/story.php?storyId=1244325

62 http://en.wikipedia.org/wiki/Northeast_blackout_of_2003

63 Hagan, N. et al., "The redistribution of protein sulfur in transgenic rice expressing a gene for a foreign, sulfur-rich protein," *Plant J.* 34 (2003): 1–11.

64 Regal, P., "Scientific Principles for Ecologically Based Risk Assessment of Transgenic Organisms," *Molecular Ecology* 3 (1994): 5-13 (The quoted sentences are from a section relevant to food safety as well as ecological safety).

65 This can occur when the transcription process starts at different locations, with the result that a single base can be part of one distinct three-base codon on one occasion and a constituent of a different codon on another, depending on where the transcription of RNA begins. In consequence, a discrete segment of DNA can participate in generating diverse segments of RNA. However, although a segment of DNA can be transcribed via alternate, overlapping reading frames, the amino acid code itself is nonoverlapping, which means that (barring errors) a single reading frame is always directly transcribed into the same RNA sequence (which can subsequently be altered via alternate splicing).

66 Segal, E. et al., "A Genomic Code for Nucleosome Positioning," *Nature* 442 (August17, 2006): 772-78. The word "superimposed" was employed by the journalist who reported the discovery in the *New York Times*: Wade, N., "Scientists Say They've Found a Code Beyond Genetics in DNA," *New York Times*, July 25, 2006: http://www.nytimes.com/2006/07/25/science/25dna.html

67 Weatheritt, R. and Babu, M., "The Hidden Codes that Shape Protein Evolution," *Science* 13: 342 (6164) (December 2013): 1325-26. The authors refer to the codes as "regulatory."

68 Stergachis, A. et al., "Exonic Transcription Factor Binding Directs Codon Choice and Affects Protein Evolution," *Science* 342(6164) (December 13, 2013): 1367-72.

69 "Scientists Discover Double Meaning in Genetic Code," *Press Release*, University of Washington December 12, 2013: http://www.washington.edu/news/2013/12/12/scientists-discover-double-meaning-in-genetic-code/

70 However, since the statement appeared in a press release, it may merely reflect the desire of the university's public relations department to hype the importance of the discovery. In any event, the resilience of ingrained presumptions to new evidence is a well-recognized phenomenon within the biological sciences, especially as regards the presumptions on which the GE venture relies.

71 Email to a group of concerned scientists, December 14, 2013.

72 Barbara A. Caulfield, executive vice president of Affymetrix, Inc., quoted in Caruso, D., "A Challenge to Gene Therapy, a Tougher Look at Biotech," *New York Times*, July 1, 2007.

73 Wilson, Stephen, "We're Not Ready for Genetic Engineering," January 15, 2011: http://lockstep.com.au/blog/2011/01/15/not-ready-for-gm

74 Ibid. Update of September 2012.

75 Ibid. Reply to a reader comment on September 11, 2012.

76 Dawkins, Richard, "Why Prince Charles is So Wrong," London Times, January 28, 2003.

77 The professorship was endowed by Charles Simonyi, who oversaw the development of Microsoft's suite of Office applications.

78 Gates, Bill, *The Road Ahead*, Penguin 1996, 228. Although Gates was referring specifically to human DNA, it's reasonable to presume that he would apply his comment to the DNA of other organisms; and it's doubtful he thinks humans have created software more advanced than the information systems of plants and animals.

79 In his speech at the World Food Prize event on October 15, 2009 he stated: "In some of our grants, we include transgenic approaches because we believe they can help address farmers' challenges faster and more efficiently than conventional breeding alone. Of course, these technologies must be subject to rigorous scientific review to ensure they are safe and effective." And it's evident he believes that adequate testing is regularly performed or readily can be. Otherwise, he would have refrained from putting so much money into developing GE crops – or at the least postponed it until after he had funded projects to develop a testing regime better suited to a technology that's altering the world's most complex information systems. Moreover, as the main text observes, if he carefully considered bioengineering in light of what's known about software engineering, he would not fund it at all and would exclusively foster sounder forms of food production.

12. Unfounded Foundational Assumptions

1 "Elements of Precaution: Recommendations for the Regulation of Food Biotechnology in Canada," The Royal Society of Canada (January 2001): 184.

2 Caruso, Denise, "A Challenge to Gene Theory, a Tougher Look at Biotech," *New York Times*, July 1, 2007: http://www.nytimes.com/2007/07/01/business/yourmoney/01frame.html?pagewanted=1&_r=2&ref=yourmoney&

3 Bernardi, Giorgio, "The Role of Chance in Evolution," in *Scientific Insights into the Evolution of the Universe and of Life*, Pontifical Academy of Sciences, Acta 20 (2009): 233. www.pas.va/content/dam/accademia/pdf/acta20/acta20-bernardi.pdf

4 Hurst, Laurence, D. et al., "The Evolutionary Dynamics of Eukaryotic Gene Order," Nature Reviews Genetics, 5 (2004): 299-310. This paper stated that "gene order has typically been assumed to be random." This assumption pertained to organisms with cellular nuclei but not bacteria, which were known to contain structures (called *operons*) in which several genes are grouped together under the control of a single promoter.

5 Bernardi, op. cit. note 3, 233.

6 Ibid.

7 Hurst, op. cit. note 4, 308.

8 Michael Antoniou, Testimony to New Zealand Royal Commission on Genetic Modification.

9 Email from Michael Antoniou. The UK GM Science Review Panel sat for two sessions. Antoniou was a member of the second panel.

10 Although the word "organic" has several denotations, an important one refers to a systematic interconnection of parts suggestive of the structure displayed by living organisms.

11 Institute of Food Technologists, *IFT Expert Report on Biotechnology and Foods*, 2000, 17.

12 Schubert, D., "A Different Perspective on GM Food," *Nature Biotechnology* 20 (October 2002): 969.

13 Beachy, R. et al., Letter to the Editor, *Nature Biotechnology* 20 (December 2002): 1195.

14 Ibid.

15 For instance, an article published in 2012 by six scientists who advocate GE foods contrasts the allegedly precise modifications made through bioengineering with the "random genetic modifications that occur in conventional breeding." Weber, Natalie et al., "Crop Genome Plasticity and Its Relevance to Food and Feed Safety of Genetically Engineered Breeding Stacks," *Plant Physiology* 160 (December 2012): 1842.

16 National Research Council and Institute of Medicine of the National Academies (NAS), "Safety of Genetically Engineered Foods: Approaches to Assessing Unintended Health Effects" (Washington D.C.: The National Academies Press, 2004), 46.

17 When sexual breeding does introduce new risk, it's usually because a wild and weedy relative that is, itself, not wholesome for humans has been crossed with a domesticated variety. So the cases requiring caution are ordinarily known. Further, in such cases the harmful substance is usually a toxin or anti-nutritive factor that already exists within the species, that's present in higher concentration in the wild varieties, and that gets expressed in the hybrid at greater levels than is normal for the cultivated varieties. Consequently, it's easier to screen for such risks because breeders know what substance to monitor.

18 All of these defects have been thoroughly discussed in Chapter 9. Moreover, (as also discussed in that chapter), most engineered plants must undergo substantial backcrossing too, and (as Chapter 5 noted) the average GE crop takes longer to develop than traditionally bred ones and, for several reasons, entails much higher cost. Consequently, there's no trade-off between increased risk and reduced cost. Instead, risks and costs are both increased.

19 Pollack, Andrew, "Panel Sees No Unique Risk From Genetic Engineering," *New York Times*, July 28, 2004.

20 Of course, one of the main reasons that the testing has been inadequate is the unrealistic belief that very little data is needed in order to demonstrate that a GE food is substantially equivalent to it's naturally produced counterpart. So faith has been to a significant degree responsible for the deficient testing.

13. The Devolution of Scientists into Spin Doctors

1 Bronowski, J., Science and Human Values, (New York: Harper & Rowe, 1965), 28.

2 Ibid., 46.

3 Ibid., 25.

4 Dutton, Diana B., *Worse than the Disease: Pitfalls of Medical Progress* (New York: Cambridge University Press, 1992), 193.

5 Ibid.

6 Ibid., 194.

7 Ibid., 194-95.

8 Ibid., 195.

9 Ibid., 193.

10 Ibid., 195.

11 Ibid.

12 As reported in *Lords of the Harvest,* Monsanto spent at least one billion dollars on research involving GE plants before it had produced even one that was marketable. Charles, Dan, *Lords of the Harvest: Biotech, Big Money, and the Future of Food* (Cambridge: Perseus, 2002), xv.

13 When I speak of modern biotechnology corporations, I'm referring to those that employ rDNA technology. And the first one of these (Genentech) was not founded until April, 1976 (by the biochemist Herbert Boyer and the venture capitalist Robert Swanson). Further, it took a few years before that company had dispelled widespread doubts about the ability of rDNA technology to produce commercially valuable products, and it did not go public until 1980, http://www.gene.com/media/company-information/chronology. Although, as Chapter 4 noted, the Cetus corporation (which had been founded by scientists in Berkeley) was pitching the promise of genetic engineering to investors in 1975, it didn't start employing that technology until well after Genentech got going: http://en.wikipedia.org/wiki/Cetus_Corporation.

14 Although Monsanto had begun research on how to produce GE plants in the late 1970's, it did not actually produce one until 1982, and it apparently did not play a significant role in the lobbying efforts mounted by the biotech industry during that decade. For instance, in her comprehensive study of the development of US biotechnology policy, Mary Ellen Jones does not mention any involvement of Monsanto until the 1980's; and her first citation of a Monsanto communication to the government is a letter it sent to the NIH sometime between 1980 and 1982 (Jones, Mary Ellen, "Politically Corrected Science: The Early Negotiation of U.S. Agricultural Biotechnology Policy," a Doctoral Dissertation in Science and Technology Studies at Virginia Polytechnic Institute [1999], n. 289, 110). Further, the index of Susan Wright's exhaustive history of the development of US biotech policy between 1972 and 1982 provides only one mention of Monsanto: in connection with an agreement it formed with the Harvard Medical School in 1974 to fund research in medically related areas of biotech (Wright, Susan, *Molecular Politics: Developing American and British Policy for Genetic Engineering 1972-1982* [Chicago: University of Chicago Press, 1994]).

15 For instance, a 2003 survey by researchers at North Dakota State University found consumers ranked university scientists as the most trustworthy source of information about GE foods (along with the US Department of Agriculture), regarding them as far more reliable than public interest groups, and, due to the

technical nature of the issue, even the clergy. Wachenheim, C. J. and W. D. Lesch, "North Dakota Shoppers Perceptions of Genetically Modified Organisms and Food: Results of a Winter 2003 Survey," Department of Agribusiness and Applied Economics, *North Dakota State University*, Agribusiness & Applied Economics Report No. 540, June 2004, p.v. (This study specifically asked for opinions about the USDA but not the FDA. Others have found that when polled about the FDA, consumers place it in the top tier of reliability as well.)

Priest, S. H., and Talbert, J., "Mass Media and the Ultimate Technological Fix: Newspaper Coverage of Biotechnology," *Southwestern Mass Communication Journal*, 10 (1), (1994): 76-85.

16 Priest and Talbert (1994), op. cit. note 15.

17 American Medical Association, Policy Statement on Biotechnology and the American Agricultural Industry, 1990.

18 For example, a report issued in 2012 contains several misleading statements. (American Medical Association, Report of the Council on Science and Public Health, CSAPH Report 2-A-12, 2012.) A few of them follow.

On p. 2, the report states: "Bioengineered foods have been consumed for close to 20 years, and during that time, no overt consequences on human health have been reported and/or substantiated in the peer-reviewed literature. . . . However, a small potential for adverse events exists." But it doesn't document that last assertion, nor does it explain how its authors arrived at the conclusion that the potential is "small." Nor does it acknowledge that many experts think the potential is significant.

On p. 4 the report denigrates the Pusztai study (that was examined in Chapter 10) by stating that "the experimental design of this study is widely regarded as flawed, with subsequent studies unable to reproduce the findings." But Chapter 10 has demonstrated why both these assertions are deceptive. Moreover, the AMA report goes on to imply that whatever differences may have been found in the rats consuming the engineered potatoes were likely caused by consumption of the lectins that were expressed by the transgene, despite the fact that (as Chapter 10 has explained) the presence of the lectins was controlled for and cannot explain the differences.

Later on p. 4, the report claims that safety assessments based on the concept of "substantial equivalence" involve "a thorough comparison" between the GE crop and its conventionally bred counterpart – a statement at odds with the opinion of numerous independent experts, including the panel that produced the 2001 report of the Royal Society of Canada.

19 Mestel, Rosie, "Scientists defend safety of genetically modified foods," *Los Angeles Times*, October 24, 2012: http://www.latimes.com/news/science/la-scigmo-food-safety-20121025,0,5914417.story?page=2

20 Although a minority of the plant's genes are attached to promoters that induce them to express in a continual manner, those promoters don't ordinarily compel the level of expression that the viral promoters do. Moreover, because those genes are essential to the plant's function, their constant expression is harmonized with the operations of the other genes. But the incessant activity of the alien genes is not.

21 For instance, besides forcing the inserted genes to hyper-express in an unregulated manner, the viral promoters that are affixed to them can directly disturb the function of some native genes. Moreover, (as Chapter 6 discussed) there's evidence that the proteins expressed by the inserted genes can gain unintended add-ons or become misfolded, either of which pose a health risk.

22 Finz, Stacy, "Biotech food measure Prop. 37 on ballot," *San Francisco Chronicle*, August 15, 2012: http://www.sfgate.com/news/article/Biotech-food measure-Prop-37-on-ballot-3788811.php

23 The first bacterium endowed by humans with recombinant DNA was created in 1973, and the first GE plant was created in 1982 (and was publicly announced in January 1983): *Timeline of Plant Tissue Culture and Selected Molecular Biology Events*, University of Florida Horticultural Sciences Department.

24 If he had merely intended to convey the idea that GE crops are just as safe as naturally produced ones, he could have said so instead of calling them "the safest crops."

25 Finz, op. cit., note 22.

26 https://www.mcdb.ucla.edu/Research/Goldberg/the_seed_institute/Biotech_exploit.pdf

27 Statement by the AAAS Board of Directors On Labeling of Genetically Modified Foods, October 20, 2012: http://www.aaas.org/sites/default/files/migrate/uploads/AAAS_GM_statement.pdf

28 http://www.newyorker.com/online/blogs/elements/2014/04/a-civildebate-over-genetically-modified-food.html

29 Cook, Guy, *Genetically Modified Language: The Discourse of Arguments for GM Crops and Food*, (London: Routledge, 2004), 2.

30 http://www.psrast.org/promplantbiot.htm

31 Dutton, op. cit. note 4, 195.

32 Quoted in Charman, K., "Brave New Nature: Spinning Science into Gold," *Sierra Club Magazine*, July/August 2001: http://www.sierraclub.org/ sierra/200107/charman.asp

33 Ibid.

34 Quoted in Ibid.

35 The information in this paragraph and the one that follows was directly communicated to me by Elaine Ingham.

36 Philip J. Regal, PhD: Declaration submitted to the US District Court, *Alliance for Bio-Integrity v. Shalala*, 1998.

37 Reported in Charman, K., op. cit. note 32.

38 Email from David Schubert.

39 Ibid.

40 One prominent GE defender who refers to scientists with whom he disagrees as "outliers" is Jon Entine. See: forbes.comhttp://www.forbes.com/sites/jonentine/2014/08/14/got-soy-milk-not-consumer-reports-which-throws-science-under-the-bus-in-warning-about-gmo-soy/.

41 European Food Safety Authority (EFSA), "*Scientific opinion: Statistical significance and biological relevance*," *EFSA J*, 9 (2011): 2372.

42 Fagan, J., Antoniou, M.C., and Robinson,C., *GMO Myths and Truths: An Evidence-Based Examination of the Claims Made for the Safety and Efficacy of Genetically Modified Crops,* 2nd Edition, version 1.0, (London: Earth Open Source, 2014), 137.

43 For a discussion of the Piltdown fraud, see Broad, W. and Wade, N., *Betrayers of the Truth: Fraud and Deceit in the Halls of Science* (New York: Simon and Schuster, 1982), 119-22.

44 van Zwanenberg, P. and Millstone, E. "'Mad Cow Disease' 1980's – 2000: How reassurances undermined precaution, " in *Late Lessons from Early Warnings: The Precautionary Principle 1896 – 2000* (Luxembourg: European Environment Agency, 2001), 161.

45 Ibid.

46 *Betrayers of the Truth,* op. cit. note 43, 191.

47 Ibid.

48 Ibid., 189.

49 Lysenko's approach promised to boost yields much faster than alternatives, and it was based on the idea that acquired characteristics can be passed on to future generations, which dovetailed with the Marxist belief that altering external surroundings can induce profound inner change.

50 *Betrayers of the Truth,* op. cit. note 43, 190.

51 Ibid., 188, 191.

52 Although the first GE plants were not created until the early 1980's, the deceptions perpetrated in order to advance the genetic engineering venture in general began at least as early as the Bethesda Conference in 1976 and were forcefully employed during the summer and fall of 1977. As discussed in Chapter 1, these deceptions set the stage for the lax regulation of GE foods and their easy entry to the US market because they quashed legislative attempts to establish sound regulation of genetic engineering and shifted the burden of proof from the technology's proponents to those who thought it should be subject to such regulation.

53 *Betrayers of the Truth,* 20.

54 Nina Fedoroff has also alleged the existence of tests that apparently never happened. On page 175 of *Mendel in the Kitchen,* she states that the Flavr Savr tomato "was subjected to $2 million-worth of testing by the FDA on top of the testing done by Calgene." However, I couldn't find a specific reference for this statement in the note pages, and the FDA records provide no indication that such testing was undertaken – and instead impart the impression that none was. Further, the agency does not ordinarily conduct tests on new additives but instead reviews those submitted by the manufacturers. Moreover, Belinda Martineau, who had extensive first-hand knowledge about the interaction between the FDA and Calgene, informed me that to the best of her understanding, the FDA had not conducted any tests on the tomato.

55 Although Lysenko propounded ideas about inheritance that were unsubstantiated and dubious, it seems that he earnestly believed them; and it appears that he did not intentionally misrepresent well-established biological processes. In contrast, scientist-proponents of bioengineering (sometimes even in government agencies) have misrepresented fundamental biological facts. For instance, they've

proffered deceptive descriptions about how promoters operate, and some have disseminated misleading accounts of what occurs in the process of grafting.

56 Khachatourians, George C., University of Saskatchewan, writing in the *AgBiotech Bulletin*, February 1998.

57 Nuffield Council on Bioethics, *The use of genetically modified crops in developing countries,* June 2003.

58 *Bioengineering of Crops,* World Bank Panel on Transgenic Crops, 1997.

59 Borlaug, Norman, "Feeding a World of Ten Billion People: The Miracle Ahead," Lecture at De Montfort University, Leicester, UK, May 6, 1997. (I am not implying that Dr. Borlaug has engaged in or advocated deception. The point is that his concerns are shared by many other scientists who have felt motivated to engage in it.)

60 *Central Constr. Co. v. Home Indemnity Co.*, 794 P.2d 595, 598 (Alaska 1990).

61 To be innocent of fraud in such circumstances, not only would a scientist have had to lack intent to confuse people, he or she would also have had to be unaware that the statements issued were incorrect – or, even if technically correct, were likely to be misleading. While there may have been more than a few scientists in this category, it seems there have been many more who do not fit within it.

14. New Directions and Expanded Horizons

1 Comstock, Gary, "Ethics and Genetically Modified Foods," SCOPE GM Food Controversy Forum (July 1, 2001).

2 Brown, Patrick, "The Promise of Plant Biotechnology – the Threat of Genetically Modified Organisms," July 2000. Available at: http://www.psrast.org/promplantbiot.htm.

3 Xue, K., "Synthetic Biology's New Menagerie," *Harvard Magazine*, September-October, 2014.

4 Ibid.

5 Ibid.

6 "Former Pro-GMO Scientist Speaks Out On The Real Dangers of Genetically Engineered Food," September 24, 2014: http://earthweareone.com/former-pro-gmo-scientist-speaks-out-on-the-real-dangers-of-genetically-engineered-food/

7 Comstock, op. cit. note 1. He emphasizes the responsibility of scientists to accurately communicate facts and says, "If scientists are dishonest, untruthful, fraudulent, or excessively self-interested, the free flow of accurate information so essential to science will be thwarted." And he adds, "The public largely trusts scientists, and scientists must in turn act as good stewards of this trust." Accordingly, because he never expresses doubt about the soundness of the information that's been disseminated by the scientist-promoters of GE foods, he apparently believes that they've been honoring their obligation – a belief that's sorely mistaken.

8 Another prominent individual who apparently shifted his position on GMOs at least in part due to misapprehension of the facts is Peter Raven. As noted in Chapter 2, during the early 1980's he shared the concerns of Ernst Mayr and Phil Regal about the risks posed by environmental releases of GMOs, and he assisted in planning a workshop at which these risks could be examined. However, he eventually

became supportive of the GE food venture; and, as is the case with so many of its scientist-supporters, he has endorsed claims about it that are inaccurate. For instance, in 2009 he participated in a study group that released a statement describing genetic engineering as the newest addition to "a long and seamless continuum of progressively more precise and predictable techniques" of plant breeding. And the statement went on to more grossly overstate the degree of precision by asserting that the genetic engineering of a plant is "accompanied by a precise analysis of the genetic and phenotypic outcomes." Among its other inaccuracies, the statement also declared that the operations of genetic engineering "affect only one or a few genes." (Previous chapters have shown that all these assertions are misaligned with reality.) Thus, Raven's association with a statement containing such flawed assertions implies that his current position on GE foods is not firmly based on the facts. (The statement referred to is: "Transgenic Plants for Food Security in the Context of Development," PAS Study Week, Vatican City, 15-19 May 2009, pp. 4, 9.)

9 Taleb, N. et al., "The Precautionary Principle (with Application to the Genetic Modification of Organisms), *Extreme Risk Initiative – NYU School of Engineering Working Paper Series*, September 4, 2014: http://nassimtaleb.org/2014/08/precautionary-principle-paper/#.VE2ocRb63mE

10 Although other legal issues would be raised as well, such as whether the manufacturers have a right not to be compelled to speak, the most compelling defense would be to demonstrate that the FDA has not merely failed to preempt the field (because its policy is admittedly one of inaction), but that it has been deliberately misrepresenting the facts and willfully violating a federal statute and its own regulations – and that this delegitimizes its determination that GE foods do not require labeling.

11 Email communication.

12 Remarks of William Jefferson Clinton, Conference of the Biotechnology Industry Organization, Chicago, Illinois, April 11, 2006.

13 Ibid.

14 Brooks, David, "The Conservative Mind," *New York Times*, September 24, 2012.

15 Thus, when I speak of traditional theism, I am not referring to a belief system in which God has created the cosmos and its laws but has not directly planned for the development of life and has instead left whether and how life would develop up to the undirected interactions of naturally occurring phenomena.

16 'The Case Against Genetic Engineering' by George Wald, in *The Recombinant DNA Debate*, Jackson and Stich (eds.), 127-28 (reprinted in *The Sciences*, September/October 1976 issue).

17 Wald, G., quoted in Kimbrell, A., *The Human Body Shop: The Engineering and Marketing of Life*, Harper Collins (1994), 159.

18 http://www.theguardian.com/news/2002/jul/02/guardianobituaries.obituaries

19 Chargaff, Edwin, *Heraclitean Fire: Sketches of a Life Before Nature*, (New York: Rockefeller University Press, 1978).

20 Declaration of Rabbi Alan Green, *Alliance for Bio-Integrity v. Shalala.*

21 *Alliance for Bio-Integrity v. Shalala,* Plaintiffs' Second Amended Complaint for Declaratory and Injunctive Relief, Paragraph 36.

22 Consumers Union, Comments to US Food and Drug Administration on *Statement of Policy: Foods Derived From New Plant Varieties*, August, 1992, 2.

23 Although these experts state that when the strict precautionary principle is employed, the proponents of an activity have the burden of proving it's safe, in the case of GMOs, it's evident from their discussion that they do not think such proof could be practically accomplished – especially as regards the risk of irreversible environmental ruin. That's because they state that GMOs should be preemptively prohibited because there is not "scientific near-certainty" about their safety; so according to their standards, any attempted proof of safety would need to achieve such near-certainty to suffice. Consequently, for practical purposes, their position amounts to full prohibition of these products.

24 Fagan, J., Antoniou, M.C., and Robinson, C., *GMO Myths and Truths: An Evidence-Based Examination of the Claims Made for the Safety and Efficacy of Genetically Modified Crops*, 2nd Edition, version 1.0, (London: Earth Open Source, 2014), Section 5.12.

25 Ibid., 284.

26 Mellon, M, and Gurian-Sherman, D., "The cost-effective way to feed the world," *The Bellingham Herald*, June 20, 2011.

27 International Assessment of Agricultural Knowledge, Science and Technology for Development (IAASTD), "Agriculture at a crossroads: Synthesis report of the International Assessment of Agricultural Knowledge, Science" and "Technology for Development: A Synthesis of the Global and Sub-Global IAASTD Reports" (Washington, DC, USA: Island Press; 2009).

28 Lean, G., "Exposed: The great GM crops myth," *The Independent*, April 20, 2008.

29 Sherman, M., Q & A: Hans Herren on "Sustainable Agriculture Solutions," *GMO Inside*, April 9, 2014: http://gmoinside.org/q-hans-herren-sustainable-agriculture-solutions/

30 Hine, R., Pretty, J. and Twarog, S., "Organic agriculture and food security in Africa," New York and Geneva: UNEP-UNCTAD Capacity-Building Task Force on Trade, Environment and Development (2008). Available at: http://bit.ly/KBCgY0

31 De Schutter, Olivier, quoted in Leahy, S., "Africa: Save climate and double food production with eco-farming," *IPS News*, March 8, 2011: http://allafrica.com/stories/201103090055.html

32 De Schutter, Olivier, quoted in "Eco-farming can double food production in 10 years, says new UN report" (press release) *United Nations Human Rights Council*, March 8, 2011: http://bit.ly/Lkfa9U

33 Drinkwater, L. E., Wagoner, P. and Sarrantonio, M., "Legume-based cropping systems have reduced carbon and nitrogen losses," Nature 396, 1998, 262–65.

34 Rodale Institute, "The Farming Systems Trials; Celebrating 30 years," 2012: http://rodaleinstitute.org/assets/FSTbooklet.pdf

35 Alteri, Miguel, "Agroecology, Small Farms, and Food Sovereignty," *Monthly Review*, vol. 61, issue 3, July/August, 2009.

36 Ikerd, John, "Family Farms of North America," in *Deep Roots*, (Rome: The Food and Agriculture Organization of the United Nations, 2014), 30-32.

37 http://www.blackwellreference.com/public/tocnode?id=g9781405184649_yr2012_chunk_g97814051846491361

38 https://en.wikipedia.org/wiki/Pure_Food_and_Drug_Act

Appendix A

1 *San Luis Obispo Mothers for Peace v. U.S. Nuclear Regulatory Comm'n*, 789 F.2d 26, 33 (D.C.Cir.1986).

2 The courts have made it clear that "general recognition" is a matter of fact. e.g. *United States v. 4680 Pails*, 725 F.2d 976, 985 (5th Cir. 1984)

3 *State Farm Mutual Automobile Insurance Co. v. Dept. of Transp.*, 680 F.2d 206, 220 (D.C. Cir. 1982).

4 Among the cases cited were *Ferro-Lac* (discussed in Chapter 5), *Natick Paperboard Corp. v. Weinberger*, 525 F.2d 1103 (1st Cir. 1975) (packaging); *U.S. v. Articles of Food . . . Pottery*, 370 F.Supp. 371 (E.D.Mi. 1974) (dinnerware).

5 Document #7 at: http://biointegrity.org/24-fda-documents (A.R. at 18960).

6 Document #19 at: http://biointegrity.org/24-fda-documents (A.R. at 18196). The FDA's dramatic shift in policy is noted in an article in the Boston College Law Review. 44 B.C.L. Rev. 733 2002-2003 at 749.

7 *SEC v. Chenery Corp.*, 332 U.S. 194 (1947) at 196.

8 116 F. 2d at 177 citing *International Fabricare Institute* v. U.S. E.P.A., 972 F.2d 384 at 389 (D.C.Cir.1992).

9 972 F.2d 384 at 390, 396, & 398.

Appendix B

1 Royal Society, "Genetically modified plants for food use and human health – an update," (2002), 6.

2 Ibid., 8.

3 Ibid., 6.

4 Institute of Food Technologists, *IFT Expert Report on Biotechnology and Foods*, 2000.

5 Page 21 of the above report states that the risks of GE are "the same in kind" as those of traditional breeding.

6 Ibid., 15.

7 Ibid., 17.

8 Although a later section on the benefits of bioengineering did note the use of viral promoters, it ignored the fact they force hyper expression of the gene affixed to them and thus avoided confronting the related safety issues.

9 Ibid., 21.

Index

Acknowledgements

I'm especially grateful for the assistance of Philip Regal, who contributed to this book in so many ways and whose heroic endeavors to align the genetic engineering venture with sound science are so interwoven with the narrative. My deep gratitude also extends to Jane Goodall for so quickly appreciating the importance of this book and contributing such an excellent foreword – and to Randall Tolpinrud, who brought it to Jane's attention and has assisted in several other significant ways. Special thanks are also due to Stephen Naylor for enhancing Chapter 3 by letting me be the first to report on the important (but as yet unpublished) research that he and Gerald Gleich conducted at the Mayo Clinic – and for devoting so much time to ensure the accuracy of my account.

I'm also very appreciative for the extensive assistance of the experts who, in addition to Dr. Regal, carefully reviewed the chapters, made many valuable comments, and answered my many questions: David Schubert, Allison Wilson, Belinda Martineau, Joseph Cummins, Joan Levin, Jack Heinemann, John Ikerd, and Richard C. Jennings. And many thanks to Susan Wright, who thoroughly reviewed Chapter 1 and answered numerous questions about details of her research on the early history of genetic engineering. Any flaws that remain are my responsibility, not theirs.

Moreover, my profound appreciation goes to Robert Kent for cheerfully and diligently performing the copy editing and providing helpful editorial suggestions. His contribution has been exemplary. I am additionally grateful for the assistance given by William Crist, Charles Yanofsky, Michael Antoniou, John Fagan, Claire Robinson, Michael Albertsen, Jonathan Latham, Ralph Bunker, Steve Nolle, Mark Rainbow, Tom Miller, Larry Weisselberg, Gerald Gleich, Tara Cook-Littman, Ken Walton, David Fisher, Ken Roseboro, Bill Freese, Gilles-Eric Séralini, Susan Bardocz, Judy Carman, Jeffrey Smith, Elaine Ingham, Robert Merritt, Dennis Mackin, Michael Hansen, Patricia Zambryski, Paul Rheingold, Steve McClaskey, Alexandrea Barogianis, Sonja Gobec, Romy Das, Shane Zisman, Kurt and Barbara Rauscher, Lucy Sharratt, Nora Mylett, George Foster – and anyone else who assisted but has inadvertently been overlooked.

Further, I want to express deepest gratitude to my dear wife, Kathryn, who has always whole-heartedly believed in this book and given it (and me) tender support in a multitude of ways. She was usually the first person to read and comment on the chapters, and through her sensitive awareness, the book has achieved greater clarity and readability.

About the Author

Steven M. Druker is a public interest attorney who initiated a lawsuit against the US Food and Drug Administration that forced it to divulge its files on genetically engineered foods. This revealed that the agency had covered up the extensive warnings of its own scientists about the unusual risks of these foods, lied about the facts, and then ushered these products onto the market in violation of explicit mandates of federal food safety law. In organizing the suit, he founded the Alliance for Bio-Integrity and assembled an unprecedented coalition of eminent scientists and religious leaders to stand with it as co-plaintiffs – the first time scientists had sued a federal administrative agency on the grounds that one of its policies is scientifically unsound.

He is a prominent commentator on the risks of genetically engineered (GE) foods and has been a featured speaker at symposia at the British House of Commons and the National Congress of Brazil and at press conferences sponsored by the Brazilian Medical Association, the Swedish Consumers' Association, and the Green Party members of the European Parliament.

He has served on the food safety panels at conferences conducted by the National Research Council and the FDA; given lectures at numerous universities (including the Biological Laboratories at Harvard, Tel Aviv University, and the University of Copenhagen); and met with government officials world-wide, including the UK's Environmental Minister and the heads of food safety for France, Ireland, and Australia. He also conferred at the White House Executive Offices with an interagency task force of President Clinton's Council on Environmental Quality.

His articles on GE food have appeared in several respected publications, including *The Congressional Quarterly Researcher*, *The Parliament Magazine*, and *The Financial Times*.

He has extensive academic background in the history and philosophy of science and in human development and ethics. He co-authored the introductory and final chapters of *Higher Stages of Human Development*, published by Oxford University Press, and wrote a chapter on ethical development for *Transcendence and Mature Thought in Adulthood*, published by Rowman and Littlefield.

He majored in philosophy at the University of California, Berkeley, received a special award for "Outstanding Accomplishment" in that field, was elected to Phi Beta Kappa in his junior year, and graduated with "Great Distinction in General Scholarship." He also received his Juris Doctor from UC Berkeley and was elected to both the California Law Review and the Order of the Coif (the legal honor society).

3 1901 05773 6904